Advanced Principles of Applied Geophysics

Advanced Principles of Applied Geophysics

Editor: Karl Seibert

STATES
ACADEMIC PRESS
www.statesacademicpress.com

States Academic Press,
109 South 5th Street,
Brooklyn, NY 11249, USA

Visit us on the World Wide Web at:
www.statesacademicpress.com

ISBN: 978-1-63989-018-7 (Hardback)

Cataloging-in-Publication Data

Advanced principles of applied geophysics / edited by Karl Seibert.
p. cm.
Includes bibliographical references and index.
ISBN 978-1-63989-018-7
1. Geophysics. 2. Earth sciences. 3. Physics. I. Seibert, Karl.
QC806 .A67 2022
550--dc23

Table of Contents

Preface

This book has been a concerted effort by a group of academicians, researchers and scientists, who have contributed their research works for the realization of the book. This book has materialized in the wake of emerging advancements and innovations in this field. Therefore, the need of the hour was to compile all the required researches and disseminate the knowledge to a broad spectrum of people comprising of students, researchers and specialists of the field.

Geophysics is a subject of natural science that deals with the physical processes and properties of the Earth and its atmosphere using quantitative methods for their analysis. It involves the geological applications of Earth's shape, gravitation, magnetic fields, internal structure, composition, dynamics, surface expression in plate tectonics, generation of magmas, volcanism, and rock formation. It also comprises the water cycle, formation of snow and ice, fluid dynamics of the oceans and the atmosphere. The study of electricity, and magnetism in the ionosphere and magnetosphere, and solar-terrestrial physics. The problems associated with the Moon and other planets arealso undertaken within this field. The topics included in this book on applied geophysics are of utmost significance and bound to provide incredible insights to readers. It attempts to understand the multiple branches that fall under this discipline and how such concepts have practical applications. This book will provide comprehensive knowledge to the readers.

At the end of the preface, I would like to thank the authors for their brilliant chapters and the publisher for guiding us all-through the making of the book till its final stage. Also, I would like to thank my family for providing the support and encouragement throughout my academic career and research projects.

Editor

Hydrothermal system in the Tatun Volcano Group, northern Taiwan, inferred from crustal resistivity structure by audio-magnetotellurics

Shogo Komori[1,6*], Mitsuru Utsugi[2], Tsuneomi Kagiyama[2], Hiroyuki Inoue[2], Chang-Hwa Chen[1], Hsieh-Tang Chiang[3], Benjamin Fong Chao[1], Ryokei Yoshimura[4] and Wataru Kanda[5]

Abstract

The present study proposes an improved conceptual model for the hydrothermal system in the Tatun Volcano Group in northern Taiwan. In the study, audio-magnetotellurics (AMT) surveys were conducted to reveal the spatial distribution of resistivity, which is highly sensitive to fluids and hydrothermal alteration. By combining the obtained resistivity structure with other geophysical and geochemical evidence, the following hydrothermal system was inferred. Beneath Chishinshan, vapor-dominant hydrothermal fluids, supplied from a deeper part, are maintained in a low to relatively low resistivity region (5 to 20 Ωm) that is covered by a clay-rich cap, represented by an upper extremely low resistivity layer. Fluid ascent is suggested by a pressure source and clustered seismicity. Exsolved gases result in fumarolic areas, such as Siao-you-keng, while mixing of gases with shallow groundwater forms a shallow flow system of hydrothermal fluids in the Matsao area, represented by a region of less than 10 Ωm. The fumarole in the Da-you-keng area originates from vapor-dominant hydrothermal fluids that may be supplied from a deeper part beneath Cing-tian-gang, suggested by a pressure source and low to relatively low resistivity. Horizontally extended vapor-bearing regions also suggest the possibility of future phreatic eruptions. The proposed conceptual model may provide clues to detecting precursors of potential volcanic activity.

Keywords: Tatun Volcano Group; Hydrothermal system; Resistivity; Vapor-dominant region; Pressure sources; Gas and groundwater geochemistry

Background

The Tatun Volcano Group (TVG) covers an area of approximately 300 km^2 on the northern tip of the island of Taiwan, only about 10 km north of the capital city of Taipei. The group is composed of over 20 andesitic volcanic composites, cones, and domes (Chen and Wu 1971; Wang and Chen 1990), enclosed by NE-SW trending faults such as the Chinshan and Kanchiao faults, as shown in Figure 1. The magmatism of the TVG is considered to have taken place in an extensional stress field, which was associated with postcollisional processes between the Philippine Sea Plate and East Asian continent, and with the opening of the southwestern end of the Okinawa Trough (Wang et al. 1999; Chang et al. 2003; Shyu et al. 2005; Chen et al. 2010). A look at historical volcanic eruptions has suggested that volcanism has been extinct since the last major activity that occurred from 0.8 to 0.2 Ma BP (e.g., Juang 1993; Tsao 1994). However, recent studies offer suggestions that the potential for volcanic activity exists in the area: extensive hydrothermal activity is discharging a large amount of heat (e.g., Chen 1970); there are strongly acidic hot springs (pH approximately 1 to 3) (e.g., Liu et al. 2011; Ohsawa et al. 2013); magmatic contributions are found in fumarolic gases (e.g., Yang et al. 1999, 2003; Lee et al. 2005; Ohba et al. 2010); relatively young ejecta can be observed (<20 ka, Chen and Lin 2002; 6 ka, Belousov et al. 2010), and the presence of clustered seismicity and shallow pressure sources suggest fluid flows (Lin et al. 2005; Konstantinou et al. 2007, 2009; Murase

*Correspondence: komori.shogo@aist.go.jp
[1]Institute of Earth Sciences, Academia Sinica, 128 Academia Road, Section 2, Nankang, Taipei 115, Taiwan
[6]Present address: Geological Survey of Japan (GSJ), AIST, Central 7, 1-1-1 Higashi, Tsukuba, Ibaraki 305-8567, Japan
Full list of author information is available at the end of the article

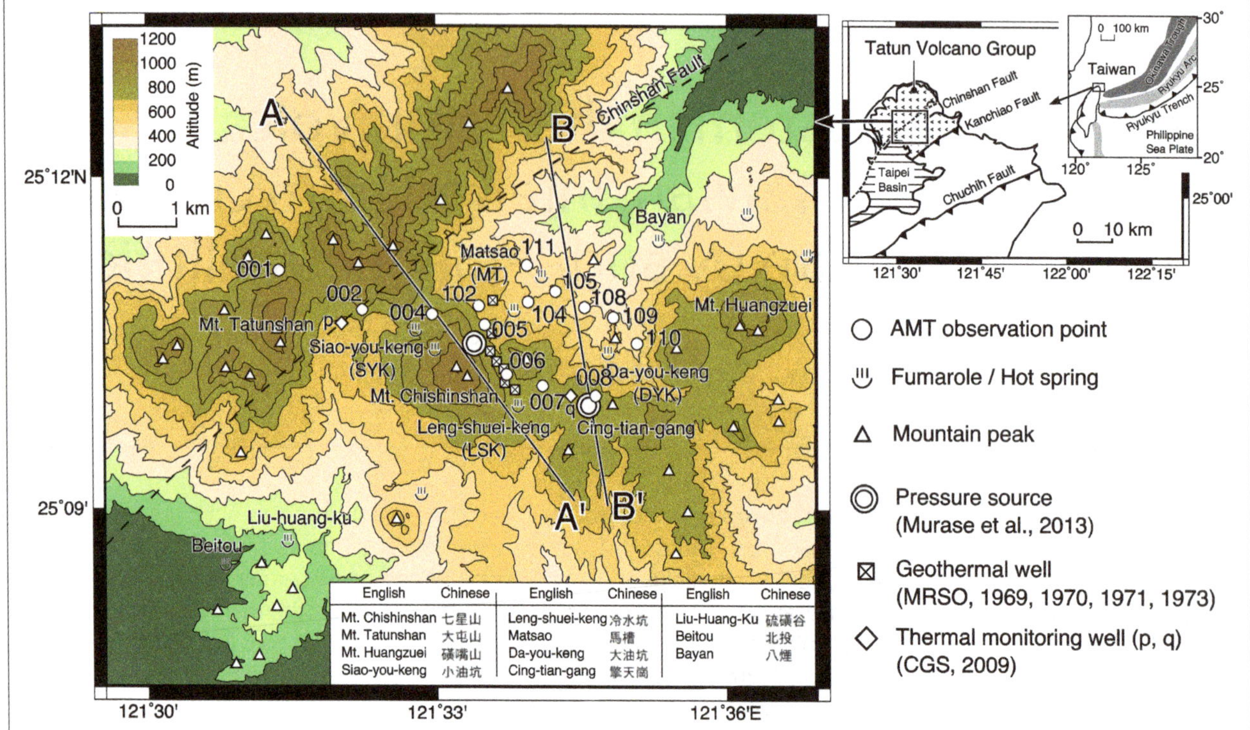

Figure 1 Location of the Tatun Volcano Group, northern Taiwan. Open circles represent the AMT observation points. Sites 001, 002, 004, 005, 006, and 007 were observed in 2011 by Utsugi et al. (2012), while sites 008, 104, 105, 108, 109, 110, and 111 were observed in 2012 for this study. Double circles represent the locations of pressure sources obtained from precise leveling surveys by Murase et al. (2013). Squares and diamonds represent the wells drilled by MRSO (1969, 1970, 1971, 1973), and CGS (2009), respectively.

et al. 2013). Based on the above studies, hazard assessments and monitoring projects are being conducted (e.g., Lee et al. 2008; Wu et al. 2009; Tsai et al. 2010; Lin et al. 2012; Rontogianni et al. 2012).

Electric resistivity is one useful physical parameter for exploring the nature of a hydrothermal system because of its great sensitivity to the existence of fluid and hydrothermal alteration. In general, bulk resistivity decreases with increases in pore water connectivity, saturation, salinity, temperature, and smectite content due to rock alteration (e.g., Revil et al. 1998, 2002; Waxman and Smits 1968). Many studies to date have found resistivity anomalies corresponding to the paths and phase states of hydrothermal fluids, and to the fluid-bearing structure in volcanic and geothermal areas (e.g., Ogawa et al. 1998; Kanda et al. 2008; Aizawa et al. 2009; Komori et al. 2010, 2013a; Revil et al. 2011). Therefore, resistivity anomalies may provide critical constraints leading to better models of hydrothermal systems. The present study proposes an integrated hydrothermal model that combines newly revealed electric resistivity structures of the crust with geophysical and geochemical evidence. We discuss implications for potential future phreatic eruptions for the TVG.

Methods and Results

AMT surveys

Electromagnetic studies at the TVG

The magnetotellurics (MT) principle (e.g., Cagniard 1953; Vozoff 1991; Simpson and Bahr 2005) is based on electromagnetic induction and has been widely used at hydrothermal and volcanic areas for estimating the crustal electric resistivity structures associated with their activity. MT was first used by Chen et al. (1998) and Chen 2009 in the TVG to provide preliminary 1-D and 2-D structures, respectively. In addition, Kagiyama et al. (2010) conducted extensive and dense VLF-MT surveys to investigate resistivity features near the surface using a singular electromagnetic frequency of 22.1 kHz. They found low resistivity anomalies (less than 30 Ωm) corresponding to the fumarole and hot spring areas. Their NE-SW trending distributions suggested that the hydrothermal activity might be controlled by polarized permeable paths developed in response to the regional stress field (Chang et al. 2003; Shyu et al. 2005; Chen et al. 2010). Utsugi et al. (2012) conducted dense audio magnetotellurics (AMT) surveys around Mt. Chishinshan to reveal a resistivity structure down to depths of 2-3 km using multiple frequencies of 1 Hz to 10 kHz. This study extended the

AMT observation points to the Matsao and Da-you-keng areas to the north, where there are also active fumaroles and hot springs (Figure 1). The present study will show the shallow crustal resistivity structure associated with the intense hydrothermal activity in the TVG in northern Taiwan. Incorporating the data from Utsugi et al. (2012), we will propose an integrated model of detailed crustal resistivity structures, particularly beneath Mt. Chishinshan, and the Matsao hot spring and Da-you-keng fumarolic areas.

It is notable that in the late 1960s to early 1970s, the Mining Research and Service Organization (MRSO 1969, 1970, 1971, 1973) conducted extensive electrical resistivity surveys (DC surveys) to obtain the spatial distribution of apparent resistivity in the TVG for the purpose of commercial use of its geothermal energy. Unfortunately, at that time, the detailed resistivity structure could not be revealed because of a lack of inversion techniques and insufficient sounding depths.

Data acquisition and processing

AMT surveys were conducted in October 2011 by Utsugi et al. (2012), and in December 2012 for the present study, using two Phoenix Geophysics MTU-5 systems. AMT pre-surveys began in 2010. There was no significant change in volcanic activity in the TVG during the time (e.g., Murase et al. 2013; Wen et al. 2013). Figure 1 shows the observation points. Those used in 2011 and 2012 were configured to intersect Mt. Chishinshan and the Matsao hot spring and Da-you-keng fumarolic areas. Four Pb-$PbCl_2$ electrodes were used to measure two orthogonal components (N-S and E-W) of the electric field, and one additional electrode was used for grounding. Three orthogonal components (N-S, E-W, and vertical) of the magnetic field were measured using three induction coils. Each consisted of a coil of copper wire wound on a core with high magnetic permeability. At each observation point, continuous data acquisition was carried out for about 4 h. To remove contamination in the data due to local noise, we used Gamble et al.'s (1979) remote reference processing method for mutual referencing within the survey area. Actually, there are noise sources like power lines and inhabited areas around the study area that seriously affected the data even after removing the noise as described above. For this study, we removed any data whose electric and magnetic fields had a coherence of less than 0.7 to maintain high quality.

Dimensionality

To examine the regional dimensionality of this study area, impedance phase tensors and skew angles (β) were calculated using the method of Caldwell et al. (2004); this method has the merit of removing distortions from original impedances due to near-surface heterogeneities. Figure 2a shows the impedance phase tensors, skew angles, and induction arrows for representative frequencies.

Skew angle β is a proxy for three-dimensional (3-D) heterogeneity in the media structure, where larger β values mean stronger effects. The β value tends to gradually increase with decreasing frequencies. In particular, relatively larger average β values were estimated at observation points near Mt. Chishinshan. Uncertainties were less than 0.1° for 100 to 10,800 Hz and less than a few degrees for 1 to 100 Hz. The resulting values were a few degrees in the range of several hertz to several hundred hertz, suggesting a relatively minor 3-D effect. In addition, the average β at the same frequencies is as small as 0.5° to 2° at the observation points off Mt. Chishinshan, so it is reasonable to infer that 3-D heterogeneity has a minor effect on a regional scale. It is also notable that the average λ values calculated using the magnitudes of the principal axes of phase tensor ellipses, which are the proxy for the one-dimensional (1-D) structure (Bibby et al. 2005), were found to be as small as 0.2 to 0.3, suggesting a relatively simple structure beneath the observation sites.

Strike estimation

The direction of the main axes of the phase tensor ellipses reflect a two-dimensional (2-D) strike (Caldwell et al. 2004). In order to estimate the regional strike, we focused on the data at lower frequencies. The tensor ellipses at frequencies of 18.8 and 8.1 Hz, shown in Figure 2a, have the following features. Most have main axes trending NW-SE at observation points around Mt. Chishinshan (sites 001, 002, 004, 102, 007) and in the Matsao area (sites 111, 104, 105). Those in the Da-you-keng area (sites 108, 109, 110, 008), however, have main axes in the N-S direction. Thus, we defined three areas, referred to as the Chishinshan, Matsao, and Da-you-keng areas, to estimate the respective direction of the major axes. Note that point 102 was at the intersect and was used in both the Chishinshan and Matsao areas.

Figure 2b shows the rose diagrams of the main axes of the phase tensor ellipses for each area. Uncertainties were lower than a few degrees for 100 to 10,800 Hz, and up to 10° for 1 to 100 Hz. The rose diagrams are rendered in discrete frequencies; this is because there are insufficient data at 530 Hz to 3.6 kHz and 40 to 79 Hz (460 Hz to 3.6 kHz, 16.2 to 116 Hz, and 1 to 6.9 Hz for the Chishinshan area). According to Caldwell et al. (2004), the dominant direction of the main axes of the phase tensor ellipses reflects the direction of either a regional strike or a conductivity discontinuity. They note that the obtained direction has a 90° ambiguity. The dominant directions of the main axes were estimated primarily

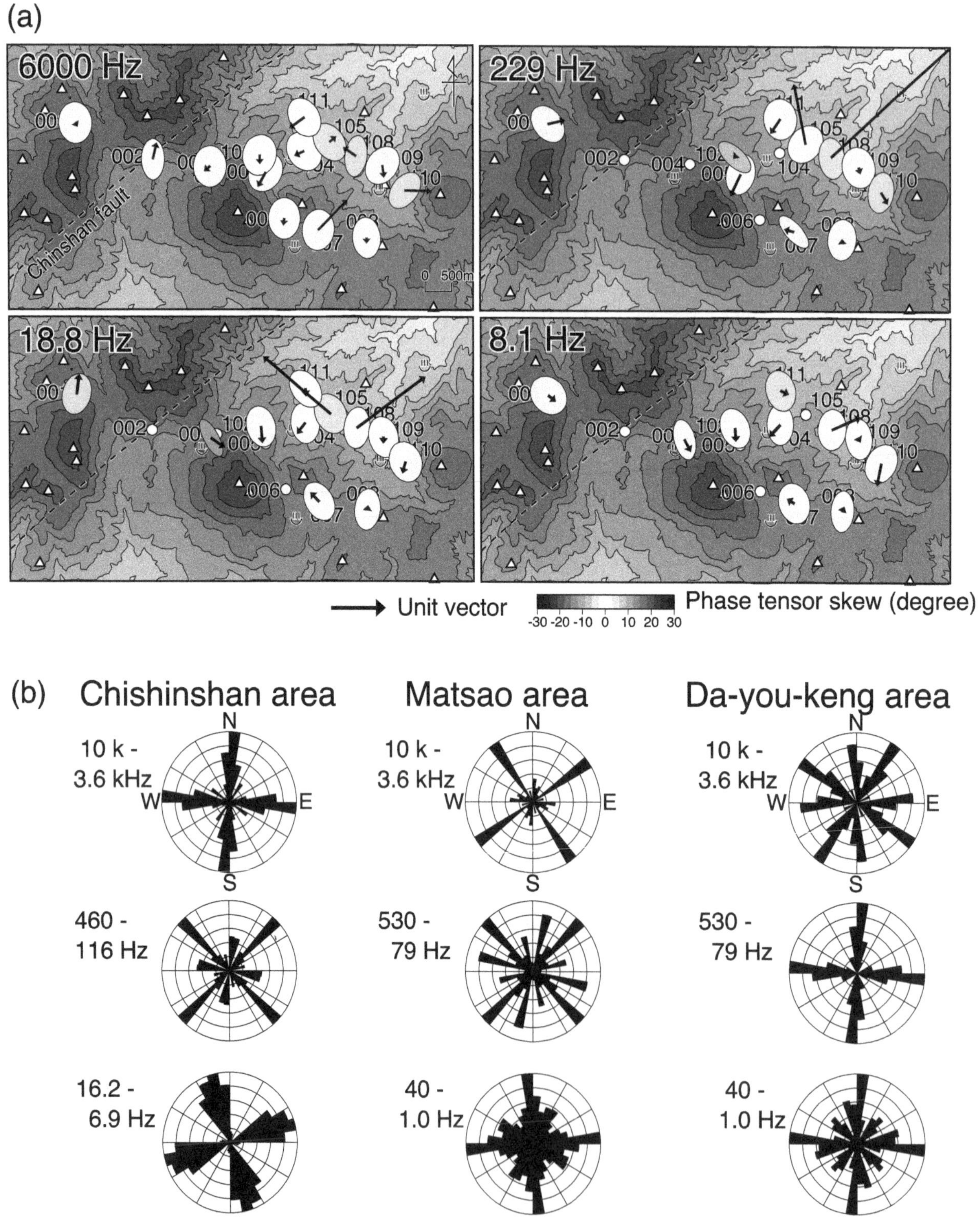

Figure 2 Data for dimensionality and strike estimation. (a) Impedance phase tensors, skew angles β, and induction arrows for representative frequencies. Note that some data are missing due to their low quality. Also note that the larger induction vectors observed at sites 007, 105, and 108 are considered to be due to noise, as these sites are located near electrical cables and roadways. **(b)** Rose diagrams of tensor axes.

from their modes in each area. The rose diagrams at several to several hundred hertz show that there are two rough types of dominant directions: NW-SE (or NE-SW) in the Chishinshan area and N-S (or E-W) at the Da-you-keng area, with the Matsao area having an intermediate direction.

As described above, northern Taiwan has regional NE-SW trending faults (approximately N60° E). Correspondingly, a NE-SW volcanic trend has developed along the Chinshan fault (e.g., Belousov et al. 2010). The TVG also has an E-W volcanic trend from Mt. Huangzuei to Mt. Tatunshan (Belousov et al. 2010), suggesting a weak extra E-W trending structure. The structural features noted above are consistent with the dominant directions of the main axes of the phase tensor ellipses. The induction vectors also trend roughly NW-SE and N-S in the Mt. Chishinshan and Da-you-keng areas, respectively, which suggests that the regional strike is perpendicular to the induction vector in each area. Based on these factors, the regional strike was estimated using the dominant directions at low- to middle-frequency ranges (below several hundred Hz) as follows: N57° E for the Chishinshan area, N65° E for the Matsao area, and and N85° W for the Da-you-keng area. This means that the 2-D cross sections are along N33° W for the Chishinshan area (line A-A' in Figure 1), N25° W for the Matsao Matsao area, and N5° E for the Da-you-keng area. The 2-D cross section of the Matsao area is close to that of the Da-you-keng area, so we combined the Matsao and Da-you-keng areas into one and assumed its 2-D cross section as N10° W (Line B-B' in Figure 1).

Accordingly, we rotated the impedance tensors to correspond to 2-D strikes oriented to N57° E for the Chishinshan area and N80° E for the Matsao-Da-you-keng area. We then performed a distortion analysis following the method of Bibby et al. (2005). This procedure estimates a distortion tensor, produced by surface heterogeneity, considering the ellipticity of a phase tensor and β angle. The effect of the distortion was removed from the rotated impedance tensor by multiplying the inversion matrix of the distortion tensor with the impedance tensor.

Soundings

Figure 3 shows the sounding curves of the apparent resistivity and phase, with open circles representing the TE mode and open rhombi for the TM mode, calculated from the impedance tensors as described above. The apparent resistivity and phase of the TM mode were calculated using electric fields perpendicular to the strike and magnetic fields parallel to the strike and the opposite for the TE-mode. For both modes, apparent resistivities have high values at high frequencies throughout and decrease with decreasing frequencies. In general, the apparent resistivity decreases with decreasing frequency for both modes.

Sounding observations at Chishinshan area, shown in Figure 3a, have the following features. Site 001 has phases less than 45° in high-frequency ranges, suggesting a high-resistivity region near the surface. At sites 001 and 002, the phase significantly decreases with decreasing frequency from a few tens of hertz to 1 Hz. On the other hand, the phase at site 004 gradually decreases compared to sites 001 and 002. This feature suggests a relatively high-resistivity body at a deeper part beneath sites 001 and 002. At site 005, the phases are greater than 60° through frequencies from 10 kHz to 10 Hz for both the TE and TM modes, and the apparent resistivities of both modes decrease to 1 to 2 Ωm at 10 Hz, a feature quite different from the other sites. This suggests that there is a significant change in the resistivity structure around site 005.

Figure 3b shows the same features for the Matsao-Da-you-keng area. At sites 111 and 105, the phase increases from 45° to 50° to 60° as frequency decreases from 10 kHz to 10 to 20 Hz for both the TE and TM modes, although their phase change patterns differ slightly. Site 108 also phase change pattern similar to sites 111 and 105. At these three sites, the phase significantly decreases to approcimately 30° as frequency decreases from 10 to 20 to 1 Hz. These features suggest a shallow conductive body and a resistive body deeper beneath the sites. At sites 109 and 110, the phase begins decreasing from a few hundred hertz, suggesting that the resistive body mentioned above could rise toward a shallower level beneath sites 109 and 110. At site 008, the phase stays higher than 60° through frequencies from 1 kHz to several hertz, suggesting a conductive body underneath.

2-D analysis

2-D resistivity structure analyses were performed using the inversion code developed by Ogawa and Uchida (1996). This code performs smoothness-constrained 2-D inversion based on the ABIC minimization method, where trade-off parameters between data misfit, model roughness, and static shift are estimated to minimize the misfit functional, assuming that static shifts follow Gaussian distributions. Error floors for apparent resistivity and phase were set to be 10% and 7% for the Chishinshan and Matsao-Da-you-keng areas, respectively. The resistivity structures were projected to lines A-A' and B-B' for the two areas, as shown in Figure 1. To represent the overall topographic change at each observation area, we defined the topography by averaging within 2 km of line A-A' for the Chishinshan area and within 1.3 km of line B-B' for the Matsao–Da-you-keng area. Note that the results thus obtained for the average topography do not differ much

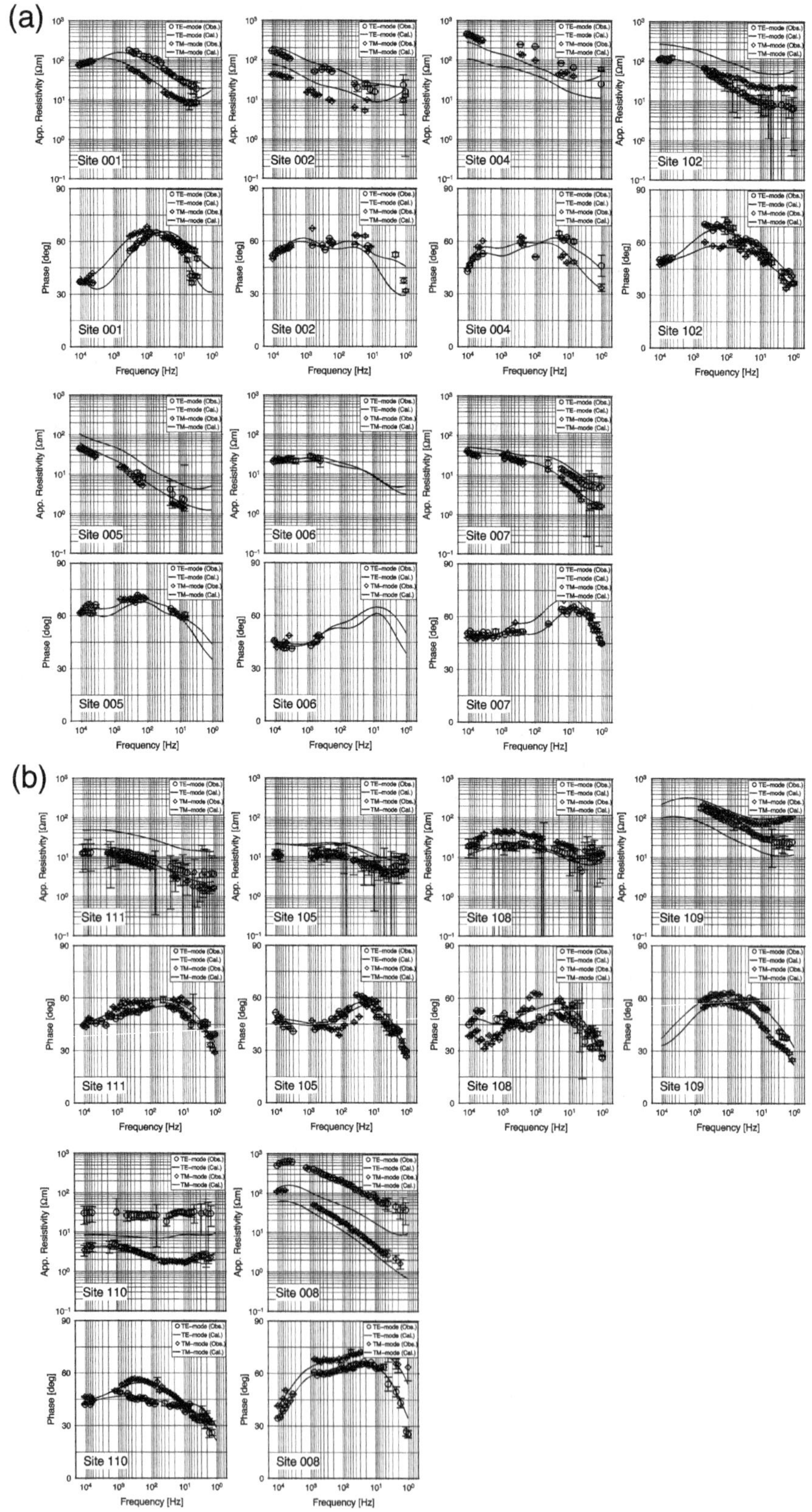

Figure 3 Sounding curves of apparent resistivity and phase for both the TM and TE modes. (a) Chishinshan area. **(b)** MatsaoŰDa-you-keng area. Circles and diamonds represent the data for the TE mode and TM mode respectively, obtained from observations. Red and Blue lines represent the data for the TE mode and TM mode respectively, calculated using the best-fit model.

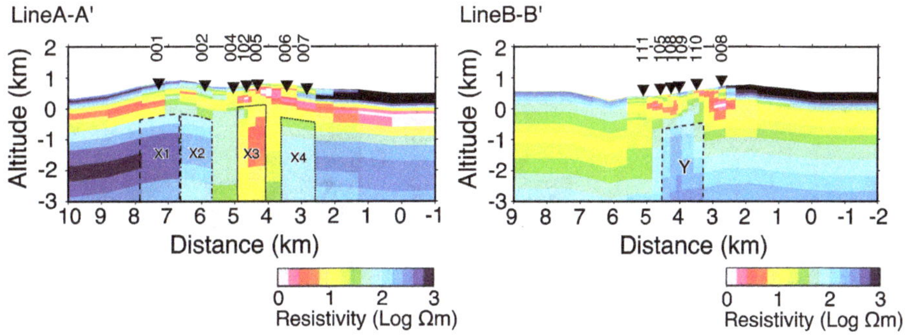

Figure 4 Best-fit models obtained from 2-D analysis. Solid inverted triangles represent observation points. Site numbers are indicated above black triangles. X1 to X4 and Y represent regions used for the sensitivity tests.

from those using the actual topography along A-A' and B-B'.

Figure 4 shows the best-fit models obtained after 42 iterations for the Chishinshan area and 31 for the Matsao-Da-you-keng area, starting from an initial model with a uniform resistivity of 100 Ωm, with both RMS misfits of 1.02. Figure 3 shows the calculated soundings for the TE mode (red line) and TM mode (blue line). The inferred models reproduce most of the apparent resistivity and phase distributions. In the Chishinshan area, there seems to be no relation between the amount of static shift and the topographic change. The TE mode at site 004 has a relatively large static shift. Fumarolic activities do sporadically occur around site 004 in the area called Siao-you-keng. This is characterized by a surface resistivity less than 30 Ωm from the VLF-MT survey (Kagiyama et al. 2010). This suggests that site 004 could have relatively strong 3-D heterogeneity near the surface, leading to the relatively large static shift. The Matsao-Da-you-keng area has a larger static shift than the Chishinshan area, presumably because most of its sites are situated amid sharp cliffs, resulting in a static shift due to the local topography (e.g., Jiracek 1990).

Sensitivity tests

Before making any interpretations on the basis of our analyses, we performed the following series of sensitivity tests:

(i) Linear sensitivity analyses based on the method of Schwalenberg et al. (2002). According to that study's authors, model sensitivity can be described as:

$$S_j = \frac{1}{\Delta_j}\sum_{i}^{N} \left\| \frac{1}{\sigma_i}\frac{\partial f_i(\mathbf{m})}{\partial m_j} \right\| \quad (1)$$

where S_j is the sensitivity of grid element j of size Δ_j, $f_i(\mathbf{m})$ is the forward solution of model $\mathbf{m}$, m_j is the resistivity change of grid element j, σ_i is the standard deviation of the data, and N is the number of elements = number of observation sites × number of frequencies × data types. This equation states that model resolution is the sum of the gradient of the forward solution to the resistivity change of a given grid element, which we normalize with regard to the standard deviation of the data, weighted by the size of each grid element. There are four types of data and forward solutions, the apparent resistivities and phases of the TE and TM modes.

Figure 5a shows the resultant model sensitivity, which decreases with depth, corresponding to the limit of the sounding frequency. The high-resistivity body situated below 1 to 1.5 km beneath sites 001 and 002 in the Chishinshan area has a variety of sensitivities ranging from 10^{-3} to 10^{-6}, as do the resistive bodies situated below 1 km beneath sites 111, 105, 108, and 109 in the Matsao–Da-you-keng area. The MT method is generally less sensitive to resistive bodies because of the properties of electromagnetic waves, which can be easily absorbed by a conductor (e.g., Cagniard 1953; Vozoff 1991; Simpson and Bahr 2005). For this reason, evaluating the reliability of the resistive body was necessary.

As shown in Figure 4, a low-resistivity, column-shaped body was found below a depth of 1 km at sites 102 and 005 in the Chishinshan area. This structural feature is represented by the low-frequency range soundings at sites 004, 102, and 005 (Figure 3). However, many data are missing because of their low quality, suggesting that such inverted deeper structures are not unique. It is also known that low-quality data or 2-D analysis of a 3-D situation may introduce inversion artifacts.

In order to confirm these regions with reliable sensitivity, the deeper resistive or conductive bodies shown by regions X1 to X4 and Y in Figures 4 and 5a were submitted to the following further sensitivity analysis.

(ii) Examination of changes in the sounding curves by making distinctive changes in the model.

Case 1: region X1 to X4

Regions X1, X2, and X4 have high resistivities with values more than 100 Ωm. For these cases, changes in

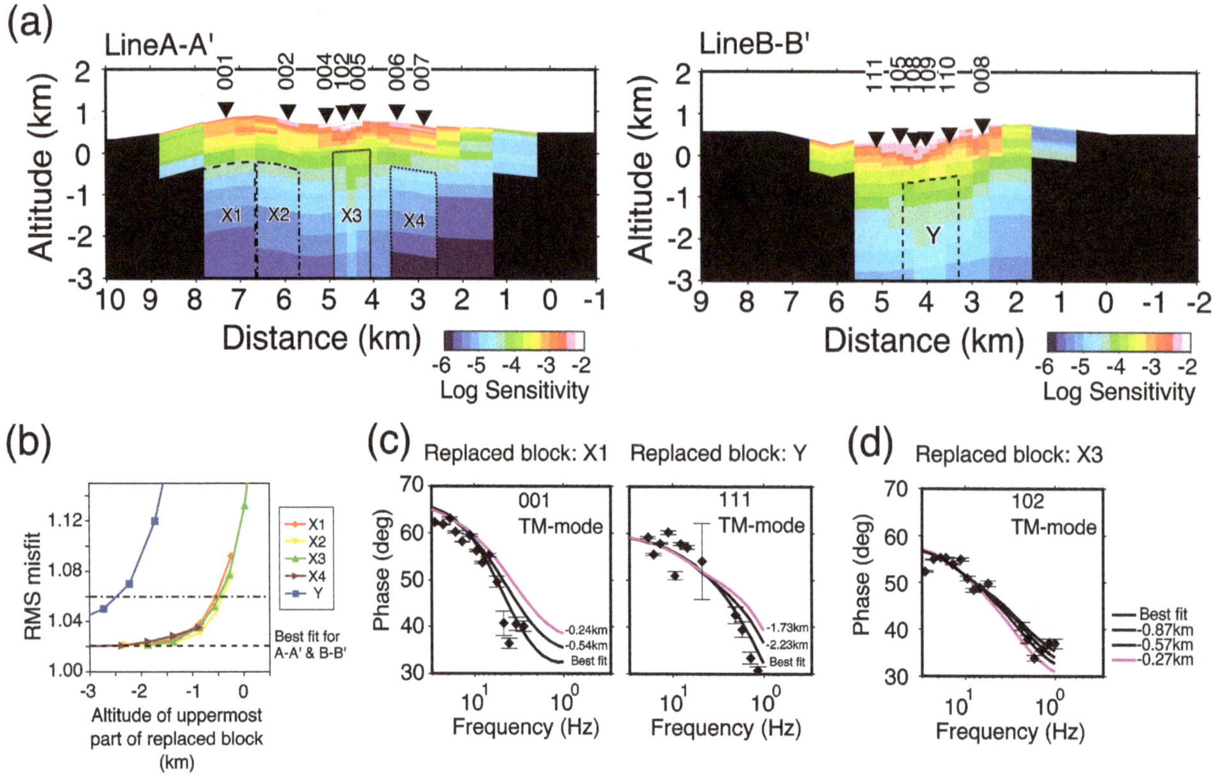

Figure 5 Sensitivity tests. (a) Results of linear sensitivity analysis. Site numbers are indicated above black triangles. **(b)** Variation in RMS misfits by further sensitivity analyses for each case. **(c)** Sensitivities of the TM phase at sites 001 and 111 by further sensitivity analysis for the cases where blocks X1 and Y are replaced. **(d)** Sensitivities of the TM phase at site 102 by further sensitivity analysis for the case where block X2 is replaced.

RMS misfits were examined by replacing the data for the regions with a conductive body. Region X3, which has low resistivities with values less than 10 Ωm, was replaced with a resistive body. Figure 5b shows the RMS misfits when a 10-Ωm body was substituted for X1, X2, and X4 and a 100-Ωm body was substituted for X3. The horizontal axis indicates the altitude of the uppermost part of the replacing body. The RMS misfits increase with the elevation of the substitute bodies, except in region X4 where there was no significant increase of the RMS misfit, suggesting that region X4 has little sensitivity. The change of TM phase at site 001 (Figure 5c) shows that the calculated sounding curve does not fit well within 3 to 7 Hz at an altitude of -0.54 km, as the RMS misfit is approximately 1.06. This is indicated by the dotted and dashed line in Figure 5b. Thus, RMS misfits greater than 1.06 indicate significant discrepancies from the best-fit model. We thus concluded that computational blocks with sensitivities of at least $10^{-4.5}$ can be supported by the observed data.

Region X3 has relatively high sensitivities at deeper depths compared to the other regions (Figure 5a). Its RMS misfit increased significantly when the block was replaced at the same altitude as the other cases (−0.27 km; Figure 5b). This occurred because of the unstable soundings at low frequencies at site 102. Figure 5d shows the change in the TM phase at site 102 that occurs when region X3 is replaced with a 100-Ωm body. Although the modified soundings show rough changes over the range of the uncertainty of the data for altitudes between −0.87 km and −0.57 km, they are still unstable. Taking into account the changes in RMS misfits, we consider the structure beneath the shallow conductive layer at sites 004, 102, and 005 to be valid for the computational block at altitudes from −0.27 to −0.57 km.

Case 2: region Y

Region Y's resistivity is high, being more than 100 Ωm. Figures 5b and c show the changes in RMS misfit and the phase sensitivities of the phase when region Y is replaced with a 10 Ωm body. As the elevation of the replacing body increases, the RMS misfit increases significantly, and the fit of the calculated to observed soundings worsens at 1 to 3 Hz, as described in case 1. Given this information, we consider that computational blocks with sensitivities of at least $10^{-4.5}$ can be supported.

2-D resistivity structure

Figure 6 shows the resistivity structures of the Chishinshan area (line A-A') and the Matsao–Da-you-keng area

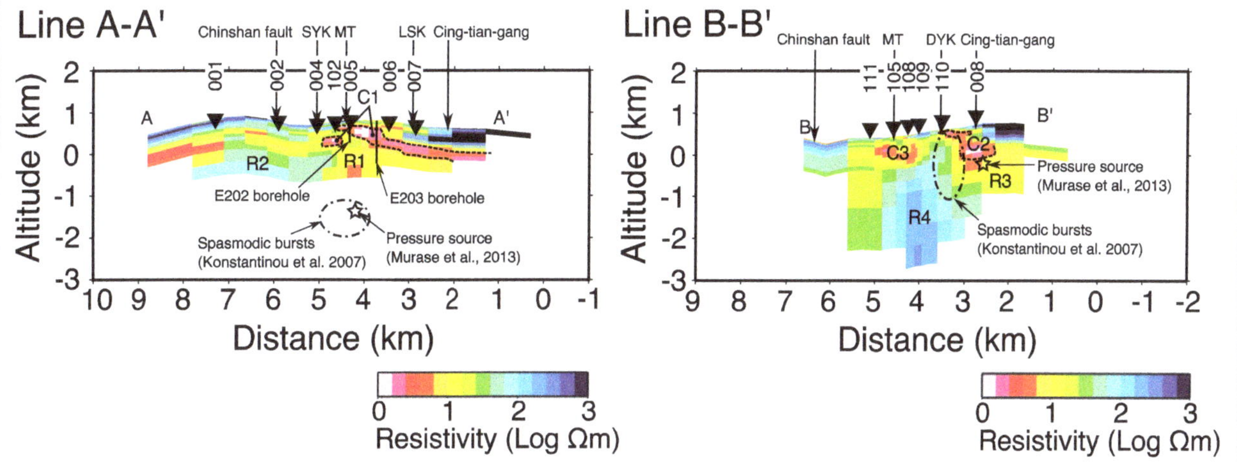

Figure 6 Resistivity structures along lines A-A' and B-B', after removal of the low-sensitivity part. Site numbers are indicated above black triangles. Locations of pressure sources (Murase et al. 2013) and spasmodic bursts (Konstantinou et al. 2007) are shown respectively as stars and ellipses in the structure along lines A-A' and B-B'. The solid lines represent the trajectories of boreholes E202 and E203 drilled by MRSO (1970).

(line B-B') after removing the low-sensitivity regions defined in the previous section.

The northern half of line A-A' has a resistive surface of several hundred to one thousand ohm meters with a thickness of one hundred to several hundreds of meters. The southern half has a surface with a resistivity of 10 to 30 Ωm, except at its southern end, where the resistivity is greater than 100 Ωm. These features are consistent with the results from previous electrical resistivity surveys (DC surveys) by the MRSO (1969, 1970, 1971, 1973) and VLF-MT surveys by Kagiyama et al. (2010). Two discrete conductive regions with resistivities less than 3 Ωm are present at depths of several hundred meters beneath Mt. Chishinshan (C1 in Figure 6). These were also suggested by the apparent resistivity distribution obtained from the MRSO DC surveys (1969, 1970, 1971, 1973).

Figure 7a shows the resistivities along borehole E203, shown in Figure 6, one of the MRSO drilling sites (1970). Red lines represent the resistivities obtained in the present study, while black represent those from well logging (MRSO 1970). The latter shows resistivities of a few Ωm at depths from 200 to 670 m with some fluctuations, increasing to several tens of ohm meters at depths of 670 to 1,000 m. The inferred resistivities around the borehole can reproduce the low resistivities from well logging quite well, although the inferred values for the resistive section of 670 to 1,000 m in depth are slightly lower than those from well logging. From these results, we determined that our method produces a reliable resistivity structure for the Chishinshan area from the surface to a depth of approximately 1 km.

Line B-B', shown in Figure 6, has a surface with a resistivity of generally less than a few tens of ohm meters . In particular, the surface at site 110 has a resistivity of less than 3 Ωm, corresponding to the Da-you-keng fumarolic area. There are two conductive regions at depths of a few hundred meters to 1 km beneath the Matsao hot spring area (C3 in Figure 6) and Cing-tian-gang (C2). They are connected to the conductive surface around the Da-you-keng fumarolic area, overlying a resistive body of about a few hundred ohm meters . Two separate conductive regions are suggested by the distribution of apparent resistivity from the MRSO DC survey (1969, 1970, 1971, 1973). Thus, by comparing the inferred results with previous work, the resistivity structure obtained along line B-B' is likewise believed to be reliable.

Discussion

This section attempts to interpret the resistivity structures obtained in this study, based mainly on reports from geothermal drillings by the MRSO (1969, 1970, 1971, 1973) and evidence from geochemistry (Ohba et al. 2010; Ohsawa et al. 2013), seismology (Konstantinou et al. 2007), and leveling (Murase et al. 2013). For easy reference, we define the following ranges of resistivity: extremely low resistivity (less than 3 Ωm), low resistivity (3 to 10 Ωm), relatively low resistivity (10 to 30 Ωm), relatively high resistivity (30 to 100 Ωm), and high resistivity (greater than 100 Ωm).

Chishinshan area

Low-permeability cap inferred from extremely low resistivity upper layer

As described in the previous section, the foot of Mt. Chishinshan has an extremely low resistivity layer a few hundred meters thick near the surface, as shown by C1 in Figure 6. This layer corresponds roughly to a region with a temperature of 100°C to 200°C, as shown in Figure 7b,

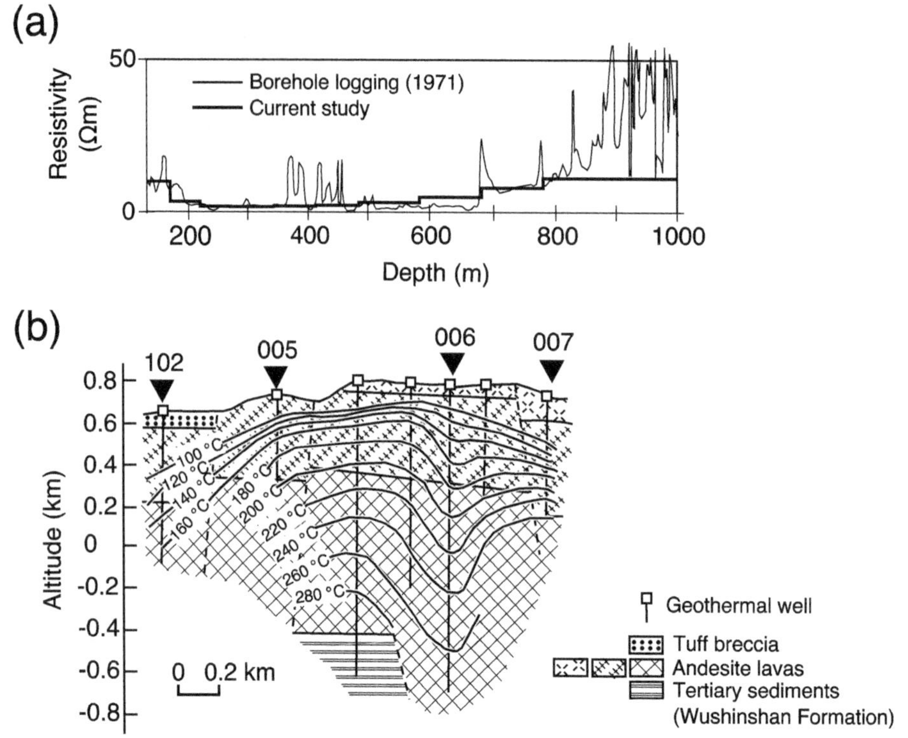

Figure 7 Comparison between inverted results and in-situ measurement. (a) Resistivities along borehole E203, shown in Figure 6. Red and black lines represent the resistivities obtained by this study and by well logging (MRSO 1970), respectively. **(b)** Geologic section along sites 102, 005, 006, and 007 at the foot of Mt. Chishinshan (MRSO 1973). Red lines represent the isotherm obtained from temperature logging of the boreholes (MRSO 1973). The locations of the boreholes are shown by the squares in Figure 1.

as obtained from temperature loggings (MRSO 1973). According to the drilling report, the 100°C to 200°C region is rich in smectite, but smectite is not present below depths corresponding to temperatures greater than 220°C (MRSO 1973). In general, smectite is stable at temperatures less than 180°C to 200°C (e.g., Pytte and Reynolds 1989; Anderson et al. 2000; Komori et al. 2013b). This significantly decreases both the resistivity and permeability of rock matrices (e.g., Ogawa et al. 1998; Revil and Cathles 1999; Revil et al. 2002; Komori et al. 2013b). Based on these earlier studies, layer C1 appears to be rich in smectite and could behave as a low-permeability cap to prevent hydrothermal fluids from easily discharging.

Layer C1 overlies the low to relatively low resistivity region R1. Beneath this region, there are 'spasmodic bursts' of clustered seismicity inferred by the seismological observations of Konstantinou et al. (2007) and a pressure source inferred by Murase et al. (2013)'s precise leveling surveys, as shown in Figure 6. The authors consider the seismic activity and crustal deformation to be associated with the flow of hydrothermal fluids. Unfortunately, the present study could not reveal a corresponding resistivity structure with high precision. We can, at least, consider that upwelling hydrothermal fluids, supplied from a deeper part, are maintained in region R1, covered by the low-permeability cap C1. This is a typical interpretation of resistivity structures obtained for shallow hydrothermal systems (e.g., Ogawa et al. 1998; Nurhasan et al. 2006; Komori et al. 2013a). In addition, the MRSO (1973) performed well tests at borehole E202, which reaches to R1 as shown in Figure 6 and extracted hydrothermal fluids with a steam ratio greater than 80%. This suggests that region R1 is dominated by vapors. This structure is considered to extend from Mt. Chishinshan toward Leng-shuei-keng and Cing-tian-gang.

Hydrothermally inactive area in the northern half of the Chishinshan area

In contrast to the area beneath Mt. Chishinshan, the northern half of line A-A' has no extremely low-resistivity layer like layer C1. This suggests that there is no structural feature that confines vapor-rich fluids as described previously. Actually, there are neither fumaroles nor temperature anomalies in this area. One borehole penetrated to a depth of 200 m (shown by rhombus 'p' in Figure 1). It has a bottom-hole temperature almost identical to the surface temperature (17°C: CGS 2009). Based on these factors, no hydrothermal activity appears to be developed in the northern half of the Chishinshan area.

This area does have two relatively high to high resistivity regions at the surface and at greater depths (R2). The high-resistivity surface corresponds to lava flows with little or no hydrothermal alteration (MRSO 1970; CGS 2009). MRSO (1970) reported that tertiary sediments are exposed to the surface around Beitou and Liu-huang-ku, a few kilometers southwest of this area. MRSO's comparison of the results of drilling and a DC survey in the above areas (MRSO 1969) indicates that weakly altered sediments correspond to an apparent resistivity of approximately 100 Ωm, while strongly altered sediments show resistivities less than 10 Ωm. This supports R2 as consisting of weakly altered tertiary sediments. Alternatively, R2 might be solidified igneous rocks, as suggested by the quite high resistivity of weakly altered lava observed at the surface. At a minimum, we can consider that resistive body R2 corresponds to weakly altered materials. There is also a possibility that region R2 might represent an extensive dry region, due to an absence of conductive fluids (e.g., Rinaldi et al. 2011), but this can be rejected based on the lack of a significant Bouguer anomaly, which is indicative of low-density material (Hsieh et al. personal communication).

Da-you-keng area

The Da-you-keng area has an intense fumarole with sulfur precipitation and hot spring water with a temperature of 100°C to 120°C (CGS 2009; Lee et al. 2005, 2008). extremely low-resistivity region, shown as C2 in Figure 6, reaches approximately 1 km near the surface of this area, extending relatively deeper beneath Cing-tian-gang. At a depth of 500 m beneath Cing-tian-gang, corresponding to the uppermost part of region C2, a temperature of 100°C has been recorded by a thermal-monitoring well, shown by rhombus 'q' in Figure 1 (CGS 2009). Based on considerations described in the section on the Chishinshan area, region C2 may correspond to a low-permeability cap due to a large amount of smectite.

The underlying low to relatively low resistivity region R3 beneath Cing-tian-gang corresponds to a pressure source inferred by Murase et al. (2013). Based on the above previous studies, hydrothermal fluids are considered to be maintained in region R3, and these fluids could ascend toward Da-you-keng along paths controlled by the low-permeability cap. The ascent of fluids could induce the seismic burst events near the Da-you-ken-Cing-tian-gang area as shown in Figure 6 (Konstantinou et al. 2007). According to Ohba et al. (2010), fumarolic gases from Da-you-keng originate in as a mixture of magmatic vapors and meteoric vapors, which suggests that the hydrothermal reservoir maintains vapor-rich fluids. Therefore, we consider R3 to be dominated by vapors, like region R1 of the Chishinshan area.

This study also shows that the Da-you-keng fumarolic area has a relatively thin conductive surface, compared to the Chishinshan area. As mentioned previously, smectite composing the conductive layer is stable at temperatures of less than 180°C to 200°C. The lack of a thick conductive layer is considered to be due to a high-temperature gradient, suggesting the presence of high-temperature fluids in the shallow subsurface of the fumarolic area. This is consistent with the dominance of vapor-rich fluids suggested by Ohba et al. (2010).

Matsao area

In the Matsao area, hot spring water with a temperature of 30°C to 50°C discharges CGS 2009; Ohsawa et al. 2013), and there is no fumarolic area around the AMT observation points. According to sulfur isotope analysis of hot spring water from the TVG (Ohsawa et al. 2013), hot springs in the Matsao area are formed by mixing meteoric water with vapors separated from deep magmatic hydrothermal fluids, which then flow at shallow depths from Mt. Chishinshan along its topography. The Matsao area has an upper region with resistivities of less than 10 Ωm at depths of 0 to 1 km, shown in C3, and an underlying high-resistivity region R4. Based on Ohsawa et al. (2013), region C3 could correspond to the shallow flow system of the Matsao area's hydrothermal fluid. Region R4 could correspond to less-altered materials, such as sediments or solidified igneous rocks, as described in the previous section. This is supported by the lack of a significant Bouguer anomaly around this area (Hsieh et al. personal communication).

Inferred hydrothermal system of the TVG and its implication for potential volcanic hazards

We discussed the relation between the resistivity structure and other geophysical and geochemical data above. On basis of the insights thus obtained, in this section, we propose a better description of the hydrothermal system of the TVG, as shown in Figure 8. Beneath Mt. Chishinshan, vapor-rich hydrothermal fluids are supplied from the deeper part. The fluids are maintained below the clay-rich, low-permeability cap, extending horizontally beneath the foot of Mt. Chishinshan. Some vapors are discharged through breaks or along the end of the clay-rich layer, developing fumarolic areas like Siao-you-keng. Further on, the vapor mixes with groundwater at the shallow subsurface, forming so-called 'steam-heated thermal waters', which develop the shallow flow system of hydrothermal fluids in the Matsao area. The fumaroles at Da-you-keng area, on the other hand, may be supplied from the deeper part beneath Cing-tian-gang and vapor-rich fluids maintained in the region, covered by a clay-rich cap.

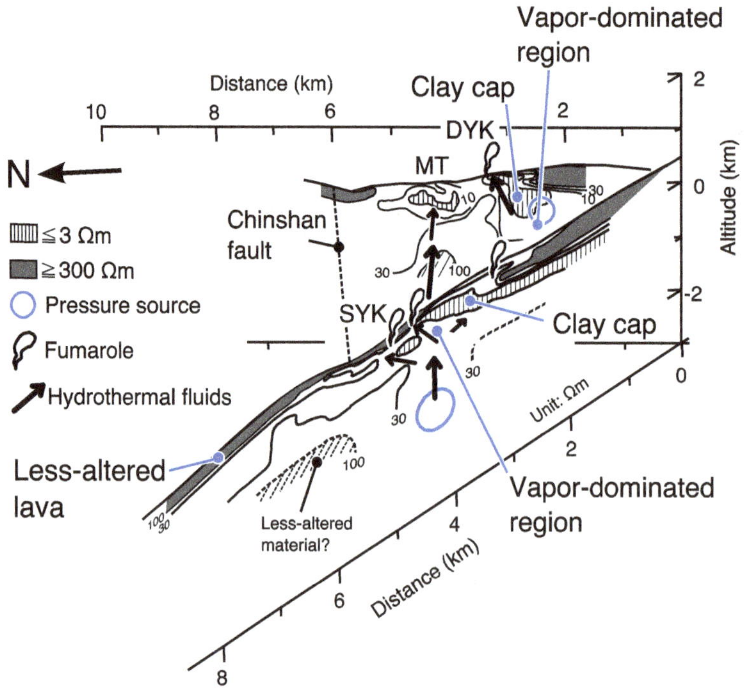

Figure 8 Conceptual model of the hydrothermal system of the Tatun Volcano Group. As inferred by interpretation of the resistivity structure and other geochemical and geophysical evidence.

We estimate that an extensive vapor-bearing region could be situated beneath the foot of Mt. Chishinshan, Da-you-keng, and Cing-tian-gang, suggesting the potential for future phreatic explosions caused by a rapid increases in the internal fluid pressure due to intense injection of volcanic fluids (e.g., Voight and Elsworth 1997; Nakada et al. 1999). A phreatic eruption actually occurred as the latest activity at Mt. Chishinshan *circa* 6,000 years ago (Belousov et al. 2010). The data also suggest that the vapor may have been maintained for several thousand years. These factors point to the need for future work to detect signals related to phreatic explosions. Precursors might be detected by the following ongoing observations: temporal variations of chemical and isotopic composition (Lee et al. 2008), microgravity changes (Hwang et al. 2012), changes in precise leveling surveys (Murase et al. 2013), and temporal changes in tiltmeter readings (Lin, personal communication).

This study, however, did not reveal magmas suggested by a deeply distributed aseismic region as in Konstantinou et al. (2007), and their interrelation with the two pressure sources, due to a lack of data with low-frequency ranges. In the future, broadband MT surveys will reveal the deeper structural features to further elucidate the conditions from the viewpoint of hydrothermal fluid flow (i.e., Aizawa et al. 2013).

Conclusions

This study revealed the resistivity structure associated with hydrothermal activity in the Tatun Volcano Group using AMT surveys to investigate the present state of this system. The structure obtained here can be consistently explained by hot spring water and fumarolic gas formation processes as indicated by geochemical work. Its association with seismicity and pressure sources supports the presence of paths of hydrothermal fluid flow. This study also inferred the presence of an extensive vapor-bearing region covered by a low-permeability cap beneath Mt. Chishinshan and Da-you-keng-Cing-tian-gang. This suggests the possibility of future phreatic explosions around the area. The results indicate that an integrated conceptual model for the hydrothermal system at the TVG can provide reliable constraints for its potential volcanic hazards.

Competing interests

This study was partly supported by Grants-in-Aid for Scientific Research (No. 21403003 and No. 23310120, T. Kagiyama) from the Ministry of Education, Culture, Sports, Science and Technology of Japan, and by a thematic research program application (AS-101-TP-A05, B.F. Chao) from the Academia Sinica, Taiwan. The authors declare that they have no competing interests.

Authors' contributions

SK organized the AMT surveys in 2012, participated in the surveys in the Mt. Chishinshan area, and carried out all of the data analyses. MU organized the surveys in 2011 and dealt with instrument logistics. RY and WK participated in the surveys in 2011 and were involved in interpreting the inverted results. RY was also involved with the data processing. TK, HI, and HTC participated in the surveys in 2012 and were involved in interpreting the inverted results. CHC and BFC arranged the entire survey plan in cooperation with Yanminshan National Park, and contributed to improving the manuscript. All authors have read and approved the final manuscript.

Acknowledgments
We would like to thank Yangmingshan National Park for arranging the surveys and M. Murase (Nihon University) for providing data about leveling surveys. We also thank Y.H. Lee, R. Sih, G.S.K. Ma, C.C. Lin, W. Minju (Academia Sinica), P. Pong (National Taiwan University), S. Yoshikawa, T. Asano, and N. Tokumoto (Kyoto Univ.) for supporting the observations. We thank the Central Geological Survey for providing data about the thermal monitoring project. We appreciate T. Lee, T. F. Yui, D. C. Lee, K. L. Wang, Y Iizuka, C. Y. Chen, C. H. Lin, X. E. Lei (Academia Sinica), C. Hwang (National Chiao Tung University), T. F. Yang, S.R. Song (National Taiwan University), K. Takemura, S. Ohsawa, and K. Matsuo (Kyoto University) for their valuable discussions. We also appreciate G. Zellmer (Massey University) for valuable comments and for improving the manuscript. The manuscript was critically reviewed by two anonymous reviewers. We appreciate H. Shimizu for the editorial support. Some figures were made using the GMT program (Wessel and Smith 1998).

Author details
[1]Institute of Earth Sciences, Academia Sinica, 128 Academia Road, Section 2, Nankang, Taipei 115, Taiwan. [2]Aso Volcanological Laboratory, Kyoto University, Minamiaso, Kumamoto 869-1404, Japan. [3]Institute of Oceanography, National Taiwan University, No. 1, Sec. 4, Roosevelt Road, Taipei 10617, Taiwan. [4]Research Center for Earthquake Prediction, Disaster Prevention Research Institute, Kyoto University, Gokasho, Uji, Kyoto 611-0011, Japan. [5]Volcanic Fluid Research Center, Tokyo Institute of Technology, 2-12-1 Ookayama, Meguro, Tokyo 152-8551, Japan. [6]Present address: Geological Survey of Japan (GSJ), AIST, Central 7, 1-1-1 Higashi, Tsukuba, Ibaraki 305-8567, Japan.

References
Aizawa K, Ogawa Y, Ishido T (2009) Groundwater flow and hydrothermal systems within volcanic edifices: delineation by electric self-potential and magnetotellurics. J Geophys Res 114(B1):B01208. doi:10.1029/2008JB005910
Aizawa K, Koyama T, Hase H, Uyeshima M, Kanda W, Utsugi M, Yoshimura R, Yamaya Y, Hashimoto T, Yamazaki K, Komatsu S, Watanabe A, Miyakawa K, Ogawa Y (2013) Three-dimensional resistivity structure and magma plumbing system of the Kirishima Volcanoes as inferred from broadband magnetotelluric data. J Geophys Res 119:198-215. doi:10.1002/2013JB010682
Anderson E, Crosby D, Ussher G (2000) Bull's-eye!-simple resistivity imaging to reliably locate the geothermal reservoir. In: Proceedings of the world geothermal congress 2000. Beppu-Morioka
Belousov A, Belousova M, Chen C-H, Zellmer GF (2010) Deposits, character and timing of recent eruptions and gravitational collapses in Tatun Volcanic Group, Northern Taiwan: Hazard-related issues. J Volcanol Geotherm Res 191:205-221. doi:10.1016/j.jvolgeores.2010.02.001
Bibby HM, Caldwell TG, Brown C (2005) Determinable and non-determinable parameters of galvanic distortion in magnetotellurics. Geophys J Int 163:915-930. doi:10.1111/j.1365-246X.2005.02779.x
Cagniard L (1953) Basic theory of the magnetotelluric method of geophysical prospecting. Geophysics 18:605-635. doi:10.1190/1.1437915
Caldwell TG, Bibby HM, Brown C (2004) The magnetotelluric phase tensor. Geophys J Int 158:457-469. doi:10.1111/j.1365-246X.2004.02281.x
Central Geological Survey (CGS) (2009) 2nd period of investigation and monitoring of particular geological hazard at Taipei district: investigation and monitoring of geological hazard and volcanic activity (2/4). Central Geological Survey, Taipei (in Chinese)
Chang C-P, Chang T-Y, Angelier J, Kao H, Lee J-C, Yu S-B (2003) Strain and stress field in Taiwan oblique convergent system: constraints from GPS observation and tectonic data. Earth Planet Sci Lett 214:115-127. doi:10.1016/S0012-821X(03)00360-1
Chen C-H (1970) Geology and geothermal power potential of the Tatun volcanic region. Geothermics 2:1134-1143. http://dx.doi.org/10.1016/0375-6505(70)90425-6
Chen C-S (2009) Investigation of potential magma chambers beneath the Tatun volcanic area: preliminary results of magnetotelluric survey. Journal of National Park 19:1-10. (in Chinese with English abstract)
Chen C-H, Lin S-B (2002) Eruptions younger than 20 ka of the Tatun Volcano Group as viewed from the sediments of the Sungshan Formation in Taipei Basin. West Pac Earth Sci 2:191-204
Chen C-H, Wu Y-J (1971) Volcanic geology of the Tatun geothermal area, northern Taiwan. Proc Geological Soc China 14:5-20
Chen C-S, Chen C-C, Chou K (1998) Deep electrical structure of Taiwan as inferred from magnetotelluric observations. Terr Atmos Ocean Sci 9:51-68
Chen C-T, Lee J-C, Chan Y-C, Lu C-Y (2010) Growth normal faulting at the western edge of the metropolitan Taipei basin since the Last Glacial Maximum, northern Taiwan. Terr Atmos Ocean Sci 21:409-428. doi:10.3319/TAO.2009.11.13.01(TH)
Gamble TD, Goubau WM, Clarke J (1979) Magnetotellurics with a remote magnetic reference. Geophysics 44:53-68. doi:10.1190/1.1440923
Hwang C, Chao B, Chen C-C, Lee C-W, Hsieh W-C, Huang J-F, Kao R (2012) Static and time-varying gravity observations at the Tatun Volcano Group in northern Taiwan. In: Abstracts of AOGS-AGU (WPGM) joint assembly. Resorts world convention centre, Sentosa Island
Jiracek GR (1990) Near-surface and topographic distortions in electromagnetic induction. Surv Geophys 11:163-203
Juang W-S (1993) Diversity and origin of quaternary basaltic magma series in northern Taiwan. Bull Natl Museum Nat Sci 4:125-165
Kagiyama T, Utsugi M, Chen C-H, Miyabuchi Y (2010) Conductivity distribution of the surface layer around Tatun Volcanic group, Taiwan. In: Abstracts of Japan geoscience union meeting 2010, Makuhari Messe international conference hall, Chiba
Kanda W, Tanaka Y, Utsugi M, Takakura S, Hashimoto T, Inoue H (2008) A preparation zone for volcanic explosions beneath Naka-dake crater, Aso volcano, as inferred from magnetotelluric surveys. J Volcanol Geotherm Res 178(1):32-45. doi:10.1016/j.jvolgeores.2008.01.022
Komori S, Kagiyama T, Hoshizumi H, Takakura S, Mimura M (2010) Vertical mapping of hydrothermal fluids and alteration from bulk conductivity: simple interpretation on the USDP-1 site, Unzen Volcano, SW Japan. J Volcanol Geotherm Res 198:339-347. doi:10.1016/j.jvolgeores.2010.09.019
Komori S, Kagiyama T, Utsugi M, Inoue H, Azuhata I (2013a) Two-dimensional resistivity structure of Unzen Volcano by AMT and MT surveys. Earth Planets Space 65:759–766. doi:10.5047/eps.2012.10.005
Komori S, Kagiyama T, Takakura S, Ohsawa S, Mimura M, Mogi T (2013b) Effect of the hydrothermal alteration on the surface conductivity of rock matrix: comparative study between relatively-high and low temperature hydrothermal systems. J Volcanol Geotherm Res 264:164-171. doi:10.1016/j.jvolgeores.2013.08.009
Konstantinou KI, Lin C-H, Liang W-T (2007) Seismicity characteristics of a potentially active quaternary volcano: The Tatun Volcano Group, northern Taiwan. J Volcanol Geotherm Res 160:300-318. doi:10.1016/j.jvolgeores.2006.09.009
Konstantinou KI, Lin C-H, Liang W-T, Chan Y-C (2009) Seismogenic stress field beneath the Tatun Volcano Group, northern Taiwan. J Volcanol Geotherm Res 187:261-271. doi:10.1016/j.jvolgeores.2009.09.011
Lee H-F, Yang TF, Lan T-F, Song S-R, Tsao S (2005) Fumarolic gas composition of the Tatun Volcano Group, northern Taiwan. Terr Atmos Ocean Sci 16:843-864
Lee H-F, Yang TF, Lan T-F, Chen C-H, Song S-R, Tsao S (2008) Temporal variations of gas compositions of fumaroles in the Tatun Volcano Group, northern Taiwan. J Volcanol Geotherm Res 178:624-635. doi:10.1016/j.jvolgeores.2008.06.005
Lin C-H, Konstantinou KI, Liang W-T, Pu H-C, Lin Y-M, You S-H, Huang Y-P (2005) Preliminary analysis of volcanoseismic signals recorded at the Tatun Volcano Group, northern Taiwan. Geophys Res Lett 32:L10313. doi:10.1029/2005GL022861
Lin C-H, Yang TF, Leu P-L, Chang W-Y (2012) Establishment of Taiwan Volcano Observatory at Tatun. In: Abstracts of AOGS-AGU (WPGM) joint assembly. Resorts world convention centre, Sentosa Island
Liu C-M, Song S-R, Chen Y-L, Tsao S (2011) Characteristics and origins of hot springs in the Tatun Volcano Group in northern Taiwan. Terr Atmos Ocean Sci 22:475-489. doi:10.3319/TAO.2011.05.25.01(TT)
Mining Research and Service Organization (MRSO) (1969) MRSO Report 90: the geothermal exploration of the, Tatun Volcano Group (I). Mining Research and Service Organization, Taipei (in Chinese)

Mining Research and Service Organization (MRSO) (1970) MRSO Report 102: the geothermal exploration of the Tatun Volcano Group (II). Mining Research and Service Organization, Taipei (in Chinese)
Mining Research and Service Organization (MRSO) (1971) MRSO Report 111: the geothermal exploration of the, Tatun Volcano Group (III). Mining Research and Service Organization, Taipei (in Chinese)
Mining Research and Service Organization (MRSO) (1973) MRSO Report 126: the geothermal exploration of the, Tatun Volcano Group (IV). Mining Research and Service Organization, Taipei (in Chinese)
Murase M, Lin C-H, Kimata F, Mori H, Suzuki A (2013) Estimated pressure source and vertical deformation in Tatun volcano group, Taiwan, detected by precise leveling during 2006-2013. In: Abstracts of IAVCEI 2013 scientific assembly, Kagoshima prefectural citizens exchange center & Kagoshima citizens social support plaza, Kagoshima
Nakada S, Shimizu H, Ohta K (1999) Overview of the 1990-1995 eruption at Unzen volcano. J Volcanol Geotherm Res 89:1-22. http://dx.doi.org/10.1016/S0377-0273(98)00118-8
Nurhasan, Ogawa Y, Ujihara N, Tank SB, Honkura Y, Onizawa S, Mori T, Makino M (2006) Two electrical conductors beneath Kusatsu-Shirane volcano, Japan, imaged by audiomagnetotellurics, and their implications for the hydrothermal system. Earth Planets Space 58: 1053-1059
Ogawa Y, Uchida T (1996) A two-dimensional magnetotelluric inversion assuming Gaussian static shift. Geophys J Int 126:69-76. doi:10.1111/j.1365-246X.1996.tb05267.x
Ogawa Y, Matsushima N, Oshima H, Takakura S, Utsugi M, Hirano K, Igarashi M, Doi T (1998) A resistivity cross-section of Usu volcano, Hokkaido, Japan, by audiomagnetotellurics soundings. Earth Planets Space 50:339-346
Ohba T, Sawa T, Taira N, Yang TF, Lee H-F, Lan T-F, Ohwada M, Morikawa N, Kazahaya K (2010) Magmatic fluids of Tatun volcanic group, Taiwan. Appl Geochem 25:513-523. http://dx.doi.org/10.1016/j.apgeochem.2010.01.009
Ohsawa S, Lee H-F, Liang B, Komori S, Chen C-H, Kagiyama T (2013) Geochemical characteristics and origins of acid hot spring waters in Tatun Volcanic Group, Taiwan. J Hot Spring Sci 62:282-293. (in Japanese with English abstract)
Pytte AM, Reynolds RC (1989) The thermal transformation of smectite to illite. In: Naeser ND, McCulloh TH (eds) Thermal history of Sedimentary Basins: Methods and Case Histories. Springer-Verlag, New York, pp 133-140
Revil A, Cathles III LM (1999) Permeability of shaly sands. Water Resour Res 35(3):651-662. doi:10.1029/98WR02700
Revil A, Cathles III LM, Losh S (1998) Electrical conductivity in shaly sands with geophysical applications. J Geophys Res 103(B10):23925-23936. doi:10.1029/98JB02125
Revil A, Hermitte D, Spangenberg E, Cochemé JJ (2002) Electrical properties of zeolitized volcaniclastic materials. J Geophys Res 107(B8):2168. doi:10.1029/2001JB000599
Revil A, Finizola A, Ricci T, Delcher E, Peltier A, Barde-Cabusson S, Avard G, Bailly T, Bennati L, Byrdina S, Colonge J, Di Gangi F, Douillet G, Lupi M, Letort J, Tsang Hin Sun E (2011) Hydrogeology of Stromboli volcano, Aeolian Islands (Italy) from the interpretation of resistivity tomograms, self-potential, soil temperature and soil CO_2 concentration measurements. Geophys J Int 186:1078-1094. doi:10.1111/j.1365-246X.2011.05112.x
Rinaldi AP, Todesco M, Vandemeulebrouck J, Revil A, Bonafede M (2011) Electrical conductivity, ground displacement, gravity changes, and gas flow at Solfatara crater (Campi Flegrei caldera, Italy): results from numerical modeling. J Volcanol Geotherm Res 207:93-105. doi:10.1016/j.jvolgeores.2011.07.008
Rontogianni S, Konstantinou KI, Lin C-H (2012) Multi-parametric investigation of the volcano-hydrothermal system at Tatun Volcano Group, Northern Taiwan. Nat Hazards Earth Syst Sci 12:2259-2270. doi:10.5194/nhess-12-2259-2012
Schwalenberg K, Rath V, Haak V (2002) Sensitivity studies applied to a two-dimensional resistivity model from the Central Andes. Geophys J Int 150:673-686. doi:10.1046/j.1365-246X.2002.01734.x
Shyu JBH, Sieh K, Chen Y-G, Liu C-S (2005) Neotectonic architecture of Taiwan and its implications for future large earthquakes. J Geophys Res 110:B08402. doi:10.1029/2004JB003251
Simpson F, Bahr K (2005) Practical magnetotellurics. Cambridge University Press, Cambridge
Tsai Y-W, Song S-R, Chen H-F, Li S-F, Lo C-H, Lo W, Tsao S (2010) Volcanic stratigraphy and potential hazards of the Chihsingshan volcano subgroup in the Tatun Volcano Group, northern Taiwan. Terr Atmos Ocean Sci 21:587-598. doi:10.3319/TAO.2010.02.22.03(TH)
Tsao S (1994) K-Ar age determination of volcanic rocks from the Tatun Volcano Group. Bull Central Geological Surv 9:137-154. (in Chinese)
Utsugi M, Kagiyama T, Chen C-H, Kanda W, Yoshimura R, Asano T, Tokumoto N, Inoue H, Yoshikawa S (2012) Audio frequency magneto-telluric survey on Tatun Volcanic Group, Taiwan In: Abstracts of Japan geoscience union meeting 2012, Makuhari Messe international conference hall, Chiba
Voight B, Elsworth D (1997) Failure of volcano slopes. Géotechnique 47:1-31
Vozoff K (1991) The magnetotelluric method. In: Nabighian MN (ed) Electromagnetic Methods in Applied Geophysics, Society of Exploration Geophysicists, Oklahoma, pp 641-711
Waxman MH, Smits LJM (1968) Electrical conductivities in oil-bearing shaly sands. Soc Pet Eng J 8:107-122
Wang W-H, Chen C-H (1990) The volcanology and fission track age dating of pyroclastic deposits in Tatun Volcano Group, northern Taiwan. Acta Geological Taiwan 28:1-30
Wang K-L, Chung S-L, Chen C-H, Shinjo R, Yang TF, Chen C-H (1999) Post-collisional magmatism around northern Taiwan and its relation with opening of the Okinawa Trough. Tectonophysics 308:363-376
Wen H-Y, Yang TF, Lan T-F, Lee H-F, Lin C-H (2013) Spatial and temporal variations of soil CO2 flux in geothermal areas of the Tatun Volcano Group, Northern Taiwan. In: Abstracts of IAVCEI 2013 scientific assembly, Kagoshima prefectural citizens exchange center & Kagoshima citizens social support plaza, Kagoshima
Wessel P, Smith WHF (1998) New, improved version of generic mapping tools released. Eos Trans AGU 79(47):579
Wu C-C, Cheng C-S, Lee M-Y, Hong C-H (2009) Numerical simulation of volcanic ash dispersal and fallout from Datun volcano and Kuei-shan-tao by using PUFF model. Meteorol Bull 48:83-99. (in Chinese with English abstract)
Yang TF, Sano Y, Song S-R (1999) $^3He/^4He$ ratios of fumaroles and bubbling gases of hot springs in Tatun Volcano Group, North Taiwan. Il Nuovo Cimento C 22:281-286
Yang TF, Chen C-H, Tien R-L, Song S-R, Liu T-K (2003) Remnant magmatic activity in the Coastal Range of East Taiwan after arc-continent collision: fission-track data and $^3He/^4He$ ratio evidence. Radiation Measurements 36:343-349

2

Rheology and stress in subduction zones around the aseismic/seismic transition

John P. Platt[1*], Haoran Xia[1,2] and William Lamborn Schmidt[1]

Abstract

Subduction channels are commonly occupied by deformed and metamorphosed basaltic rocks, together with clastic and pelagic sediments, which form a zone up to several kilometers thick to depths of at least 40 km. At temperatures above ~ 350 °C (corresponding to depths of > 25–35 km), the subduction zone undergoes a transition to aseismic behavior, and much of the relative motion is accommodated by ductile deformation in the subduction channel. Microstructures in metagreywacke suggest deformation occurs mainly by solution-redeposition creep in quartz. Interlayered metachert shows evidence for dislocation creep at relatively low stresses (8–13 MPa shear stress). Metachert is likely to be somewhat stronger than metagreywacke, so this value may be an upper limit for the shear stress in the channel as a whole. Metabasaltic rocks deform mainly by transformation-assisted diffusional creep during low-temperature metamorphism and, when dry, are somewhat stronger than metachert. Quartz flow laws for dislocation and solution-redeposition creep suggest strain rates of ~ 10^{-12} s^{-1} at 500 °C and 10 MPa shear stress: this is sufficient to accommodate a 100 mm/yr. convergence rate within a 1 km wide ductile shear zone.

The up-dip transition into the seismic zone occurs through a region where deformation is still distributed over a thickness of several kilometers, but occurs by a combination of microfolding, dilational microcracking, and solution-redeposition creep. This process requires a high fluid flux, released by dehydration reactions down-dip, and produces a highly differentiated deformational fabric with alternating millimeter-scale quartz and phyllosilicate-rich bands, and very abundant quartz veins. Bursts of dilational microcracking in zones 100–200 m thick may cause cyclic fluctuations in fluid pressure and may be associated with episodic tremor and slow slip events. Shear stress estimates from dislocation creep microstructures in dynamically recrystallized metachert are ~ 10 MPa.

Keywords: Dynamically recrystallized grain size, Dislocation creep, Pressure solution, Tremor, Slow slip

Introduction

Seismicity along the subduction zone interface at shallow depths transitions downwards into a zone of aseismic creep at depths of 25–40 km (Tichelaar and Ruff 1993). The character and properties of the creeping zone are poorly known: many analyses assume that it is simply a zone of stable slip along the interface (e.g., Holtkamp and Brudzinski 2010), but exhumed rocks from these depths suggest there may be a so-called subduction channel containing metasedimentary and volcanic rocks, which take up much or all of the displacement (Gerya 2002; Warren et al. 2008; Beaumont et al. 2009; Blanco-Quintero et al. 2011; Behr and Platt 2013). The transition between the seismic and aseismic zones is particularly interesting, as in a number of subduction zones this is the source of tectonic tremor and slow slip events (Hirose et al. 1999; Dragert et al. 2001; Obara 2002), the origin of which is uncertain (Fig. 1).

Slow slip events involve geodetically determined surface displacements of a few centimeters over periods of a few days to ~ 1 year (Peng and Gomberg 2010; Gao et al. 2012); these appear to reflect the release of elastic strain by displacement within the subduction zone at rates that are sub-seismic, but one to two orders of magnitude faster than plate motion rates. Slow slip is commonly accompanied by tremor, a non-impulsive seismic signal that may last for periods up to the duration of the slow slip event (Obara 2002; Rogers and Dragert 2003; Schwartz and Rokosky 2007). Tremor is now thought to be made up of concatenated low-frequency earthquakes (LFEs)—low-

* Correspondence: jplatt@usc.edu
[1]Department of Earth Sciences, University of Southern California, Los Angeles, CA 90089-0740, USA
Full list of author information is available at the end of the article

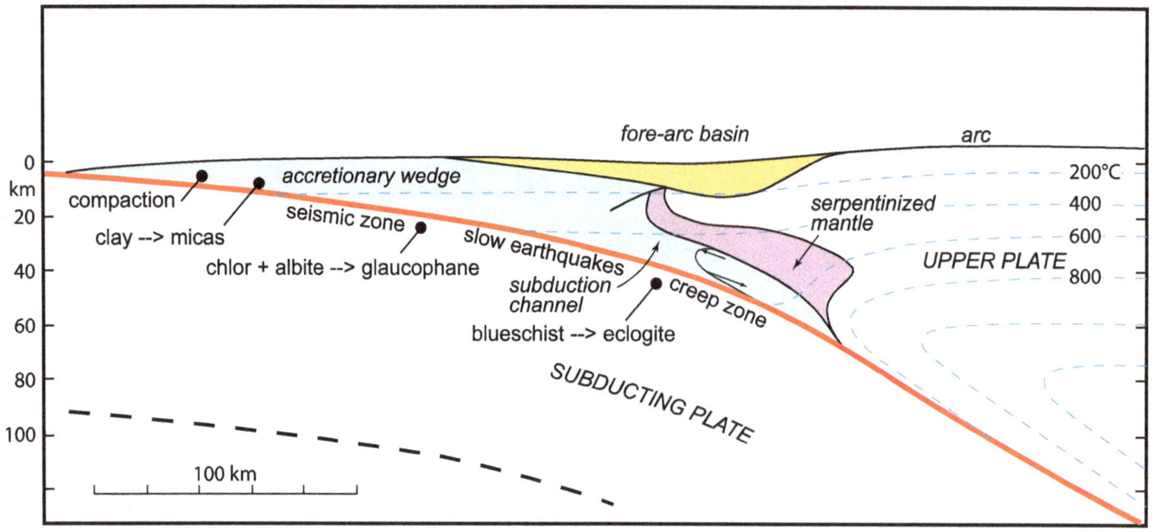

Fig. 1 The seismic-aseismic transition in a subduction zone. Isotherms after Peacock et al. (2011). Slow earthquakes: source region of tectonic tremor and slow slip events. Dehydration processes contributing water to the subduction zone in green. Distribution of serpentinized mantle is schematic

magnitude events with unusually long durations and low frequencies (Shelly et al. 2007; Frank et al. 2016). In this paper, we refer to the spectrum of slow phenomena, such as LFEs, tectonic tremor, and aseismic slow slip events, as slow earthquakes.

Controls on the location of the transition zone remain uncertain. There is general agreement that it does not correspond to a specific temperature (Peacock 2009; Boyarko and Brudzinski 2010), and that pore-fluid pressure is likely to be important (Audet et al. 2009; Peng and Gomberg 2010; Peacock et al. 2011). The seismic-aseismic transition generally occurs in the depth range 25–40 km, corresponding to temperatures in the range 350–500 °C (Peacock et al. 2011), above the lower temperature limit for crystal plasticity in quartz (Hirth et al. 2001).

A key factor controlling the rheology and response of subduction zones is water. Water is released from hydrated mantle in the subducted plate, the subducted ocean floor, and its sedimentary cover; initially by compaction, and then at progressively increasing depth and temperature by metamorphic dehydration reactions, such as the breakdown of clay minerals to micas and chlorite, chlorite and albite to form glaucophane, serpentine minerals to talc and forsterite (Audet et al. 2009), and blueschist-facies assemblages including sodic amphibole and lawsonite to eclogite-facies assemblages including sodic clinopyroxene (omphacite) and garnet (Peacock et al. 2011). Water can facilitate brittle fracture at depths where the confining pressure would normally inhibit it, by reducing the effective pressure and inducing hydraulic fracture. This process is limited, however, because fracture generally induces dilatancy, which in turn reduces the fluid pressure (Peng and Gomberg 2010); this negative feedback may provide an explanation for slow slip events (Segall et al. 2010). Water also reduces the stress required for crystal plasticity (e.g., Holyoke and Kronenberg 2013) and facilitates metamorphic reactions, producing weak hydrous phases.

Interseismic landward motion recorded by continuous GPS above subduction zones is interpreted as the accumulation of elastic stress that is subsequently released either by true seismic events or by slow earthquakes (Dragert et al. 2001). Slow earthquakes are interpreted by most workers to occur on the plate interface (e.g., Beroza and Ide 2010), and the interpreted focal mechanisms for LFEs are consistent with slip along a gently dipping thrust fault (Ide et al. 2007a, b, Shelly et al. 2007, Frank et al. 2013). Locational uncertainties for LFEs are of the order of 5 km, however (Shelly et al. 2007; Brown et al. 2009; Peng and Gomberg 2010), leaving open the possibility that the sources may not be confined to a single discrete slip surface.

The transition from seismic slip to aseismic creep along the subduction zone is likely to involve a progressive increase in off-fault deformation with depth. The role of this off-fault deformation in accommodating plate convergence at the depth of slow earthquakes is neglected in the above interpretations, and there is an open question as to whether off-fault deformation is responsible for slow earthquakes, or for some component of the interseismic landward motion. The discovery by Ide et al. (2007a, b) of separate moment/duration scaling relationships for normal seismic events and slow earthquakes suggests a fundamentally different mechanism for slow events. This may involve a component of hydraulic fracture, as suggested by the fact that they occur in a zone characterized by high Vp/Vs ratios (e.g., Beroza and Ide 2010), indicating high water content. Tremor and slow slip events have also been attributed to fracture of competent rock lenses in otherwise ductile shear zones or melange zones that may be hundreds of meters to kilometers thick (Skarbek et al. 2012; Fagereng et al. 2014; Hayman and Lavier 2014; Behr et al. in press).

The purpose of this paper is to present field and microstructural data from two terranes in California that represent rocks exhumed from the subduction channel developed in the Late Mesozoic to Early Tertiary subduction zone along the western margin of the North American plate. One of these is the Pelona schist in the San Gabriel Mountains of southern California; this is a body of volcanic and sedimentary rocks derived from the underthrust oceanic plate that was metamorphosed at a temperature of ~ 500 °C and a depth of ~ 39 km during the latest Cretaceous Laramide flat-slab subduction event (Xia and Platt 2017). The other is the South Fork Mountain Schist in the Coast Ranges of northern California, which was metamorphosed at ~ 350 °C and a depth of ~ 30 km along the subduction zone interface during early Cretaceous subduction (Broecker and Day 1995; Cooper et al. 2011; Schmidt and Platt 2018). The Pelona Schist appears to represent the products of aseismic creep; the conditions under which the South Fork Mountain Schist formed correspond well to the transition zone, hence with a possible source region of slow earthquakes. These two exhumed terrains give us a unique insight into the nature of the seismic-aseismic transition in a subduction zone.

The Pelona Schist: aseismic creep in the subduction zone

The Pelona Schist in southern California is a body of oceanic metavolcanic and sedimentary rocks exposed in a series of tectonic windows, where it lies beneath Proterozoic crystalline rocks attributed to the North American continent and Mesozoic granitoids belonging to the subduction-related magmatic arc inboard from the trench (Fig. 2). The Pelona Schist is thought to have been thrust beneath the margin during the Laramide flat-slab subduction event (e.g., Jacobson et al. 2007), which truncated the roots of the arc and removed the subcontinental lithosphere, emplacing Laramide oceanic lithosphere and superjacent sedimentary sequences in its place. The situation at the end of the Cretaceous along the western margin of North America at the latitude of California probably closely resembled the present-day subduction zone near Guerrero on the Mexican margin, where flat-slab subduction at ~ 45 km depth is on-going at present (Pérez-Campos et al. 2008; Frank et al. 2014).

The Pelona Schist in the San Gabriel Mountains of California is directly overlain by a 500–800 m thick mylonite zone that was originally thought to represent the subduction zone interface. Recent structural and petrological investigations have shown that this ductile shear zone is part of a normal fault system that exhumed

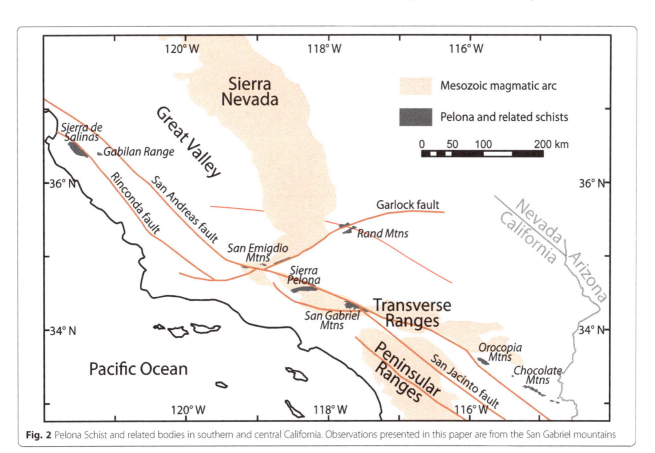

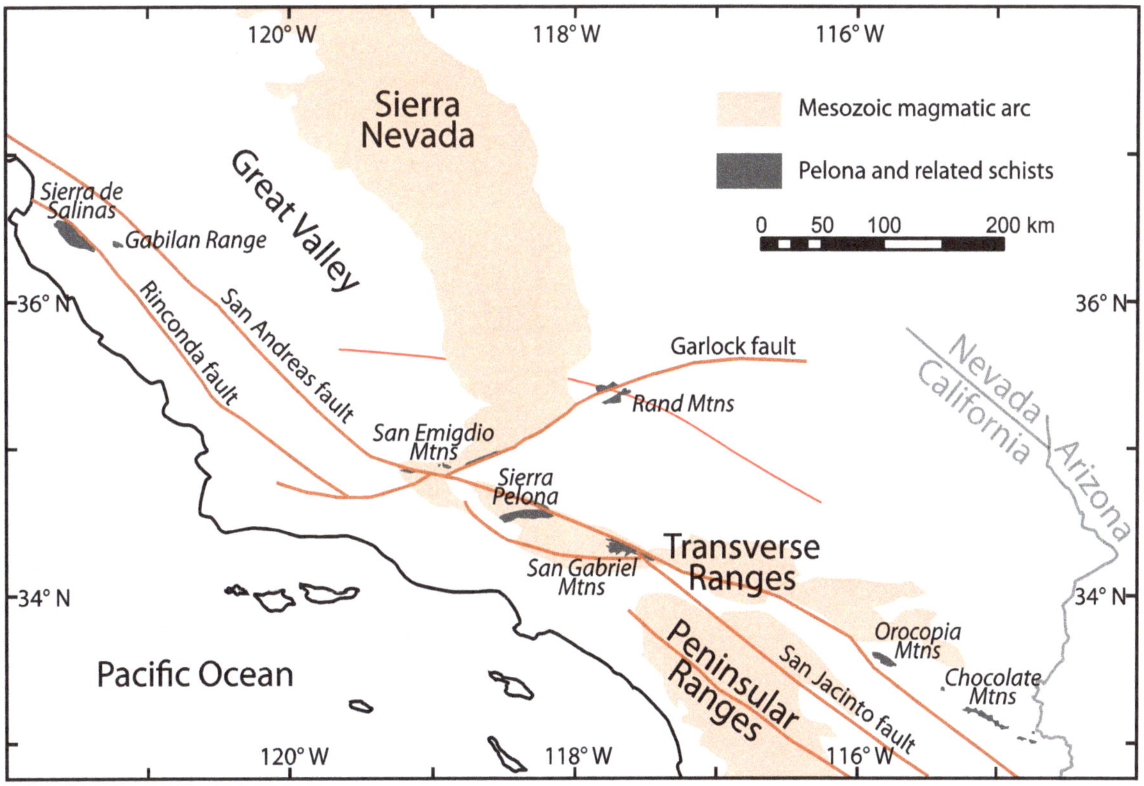

Fig. 2 Pelona Schist and related bodies in southern and central California. Observations presented in this paper are from the San Gabriel mountains

the Pelona Schist during early Cenozoic time and that the subduction zone interface is no longer preserved (Xia and Platt, in review). The Pelona Schist itself shows a reversal of shear sense over its exposed ~4 km structural depth. This spatial reversal was established during the early stages of exhumation, associated with retrograde metamorphism immediately after peak temperature and pressure; it is interpreted by Xia and Platt (2017) as indicating that the schist was deformed and exhumed in a subduction channel, probably 5–10 km thick, the lower part of which is not exposed. Metamorphic conditions during this process were around 500 °C and 1 GPa, consistent with a depth of ~ 39 km in the subduction zone (Xia and Platt 2017). The Pelona Schist was subsequently exhumed along a major normal fault system, accompanied by mylonitization of both the hanging wall gneisses and granitoids, and the uppermost part of the Pelona Schist itself, under conditions of decreasing pressure and temperature, as is observed for other outcrops of Pelona-type schist (Jacobson et al. 2007).

The bulk of the Pelona Schist consists of metagreywacke, with lesser amounts of metabasaltic rocks (now mafic greenschist) and metachert. Deformation is intense, with complete transposition of bedding and early deformational fabrics into the plane of the dominant foliation, which formed in the subduction channel during exhumation and decompression (Fig. 3a). No discontinuities have been identified associated with this deformation, which was entirely ductile and distributed across at least the 4 km structural thickness of the schist that is exposed. For this reason, we interpret the deformation in the schist as representing the creeping section of the subduction zone. Deformation in the metagreywacke was accomplished predominantly by pressure solution, which created a differentiated fabric, with laminae rich in sheet silicates (chlorite and white mica) alternating at the millimeter-scale with laminae rich in quartz (Fig. 3b). Rigid porphyroblasts of

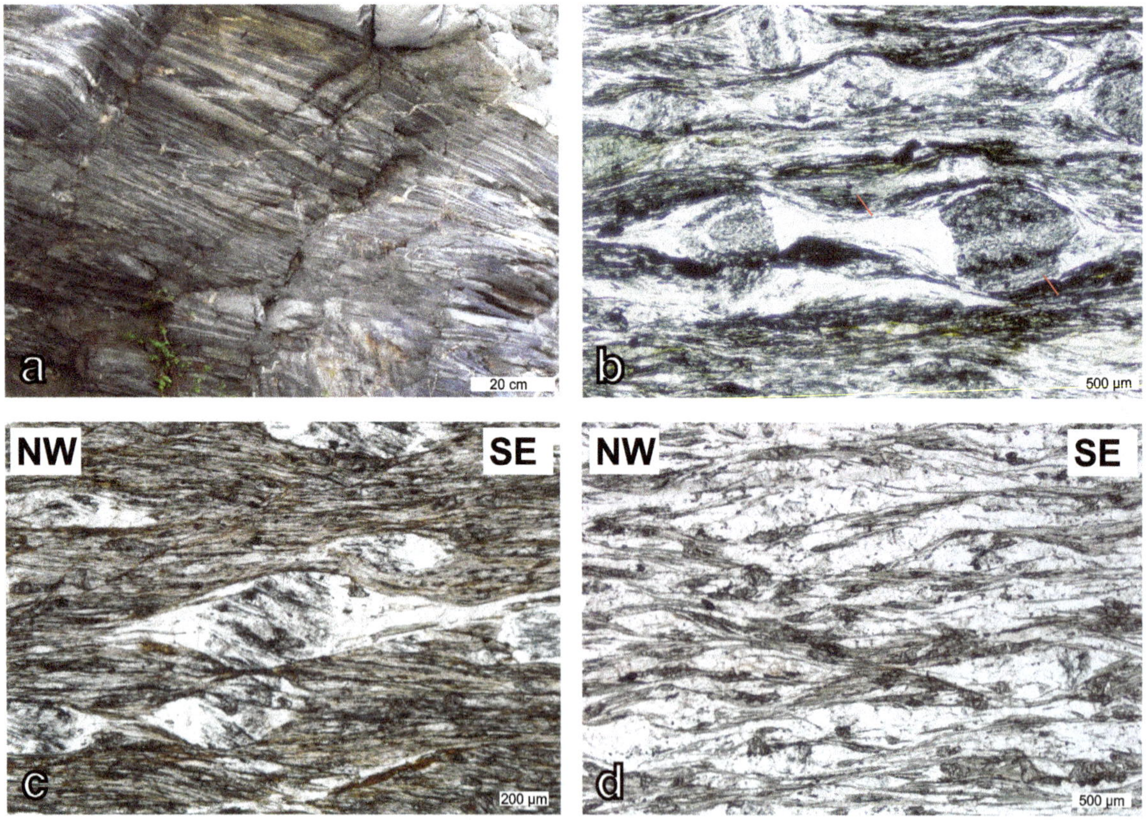

Fig. 3 Structural and microstructural characteristics of the Pelona Schist. **a** Field photograph of thinly bedded metagreywacke with tight folds formed during return flow. This illustrates the intense ductile deformation characteristic of the schist. **b** Albite porphyroblasts formed at close to peak metamorphic conditions were pulled apart during return flow, and quartz was precipitated by pressure solution between them. The alternating bands of quartz and sheet silicates (white mica and chlorite) are a characteristic microstructure formed by pressure solution. **c** Quartz precipitated in pressure shadows around albite porphyroblasts in metagreywacke has an asymmetric shape indicative of top to the left (top NW) shear. The sample comes from the deepest part of the schist, and the microstructures represent the lower part of the subduction channel. **d** Asymmetric pressure shadows around albite porphyroblasts from the upper part of the subduction channel have an asymmetry indicating top SE shear sense

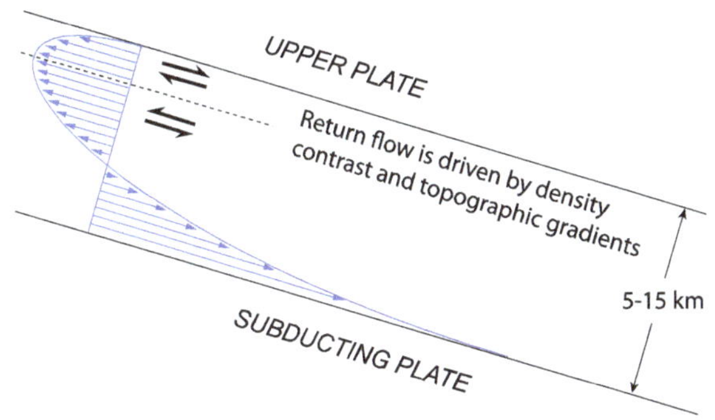

Fig. 4 Pattern of flow in a subduction channel. This is a combination of Couette flow (simple shear) driven by the subducting plate, and Poiseulle flow (channelized flow) driven by a pressure gradient produced by the buoyancy of the subducted sediment and the arc-trench topographic gradient. Dashed black line indicates the locus of maximum exhumation rate, across which the sense of shear changes, as shown. Thickness of channel estimated for the Pelona Schist from seismic tomographic data in southern California (Lee et al. 2014)

albite are surrounded by pressure shadows filled with quartz. Inclusion trails within albite define an earlier differentiated fabric that formed during subduction, suggesting that pressure solution was the dominant deformation mechanism throughout the process of subduction and return flow. This is consistent with the widely reported occurrence of pressure solution as the dominant deformation mechanism in many subduction complexes (e.g., Bolhar and Ring 2001; Gratier et al. 2011; Behr and Platt 2013; Wassmann and Stöckhert 2013).

The pressure shadows around albite porphyroblasts are commonly asymmetric and indicate the sense of shear during flow. The sense is top to WNW (in present coordinates) in the lower part of the exposed schist (Fig. 3c) and top to ENE in the top 1000 m (Fig. 3d). Xia and Platt (2017) interpret this in terms of return flow in a subduction channel ~ 5–10 km thick (Fig. 4), which is consistent with the inferred thickness of the Pelona Schist beneath southern California (Lee et al. 2014).

Sparsely distributed metachert within the Pelona Schist shows isoclinal folds and sheath folds, demonstrating that it participated fully in the ductile deformation. The microstructure of the metachert suggests dislocation creep and dynamic recrystallization by the subgrain rotation mechanism (Fig. 5), and this is supported by the presence of a crystallographic preferred orientation of the quartz, with an asymmetric single girdle of c-axes. The dynamically recrystallized grain size in the quartz from metachert on both the ESE- and WNW-directed "limbs" of the return flow system lies in the range 49–85 μm, which indicates a shear stress during deformation of 8–13 MPa, using the paleopiezometer of Stipp and Tullis (2003), with corrections as suggested by Holyoke and Kronenberg (2010). Differential stress inferred from the piezometer has here been converted to shear stress, appropriate to bulk simple shear flow, by dividing it by $\sqrt{3}$ (Behr and Platt 2013).

The shear stress inferred from the metachert is consistent with dislocation creep (Fig. 6), as is also suggested by the microstructure, but it appears to be inconsistent with the fact that the metagreywackes deformed predominantly

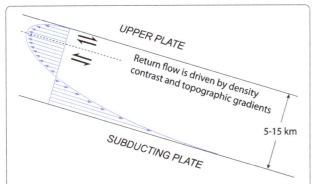

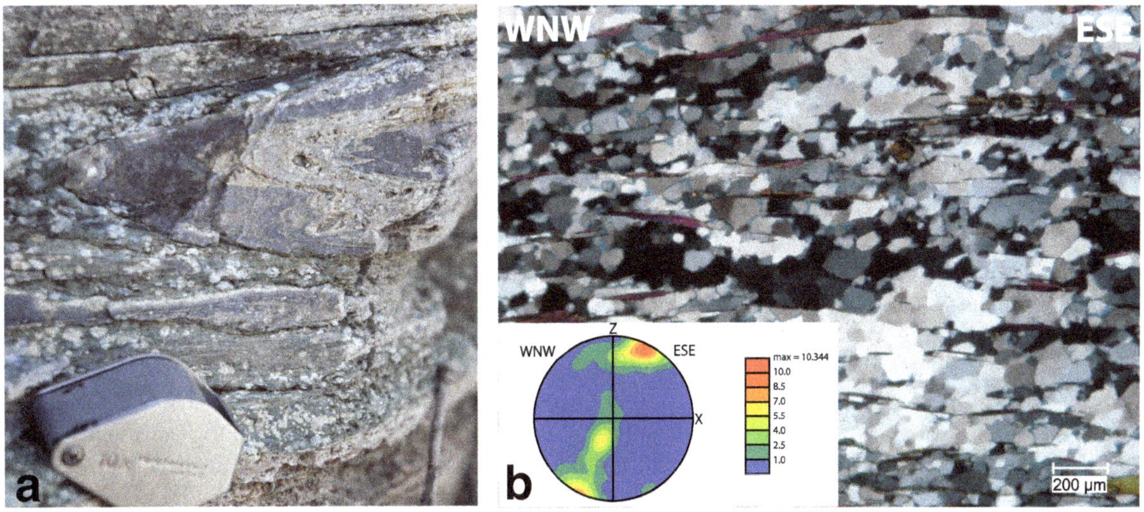

Fig. 5 Paleopiezometry on the Pelona Schist. **a** Isoclinally folded metachert layer with sheath folds demonstrates that the metachert participates in the same intense ductile deformation as the surrounding metagreywacke. **b** The metachert shows microstructural evidence for dislocation creep. Dynamically recrystallized grains define an oblique shape fabric (inclined left), indicating SE shear sense in the upper part of the subduction channel. This is confirmed by the oblique single girdle of quartz c-axes measured by EBSD (inset). X and Z are finite strain axes defined by the macroscopic foliation and lineation. The dynamically recrystallized grain size lies in the range 49–85 μm, which corresponds to a shear stress of 8–13 MPa (see text for details)

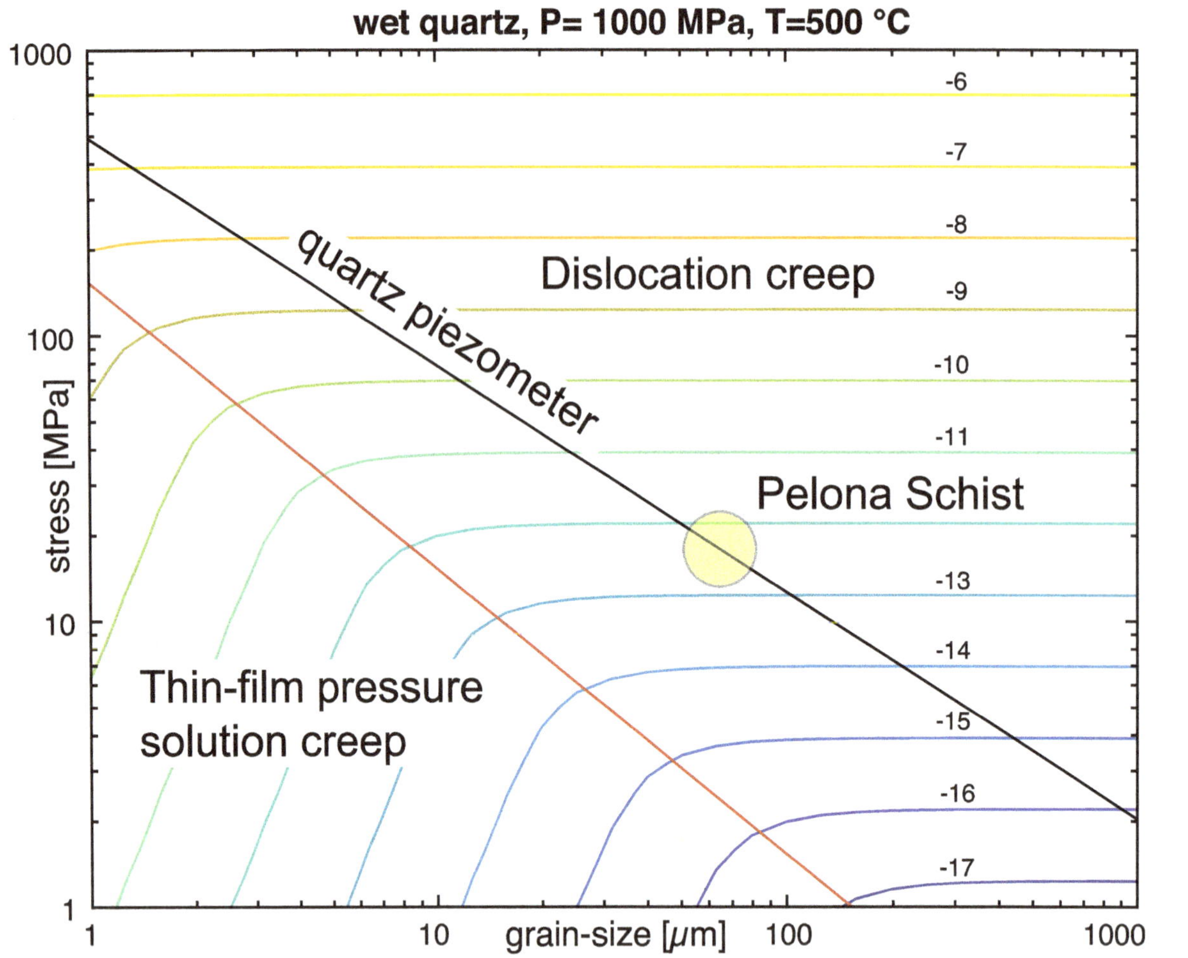

Fig. 6 Stress—grain-size map for quartz at 1000 MPa pressure and 500 °C, showing strain-rate contours in sec^{-1}, the boundary (red) between the dislocation creep and thin-film pressure solution fields, and the location of the Pelona Schist metacherts. Pressure solution flow law from Den Brok (1998), dislocation creep flow law from Hirth et al. (2001) assuming water activity of one, and piezometric line from Stipp and Tullis (2003), with correction from Holyoke and Kronenberg (2010)

by pressure solution, which in pure quartz would require lower stress or smaller grain size, or both (Fig. 6). Quartz in metagreywacke has recrystallized grain sizes in the range 100–200 μm, but metachert and metagreywacke appear to have been deformed together to comparable strains, so Xia and Platt (2017) suggest that the rate of pressure solution in the greywacke was enhanced by the presence of abundant sheet silicates, which facilitated the diffusion of aqueous fluids.

Metabasaltic rocks in the Pelona schist were metamorphosed under high-pressure greenschist facies conditions, and contain mineral assemblages dominated by albite, epidote, and Ca-amphibole. During metamorphism these rocks are likely to have deformed by transformation-assisted diffusion creep, and may have had very low strength. Once the new mineral assemblage was formed, however, the rock is likely to have been quite strong. Currently available experimental data do not allow us to quantify this, but the field relationships suggest that the rock was at least as strong as the intercalated metachert.

The strain rate in the metachert inferred from the stress and temperature is $\sim 10^{-12}\ s^{-1}$, using the quartz flow law of Hirth et al. (2001). This is sufficient to accommodate a subduction rate of 100 mm/year, appropriate to the late Cretaceous subduction zone off California, in a zone of simple shear < 1 km thick. Given that the subduction channel in which the Pelona Schist was deformed was at least 5 km thick, this appears to justify our assumption that it represents a zone of distributed ductile shear without the need for any discontinuities.

The South Fork Mountain Schist: a possible slow earthquakes source

The South Fork Mountain Schist (SFMS) is a body of blueschist facies metasedimentary and metavolcanic rocks of oceanic origin that extends for 250 km along

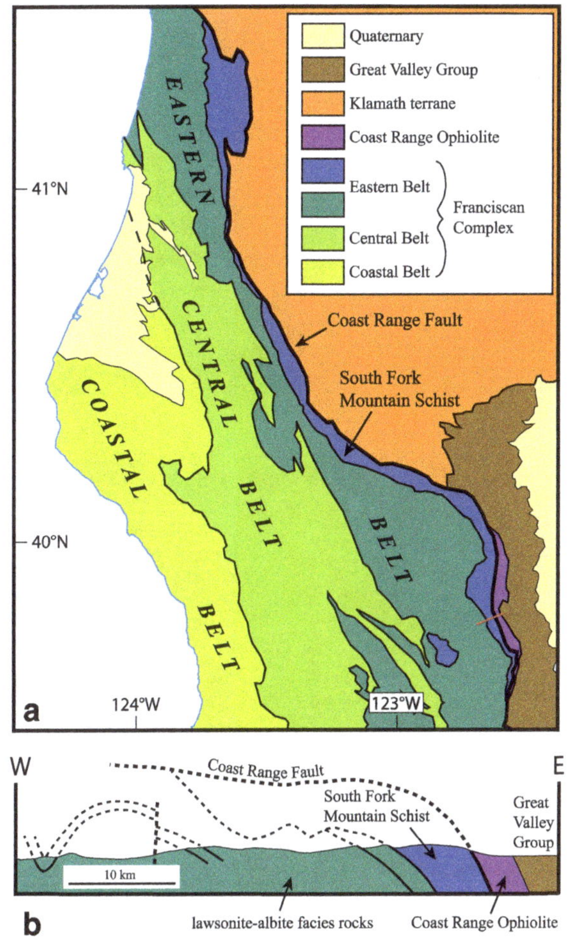

Fig. 7 **a** Tectonic sketch map of the Franciscan Complex in the Coast Ranges of northern California, showing the extent of the South Fork Mountain Schist, based on a map compiled from numerous sources by Dumitru et al. (2010). Observations presented in this paper are from the Thomes Creek transect (red line). **b** Schematic profile across the eastern part of the Franciscan Complex in the region of our transect, to show the large-scale tectonic relationships

the eastern margin of the Franciscan accretionary complex in northern California (Fig. 7) (Jayko and Blake 1989). Limited detrital zircon ages suggest that the youngest sedimentary components are ~ 137 million years old, and Ar-Ar ages on metamorphic mica indicate crystallization or cooling at ~ 121 Ma (Dumitru et al. 2010), which suggests that the SFMS is the oldest large-scale coherent body of metamorphic rock within the Franciscan Complex. This and its structural position immediately beneath the Coast Range ophiolite, which represents the upper plate of the subduction zone, suggest that it formed along the subduction zone interface in Early Cretaceous time (Schmidt and Platt 2018). The unit is ~ 3.5 km thick and consists largely of pelitic schist, with subordinate amounts of metabasaltic rocks and metachert.

The SFMS was metamorphosed at around 350 °C and 800 MPa, corresponding to a depth of ~ 30 km in the subduction zone (Broecker and Day 1995; Cooper et al. 2011). This places it at a position equivalent to the transition from seismic to aseismic creep on present-day active margins such as the Nankai trough (Peacock 2009) or the Cascadia margin (Peacock et al. 2011), and in a temperature regime similar to that calculated for tremor regions beneath the Kii Peninsula (Peacock 2009). It is intensely deformed, with a strong differentiated fabric composed of alternating millimeter-scale laminae of quartz and sheet silicates, formed by solution-redeposition processes (Schmidt and Platt 2018). This fabric has been redeformed by a predominantly W-vergent set of folds, ranging in scale from millimeter-scale crenulations to folds several hundred meters in wavelength. These folds are accompanied by a variably developed axial-planar crenulation cleavage (Fig. 8a, b). A distinctive aspect of these folds is that they were accompanied by dilational microcracking on a range of scales. The most common expression of this is dilational arcs in the hinges of the millimeter-scale crenulations, which were progressively opened by dilational microcracking and then healed by infillings of quartz (Figs. 8c, d and 9a). Quartz veins also formed on a range of scales at all stages in the deformation history (Fig. 10b). These features form either as hydraulic fractures or as hydraulically assisted shear fractures and indicate fluid pressures approaching lithostatic, at least transiently. Given the low solubility of silica in water at 300 °C, it is likely that these hydraulic fractures opened repeatedly to produce the present-day geometry.

Quartz precipitated in veins and dilational fractures shows microstructural evidence for limited crystal plastic deformation, in the form of subgrains, grain-boundary suturing (bulging), and some dynamic recrystallization (Fig. 10b). Plastic deformation alternated with hydraulic fracture, as younger veins show lesser degrees of plastic deformation than older ones (Schmidt and Platt 2018). The size of the dynamically recrystallized grains (~ 40 μm) indicates shear stress of the order of 10 MPa, but the evidence for alternating phases of plastic deformation and hydraulic fracture suggests that the shear stress and fluid pressure may have fluctuated repeatedly.

An important aspect of the deformation in the SFMS is that the crenulation trains have the geometry and mechanical characteristics of kink-bands (Fig. 10a). Kink-bands are characteristic of the deformation of strongly anisotropic materials, such as schists (Donath 1968), and they tend to propagate rapidly parallel to the direction of maximum rate of shear strain in the deforming medium (Cobbold et al. 1971; Gay and Weiss 1974). Kinking is greatly facilitated by dilation in the hinge area of the kink,

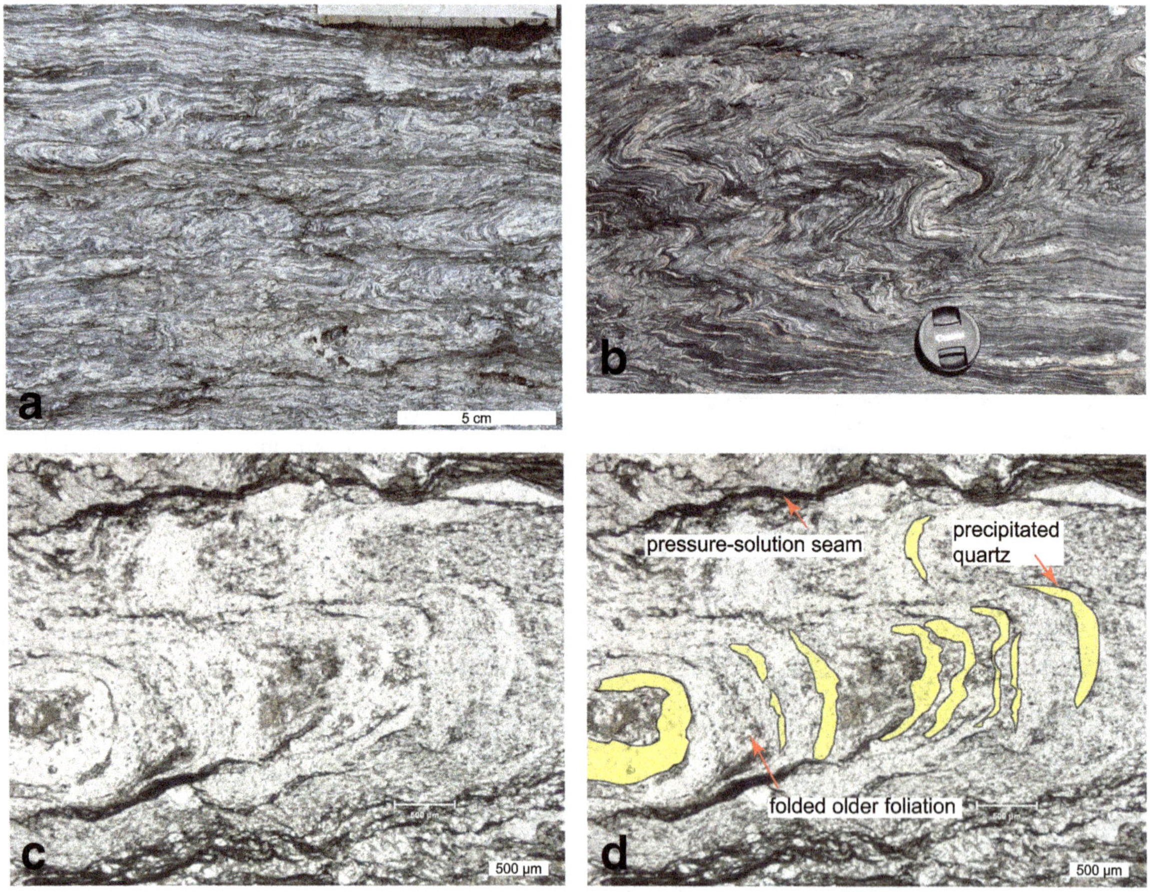

Fig. 8 Structural and microstructural characteristics of the South Fork Mountain Schist. **a** The dominant foliation in pelitic schist is a transposed crenulation cleavage, formed by microfolding an earlier differentiated fabric, accompanied by pressure solution. **b** Mesoscopic folds deforming a differentiated fabric, with a new crenulation cleavage forming parallel to the axial planes. **c**, **d** Photomicrograph showing the dilational arcs and pressure-solution seams that form in conjunction with the dominant crenulation cleavage

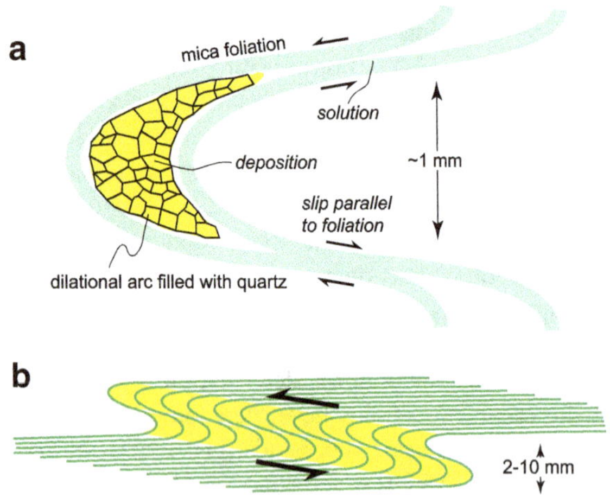

Fig. 9 **a** Relationship between a microfold (crenulation), the dilational arc in the hinge, and pressure-solution seams along the limbs. **b** A microfold train is a dilatant shear zone. Incremental growth of these microfold trains can produce displacement and equivalent seismic moment comparable to low-frequency earthquakes

as this allows it to develop without components of extension or shortening parallel to the rotating foliation. The crenulations in the SFMS show characteristic dilation in the crenulation arcs, suggesting that high fluid pressure facilitated their formation. The dilational cracks are infilled with quartz, resulting from solution and redeposition of silica during the deformation (Fig. 8). The source of the silica is likely to be the limbs of the crenulations, as these commonly show evidence for pressure solution, such as depletion in quartz and concentration of mica and insoluble materials such as graphite (Schmidt and Platt 2018). A train of these crenulations, such as that illustrated in Fig. 10a, is therefore likely to approximate closely to a zone of simple shear (Fig. 9b), accommodating displacement at a rate that is controlled on short time-scales by hydraulic fracture, and on long time-scales by solution and redeposition of silica.

At 800 MPa pressure and 350 °C, the molar concentration of SiO_2 in water is 0.0444 mol/kg, and the mole fraction is close to 8×10^{-4} (Fournier and Potter 1982; den

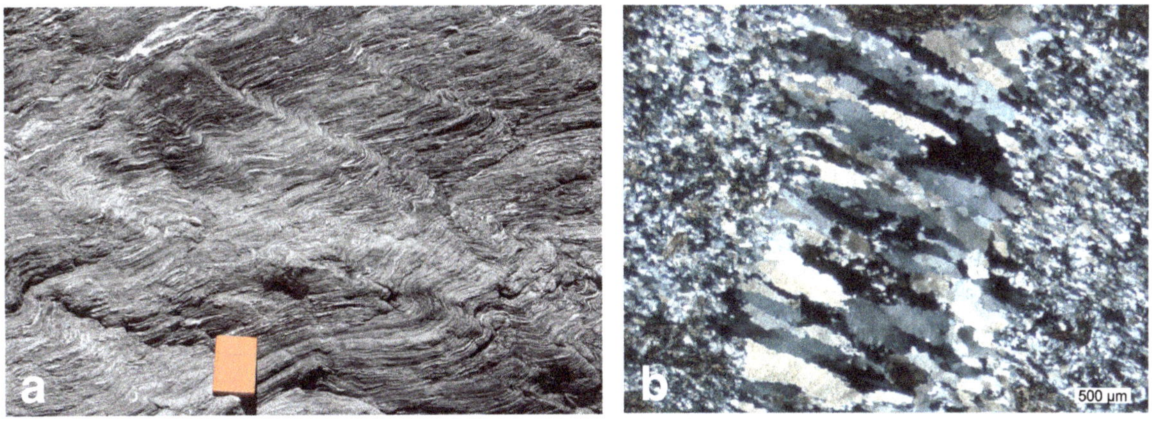

Fig. 10 a Train of crenulations in pelitic schist, showing the kink-like geometry of the microfolds. **b** Dynamically recrystallized metachert deformed by dislocation creep, cut by a quartz vein produced by hydraulic fracture. The vein shows cross-fiber geometry, indicating the direction of opening, and grain-boundary bulging and dynamic recrystallization indicating renewed dislocation creep after emplacement

Brok 1998). Hence, a quartz-filled dilation site needs about 1250 times its volume of SiO_2-saturated water to pass through it in order to fill it. This could have been achieved by repeated cycles of opening and closing of the void, driven by fluctuations in fluid pressure. Evidence for this type of repeated opening and closing comes from some quartz veins, which show crack-seal type microstructures (Ramsay 1980) indicative of this, but we have not seen evidence for this process in the crenulation arcs. Alternatively, these relatively small voids may have stayed open long enough for sufficient fluid to flow through them to fill them with quartz. Geophysical evidence for large-scale fluid flow during slow slip events has been documented by Frank et al. (2015), Skarbek and Rempel (2016) and Taetz et al. (2018).

We postulate that propagation of a train of crenulations is likely to be triggered by high-fluid pressure, rapidly opening up dilational arcs, each of ~1 mm dimension, in a coordinated fashion to form a shear band with in-plane dimensions of the order of one to a few tens of meters, and an effective shear displacement of ~1 mm (Fig. 9). The equivalent seismic moment is the slip × area × elastic modulus. Assuming a shear modulus of 3×10^{10} Pa, and a linear dimension of 30 m, giving an in-plane area of 10^3 m^2, the equivalent moment is 3×10^{11} N m, which is the characteristic size of the LFEs making up tremor bursts (Gao et al. 2012; Frank et al. 2016). The crenulation bands seen in outcrop mostly have linear dimensions of the order of 1 m, but some of them are likely to have extended for

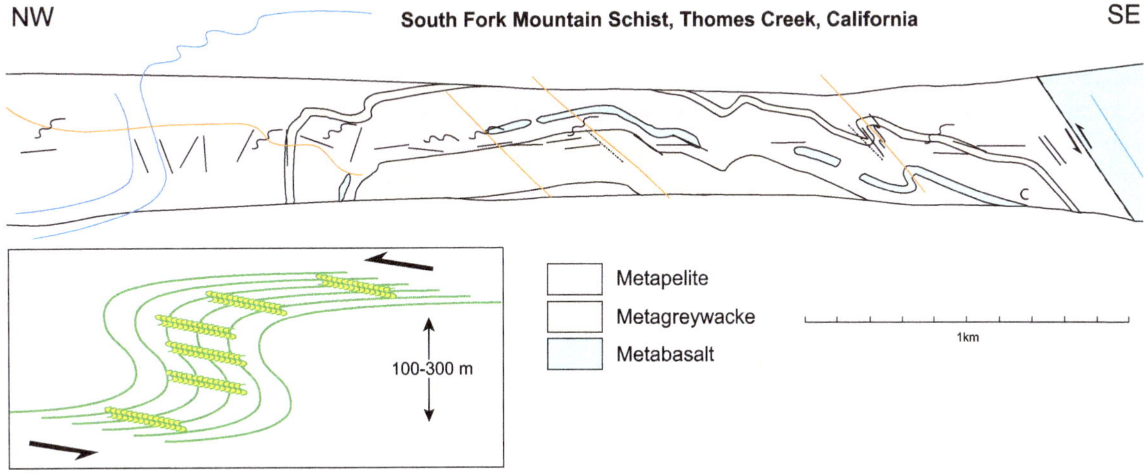

Fig. 11 Section across part of the South Fork Mountain Schist in Thomes Creek, California, showing the distinctive pattern of large-scale folds in revealed by structural analysis. Inset (left) shows schematically how these folds develop by the successive initiation and propagation of crenulation bands. Incremental growth of folds at scales of meters to hundreds of meters can produce displacement and equivalent seismic moment comparable to tremor bursts and slow slip events

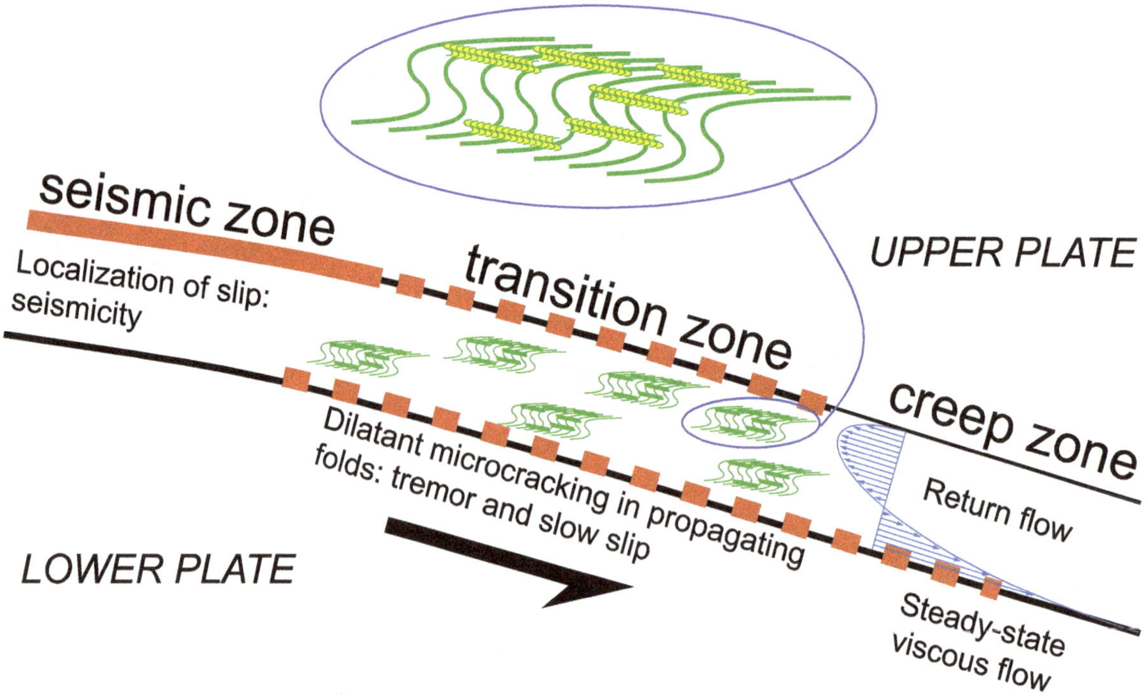

Fig. 12 Relationship between the seismic, transition, and creep zones in a subduction zone, and the processes associated with each. The transition zone is the source of tectonic tremor and slow slip and is also the likely locus of underplating, where the metamorphosed sedimentary and volcanic cover of the subducting plate is progressively accreted to the upper plate. This produces significant off-fault deformation, and the active slip surface migrates downwards. In the creep zone, no discrete displacement surface exists; ductile deformation is distributed in a subduction channel between the two plates. Inset shows folds with axial-planar cleavage of the type shown in Fig. 11; we suggest that the hierarchy of structures at different scales can explain LFEs, tremor bursts, and slow slip events

tens of meters. Given that the cumulative moment release in LFEs is only a small proportion of the total moment release in slow slip events (Peng and Gomberg 2010), we suggest that the largest crenulation bands may be good candidates for LFE sources.

What we cannot determine from the field relationships or the microstructure is the time-scale on which these crenulation bands propagate. The rate-limiting step is the rate at which fluid can permeate from the surrounding rock into the crenulation arcs, which depends on the permeability of the rock and the distance over which the fluid must migrate. At present, we have too few constraints to estimate this.

The formation of these crenulation bands must occur repeatedly to produce the pervasive sets of crenulation cleavages in the SFMS, and the scale of the crenulation arcs is likely to be consistent, as it is controlled by the thickness of the foliation bands making up the earlier fabric that is being crenulated. The process we describe is therefore likely to produce repeating events with very similar characteristics, which is what is observed for LFEs (Shelly et al. 2007; Frank et al. 2014).

The formation of crenulation bands is clearly linked geometrically and mechanically to larger scale folds. These folds were produced by the deformation of strongly layered sandstone-shale sequences, and hence also have the geometry of kink-bands. The folds form a hierarchy of scales, from centimeter scale to hundreds of meters (Schmidt and Platt 2018), controlled by the scale lengths of mechanical layering in the rocks, which varies from individual beds, with thicknesses of centimeters to tens of centimeters, to packages of beds that may have formed in channels or lobes within a submarine fan, with thicknesses ranging from tens of meters to kilometers (Pickering and Hiscott 2016). Buckle folds in deformed sand-shale sequences typically have wavelengths around one order of magnitude larger than the layer thickness, so that scales of folding in any particular sequence will form a hierarchy (Ramberg 1964; Hudleston 1973). Amplification of the larger scale structures is accommodated by the accumulation of displacements on smaller scale structures. This corresponds well to what is observed in slow slip events: LFEs are concatenated to form tremor bursts, with equivalent moments of ~ 10^{14} N m: these might correspond to the amplification of a meter-scale fold train, with areal dimensions of 10^5 m^2, in which the activity of a series of crenulation bands in a burst leads to a net displacement of 10 mm (Fig. 8b). Development of folds on scales of 10–100 m wavelength and corresponding larger areal dimensions could lead in the same way to displacements of tens of centimeters (Fig. 11), corresponding to slow slip events with moment equivalents of up to 5×10^{18} N m, which might last for several weeks. Each of these events would involve a concatenated series of amplifications of individual smaller scale folds and crenulation bands, as described above. Initiation of these events would correspond to a peak in fluid pressure, and they would be self-limited by the drop in fluid pressure caused by the opening of the dilational cracks (Segall et al. 2010).

Conclusions

Field and microstructural observations from the Pelona Schist, an exhumed subduction complex in southern California, suggest that the creeping section of the subduction zone interface, below 30–40 km depth, is characterized by pervasive ductile flow at a temperature of ~ 500 °C, including components of shear driven by subduction (Couette flow) and return flow driven by buoyancy and topographic gradients (Poiseuille flow). The predominant deformational mechanism in meta-greywacke is thin-film pressure solution, accompanied by dislocation creep in quartz-rich domains including metachert. The dynamically recrystallized grain-size piezometer suggests shear stresses of 8–13 MPa. Inferred strain rates of 10^{-12} s^{-1} are sufficient to accommodate all the displacement without the need for a discrete slip surface.

At around 30 km depth and a temperature of 350 °C, there is a transition into zone of seismic slip. The transition zone itself is the source of episodic tremor and slow slip events. Field observations from the South Fork Mountain Schist in the northern California Coast Ranges suggest that this is a zone of distributed deformation ~ 3.5 km thick, characterized by intensive microfolding, dilatant fracturing in microfold hinges, and solution-redeposition of quartz, producing multiple differentiated foliations. The dilational microcracks indicate high fluid pressures and fluid content, as is expected in a zone of low seismic velocity and high Vp/Vs ratio. Microstructural evidence shows that these processes alternated with dislocation creep, and dynamically recrystallized grain sizes indicate peak shear stresses of the order of 10 MPa.

The crenulation trains are kinematically equivalent to shear bands with a reverse shear sense, which can propagate rapidly both up-dip and laterally, and which are capable of producing displacements and seismic moment comparable to low-frequency earthquakes. Because these shear bands accommodate subduction-driven thrust motion, their activity could produce a focal mechanism consistent with that of a gently dipping thrust fault. Development of and amplification of mesoscopic folds by the coordinated development of crenulation trains could be responsible for tremor bursts. The development of larger scale folds with wavelengths of tens to hundreds of meters by the same mechanism could be responsible for slow slip events. The different scales of deformation can therefore be related to the different scales of slow earthquake phenomena. In each case, reasonable estimates of the incremental displacements produced and the equivalent seismic moment are comparable to the observed geophysical phenomena. The distribution of all these processes along the subduction zone interface is summarized in Fig. 12.

Abbreviations

LFE: Low-frequency earthquake; SFMS: South Fork Mountain Schist

Acknowledgements

JPP gratefully acknowledges a travel grant from Japan Geoscience Union to attend the joint JpGU/AGU meeting held in Chiba, Japan. We thank the reviewers for their comments, which helped us to improve the paper; Simon Wallis for editorial handling; and William Frank for assistance with terminology and referencing.

Funding

This research was supported in part by NSF grant EAR-1250128 awarded to J.P. Platt.

Authors' contributions

Work on the Pelona Schist was a PhD project, conceived by JPP, and developed and carried out by HX. Work on the SFMS is an ongoing PhD project, conceived by JPP and being developed and carried out by WLS. Interpretations were developed by all authors in collaboration. All authors read and approved the final manuscript.

Competing interests

The authors declare that they have no competing interests.

Author details

[1]Department of Earth Sciences, University of Southern California, Los Angeles, CA 90089-0740, USA. [2]Present Address: Chevron Energy Technology Company, 1500 Louisiana St, Houston, TX 77002, USA.

References

Audet P, Bostock MG, Christensen NI, Peacock SM (2009) Seismic evidence for overpressured subducted oceanic crust and megathrust fault sealing. Nature 457:76–78

Beaumont C, Jamieson RA, Butler JP, Warren CJ (2009) Crustal structure: a key constraint on the mechanism of ultra-high-pressure rock exhumation. Earth Planet Sci Lett 287(1-2):116–129

Behr W, Kotowski A, Ashley K (in press) Dehydration induced rheological heterogeneity and the deep tremor source in subduction zones. Geology

Behr WM, Platt JP (2013) Rheological evolution of a Mediterranean subduction complex. J Struct Geol 54:136–155

Beroza GC, Ide S (2010) Slow earthquakes and nonvolcanic tremor. Annu Rev Earth Planet Sci 39:271–296

Blanco-Quintero I, García-Casco A, Gerya T (2011) Tectonic blocks in serpentinite mélange (eastern Cuba) reveal large-scale convective flow of the subduction channel. Geology 39:79–82

Bolhar R, Ring U (2001) Deformation history of the Yolla Bolly terrane at Leech Lake Mountain, Eastern belt, Franciscan subduction complex, California Coast Ranges. Geol Soc Am Bull 113:181–195

Boyarko DC, Brudzinski MR (2010) Spatial and temporal patterns of nonvolcanic tremor along the southern Cascadia subduction zone. J Geophys Res Solid Earth 115:B8

Broecker M, Day HW (1995) Low-grade blueschist facies metamorphism of metagreywackes, Franciscan Complex, Northern California. J Metamorph Geol 13:61–78

Brown JR, Beroza GC, Ide S, Ohta K, Shelly DR, Schwartz SY, Rabbel W, Thorwart M, Kao H (2009) Deep low-frequency earthquakes in tremor localize to the plate interface in multiple subduction zones. Geophys Res Lett 36:L19306

Cobbold PR, Cosgrove JW, Summers JM (1971) The development of internal structures in deformed anisotropic rocks. Tectonophysics 12:23–53

Cooper FJ, Platt JP, Anczkiewicz R (2011) Constraints on early Franciscan subduction rates from 2-D thermal modeling. Earth Planet Sci Lett 312(1-2):69–79

den Brok SWJ (1998) Effect of microcracking on pressure-solution strain rate: the Gratz grain boundary model. Geology 26:915–918

Donath FA (1968) Experimental study of kink-band development in strongly anisotropic rock. In: A. J. Ball and D. K. Norris Eds. Conference on Research in Tectonics. Geol Surv Can Pap 68-52:255–288

Dragert H, Wang K, James TS (2001) A silent slip event on the deeper Cascadia subduction interface. Science 292(5521):1525–1528

Dumitru T, Wakabayashi J, Wright JE, Wooden JL (2010) Early cretaceous (ca. 123 Ma) transition from nonaccretionary behavior to strongly accretionary behavior within the Franciscan subduction complex. Tectonics 29:TC5001

Fagereng Å, Hillary GWB, Diener JFA (2014) Brittle-viscous deformation, slow slip, and tremor. Geophys Res Lett 41:4159–4167

Fournier RO, Potter RW (1982) An equation correlating the solubility of quartz in water from 25 to 900 C at pressures up to 10,000 bars. Geochim Cosmochim Acta 46(10):1969–1973

Frank WB, Shapiro NM, Husker AL, Kostoglodov V, Bhat HS, Campillo M (2015) Along-fault pore-pressure evolution during a slow-slip event in Guerrero, Mexico. Earth Planet Sci Lett 413:135–143

Frank WB, Shapiro NM, Husker AL, Kostoglodov V, Gusev AA, Campillo M (2016) The evolving interaction of low-frequency earthquakes during transient slip. Sci Adv 2:e1501616

Frank WB, Shapiro NM, Husker AL, Kostoglodov V, Romanenko A, Campillo M (2014) Using systematically characterized low-frequency earthquakes as a fault probe in Guerrero, Mexico. J Geophys Res Solid Earth 119:7686–7700

Frank WB, Shapiro NM, Kostoglodov V, Husker AL, Campillo M, Payero JS, Prieto GA (2013) Low-frequency earthquakes in the Mexican sweet spot. Geophys Res Lett 40:2661–2666

Gao H, Schmidt DA, Weldon RJ (2012) Scaling relationships of source parameters for slow slip events. Bull Seismol Soc Am 102(1):352–360

Gay NC, Weiss LE (1974) The relationship between principal directions and the geometry of kinks in foliated rocks. Tectonophysics 21:287–300

Gerya TV (2002) Exhumation rates of high pressure metamorphic rocks in subduction channels: the effect of rheology. Geophys Res Lett 29(8):102-1–102-4

Gratier JP, Richard J, Renard F, Mittempergher S, Doan ML, Di Toro G, Hadizadeh J, Boullier AM (2011) Aseismic sliding of active faults by pressure solution creep: evidence from the San Andreas Fault Observatory at Depth. Geology 39:1131–1134

Hayman NW, Lavier LL (2014) The geologic record of deep episodic tremor and slip. Geology 42:195–198

Hirose H, Hirahara K, Kimata F, Fujii N, Miyazaki S (1999) A slow thrust slip event following the two 1996 Hyuganada earthquakes beneath the Bungo Channel, southwest Japan. Geophys Res Lett 26:3237–3240

Hirth G, Teyssier C, Dunlap WJ (2001) An evaluation of quartzite flow laws based on comparisons between experimentally and naturally deformed rocks. Int J Earth Sci 90:77–87

Holtkamp S, Brudzinski MR (2010) Determination of slow slip episodes and strain accumulation along the Cascadia margin. J Geophys Res 115:B00A17

Holyoke CW, Kronenberg AK (2010) Accurate differential stress measurement using the molten salt cell and solid salt assemblies in the Griggs apparatus with applications to strength, piezometers and rheology. Tectonophysics 494:17–31

Holyoke CW, Kronenberg AK (2013) Reversible water weakening of quartz. Earth Planet Sci Lett 374:185–190

Hudleston PJ (1973) Fold morphology and some geometrical implications of theories of fold development. Tectonophysics 16:1–46

Ide S, Beroza C, Shelly DR, Uchide T (2007a) A scaling law for slow earthquakes. Nature 447:76–79

Ide S, Shelly DR, Beroza GC (2007b) Mechanism of deep low frequency earthquakes: further evidence that deep non-volcanic tremor is generated by shear slip on the plate interface. Geophys Res Lett 34:2191–2195

Jacobson C, Grove M, Vucic A, Pedrick J, Cloos M, Carlson W, Gilbert M, Liou J, Sorensen S (2007) Exhumation of the Orocopia Schist and associated rocks of southeastern California: relative roles of erosion, synsubduction tectonic denudation, and middle Cenozoic extension, pp 1–37

Jayko AS, Blake MC (1989) Deformation of the Eastern Franciscan Belt, northern California. J Struct Geol 11: 375–390

Lee E-J, Chen P, Jordan TH, Maechling PB, Denolle MA, Beroza GC (2014) Full-3-D tomography for crustal structure in Southern California based on the scattering-integral and the adjoint-wavefield methods. J Geophys Res Solid Earth 119:6421–6451

Obara K (2002) Nonvolcanic deep tremor associated with subduction in southwest Japan. Science 296(5573):1679–1681

Peacock SM (2009) Thermal and metamorphic environment of subduction zone episodic tremor and slip. J Geophys Res Solid Earth 114:B8

Peacock SM, Christensen NI, Bostock MG, Audet P (2011) High pore pressures and porosity at 35 km depth in the Cascadia subduction zone. Geology 39: 471–474

Peng Z, Gomberg J (2010) An integrated perspective of the continuum between earthquakes and slow-slip phenomena. Nat Geosci 3:599–607

Pérez-Campos X, Kim Y, Husker A, Davis PM, Clayton RW, Iglesias A, Pacheco JF, Singh SK, Manea VC, Gurnis M (2008) Horizontal subduction and truncation of the Cocos Plate beneath central Mexico. Geophys Res Lett 35:L18303

Pickering KT, Hiscott RN (2016) Deep marine systems: processes, deposits, environments, tectonics and sedimentation. Wiley, Oxford, 657 pp.

Ramberg H (1964) Selective buckling of composite layers with contrasted physical properties : a theory for simultaneous formation of several orders of folds. Tectonophysics 1:307

Ramsay JG (1980) The crack-seal mechanism of rock deformation. Nature 284:135–139

Rogers G, Dragert H (2003) Episodic tremor and slip on the Cascadia subduction zone: the chatter of silent slip. Science 300(5627):1942–1943

Schmidt, W. L., and J. P. Platt (2018), Subduction, accretion, and exhumation of coherent Franciscan blueschist-facies rocks, northern coast ranges, California, Lithosphere

Schwartz SY, Rokosky JM (2007) Slow slip events and seismic tremor at circum-Pacific subduction zones. Rev Geophys 45(3):RG3004

Segall P, Rubin AM, Bradley AM, Rice JR (2010) Dilatant strengthening as a mechanism for slow slip events. J Geophys Res 115:B12305

Shelly DR, Beroza GC, Ide S (2007) Non-volcanic tremor and low-frequency earthquake swarms. Nature 446:305–307

Skarbek RM, Rempel AW (2016) Dehydration induced porosity waves and episodic tremor and slip. Geochem Geophys Geosyst 17:442–469

Skarbek RM, Rempel AW, Schmidt DA (2012) Geologic heterogeneity can produce aseismic slip transients. Geophys Res Lett 39, L21306

Stipp M, Tullis J (2003) The recrystallized grain size piezometer for quartz. Geophys Res Lett 30:2088 3-1 - 3-5

Taetz S, John T, Bröcker M, Spandler C, Stracke A (2018) Fast intraslab fluid-flow events linked to pulses of high pore fluid pressure at the subducted plate interface. Earth Planet Sci Lett 482:33–43

Tichelaar BW, Ruff LJ (1993) Depth of seismic coupling along subduction zones. J Geophys Res 98(B2):2017–2037

Warren C, Beaumont C, Jamieson RA (2008) Deep subduction and exhumation: role of crustal strength and strain weakening in continental subduction and ultrahigh-pressure rock exhumation. Tectonics 27:TC6002

Wassmann S, Stöckhert B (2013) Rheology of the plate interface—dissolution precipitation creep in high pressure metamorphic rocks. Tectonophysics 608:1–29

Xia H, Platt JP (2017) Structural and rheological evolution of the Laramide subduction channel in southern California. Solid Earth 8:379–403

3

Dynamic responses of the Earth's outer core to assimilation of observed geomagnetic secular variation

Weijia Kuang[1]* and Andrew Tangborn[2]

Abstract

Assimilation of surface geomagnetic observations and geodynamo models has advanced very quickly in recent years. However, compared to advanced data assimilation systems in meteorology, geomagnetic data assimilation (GDAS) is still in an early stage. Among many challenges ranging from data to models is the disparity between the short observation records and the long time scales of the core dynamics. To better utilize available observational information, we have made an effort in this study to directly assimilate the Gauss coefficients of both the core field and its secular variation (SV) obtained via global geomagnetic field modeling, aiming at understanding the dynamical responses of the core fluid to these additional observational constraints. Our studies show that the SV assimilation helps significantly to shorten the dynamo model spin-up process. The flow beneath the core-mantle boundary (CMB) responds significantly to the observed field and its SV. The strongest responses occur in the relatively small scale flow (of the degrees $L \approx 30$ in spherical harmonic expansions). This part of the flow includes the axisymmetric toroidal flow (of order $m = 0$) and non-axisymmetric poloidal flow with $m \geq 5$. These responses can be used to better understand the core flow and, in particular, to improve accuracies of predicting geomagnetic variability in the future.

Keywords: Geodynamo, Geomagnetic field, Secular variation, Core flow, Data assimilation

Background

Geomagnetic field observed at the Earth's surface varies significantly in time: its temporal scales range from minutes to geological time scales. Though it was first noticed by mankind over 5000 years ago (Roberts 1992), and its origin was sought as early as 800 years ago (Dibner Library 1980), the modern theory that the geomagnetic field is generated and maintained by convective flow in the Earth's outer core (geodynamo) was originated from the seminal work of Larmor (1919). Successful numerical simulation of the geodynamo was first carried out by Glatzmaier and Roberts (1995), and then followed by Kageyama and Sato (1997) and by Kuang and Bloxham et al. (1997). Christensen et al. (2010) provided a comprehensive summary of numerical geodynamo solutions and their relevances to geomagnetic observations.

*Correspondence: weijia.kuang-1@nasa.gov
[1] Planetary Geodynamics Laboratory, NASA Goddard Space Flight Center, 8800 Greenbelt Road, Greenbelt, MD 20771, USA
Full list of author information is available at the end of the article

Assimilation of geomagnetic observations with numerical geodynamo models started less than a decade ago. Sun et al. (2007) and Fournier et al. (2007) used simplified magnetohydrodynamic (MHD) systems, and synthetic data tested the applicability of assimilation of sparse magnetic data. Liu et al. (2007) first used observation system simulation experiments (OSSEs) with a full dynamo model and demonstrated clearly that one could use assimilation of magnetic field at the surface to estimate the dynamo state deep in the fluid core. Kuang et al. (2008) published the first working geomagnetic data assimilation system MoSST_DAS in which the Gauss coefficients of various geomagnetic and paleomagnetic field models are assimilated with their MoSST geodynamo model (Kuang and Chao 2003; Jiang and Kuang 2008) for estimation of the core state and prediction of geomagnetic field variation. Kuang et al. (2009) then used this assimilation system and 100 years of the Gauss coefficients from *gufm1* (Jackson et al. 2000) and CM4 (Sabaka et al. 2004) to understand the responses of the core state to surface geomagnetic observations, and their implications to core state estimation and secular variation (SV) prediction. We refer the reader to Fournier et al. (2010) for a comprehensive review of the data assimilation algorithms for geomagnetic data assimilation (GDAS) and some of the early results.

Rapid advances have occurred in multiple facets of GDAS. Several independent assimilation systems have been developed to understand better the core dynamical state. For example, Aubert and Fournier (2011) and Fournier et al. (2011, 2013) carried out OSSEs with synthetic observations and numerical dynamo models to examine possibilities of core state determination. Fournier et al. (2011, 2013) also tested their approach with a geomagnetic field model. Aubert (2013, 2014) investigated possibilities of inverting core state properties using the observed field and SV. In addition to the sequential data assimilation systems mentioned above, there are also efforts in developing GDAS systems based on variational data assimilation techniques. For example, Li et al. (2011, 2014) have been continuing their effort on a new combined forward and adjoint system towards a full geodynamo model. Encompassed application is the contributions of assimilation results to international geomagnetic reference field (IGRF) (Kuang et al. 2010) and efforts to determine field model error statistics (Gillet et al. 2013).

Despite these advances, GDAS is still in an early stage similar to that of early numerical weather prediction (NWP) (for a more comprehensive review, see, e.g., Kalnay 2003). Many important questions are still to be fully answered, such as comprehensive assessment of numerical dynamo model biases, observation and core state covariances and error statistics, and the dynamic responses of dynamo state to the observed geomagnetic field. The latter is of particular importance to the spin-up processes of the numerical models which is characterized by the difference between the observation and the forecast, often called ($\mathcal{O}$-$\mathcal{F}$). These, in turn, determine how fast and how close the numerical solutions can be pulled to the true state of the core.

Concerns on the spin-up of the numerical models can be examined from the time scales of the observed field and of the numerical models. Global field model results from the past 400 years of geomagnetic data (e.g., Jackson et al. 2000; Sabaka et al. 2004, 2015; Olsen et al. 2006, 2014) show that the typical time scales τ_l of the degree l components (Stacey 1992; Hulot and Le Mouël 1994; Olsen et al. 2006)

$$\tau_l = \left[\frac{\sum_m \left(g_l^m\right)^2 + \left(h_l^m\right)^2}{\sum_m \left(\dot{g}_l^m\right)^2 + \left(\dot{h}_l^m\right)^2} \right]^{1/2} \tag{1}$$

varies from over 1000 years for the dipole ($l = 1$) to less than 100 years for higher degrees. In (1), $\left(g_l^m, h_l^m\right)$ are the Gauss coefficients of the field, and $\left(\dot{g}_l^m, \dot{h}_l^m\right)$ are their first order time derivatives, i.e., the Gauss coefficients of the SV. Currently, the longest record for low-degree ($l \leq 5$) field coefficients is from the paleo/archeo magnetic data (e.g., Korte et al. 2011; Nilsson et al. 2014). The high-quality coefficients for up to degree $l \leq 8$ could be obtained from historical and observatory data (Jackson et al. 2000). Very high quality coefficients for degrees $l \leq 13$ are obtained in the past 50 years with satellite magnetic data (Sabaka et al. 2004, 2015; Olsen et al. 2006, 2014). In summary, the data record is no more than 10 times of the typical time scales of the geomagnetic field. This brings the very concern on whether the observational record is sufficient to spin up numerical dynamo models. OSSEs results also suggest long spin-up time in geomagnetic data assimilation. For example, Liu et al. (2007) showed, via their OSSEs with synthetic magnetic data and a fully nonlinear dynamo model, that the difference between the forecast and the truth reach the minimum in approximately 40 % of the magnetic free decay time, or 8000 years, if only the poloidal field (of the first eight spherical harmonic degree coefficients) is assimilated (see Figure four in Liu et al. 2007). Therefore, the model spin-up also has direct consequence on estimation of the core state.

How could we improve geomagnetic data assimilation systems within the observational limit? There are several areas for improvements. For example, improvements in global geomagnetic field modeling are needed since the Gauss coefficients from various field models have been used in most of the previous GDAS studies. Currently, there are many field models covering different epochs (e.g., Jackson et al. 2000; Korte et al. 2011; Gillet et al. 2013; Olsen et al. 2014; Sabaka et al. 2015). A unified field model covering the longest possible period could certainly reconcile differences in these models and thus help greatly GDAS systems. There is an ongoing effort on constructing a unified global field model for the past millennium (private communication with Korte). The field model error statistical information of such unified field models, such as those in Gillet et al. (2013), is also necessary for GDAS.

Improvement in the assimilation algorithms could also help data utilization. Some efforts were made by Kuang et al. (2010) in which a subset of the Gauss coefficients (of lower degrees) with much longer records are assimilated first to spin up the model, followed by assimilating those of higher degrees for the past 100 years. Tangborn and Kuang (2015) showed, via a set of experiments, that such assimilation methodology can have positive impact on core state and improve accuracies of predicting the subset of the Gauss coefficients not assimilated. Another example is employment of ensemble Kalman filtering (EnKF) approach (Evensen 1994). Fournier et al. (2011, 2013) used OSSEs to show the potential to speed up the transfer of information from geomagnetic data to the core state. But such speedy transfer depends on model errors (that are in general very large due to limitations of numerical dynamo models) not considered in their studies. It should also point out that GDAS is computationally very expensive. Such expense needs to be considered in the algorithm improvement.

Another improvement is on exploiting and utilizing further geodynamic information embedded in surface geomagnetic measurements. An immediate candidate for such exploitation is the geomagnetic secular variation (SV), described by the first-order time derivative $\left(\dot{g}_l^m, \dot{h}_l^m\right)$ of the Gauss coefficients since, as we will describe in the next section, they provide additional constraints on the core flow beneath the core-mantle boundary (CMB) and on the radial variation of the magnetic field. The former is not new, as there is a long history of, started from Roberts and Scott (1965), core flow inversion from observed SV at the Earth's surface via the "frozen-flux" approximation (in which the Ohmic dissipation beneath the CMB is ignored). However, this approximation comes with the price: the core flow cannot be uniquely inverted (e.g., Roberts and Scott 1965; Backus 1968). Thus, additional constraints on core flow properties are necessary in such core flow inversion studies (for more complete reviews, please read, e.g., Holme 2007; Kuang and Tangborn 2011). If the Ohmic dissipation is retained (no "frozen flux" approximation), then the observed SV imposes the constraints on the radial variation of the field in the core, as the latter is part of the magnetic induction. Since both field advection and Ohmic dissipation are included in geodynamo modeling, both kinds of constraints can be examined in MoSST_DAS or any other GDAS system without mathematical difficulties.

Therefore, a natural expansion of data utilization in GDAS is to assimilate both the field and its SV, so that the embedded geodynamic constraints can be used to make more optimal analysis, thus speeding up the transport of information from the surface geomagnetic observations to the dynamical state in the outer core. Since the SV is not included in the state vector of numerical geodynamo models, it will be connected through a non-linear observation operator, $\mathcal{H}$, which transforms the model state space to the observations space. Obviously, $\mathcal{H}$ will depend on, among others, fundamental physical properties of the magnetic field.

It should be pointed out here that assimilating the rate of change of geodynamic observables has been routinely used in NWP. For example, precipitation rate measured from a variety of satellite instruments is assimilated, despite not being a state variable in a GCM (Hou et al. 2000).

There is also a long history of attempting to invert core state from surface observations. In addition to core flow inversion initiated by Roberts and Scott (1965), Zatman and Bloxham et al. (1997) extended further the inversion of the poloidal magnetic field in the outer core. More recently, Aubert (2013, 2014) attempted a new approach to invert core dynamical state with the observed SV and the dynamo models. These inversions, however, are not data assimilation. However, the inversion results could benefit geomagnetic data assimilation. For example, the inversion by Aubert (2013, 2014) could be utilized for making "analysis" in a geomagnetic assimilation system.

In this paper, we describe in detail the results from our recent effort on assimilation of both the field and its SV. These results, from a series of experiments, will demonstrate the improvement in prediction and knowledge on core flow responses to the SV assimilation. The results also provide valuable information for further development in this direction.

This paper is organized as follows: the numerical model details and the mathematical formulation for SV assimilation will be given in the next section. Followed are the experimental results we have with this assimilation approach. Discussions and plans for further improvements are presented in the last section.

Methods

The mathematical formulation for SV assimilation depends on the numerical geodynamo models and the assimilation algorithms, in addition to the physics controlling the time variation of the magnetic field. In this section, we provide the mathematical methodologies used in MoSST_DAS employed in this study (Kuang et al. 2008; Sun and Kuang 2015). But, with some modifications, they can be applied to other GDAS systems.

Dynamo state vector and geomagnetic observation

MoSST_DAS utilizes the MoSST core dynamics model for time integration of the magnetic field (Kuang et al. 2008; Sun and Kuang 2015). In this system, the state vector $\mathbf{x}$

$$\mathbf{x} = (\mathbf{v}, \mathbf{B}, \delta\varrho)^T \tag{2}$$

includes the velocity field $\mathbf{v}$ and the density anomaly $\delta\varrho$ in the outer core $r_i \leq r \leq r_c$ (r_i and r_c are the mean radii of the inner core boundary (ICB) and CMB, respectively); and the magnetic field $\mathbf{B}$ in the outer core, the electrically conducting inner core $r \leq r_i$, and the D''-layer $r_c \leq r \leq r_d$ (r_d is the mean radius at the top of the layer). The superscript "T" in (2) implies the transpose. The solid mantle above the D''-layer $r_d \leq r \leq r_s$ (r_s is the mean radius of the Earth's surface) is electrically insulating. The whole system is defined in the reference frame fixed with the solid mantle.

The velocity field **v** and the magnetic field **B** are decomposed into the poloidal and toroidal components, with the scalars described via spherical expansions

$$(\mathbf{v}, \mathbf{B})^T = \nabla \times \left[(T_v, T_b)^T \mathbf{1}_r\right] + \nabla \times \nabla \times \left[(P_v, P_b)^T \mathbf{1}_r\right], \tag{3}$$

$$(P_v, T_v, P_b, T_b, \delta\varrho)^T = \sum_{0 \le m \le l}^{L_M} \left(v_l^m, \omega_l^m, b_l^m, j_l^m, \vartheta_l^m\right)^T \times Y_l^m(\theta, \phi) + C.C., \tag{4}$$

where $\mathbf{1}_r$ is the unit radial vector, θ is the co-latitude, ϕ is the longitude, Y_l^m are the fully normalized spherical harmonic functions of degree l and order m, L_M is the truncation order, and $C.C.$ implies the complex conjugate part. P and T in (3) are called the poloidal and toroidal scalars. It is therefore convenient to write

$$\mathbf{x} = \left(\mathbf{x}_v, \mathbf{x}_\omega, \mathbf{x}_b, \mathbf{x}_j, \mathbf{x}_\rho\right)^T, \tag{5}$$

where the subsets are defined with the relevant spectral coefficients in (4), e.g.,

$$\mathbf{x}_b = \left\{b_l^m(r_k) \mid 0 \le r_k \le r_d;\ 0 \le m \le l \le L_M\right\}^T \tag{6}$$

for the poloidal magnetic field. (5) and (6) can be different for other dynamo models.

In geomagnetic field modeling, geomagnetic measurements are used to obtain the magnetic field $\mathbf{B}^o$ originated from the core (simply called the geomagnetic field hereafter) that is described as

$$\mathbf{B}^o = -\nabla\Psi, \tag{7}$$

$$\Psi = r_s \sum_{0 \le m \le l}^{L_o} \left(\frac{r_s}{r}\right)^{l+1} \left(g_l^m \cos m\phi + h_l^m \sin m\phi\right) P_l^m(\theta) \tag{8}$$

where P_l^m is the Schmidt normalized associate Legendre polynomial of degree l and order m, $\left(g_l^m, h_l^m\right)$ are the Gauss coefficients (slightly different from the standard notation), and L_o is the maximum degree ($L_o \le 13$ in general). Since these Gauss coefficients $\left(g_l^m, h_l^m\right)$ are provided by different field models over the past 10,000 years (e.g., Jackson et al. 2000; Korte et al. 2005, 2011; Gillet et al. 2013; Olsen et al. 2014; Sabaka et al. 2015), they are used as the "observations" in our study.

By (3), (4), (7), and (8), we can obtain the relationship between $\left(g_l^m, h_l^m\right)$ in (8) and the observed $b_l^{m(o)}$ in the form of (4) via the radial component B_r of the magnetic field **B**

$$\begin{aligned} B_r^o &= -\frac{\partial\Psi}{\partial r} = \sum_{0 \le m \le l}^{L_o} (l+1) \left(\frac{r_s}{r}\right)^{l+2} \\ &\quad \times \left(g_l^m \cos m\phi + h_l^m \sin m\phi\right) P_l^m(\theta) \\ &= -\frac{\hat{L}}{r^2} P_b^o = \sum_{0 \le m \le l}^{L_o} \frac{l(l+1)}{r^2} b_l^{m(o)} Y_l^m + C.C. \end{aligned} \tag{9}$$

where $\hat{L}$ is the angular momentum operator. With the definitions of Y_l^m and P_l^m, (9) requires that

$$\begin{aligned} b_l^{m(o)}(r) &= \frac{r_s^2}{l} \left(\frac{r_s}{r}\right)^l G_m \left(g_l^m - i h_l^m\right), \\ G_m &= \left[\frac{2\pi(1 + \delta_{m0})}{2l+1}\right]^{1/2} \end{aligned} \tag{10}$$

for $r_d \le r \le r_s$. The spectral coefficients of the SV are the time derivatives of (10):

$$\dot{b}_l^{m(o)}(r) = \frac{r_s^2}{l} \left(\frac{r_s}{r}\right)^l G_m \left(\dot{g}_l^m - i \dot{h}_l^m\right) \quad \text{for} \quad r_d \le r \le r_s, \tag{11}$$

where $(\dot{\ })$ means the time derivative.

SV and core state

Geomagnetic observations only provide the time series of $\left(g_l^m, h_l^m\right)$. The SV coefficients $\left(\dot{g}_l^m, \dot{h}_l^m\right)$ are actually derived. Assimilation of the SV thus raises two major concerns: could the SV be approximated as "instantaneously" measured and whether it is redundant to the assimilation of the field?

Answers to the first concern depend on the significance of numerical errors in SV calculation. Consider, for example, a central difference scheme is used:

$$\dot{g}_l^m(t) \approx \frac{g_l^m(t + \delta t) - g_l^m(t - \delta t)}{2\delta t}.$$

Then, the relative numerical error can be estimated as

$$\frac{g_l^m(t + \delta t) - g_l^m(t - \delta t)}{2\delta t} = \dot{g}_l^m(t) + \mathcal{O}\left[\frac{\partial^3 g_l^m}{\partial t^3} (\delta t)^2\right]$$

Since $\partial^3 g_l^m / \partial t^3 \approx \dot{g}_l^m / \tau_l^2$ by (1) and $\delta t \approx t_o$ (the typical time intervals of data series), we have

$$\begin{aligned} \frac{g_l^m(t + \delta t) - g_l^m(t - \delta t)}{2\delta t} &= \dot{g}_l^m(t) \left[1 + \mathcal{O}\left(\tau_o/\tau_l\right)^2\right] \\ &\equiv \dot{g}_l^m(t) \left(1 + \epsilon_n\right) \end{aligned}$$

In general, $\tau_o \le 1$ month in the field models using modern observatory and satellite data (e.g., Sabaka et al. 2004, 2015; Olsen et al. 2006, 2014), while τ_l can be as short as 10 years (Christensen et al. 2012) for high-degree coefficients. Thus, $\epsilon_n \approx 10^{-6}$, which leads to an order

10^{-4} nT/year error in SV. On the other hand, the external field is several tens of nanotesla at the Earth's surface (Sabaka et al. 2015) on the solar cycle (~11 years) and shorter time scales. Thus, ϵ_n is negligible compared to those arising from, e.g., separation of the external and the internal magnetic signals. One could then argue that both the field and its SV are "concurrently" measured.

The redundancy is not an issue because the observed SV brings different knowledge of the core state $\mathbf{x}$ compared to the observed field. To see this, let us consider the magnetic induction of the poloidal magnetic field beneath the impenetrable and "free-slip" CMB ($r = r_c^-$)

$$\dot{b}_l^m = -\frac{r^2}{l(l+1)}\left[\nabla_h \cdot (\mathbf{v}_h B_r)\right]_l^m + \eta\left[\frac{\partial^2}{\partial r^2} - \frac{l(l+1)}{r^2}\right] b_l^m, \tag{12}$$

and in the D''-layer

$$\dot{b}_l^m = \eta_d\left[\frac{\partial^2}{\partial r^2} - \frac{l(l+1)}{r^2}\right] b_l^m. \tag{13}$$

In (12), the subscript "h" implies the horizontal components of the velocity field $\mathbf{v}$ and η is the magnetic diffusivity of the outer core fluid; η_d in (13) is the magnetic diffusivity of the D''-layer ($\eta \le \eta_d$ in general). These two equations show clearly that the observed $\dot{b}_l^{m(o)}$ will impose the constraint on $\mathbf{v}$ and on the non-potential part of the poloidal field.

The latter, i.e., (13), implies that, at the top of the D''-layer ($r = r_d$), a potential poloidal field $b_l^{m(p)}$ can fully recover the observed field $b_l^{m(o)}$. However, it cannot recover the observed SV $\dot{b}_l^{m(o)}$ since

$$\frac{\partial^2 b_l^{m(p)}}{\partial r^2} - \frac{l(l+1)}{r^2} b_l^{m(p)} = 0.$$

In other words, the observed SV provides the information on the non-potential part of the field that is missing in the observed field $b_l^{m(o)}$. Therefore, SV assimilation is not redundant to the field assimilation.

Indeed, our assimilation results (in Fig. 2) demonstrate clearly that assimilation of $b_l^{m(o)}$ could not reduce the differences between the forecast SV $\dot{b}_l^{m(f)}$ and the observed SV $\dot{b}_l^{m(o)}$, called ($\mathcal{O}$-$\mathcal{F}$) of the SV, although that of the field is reduced very rapidly in the first few analysis cycles, a strong indication for the need of SV assimilation.

New assimilation approach

We have been using the sequential assimilation approach in MoSST_DAS (e.g., Kuang et al. 2008; Sun and Kuang 2015). It can be summarized as follows: at the analysis time t_a when the observation $\mathbf{y}$ is made, a new initial condition $\mathbf{x}^a$ (called the "analysis") is made from the forecast $\mathbf{x}^f$ and the observation $\mathbf{y}$, future forecast for $t > t_a$ can then be made with the following initial value system:

$$\frac{\partial \mathbf{x}^f}{\partial t} = \mathbf{M}\left(\mathbf{x}^f\right), \qquad \mathbf{x}^f(t_a) = \mathbf{x}^a. \tag{14}$$

If there is a linear observation operator $\mathbf{H}$ that projects $\mathbf{x}$ to the observation space (where $\mathbf{y}$ is defined), then the analysis $\mathbf{x}^a$ is of the form

$$\mathbf{x}^a = \mathbf{x}^f + \mathbf{K}\left(\mathbf{y} - \mathbf{H}\mathbf{x}^f\right) \tag{15}$$

$$\mathbf{K} = \mathbf{P}^f\mathbf{H}^T\left(\mathbf{H}\mathbf{P}^f\mathbf{H}^T + \mathbf{R}\right)^{-1} \tag{16}$$

where $\mathbf{K}$ is called the gain matrix, $\mathbf{P}^f$ and $\mathbf{R}$ are the error covariance matrices of the forecast $\mathbf{x}^f$ and of the observation $\mathbf{y}$, respectively. (15) is obtained to minimize the error $\left|\mathbf{H}\cdot(\mathbf{x}^t - \mathbf{x}^a)\right|^2$ between the analysis $\mathbf{x}^a$ and the truth $\mathbf{x}^t$.

In our assimilation system MoSST_DAS, the error covariance $\mathbf{P}^f$ is calculated via three different approaches: an ensemble-based covariance analysis (Sun et al. 2007; Sun and Kuang 2015), an empirical covariance based on forecast solution properties (Tangborn and Kuang 2015), and an optimal interpolation (OI) scheme with a predefined time-invariant covariance (Kuang et al. 2009). The first approach is computationally very expensive, in which an ensemble N initial states are created $\mathbf{x}_i = \mathbf{x}^f + \epsilon_i$ (ϵ_i are random white noise perturbations). Their free running model solutions at a later time (often a fraction of the magnetic free decay time) are then used to calculate the covariance $\mathbf{P}^f = \langle(\mathbf{x}_i - \overline{\mathbf{x}})(\mathbf{x}_i - \overline{\mathbf{x}})^T\rangle$ ($\overline{\mathbf{x}} = \sum \mathbf{x}_i/N$ is the mean). The empirical covariance is assumed diagonal (no cross covariance between different degrees l and orders m) and is determined by the forecast error standard deviations and an exponentially decaying spatial correlation function. This approach can be updated at analysis time without much computational effort. The OI-type error covariance is similar to the empirical one, except that the forecast error standard deviations are assumed time-invariant (i.e., constant throughout the assimilation) and that spatial correlation does not decay exponentially with the distance. We simply use the OI-type error covariance in this study.

If only the observed field is assimilated, then

$$\mathbf{y} = \left\{b_l^{m(o)}(r_d)\middle| 0 \le m \le l \le L_o\right\}^T \equiv \mathbf{y}_b. \tag{17}$$

By (2) and (5), $\mathbf{H}$ is linear and very simple

$$\mathbf{H} = (\mathbf{0}, \mathbf{0}, \mathbf{H}_b, \mathbf{0}, \mathbf{0})^T, \tag{18}$$

where $\mathbf{H}_b$ corresponds to the subset $\mathbf{x}_b$ and has only nonzero entries for $b_l^m(r_d)$ with $0 \le m \le l \le L_o$. If the observed SV is also assimilated, then

$$\mathbf{y} = \left(\mathbf{y}_b, \mathbf{y}_{\dot{b}}\right)^T \tag{19}$$

$$\mathbf{y}_{\dot{b}} \equiv \left\{\dot{b}_l^{m(o)}(r_d) \mid 0 \le m \le l \le L_o\right\}^T, \tag{20}$$

However, by (12) and (13), transformation between $\mathbf{y}_b$ and $\mathbf{x}^f$ is a differential-functional projection and is denoted as $\mathcal{H}(\mathbf{x}^f)$. One could of course construct an independent projection system which evaluates $\mathcal{H}(\mathbf{x}^f)$ directly (see, e.g., Remarks 5.3.1 in Kalnay 2003). Alternatively, a linearization approximation $\mathcal{H}(\mathbf{x}^f) \approx \mathbf{H} \cdot \mathbf{x}^f$ could be made so that (15) can still be used.

There are different means to linearize $\mathcal{H}(\mathbf{x}^f)$. In our current study, we create an effective observed field $\widetilde{b}_l^{m(o)}$ defined in the D''-layer that matches both $b_l^{m(o)}$ and $\dot{b}_l^{m(o)}$. In this approach, $\widetilde{b}_l^{m(o)}$ comprises of a potential field that accounts for $b_l^{m(o)}$ and a non-potential field for $\dot{b}_l^{m(o)}$:

$$\widetilde{b}_l^{m(o)}(r) = \left(\frac{r_d}{r}\right)^l b_l^{m(o)}(r_d) + \frac{1}{2\eta_d} \times (r - r_d)^2 \dot{b}_l^{m(o)}(r_d) \quad \text{for} \quad r_c \le r \le r_d. \tag{21}$$

Obviously, at the top of the D''-layer $r = r_d$,

$$\widetilde{b}_l^{m(o)} = b_l^{m(o)}, \quad \frac{\partial \widetilde{b}_l^{m(o)}}{\partial r} = \frac{\partial b_l^{m(o)}}{\partial r} = -\frac{l}{r_d} b_l^{m(o)}, \quad \dot{\widetilde{b}}_l^{m(o)} = \dot{b}_l^{m(o)}.$$

The coefficients $b_l^m(r)$ in the D''-layer can be approximated via Taylor series expansion

$$b_l^m(r) = b_l^m(r_d) + b_l^{m\prime}(r_d)(r - r_d) + \frac{1}{2} b_l^{m\prime\prime} \times (r_d)(r - r_d)^2 + \mathcal{O}\left[b_l^{m\prime\prime\prime}(r - r_d)^3\right],$$

where $'$ implies the radial derivative. Since the approximation (21) satisfies the first three terms in the above expansion, the errors are given by the last term. Since $b_l^{m\prime\prime\prime} \sim b_l^m / r^3$ in D''-layer, the relative errors are of the order $[(r_d - r_c)/r_c]^3$. For a 20 km layer thickness, it is smaller than 10^{-6}. (21) allows us to extend the surface observations to the CMB. The observation vector $\mathbf{y}$ is now of the form

$$\mathbf{y} = \left\{\widetilde{b}_l^{m(o)}(r) \mid 0 \le m \le l \le L_o;\ r_c \le r \le r_d\right\}^T \equiv \mathbf{y}_{\widetilde{b}}. \tag{22}$$

The observation projection is again linear:

$$\mathcal{H}(\mathbf{x}^f) = \mathbf{H} \cdot \mathbf{x}^f \tag{23}$$

with $\mathbf{H}$ defined in (18). However, $\mathbf{H}_b$ now includes non-zero entries on all grid points in the D''-layer $r_c \le r \le r_d$.

We can use this approach to further construct an effective observed velocity field $\widetilde{\mathbf{v}}^o$ beneath the CMB $(r = r_c^-)$. Since $\dot{b}_l^m$ is continuous across the CMB, by (12), (13) and (21), we have

$$-\frac{r_c^2}{l(l+1)}\left[\nabla_h \cdot (\widetilde{\mathbf{v}}_h^o \widetilde{B}_r^o)\right]_l^m + \eta\left[\frac{\partial^2}{\partial r^2} - \frac{l(l+1)}{r_c^2}\right]\widetilde{b}_l^{m(o)} = \dot{\widetilde{b}}_l^{m(o)}(r_c) = \dot{b}_l^{m(o)}(r_d)\left[1 - \frac{l(l+1)}{2r_c^2}(r_c - r_d)^2\right] \tag{24}$$

For those geodynamo models without an electrically conducting D''-layer, one can simply replace $\dot{\widetilde{b}}_l^{m(o)}(r_c)$ in the above equation with $\dot{b}_l^{m(o)}(r_c)$ continued downward from the surface observation.

Obviously, (24) is an underdetermined system, since both $\widetilde{b}_l^{m(o)}$ and $\widetilde{\mathbf{v}}^o$ are unknown at r_c^-. But one can find the "best-fit" $\widetilde{\mathbf{v}}^o$ and $\widetilde{b}_l^{m(o)}$ via minimizing the following difference

$$\min_{\widetilde{\mathbf{v}}^o, \widetilde{b}_l^{m(o)}} \left| \dot{\widetilde{b}}_l^{m(o)}(r_c) + \frac{r_c^2}{l(l+1)}\left[\nabla_h \cdot (\mathbf{v}_h B_r)\right]_l^m - \eta\left[\frac{\partial^2}{\partial r^2} - \frac{l(l+1)}{r_c^2}\right] b_l^m \right|^2. \tag{25}$$

We provide here only a sketch of the minimization procedure: denoting by $\mathbf{x}_b$ and $\mathbf{x}_v$ the vectors of the spectral coefficients of B_r and of $\mathbf{v}_h$ in (25), respectively, and then (25) is equivalent to minimize the (inner) product

$$[\dot{\mathbf{x}}_b + \mathbf{A} \cdot \mathbf{x}_b + \mathbf{B}(\mathbf{x}_b) \cdot \mathbf{x}_v]^T \cdot [\dot{\mathbf{x}}_b + \mathbf{A} \cdot \mathbf{x}_b + \mathbf{B}(\mathbf{x}_b) \cdot \mathbf{x}_v] \tag{26}$$

with respect to $\mathbf{x}_v$. Note that $\mathbf{A} \cdot \mathbf{x}_b$ and $\mathbf{B}(\mathbf{x}_b) \cdot \mathbf{x}_v$ in (26) describe the diffusion and the advection terms in (25), respectively. Minimizing the product (26) leads to the solution

$$\left(\mathbf{B}^T \cdot \mathbf{B}\right) \cdot \tilde{\mathbf{x}}_v = -\mathbf{B}^T \cdot [\dot{\mathbf{x}}_b + \mathbf{A} \cdot \mathbf{x}_b] \tag{27}$$

which provides the spectral coefficients of the effective observed velocity field $\widetilde{\mathbf{v}}^o(r_c^-)$ beneath the CMB. If $\widetilde{\mathbf{v}}^o(r_c^-)$ is included, then the observation vector $\mathbf{y}$ is

$$\mathbf{y} = \left(\mathbf{y}_{\widetilde{v}}, \mathbf{y}_{\widetilde{\omega}}, \mathbf{y}_{\widetilde{b}}\right)^T \tag{28}$$

where $\mathbf{y}_{\widetilde{\omega}}$ includes, as shown in (3) and (4), the spectral coefficients $\widetilde{\omega}_l^{m(o)}$ of $\widetilde{\mathbf{v}}^o$ at r_c^-. Again, the linearized observation projection (23) is achieved. However, $\mathbf{H}$ includes additional subsets:

$$\mathbf{H} = (\mathbf{H}_v, \mathbf{H}_\omega, \mathbf{H}_b, \mathbf{0}, \mathbf{0})^T, \tag{29}$$

where $\mathbf{H}_v$ and $\mathbf{H}_\omega$ include only non-zero entries for $\widetilde{v}_l^{m(o)}$ and $\widetilde{\omega}_l^{m(o)}$ at $r = r_c^-$, respectively.

Effective observation error covariance

Since the gain matrix $\mathbf{K}$ in (16) depends on the observation error covariance $\mathbf{R}$, we need to determine the effective error covariance $\widetilde{\mathbf{R}}$ for $\widetilde{b}_l^{m(o)}$ which can be calculated from those of the Gauss coefficients g_l^m and h_l^m. In this section, we only describe a formal procedure without going into the details.

In geomagnetic field modeling (Jackson et al. 2000; Sabaka et al. 2004; Korte et al. 2005; Olsen et al. 2006; Gillet et al. 2013), the Gauss coefficients, e.g. g_l^m, can be described in general as

$$g_l^m = \mathbf{S}^T(t) \cdot \boldsymbol{\alpha}_{lm}, \tag{30}$$

where $\mathbf{S}$ is the vector describing deterministic, model-specific base functions in the time domain, e.g., B-spline functions, and $\boldsymbol{\alpha}$ is the coefficient vector which includes the observation error statistics.

For illustrative purpose, we use the simplest error statistics for our derivation. Assume that geomagnetic observations (and thus $\boldsymbol{\alpha}$) are unbiased and with known error covariances:

$$\boldsymbol{\alpha}_{lm} = \boldsymbol{\alpha}_{lm}^t + \boldsymbol{\epsilon}_\alpha, \quad \langle \boldsymbol{\epsilon}_\alpha \rangle = \mathbf{0}, \quad \left\langle \boldsymbol{\epsilon}_\alpha \boldsymbol{\epsilon}_\alpha^T \right\rangle = \mathbf{C}_\alpha, \tag{31}$$

where $\boldsymbol{\alpha}_{lm}^t \equiv \langle \boldsymbol{\alpha}_{lm} \rangle$ is the truth (expectation) and $\mathbf{C}_\alpha$ is the observation error covariance matrix of $\boldsymbol{\alpha}_{lm}$. Thus, by (30),

$$\begin{aligned} &g_l^m = g_l^{m(t)} + \epsilon_g, \quad g_l^{m(t)} = \mathbf{S}^T \cdot \boldsymbol{\alpha}_{lm}^t, \quad \epsilon_g = \mathbf{S}^T \cdot \boldsymbol{\epsilon}_\alpha, \\ &\left\langle \epsilon_g^2 \right\rangle = \mathbf{S}^T \cdot \mathbf{C}_\alpha \cdot \mathbf{S} \equiv R_g^{lm}. \end{aligned} \tag{32}$$

Similar formulation applies to h_l^m, but with the error ϵ_h. By (10) and (32), we have

$$b_l^{m(o)}(r) = b_l^{m(t)}(r) + \epsilon_b(r), \tag{33}$$

$$b_l^{m(t)}(r) = \frac{r_s^2}{l}\left(\frac{r_s}{r}\right)^l G_m\left(g_l^{m(t)} - i h_l^{m(t)}\right), \tag{34}$$

$$\epsilon_b(r) = \frac{r_s^2}{l}\left(\frac{r_s}{r}\right)^l G_m\left(\epsilon_g - i\epsilon_h\right) \tag{35}$$

This leads to

$$\langle \epsilon_b \epsilon_b^* \rangle = \left(\frac{r_s^2}{l}\right)^2 \left(\frac{r_s}{r}\right)^{2l} G_m^2 \left[\left(R_g^{lm}\right)^2 + \left(R_h^{lm}\right)^2\right] \tag{36}$$

One can use this equation to evaluate the covariance at any location in the mantle, including $r = r_d$ the top of the D''-layer. If $\mathbf{S}$ in (32) is replaced by $\dot{\mathbf{S}}$, then we can obtain the covariance $R_{\dot{g}}^{lm}$ of the SV

$$R_{\dot{g}}^{lm} = \dot{\mathbf{S}}^T \cdot \mathbf{C}_\alpha \cdot \dot{\mathbf{S}},$$

and therefore the variance of $\dot{b}_l^{m(o)}$

$$\dot{b}_l^{m(o)}(r) = \dot{b}_l^{m(t)}(r) + \epsilon_{\dot{b}}(r), \tag{37}$$

$$\left\langle \epsilon_{\dot{b}} \epsilon_{\dot{b}}^* \right\rangle(r) = \left(\frac{r_s^2}{l}\right)^2 \left(\frac{r_s}{r}\right)^{2l} G_m^2 \left[\left(R_{\dot{g}}^{lm}\right)^2 + \left(R_{\dot{h}}^{lm}\right)^2\right]. \tag{38}$$

The full error covariance of $\widetilde{b}_l^{m(o)}(r)$ can then be determined from (21), (36), and (38).

Results

In this study, we focus only on (22), i.e., assimilation of the effective observed field $\widetilde{b}_l^{m(o)}$ which matches both the observed field $b_l^{m(o)}$ and the observed SV $\dot{b}_l^{m(o)}$ at the top of the D''-layer, mainly for two goals: to explore improvements of the assimilation system with the observed SV, such as the model spin-up process and *rms* of the observed minus forecast ($\mathcal{O}$-$\mathcal{F}$) of the magnetic field; and to understand responses of the core state $\mathbf{x}$ to the observed SV, in particular changes of the velocity field $\mathbf{v}$ beneath the CMB. Both are critical for determination of the effective velocity field $\widetilde{\mathbf{v}}$ in (25) and thus for implementation of the more comprehensive observation (28).

The baseline geodynamo model is the MoSST core dynamics model (Kuang and Chao 2003; Jiang and Kuang 2008) for the thermal convection of a Boussinesq electrically conducting fluid in the (rapidly rotating) outer core, confined between the electrically conducting inner core and the D''-layer. The non-dimensional parameters include the Ekman number E (for viscosity), the magnetic Rossby number R_o (for magnetic diffusivity), the modified Prandtl number q_κ (for thermal conductivity), and the modified Rayleigh number R_{th} (for buoyancy force). In our assimilation, the ICB and the CMB are assumed impenetrable, stress-free, and fixed heat fluxes. We also select the following parameter values:

$$E = R_o = 1.25 \times 10^{-6}, \quad R_o = 1.0 \quad R_{th} = 15 R_{th}^c, \tag{39}$$

where R_{th}^c is the critical Rayleigh number for purely thermal convection. In our assimilation, the numerical truncation order is $L_M = 96$. There are 20 grid points in D''-layer (which is 20 km thick with $\eta_d = 20\,\eta$), 80 grid points in the outer core, and 40 grid points in the inner core. With this set of the parameter values, the mean time scale of the dipole field is approximately 0.7 % of the magnetic diffusive time τ_η (which is used for the time scaling of the dynamo model) and those for non-dipole components are more than an order of magnitude shorter. These are very similar to those derived from satellite magnetic data. Therefore, in our assimilation, we choose $\tau_\eta = 200,000$ years to convert non-dimensional dynamo time to years (with this conversion, the mean time scale of the dipole field is approximately 1400 years).

We consider only the observations for the time period 1900 – 2000 simply because modern observatory and satellite data provide very high quality $\left(g_l^m, h_l^m\right)$ and $\left(\dot{g}_l^m, \dot{h}_l^m\right)$. These coefficients are from *gufm1* (Jackson et al. 2000) for 1900–1962 and CM4 (Sabaka et al. 2004) for 1962–2000. We also set $L_o = 8$, lower than the highest degrees of the two models. For our research purposes, we carry out three distinct experiments:

Case I: Free-running model (no assimilation)
Case II: Assimilation of $b_l^{m(o)}$ with (17)
Case III: Assimilation of $\widetilde{b}_l^{m(o)}$ with (22)

Except the differences in the data $\mathbf{y}$ in analysis, everything else is identical in the experiments, including the original initial state at 1900. The analysis cycle is $\Delta t = 5$ years. By this design, we can identify exactly the causes of changes in the dynamo state $\mathbf{x}$: the differences between the solutions of case I and case II are due to assimilation of the observed field $b_l^{m(o)}$ and the differences between the solutions of case II and case III are due to the assimilation of the observed SV $\dot{b}_l^{m(o)}$. These allow us to understand clearly the responses of the core state to surface observations and their dynamical consequences.

We use a modeled observation error covariance, since the actual error covariances of the field models are not yet available. The model error covariance $\mathbf{R}$ is assumed diagonal, with the diagonal elements defined as

$$R_{lm} = \left|\epsilon_R(l) b_l^m\right|^2, \quad \epsilon_R(l) = \epsilon_0(t) + [\epsilon_1(t) - \epsilon_0(t)] \frac{l-1}{L_o - 1}, \tag{40}$$

where ϵ_0 and ϵ_1 decreases linearly in time: ϵ_0 decreases from 0.01 in 1900 to 0.001 in 2000 and ϵ_1 decreases from 0.3 in 1900 to 0.1 in 2000. These imply that the relative errors in (40) decreases in time, but increases with the degree l.

We would like to point out here that Gillet et al. (2013) provided a global field model which includes a full error covariance of the Gauss coefficients. This model and any future model with specified error statistic knowledge are more appropriate for GDAS. However, we conjecture that (40) is sufficient for our current objectives.

Responses of the magnetic field to SV assimilation

The quantities used to understand the responses of the magnetic field are the $(\mathcal{O}\text{-}\mathcal{F})$ of the radial magnetic field B_r and its SV $\dot{B}_r$. Instead of using traditional $(\mathcal{O}\text{-}\mathcal{F})$, we prefer the following modified definition

$$(\mathcal{O} - \mathcal{F})_B^2 = \sum_{1 \le l}^{L_o} \left\{ \left[\sum_{0 \le m \le l} \left| \frac{b_l^{m(o)}}{b_1^{0(o)}} - \frac{b_l^{m(f)}}{b_1^{0(f)}} \right|^2 \right] \times \left[\sum_{0 \le m \le l} \left| \frac{b_l^{m(o)}}{b_1^{0(o)}} \right|^2 \right]^{-1} \right\} \tag{41}$$

at $r = r_d$, i.e., the misfit normalized by the observed field strength for a given degree l. Replacing b_l^m by $\dot{b}_l^m$ in (41), we have $(\mathcal{O}\text{-}\mathcal{F})_{\dot{B}}$ of the SV. This modified $(\mathcal{O}\text{-}\mathcal{F})$ can tell us more accurately how close is the forecast to observation because it eliminates the effect of changes in the magnitude of the individual spectral coefficients. Our modified $(\mathcal{O}\text{-}\mathcal{F})$ in (41) is different from other quantities used for measuring the difference (or misfit) between the observation $\mathbf{y}$ and the forecast $\mathbf{H} \cdot \mathbf{x}^f$, such as the misfit normalized by the error covariances (e.g., Aubert 2014). Since, in our assimilation, the modeled error covariances are proportional in magnitude to the observed field strength; (41) is actually very similar to the normalized misfit.

Figure 1 is the $(\mathcal{O}\text{-}\mathcal{F})_B$ of Case II (dashed lines) and Case III (solid lines). From this figure, we can observe clearly that their magnitudes in Case III are approximately 30 % smaller than those in Case II over the entire assimilation period, demonstrating a substantial improvement in forecast accuracies with the SV assimilation (21) and (22). Similar improvement can be also observed from $(\mathcal{O}\text{-}\mathcal{F})_{\dot{B}}$ shown in Fig. 2.

The SV assimilation also helps accelerate the dynamo model spin-up process. For example, we can observe from Fig. 1 that the time variations of $(\mathcal{O}\text{-}\mathcal{F})_B$ are nearly identical in both cases: they decay nearly monotonically over much of the assimilation period before leveling off in the last 20 years (from 1980 to 2000). But $(\mathcal{O}\text{-}\mathcal{F})_{\dot{B}}$, as shown in Fig. 2, are very different in the two cases: in Case II, it increases first from 1900 to 1940 and only starts to decay continuously in the last 20 years. In Case III, however, $(\mathcal{O}\text{-}\mathcal{F})_{\dot{B}}$ decays almost monotonically in time, except two small surges around 1940 and 1980. This implies that the dynamo core state $\mathbf{x}^f$ responds stronger to the SV assimilation. In other words, the SV assimilation helps to accelerate the model spin-up process.

To better understand how do the forecasts $b_l^{m(f)}$ and $\dot{b}_l^{m(f)}$ respond to the observations $\mathbf{y}_b$ in (17) and $\mathbf{y}_{\widetilde{b}}$ in (22), we examine first the $(\mathcal{O}\text{-}\mathcal{F})$ for individual degrees. In Fig. 3 are $(\mathcal{O}\text{-}\mathcal{F})_B$ for the degrees $l \le 6$. Improvements are clearly shown in all six degrees, as all values are smaller in Case III than those in Case II. But we can also observe different patterns among these degrees. For example, $(\mathcal{O}\text{-}\mathcal{F})_B$ for the odd degrees ($l = 1, 3, 5$) increase in magnitude again from around 1980 to the end of the assimilation period. But there is no such clear reversing trend in those for the even degrees ($l = 2, 4, 6$): it decreases monotonically for $l = 2$; for $l = 4$, it oscillates with a damping amplitude; and the $(\mathcal{O}\text{-}\mathcal{F})$ for $l = 4$ remains nearly constant after the rapid decay in the first two analysis cycles.

As shown in Fig. 4, the difference between the odd and even degrees of $(\mathcal{O}\text{-}\mathcal{F})_{\dot{B}}$ is even more significant. There is still a strong surge in magnitude for $l = 3$ around 1980 in the both cases. But the reduction for $l = 5$ is minimal.

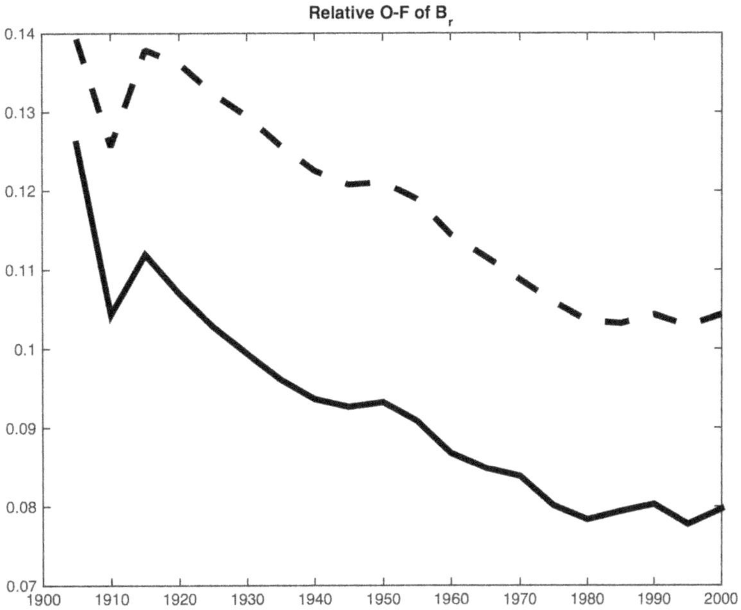

Fig. 1 The *rms* $(\mathcal{O}\text{-}\mathcal{F})_B$ of the magnetic field in case II (*dashed line*) and case III (*solid line*). In both cases, $(\mathcal{O}\text{-}\mathcal{F})_B \ll 1$ and decays monotonically after the first three analysis cycles and then levels off in the last 20 years. This shows the continuing improvement in the forecast accuracies. In addition, the $(\mathcal{O}\text{-}\mathcal{F})$ results in case III (with the assimilation of $b_l^{m(o)}$ and $\dot{b}_l^{m(o)}$) are in general more than 20 % smaller than in case II (with only the assimilation of $b_l^{m(o)}$), showing a clear improvement in forecast accuracies

In particular, it does not decay monotonically in time in either case. These differences may indicate potential inconsistencies between the core dynamics of the model and the time variation of the Gauss coefficients. We will discuss this again later in this paper.

Responses of the velocity field to SV assimilation

Why does the dynamo model respond faster and stronger in case III than in case II? We can find at least partial answers from the difference between the free-running model solutions $\mathbf{x}^M$ (case I) and the forecasts

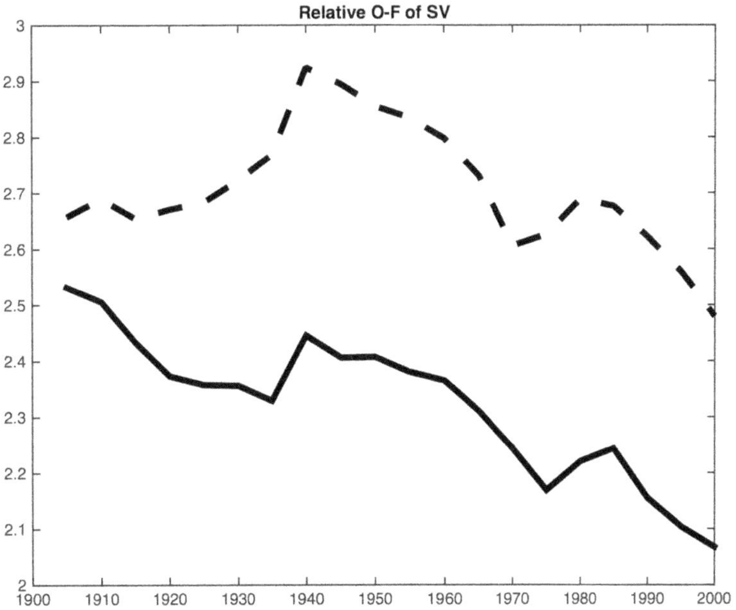

Fig. 2 Similar to Fig. 1, but for $(\mathcal{O}\text{-}\mathcal{F})_{\dot{B}}$ of the SV. In case II (*dashed line*), $(\mathcal{O}\text{-}\mathcal{F})_{\dot{B}} = \mathcal{O}(1)$ for much of the assimilation period before decays gradually in the last 20 years, implying that there is no similarity between the forecasted SV $\dot{b}_l^{m(f)}$ and the observed SV $\dot{b}_l^{m(o)}$. But its magnitude is much smaller in case III (*solid line*), and it decays monotonically in time, indicating that the SV assimilation accelerates the spin-up process

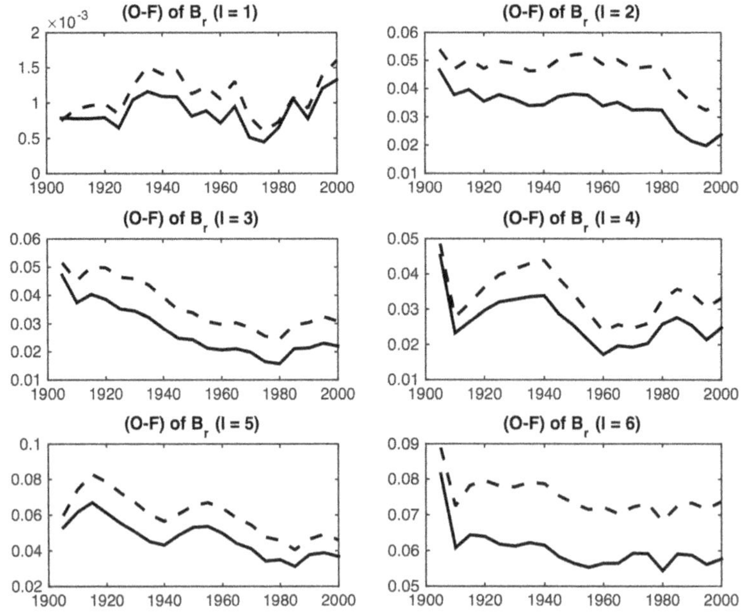

Fig. 3 The $(\mathcal{O}\text{-}\mathcal{F})_B$ of the first six spherical harmonic degrees in case II (*dashed lines*) and case III (*solid lines*)

$\mathbf{x}^f$ in cases II and III, in particular the differences in the velocity field $\mathbf{v}$ beneath the CMB, because they are the direct consequences of the magnetic induction (12). The knowledge is also very important for obtaining the "effective" observed velocity field (25) for future studies.

Since in our geodynamo model, the CMB is impenetrable and is free-slip, the radial velocity $v_r = 0$, and, by (3) and (4), the horizontal velocity $\mathbf{v}_h$ depends on $\partial v_l^m/\partial r$ and ω_l^m at $r = r_c$. Therefore, it is very convenient to examine the following two variables beneath the CMB:

$$v_r' \equiv \frac{\partial v_r}{\partial r} = \sum_{0 \leq m \leq L}^{L_M} \frac{l(l+1)}{r_c^2} \frac{\partial v_l^m}{\partial r} Y_l^m(\theta, \phi) + C.C. \quad (42)$$

$$\omega_r \equiv (\nabla \times \mathbf{v})_r = \sum_{0 \leq m \leq L}^{L_M} \frac{l(l+1)}{r_c^2} \omega_l^m Y_l^m(\theta, \phi) + C.C., \quad (43)$$

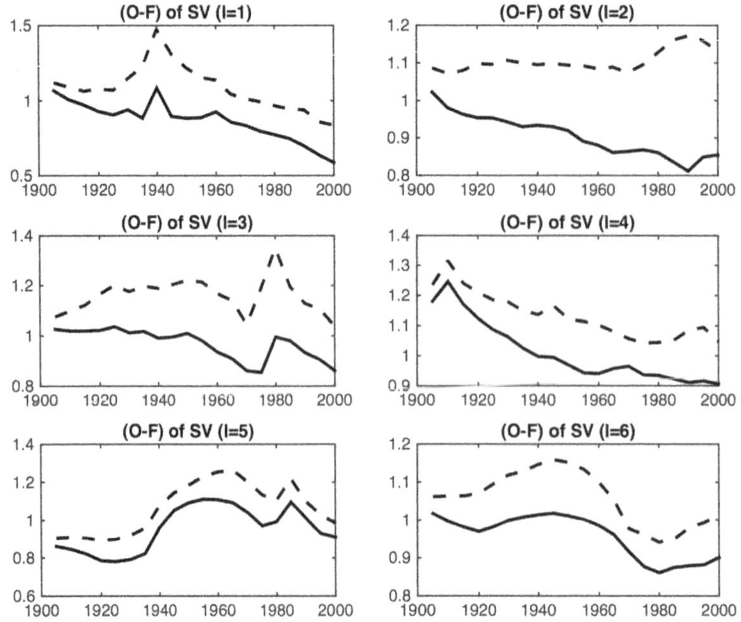

Fig. 4 Similar to Fig. 3, but for $(\mathcal{O}\text{-}\mathcal{F})_{\dot{B}}$

where v'_r is poloidal and describes the up-and-down welling and ω_r is toroidal and describes the differential rotation. The *rms* differences $(\mathcal{M}\text{-}\mathcal{F})$ between the two variables of the forecast $\mathbf{x}^f$ (cases II and III) and of the free-running model $\mathbf{x}^M$ (case I) can be used to quantify the responses of the core flow to the assimilation of surface observations:

$$(\mathcal{M}\text{-}\mathcal{F})_{v_P} \equiv \|v_r'^M - v_r'^f\|_2 = \left[\sum_{0\le m\le l}^{L_M} \frac{l^2(l+1)^2}{r_c^4}\left|\frac{\partial v_l^{m(M)}}{\partial r} - \frac{\partial v_l^{m(f)}}{\partial r}\right|^2\right]^{1/2} \quad (44)$$

$$(\mathcal{M}\text{-}\mathcal{F})_{v_T} \equiv \|\omega_r^M - \omega_r^f\|_2 = \left[\sum_{0\le m\le l}^{L_M} \frac{l^2(l+1)^2}{r_c^4}\left|\omega_l^{m(M)} - \omega_l^{m(f)}\right|^2\right]^{1/2} \quad (45)$$

In the above equations, $\|\cdot\|_2$ is the L_2–norm (or *rms*) over the CMB.

In Fig. 5 are the non-dimensional (with the scaling factor 5×10^{-6} year^{-1} for dimensional values) $\|v'_r\|_2$ (red) and $\|\omega_r\|_2$ (blue) of the free-running model (case I). As shown in the figure, v'_r increases slightly in magnitude in the assimilation period and ω_r remains flat. But, the *rms* differences $(\mathcal{M}\text{-}\mathcal{F})_{v_P}$ (shown in Fig. 6) and $(\mathcal{M}\text{-}\mathcal{F})_{v_T}$ (shown in Fig. 7) increase in time, i.e., a growing divergence between the forecast state $\mathbf{x}^f$ and the free-running model state $\mathbf{x}^M$.

From Figs. 6 and 7, we can also observe that $(\mathcal{M}\text{-}\mathcal{F})$ of case III (the solid lines) are slightly larger than those of case II (dashed lines), implying that $\mathbf{x}^f$ moves away from $\mathbf{x}^M$ faster with the SV assimilation (22), another demonstration of improved model spin-up with the SV assimilation. However, the differences are much less significant than those of the magnetic field. This suggests the need for the effective observed velocity field $\tilde{\mathbf{v}}^o$ to increase further $(\mathcal{M}\text{-}\mathcal{F})$ of the velocity field and thus to expedite the model spin-up process.

To aid the future study of determining the effective core flow from the observed SV via (25), we need to understand better the details of $(\mathcal{M}\text{-}\mathcal{F})$, e.g., their distributions in the spectral space defined by the spherical harmonic degrees l and orders m. We shall pay special attention to their distributions in l, i.e., the summation of the terms in (44-45) with $0 \le m \le l$ for a given degree l, and their distributions in m, i.e., the summation of the terms in (44-45) with $m \le l \le L_M$ for a given order m. Since, as shown in Figs. 6 and 7, the differences between the two cases are very small, we can focus only on case III without loss of generality.

In Fig. 8 is the distribution of $(\mathcal{M}\text{-}\mathcal{F})_{v_P}$ in the degree l and in Fig. 9 is its distribution in the order m. From the figures, we can find that $(\mathcal{M}\text{-}\mathcal{F})_{v_P}$ varies substantially in the spectral spaces. As shown in Fig. 8, the differences for the degrees $15 \le l \le 35$ increase the fastest in time,

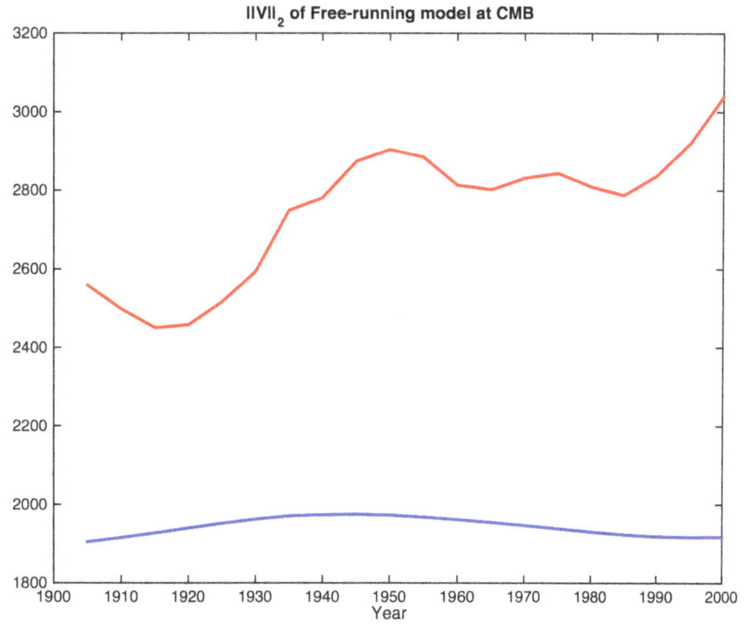

Fig. 5 The non-dimensional $\|v'_r\|_2$ (*red*) and $\|\omega_r\|_2/10$ (*blue*) beneath the CMB $r = r_c^-$ from the free-running model solutions (case I). The dimensional values can be obtained with the scaling factor 5×10^{-6} year^{-1}

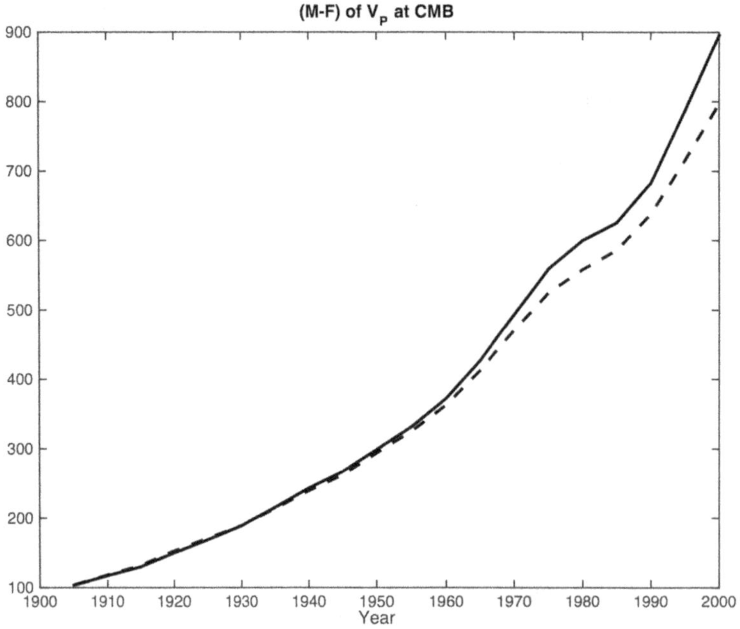

Fig. 6 The ($\mathcal{M}$-$\mathcal{F}$) of the poloidal velocity field v_r' as defined in (44). The *dashed lines* are the results without SV assimilation (case II) and the *solid lines* are those with the SV assimilation (case III)

and their magnitudes are the largest at the end of the assimilation period, with the peak at $l = 20$. The differences are much smaller and grows slower in time for the degrees $l \leq 5$ and $l \geq 40$. But, as shown in Fig. 9, the distribution in m is more broadband: the differences for $5 \leq m \leq 35$ increase rapidly in time and reach comparable values in magnitude at the end of the assimilation period. However, ($\mathcal{M}$-$\mathcal{F}$) for $m \leq 4$ are very different: they remain small and nearly unchanged throughout the entire assimilation. These suggest that the responses of the poloidal velocity is dominantly non-axisymmetric ($m > 0$).

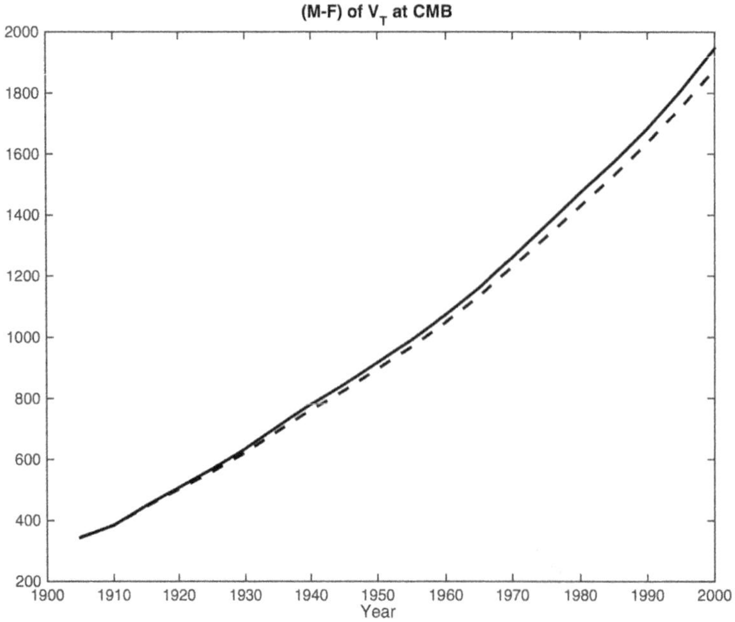

Fig. 7 Similar to Fig. 6, but for The ($\mathcal{M}$-$\mathcal{F}$) of the toroidal velocity ω_r as defined in (45)

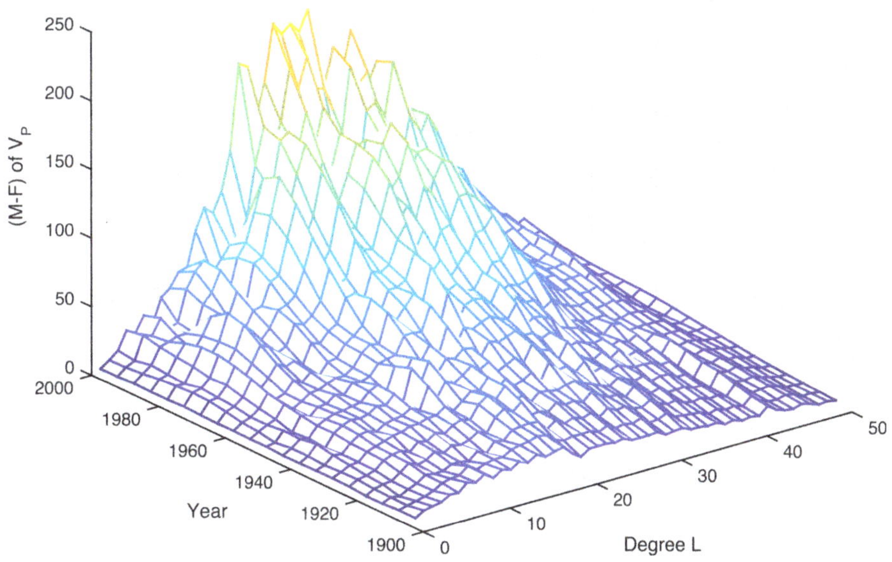

Fig. 8 The distribution of $(\mathcal{M}\text{-}\mathcal{F})_{v_P}$ in spherical harmonic degrees l with the SV assimilation (case III)

The distribution of $(\mathcal{M}\text{-}\mathcal{F})_{v_T}$ of the toroidal velocity, as shown in Figs. 10 and 11, displays both similar and distinct characteristics. Its distribution in l is very similar to that of $(\mathcal{M}\text{-}\mathcal{F})_{v_P}$, except that it peaks at a higher degree $l = 30$. But its distribution in m (Fig. 11) is very different: the differences for $m \leq 20$ remain comparable in both the magnitude and the time increasing rate. But they decay rapidly for larger m. It should be pointed out in particular that, opposite to $(\mathcal{M}\text{-}\mathcal{F})_{v_P}$ (in Fig. 9), $(\mathcal{M}\text{-}\mathcal{F})_{v_T}$ of the axisymmetric toroidal velocity ($m = 0$) remains very large, implying that the axisymmetric toroidal flow is very sensitive to the surface observations.

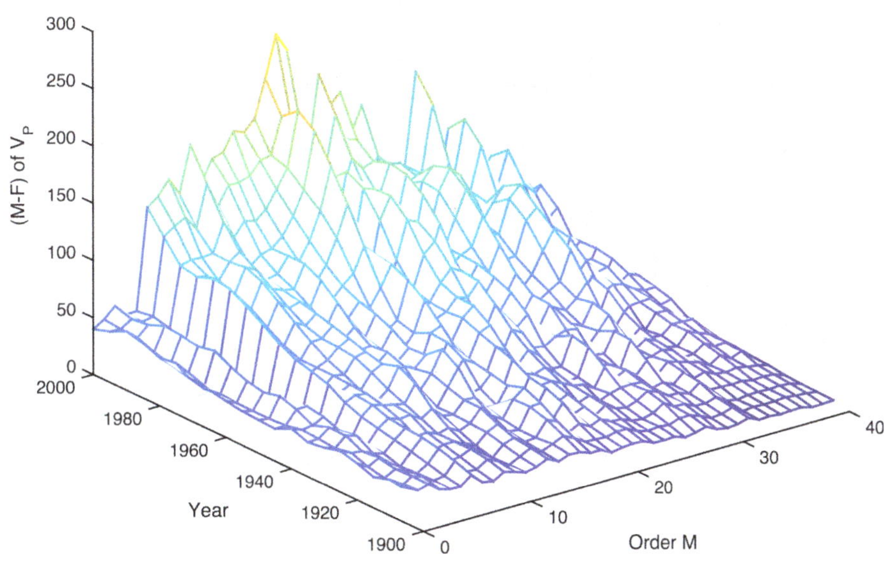

Fig. 9 Similar to Fig. 8, but for the distribution of $(\mathcal{M}\text{-}\mathcal{F})_{v_P}$ in spherical harmonic orders m

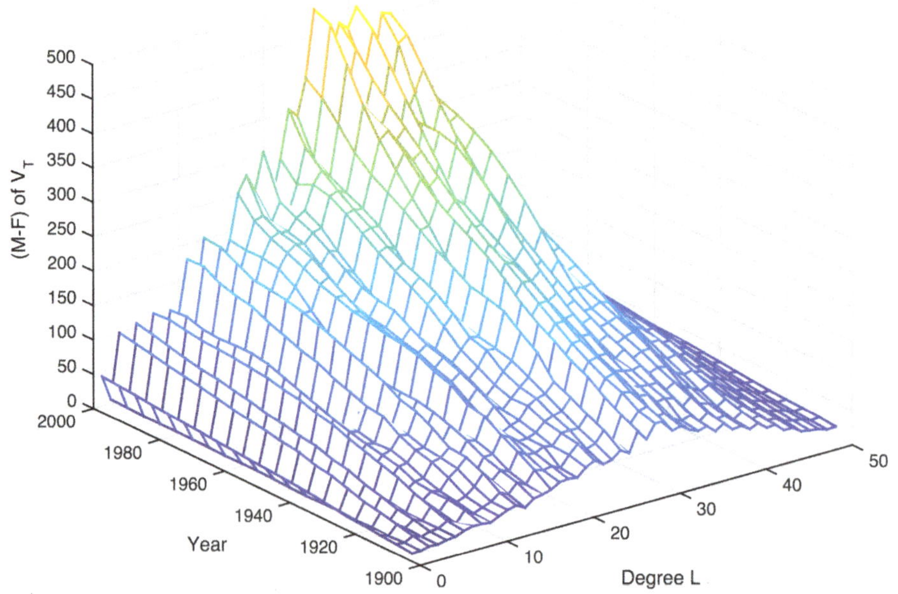

Fig. 10 Similar to Fig. 8, but for $(\mathcal{M}\text{-}\mathcal{F})_{V_T}$

Discussions and conclusions

In this study, we have examined the consequences of assimilating the observed SV on geomagnetic forecasts and on the responses of the dynamo core state. We argued that,because geomagnetic data sampling frequencies are several orders of magnitude higher than those of the SV, the geomagnetic field and its SV are concurrently measured. We further demonstrated that the observed SV provides unique knowledge of the magnetic field and the velocity field in the core. Thus, assimilations of the

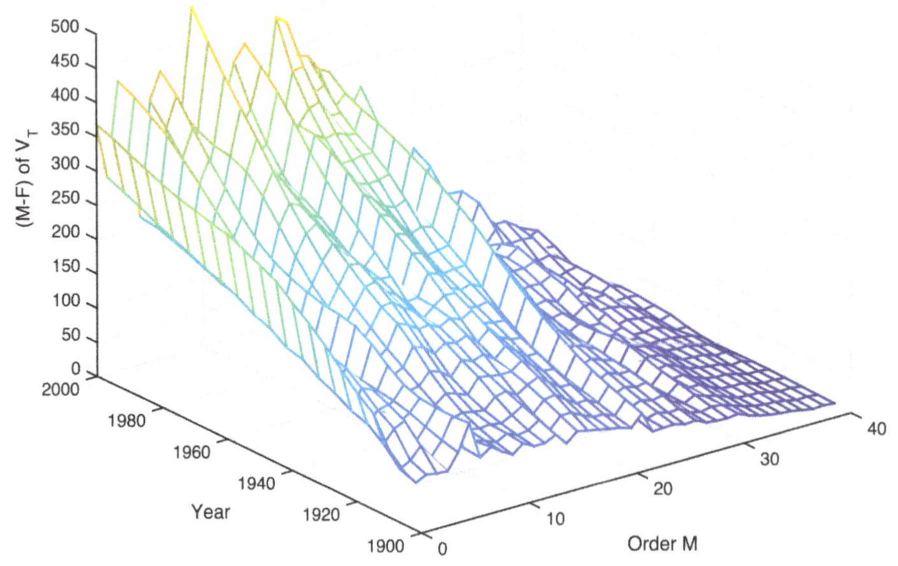

Fig. 11 Similar to Fig. 9, but for $(\mathcal{M}\text{-}\mathcal{F})_{V_T}$

observed field and of the observed SV are necessary and are not redundant.

In this study, we incorporate the observed SV into the observation vector $\mathbf{y}$ via introducing the effective poloidal field $\widetilde{b}_l^{m(o)}$ (21) in the D''-layer, which is then used in the sequential assimilation algorithm (15). We designed the three experiments to identify the impact of SV assimilation: a free-running model dynamo simulation (case I), an experiment with the assimilation of the observed field (case II), and an experiment with the assimilation of both the observed field and its SV. The relative $(\mathcal{O}\text{-}\mathcal{F})$ of the field and SV, defined in (41), at the top of the D''-layer are used to measure the forecast accuracies; the $(\mathcal{M}\text{-}\mathcal{F})$ of the poloidal velocity field (44) and of the toroidal velocity field (45) beneath the CMB are used to characterize the responses of the core state to the SV assimilation.

The results of our experiments demonstrate clearly that the SV assimilation with (21) improves significantly the geomagnetic forecast accuracies since, as shown in Figs. 1 and 2, both $(\mathcal{O}\text{-}\mathcal{F})_B$ and $(\mathcal{O}\text{-}\mathcal{F})_{\dot{B}}$ in case III are more than 20 % smaller than those in case II. In particular, the improvements occur to all degrees, as shown in Figs. 3 and 4. The nearly monotonic decay in time of $(\mathcal{O}\text{-}\mathcal{F})_{\dot{B}}$ in case III (Fig. 2) shows clearly that the SV assimilation accelerates the spin-up of the dynamo model.

The improvement by the SV assimilation can be also seen from the differences $(\mathcal{M}\text{-}\mathcal{F})_{v_P}$ and $(\mathcal{M}\text{-}\mathcal{F})_{v_T}$ between the free-running model state and those of the assimilations. As shown in Figs. 6 and 7, these differences grow rapidly in time, showing an accelerated departure of the core state with assimilation from the free-running model state. The differences in case III are slightly larger than those in case II, further demonstrating the improvement brought by the SV assimilation, though such increment is less significant that those in the $(\mathcal{O}\text{-}\mathcal{F})$ of the magnetic field (Figs. 1 and 2).

Our results have further implications. First, even with the help of (21), the dynamo model is still not fully spun up. For example, though $(\mathcal{O}\text{-}\mathcal{F})_{\dot{B}}$ decreases monotonically in time, the SV forecast is still very far away from the observations, as $(\mathcal{O}\text{-}\mathcal{F})_{\dot{B}} \approx \mathcal{O}(1)$ for all degrees (see Fig. 4). This can be shown further by the continuously growing differences $(\mathcal{M}\text{-}\mathcal{F})_{v_P}$ and $\mathcal{M}\text{-}\mathcal{F})_{v_T}$ between the forecast velocity field and that of the free-running model (see Figs. 6 and 7). In addition, the differences at the end of the assimilation period are still very small, approximately 10 % in the magnitude of the velocity field of the free-running model (Fig. 5).

These suggest that much larger velocity differences $(\mathcal{M}\text{-}\mathcal{F})_{v_P}$ and $(\mathcal{M}\text{-}\mathcal{F})_{v_T}$ are needed over a shorter assimilation period for expediting the model spin-up. Assimilation of the effective observed velocity (25) could be an answer. However, as discussed earlier, (25) is an underdetermined system, since both $\mathbf{v}_h^{(o)}$ and $\widetilde{b}_l^{m(o)}$ (more specifically, $\partial^2\widetilde{b}_l^{m(o)}/\partial r^2$) are unknown beneath the CMB. Thus, the responses of $\mathbf{v}_h$ to the SV assimilation, e.g., $(\mathcal{M}\text{-}\mathcal{F})_{v_P}$ (in Figs. 8 and 9) and $(\mathcal{M}\text{-}\mathcal{F})_{v_T}$ (in Figs. 10 and 11) are needed to determine $\mathbf{v}_h^{(o)}$. For example, as shown in the two figures, the non-axisymmetric ($m > 0$) poloidal velocity $v_l^{m\prime}$ around the degree $l = 20$ and the toroidal velocity ω_l^m around the degree $l = 30$ and order $m \leq 20$ should be given more attention, as they are most sensitive to the surface observations.

An alternative answer could be the core states inverted from the surface observations and dynamo solutions, such as those of Aubert (2013, 2014). These can be used as the analysis of the assimilation system. But cautions should be taken with this approach. For example, the inverted velocity field beneath the CMB is actually derived with the observed field and SV and the magnetic diffusion of the dynamo state (Aubert 2014). This could potentially lead to dynamical inconsistencies as well as uncertainties in error statistics.

Our results also show several new features that may have implications to field modeling and to core flow inversion. One new knowledge is from the time variation of $(\mathcal{O}\text{-}\mathcal{F})_{\dot{B}}$. As shown in Fig. 4, $(\mathcal{O}\text{-}\mathcal{F})_{\dot{B}}$ of the odd degrees ($l = 1, 3, 5$) are significantly different from those of the even degrees ($l = 2, 4, 6$): the values of the even degrees decay nearly monotonically in time; but those of the odd orders show either spikes (for $l = 1, 3$) during the assimilation or even increase over time (for $l = 5$). These even-odd degree disparities suggest inconsistencies between the model and the observations. These inconsistencies could be entirely due to numerical dynamo model which may include a magnetic induction different from those in the Earth's outer core or may include some mechanisms resulting in different symmetry properties of the core state. But the inconsistencies could also come from possible biases in the field models that are not included in the observation error covariances. For example, ionospheric ring current generated field (an external field component) contributes dominantly to the Gauss coefficients of degrees $l = 1, 3, 5$ and varies on time scales comparable to those of SV, e.g., the solar activity cycles (Sabaka et al. 2015). Model biases exist if this part of the signals is not well separated from those of the core field. This is potentially an area for application of geomagnetic data assimilation.

The core fluid flow responses, i.e., the differences $(\mathcal{M}\text{-}\mathcal{F})_{v_P}$ and $(\mathcal{M}\text{-}\mathcal{F})_{v_T}$ between the forecast velocity field $\mathbf{v}_h^f$ and the $\mathbf{v}_h^M$ of the free-running model (see Figs. 8, 9, 10, and 11), from our experiments could also help the inversion of core flow from the observed SV. For example, the different characteristics in $(\mathcal{M}\text{-}\mathcal{F})_{v_P}$ and $(\mathcal{M}\text{-}\mathcal{F})_{v_T}$ suggest that the poloidal velocity field and the toroidal velocity field could be treated separately in the core flow inversion. It should be pointed out that purely toroidal core flow approximations were used in previous studies (e.g., Bloxham et al. 2002; Olsen and Mandea 2008). The strong responses of the high-degree core flow ($l \approx 20$ for the poloidal flow and $l \approx 30$ for the toroidal flow) to the observed SV (for $l \leq 8$) indicate that higher degree velocity field should be included in the core flow inversion. For example, one would normally expect that due to nonlinear effects, i.e., the quadratic terms in Navier-Stokes equation and the induction equation, the core flow up to the degrees twice as much as that of the SV should be sufficient for the core flow inversion (as in Aubert 2013). But our results show that time evolution of the core flow leads to the strongest responses for the degrees more than triple of the maximum degree of the SV. Therefore, inversion of time-dependent core flow from the observed SV (up to degree 13) should include high-degree ($l > 40$) spectral coefficients. Of course, the core flow response may also suggest that the free-running model solutions do not provide a good description of the small-scale dynamical processes in the core. And assimilation of geomagnetic observations could substantially improve our understanding of these small-scale processes. Regardless, our results suggest that unobservable small-scale dynamical processes are very important in interpretation and prediction of fast SV observable at the Earth's surface.

Again, we should point out that our results could be improved in several areas: assimilation with other geodynamo models different from (39) and with more sophisticated assimilation algorithms and field models with more accurate error statistics. The former will build up an ensemble of assimilation results to provide more accurate statistics on the characteristics of the core flow to the observed SV. Improvements from the latter are also obvious. For example, we anticipate more accurate estimation of ($\mathcal{O}$-$\mathcal{F}$) for both the field and the SV, and better assessment of the core state responses if a full ensemble approach is used for the covariance $\mathbf{P}^f$, and a more appropriate observation error covariance, e.g., those determined by Gillet et al. (2013), than (40) used in this study. Regardless, our assimilation experiments have shown clearly the importance of SV assimilation and the improvements that the SV assimilation brought to forecast accuracies and to model spin-up processes.

Abbreviations
CMB: core-mantle boundary; GDAS: geomagnetic data assimilation; ICB: inner core boundary; SV: secular variation.

Competing interests
The authors declare that they have no competing interests.

Authors' contributions
WK proposed and designed the study and carried out the assimilation experiments. AT provided error covariance models and mathematical derivations for the sequential assimilation algorithm used in the experiments. Both authors read and approved the final manuscript.

Acknowledgements
This research is funded by grants from NASA Earth Surface and Interior Program (NNZ09AK70G) and from NSF (EAR-0757880). We also thank NASA NAS for support on high-end scientific computation.

Author details
[1]Planetary Geodynamics Laboratory, NASA Goddard Space Flight Center, 8800 Greenbelt Road, Greenbelt, MD 20771, USA. [2]Joint Center for Earth System Technologies, University of Maryland, Baltimore County, 1000 Hilltop Circle, Baltimore, MD 21250, USA.

References
Aubert J (2013) Flow through the Earth's core inverted from geomagnetic observations and numerical dynamo models. Geophys. J. Int. 192:537–56
Aubert, J (2014) Earth's core internal dynamics 1840–2010 imaged by inverse geodynamo modeling. Geophys J Int. 197:1321–34
Aubert J, Fournier A (2011) Inferring internal properties of Earth's core dynamics and their evolution from surface observations and a numerical geodynamo. Nonlin Processes Geophys. 18:657–74
Backus GE (1968) Kinematics of geomagnetic secular variation in a perfectly conducting core. Phil Trans R Soc London A263:239–66
Bloxham J, Zatman S, Dumberry M (2002) The origin of geomagnetic jerks. Nature 420:65–8
Christensen U. R, Aubert J, Hulot G (2010) Conditions for Earth-like geodynamo models. Earth Planet Sci Lett. 296:487–96
Christensen U. R, Wardingsky I, Aubert J, Lesur V (2012) Timescales of geomagnetic secular acceleration in satellite field models and geodynamo models. Geophys J Int. 190:243–54
Evensen G (1994) Sequential data assimilation with a nonlinear quasi-geostrophic model using Monte Carlo methods to forecast error statistics. J Geophys Res. 99:10143–62
Fournier A, Eymin C, Alboussiere T (2007) A case for variational geomagnetic data assimilation: insights from a one-dimensional, nonlinear, and sparsely observed MHD system. Nonlinear Process Geophys 14:163–80
Fournier A, Hulot G, Jault D, Kuang W, Tangborn A, Gillet N, et al (2010) An introduction to data assimilation and predictability in geomagnetism. Space Sci Rev. doi:10.1007/s11214-010-9669-4
Fournier A, Aubert J, Thébault E (2011) Inference on core surface flow from observations and 3-D dynamo modeling. Geophys J Int. 186:118–36
Fournier A, Nerger L, Aubert J (2013) An ensemble Kalman filter for the time-dependent analysis of the geomagnetic field. Geochem Geophys Geosys. doi:10.1002/ggge.20252
Gillet N, Jault D, Finlay CC, Olsen N (2013) Stochastic modeling of the Earth's magnetic field: inversion for covariances over the observatory era. Geochem Geophys Geosys. doi:10.1002/ggge.20041
Glatzmaier GA, Roberts PH (1995) A three-dimensional self-consistent computer simulation of a geomagnetic field reversal. Nature 377:203–6
Holme R (2007) Large-scale flow in the core. In: Olson P (ed). Core dynamics, treaties on geophysics. Elsevier, Amstedam
Hou AY, Ledvina DV, da Silva AM, Zhang SQ, Joiner J, Atlas RM, et al (2000) Assimilation of SSM/I-derived surface rainfall and total precipitable water for improving the GEOS analysis for climate studies. Mon Weath Rev. 128:509–37
Hulot G, LeMouël JL (1994) A statistical approach to the Earth's main magnetic field. Phys Earth Planet Inter. 82:167–83
Jackson A, Jonkers AR, Walker MR (2000) Our centuries of geomagnetic secular variation from historical records. Phil Trans R Soc Lond. A358:957–90
Kageyama A, Sato T (1997) Generation mechanism of a dipole field by a magnetohydrodynamic dynamo. Phys Rev E. 55:4617–26
Kalnay E (2003) Atmospheric modeling, data assimilation and predictability. Cambridge University Press, Cambridge, U.K.
Korte M, Genevey A, Constable CG, Frank U, Schnepp E (2005) Continuous geomagnetic field models for the past 7 millennia: 1. A new data compilation. Geochem Geophys Geosys. doi:10.1029/2004GC000800
Korte M, Constable CG, Donadini F, Holme R (2011) Reconstructing the Holocene geomagnetic field. Earth Planet Sci Lett. 312:497–505
Kuang W, Bloxham J (1997) An Earth-like numerical dynamo model. Nature 389:371–4
Kuang W, Tangborn A (2011) Interpretation of core field models. In: Mandea M, Korte M (eds). Geomagnetic observations and models. Springer, Heidelberg
Kuang W, Tangborn A, Jiang W, Liu D, Sun Z, Bloxham J, Wei Z (2008)

MoSST_DAS: The first generation geomagnetic data assimilation framework. Comm Comp Phys. 3:85–108
Kuang W, Tangborn A, Wei Z, Sabaka TJ (2009) Constraining a numerical geodynamo model with 100 years of surface observations. Geophys J Int. 179:1458–68
Kuang W, Wei Z, Holme R, Tangborn A (2010) Prediction of geomagnetic field with data assimilation: a candidate secular variation model for IGRF-11. Earth Planet Space. 62:775–85
Larmor J (1919) How could a rotating body such as the Sun become a magnet? Rep Br Assoc 87:159–60
Li D, Jackson A, Livermore PW (2011) Variational data assimilation for the initial value dynamo problem. Phys Rev E. doi:10.1103/PhysRevE.84.056321
Li, D, Jackson A, Livermore PW (2014) Variational data assimilation for a forced, inertia-free magnetohydrodynamic dynamo model. Geophys J Int. 199:1662–76
Liu D, Tangborn A, Kuang W (2007) Observing system simulation experiments in geomagnetic data assimilation. J Geophys Res. doi:10.1029/2006JB004691
Nilsson A, Holme R, Korte M, Suttie N, Hill M (2014) Reconstructing Holocene geomagnetic field variation: new methods, models and implications. Geophys J Int. doi:10.1093/gji/ggu120
Olsen N, Mandea M (2008) Rapidly changing flows in the Earth's core. Nature Geosci. 1:390–4
Olsen N, L uhr H, J S. T, Mandea M, Rother M, Tøffner-Clausen L, Choi S (2006) CHAOS-a model of the Earth's magnetic field derived from CHAMP, Ørsted, and SAC-C magnetic satellite data. Geophys J Int. 166:67–75
Olsen N, Lühr H, Finlay C. C, Sabaka T. J, Michaelis I, Rauberg J, Tøffner-Clausen L (2014) The CHAOS-4 geomagnetic field model. Geophys J Int. 197:815–27
Roberts PH (1992) Geomagnetism. Encyclopedia Earth Syst Sci. 2:277–94
Roberts PH, Scott S (1965) On analysis of the secular variation, 1, A hydromagnetic constraint: theory. J Geomag Geoelec. 17:137–51
Sabaka TJ, Olsen N, Purucker M (2004) Extending comprehensive models of the Earth's magnetic field with Ørsted and champ data. Geophys J Int. 159:521–47
Sabaka T. J, Olsen N, Tyler R. H, Kuvshinov A (2015) CM5, a pre-Swarm comprehensive geomagnetic field model derived from over 12 year of CHAMP, Ørsted, SAC-C and observatory data. Geophys J Int. 200:1596–626
Stacey FD (1992) Physics of the Earth. Brookfield Press, Kenmore, Australia
Sun Z, Kuang W (2015) An ensemble algorithm based component for geomagnetic data assimilation. Terr Atmos Ocean Sci. doi:10.3319/TAO.2014.08.19.05
Sun Z, Tangborn A, Kuang W (2007) Data assimilation in a sparsely observed one-dimensional modeled MHD system. Nonlin Proc Geophys. 14:181–92
Tangborn A, Kuang W (2015) Geodynamo model and error parameter estimation using geomagnetic data assimilation. Geophys J Int. doi:10.1093/gji/ggu409
Zatman S, Bloxham J (1997) Torsional oscillations and the magnetic field within the Earth's core. Nature 388:760–3

4

Towards a seamlessly diagnosable expression for the energy flux associated with both equatorial and mid-latitude waves

Hidenori Aiki[1,2*], Richard J. Greatbatch[3,4] and Martin Claus[3]

Abstract

For mid-latitude Rossby waves (RWs) in the atmosphere, the expression for the energy flux for use in a model diagnosis, and without relying on a Fourier analysis or a ray theory, has previously been derived using quasi-geostrophic equations and is singular at the equator. By investigating the analytical solution of both equatorial and mid-latitude waves, the authors derive an exact universal expression for the energy flux which is able to indicate the direction of the group velocity at all latitudes for linear shallow water waves. This is achieved by introducing a streamfunction as given by the inversion equation of Ertel's potential vorticity, a novel aspect for considering the energy flux. For ease of diagnosis from a model, an approximate version of the universal expression is explored and illustrated for a forced/dissipative equatorial basin mode simulated by a single-layer oceanic model that includes both mid-latitude RWs and equatorial waves. Equatorial Kelvin Waves (KWs) propagate eastward along the equator, are partially redirected poleward as coastal KWs at the eastern boundary of the basin, and then shed mid-latitude RWs that propagate westward into the basin interior. The connection of the equatorial and coastal waveguides has been successfully illustrated by the approximate expression of the group-velocity-based energy flux of the present study. This will allow for tropical-extratropical interactions in oceanic and atmospheric model outputs to be diagnosed in terms of an energy cycle in a future study.

Keywords: Group velocity, Model diagnosis, Equatorial Rossby waves, Equatorial mixed Rossby-gravity waves, Equatorial inertia-gravity waves, Equatorial Kelvin waves, Coastal Kelvin waves, Mid-latitude Rossby waves, Mid-latitude inertia-gravity waves, Tropical-extratropical interactions

Introduction

A feature of many phenomena in the equatorial oceans is the role played by equatorial Kelvin waves (KWs), examples being El Niño Southern Oscillation (ENSO; Philander 1989) and the so-called Atlantic Niño (Merle 1980). KWs propagate along the equator and are partially redirected into coastal KWs at the eastern boundary, where they can influence off-equatorial latitudes (e.g., Lübbecke et al. 2010) as well as excite extratropical Rossby waves (RWs) that subsequently propagate into the ocean interior (McPhaden and Ripa 1990; Isachsen et al. 2007). A striking example of this behavior is the equatorial basin mode (Cane and Moore 1981). For the gravest basin mode, the time scale is set by the time taken for an equatorial KW to propagate across the basin and for the reflected gravest long Rossby wave to return to the western boundary (that is $4L/c$ where L is the basin width and c is the phase propagation speed for KWs). In addition to waves that are trapped on the equator, equatorial basin modes also feature coastal KWs that propagate along the eastern boundary and extratropical RWs that are excited by these KWs and refocus on the equator, as described by Schopf et al. (1981). There is growing evidence that equatorial basin modes play an important role in equatorial ocean dynamics. For example, basin modes have been associated with the equatorial deep jets (Johnson and Zhang 2003; Brandt et al. 2011; Claus et al. 2016) and with the semi-annual (Thierry et al. 2004) and annual cycles (Brandt

*Correspondence: aiki@nagoya-u.jp
[1] Institute for Space-Earth Environmental Research, Nagoya University, Nagoya City, 464-8601 Aichi, Japan
[2] Application Laboratory, Japan Agency for Marine-Earth Science and Technology, Yokohama, Japan
Full list of author information is available at the end of the article

et al. 2016) in the equatorial Atlantic. However, the energy cycle associated with equatorial basin modes has received little attention and is an important factor when considering the forced/dissipative basin modes that one can relate to observations. A particularly interesting example is the upward energy propagation associated with the Atlantic equatorial deep jets (Johnson and Zhang 2003; Brandt et al. 2011; Mathiessen et al. 2015). Yet, the detailed energy cycle associated with the jets remains largely unknown.

One way to approach the energy flux is to use ray theory. However, ray theory is linked to the dispersion relation of a single type of wave and is not suitable for investigating the sequential connection of different types of waves that are associated with a basin mode. Likewise, a Fourier analysis is not suitable for the investigation of waves near the coastal boundaries of the ocean. In fact, it is only for mid-latitude inertia-gravity waves (IGWs) that the flux of wave energy has been diagnosed from oceanic model output (Cummins and Oey 1997; Niwa and Hibiya 2004; Furuichi et al. 2008). On the other hand, in the atmospheric literature, the model diagnosis of pseudomomentum (or wave activity) flux has been more popular than the model diagnosis of the energy flux (Hoskins et al. 1983; Plumb 1986; Takaya and Nakamura 1997; Nakamura and Solomon 2011).

Here, we seek a general expression that can be used to diagnose the energy flux associated with linear shallow water waves at all latitudes from model output. This manuscript is organized as follows. First we provide the theoretical background. Then, we present an analytical investigation that leads to a general expression for the energy flux that can indicate the exact profile of the group velocity times wave energy for both equatorial and mid-latitude waves. The utility of the universal expression of energy flux as a model diagnostic is illustrated for a forced/dissipative equatorial basin mode simulated by a single-layer model. The model diagnosis is achieved by introducing an inversion for the linearized version of Ertel's potential vorticity. This is a novel aspect for considering the energy flux in the presence of a coastal waveguide that connects the equatorial and mid-latitude regions.

Theoretical background

We use the shallow water equations for a single vertical normal mode (Gill 1982) appropriate to linear waves in a rotating frame of reference and in the absence of a mean flow. Let an arbitrary variable with an associated physical dimension be expressed by A^*, and let Cartesian-horizontal coordinates be labelled by the set of independent variables x^*, y^*, t^*, where each of x^*, y^* increases eastward and northward, respectively, and u^*, v^* are the corresponding horizontal components of velocity (a list of variables is given in Table 1)[1]. The equations may then be written as

$$\frac{\partial u^*}{\partial t^*} - f^* v^* + \frac{\partial p^*}{\partial x^*} = 0, \tag{1a}$$

$$\frac{\partial v^*}{\partial t^*} + f^* u^* + \frac{\partial p^*}{\partial y^*} = 0, \tag{1b}$$

$$\frac{\partial p^*}{\partial t^*} + c^{*2}\left(\frac{\partial u^*}{\partial x^*} + \frac{\partial v^*}{\partial y^*}\right) = 0, \tag{1c}$$

where $f^* = f_0^* + \beta^* y^*$ is the Coriolis parameter, $p^* = p^*(x^*, y^*, t^*)$ corresponds to the pressure[2] or geopotential, and c^* is a uniform constant representing the propagation speed of nonrotating gravity waves for a given mode. Manipulation of (1a)–(1c) yields a prognostic equation for the linearized version of Ertel's potential vorticity (hereafter EPV and symbolized as q^*) to read

$$\frac{\partial}{\partial t^*}\underbrace{\left(\frac{\partial v^*}{\partial x^*} - \frac{\partial u^*}{\partial y^*} - \frac{f^*}{c^{*2}} p^*\right)}_{\equiv q^*} + v^* \beta^* = 0, \tag{2}$$

which is applicable to waves at all latitudes, such as mid-latitude RWs, mid-latitude IGWs, and equatorial waves [i.e., equatorial RWs and IGWs, equatorial Rossby-gravity waves (RGWs, i.e., Yanai waves), and equatorial KWs; Matsuno 1966; Yanai and Maruyama 1966], understanding $f_0^* = 0$ for an equatorial β-plane and $\beta^* = 0$ for a mid-latitude f-plane. Both mid-latitude IGWs (i.e., $\beta^* =$

Table 1 List of symbols, where A^* and A are arbitrary quantities written dimensionally or non-dimensionally, respectively

$f^* = f_0^* + \beta^* y^*$	Coriolis parameter
c^*	Speed of long gravity wave
x, y, t	Cartesian coordinates wherein x and y increase eastward and northward
$\langle\langle a, b \rangle\rangle$	Horizontal vector with eastward and northward components a and b
$\mathbf{V} = \langle\langle u, v \rangle\rangle$	Horizontal velocity vector
$\nabla \equiv \langle\langle \partial_x, \partial_y \rangle\rangle$	Horizontal gradient operator
p	Pressure
$q \equiv v_x - u_y - yp$	Linearized Ertel's potential vorticity: $q^* \equiv v^*_{x^*} - u^*_{y^*} - (f^*/c^{*2})p^*$
φ	Solution of $\nabla^2\varphi - y^2\varphi - 3\varphi_{tt} = q$, see (16) & (17a)
φ^{app}	Solution of $\nabla^2\varphi^{app} - y^2\varphi^{app} = q$, see (26a) & (18a)
$(u^2 + v^2 + p^2)/2$	Wave energy: $(u^{*2} + v^{*2} + p^{*2}/c^{*2})/2$
$\theta = kx - \omega t$	Wave phase
k	Zonal wavenumber
ω	Wave frequency
$H^{(n)}$	Hermite polynomial, see endnote 1
n	Meridional mode number of free equatorial waves
$\overline{A}$	Phase average of A

0) and equatorial KWs (i.e., $v^* = 0$) are characterized by $q^* = 0$, as noted in Table 2.

On the other hand, a prognostic equation for wave energy may be derived from (1a)–(1c) as

$$\frac{\partial}{\partial t^*}\frac{1}{2}\overline{\left(u^{*2} + v^{*2} + \frac{p^{*2}}{c^{*2}}\right)} + \nabla^* \cdot \langle\!\langle \overline{u^* p^*}, \overline{v^* p^*} \rangle\!\rangle = 0, \quad (3)$$

where $\nabla^* \equiv \langle\!\langle \frac{\partial}{\partial x^*}, \frac{\partial}{\partial y^*} \rangle\!\rangle$ and the overbar symbol represents a phase-average operator (i.e., for a sinusoidal wave, $\overline{A^*} = 0$ for $A^* = u^*, v^*$, and p^*) or a low-pass time filter (for this reason, we retain the local time derivative in (3) to allow for slow time variations in the general case).

For mid-latitude IGWs in the ocean and atmosphere, the group velocity vector points in the same direction as the energy flux vector in (3):

$$\overline{\mathbf{V}^* p^*} = \langle\!\langle \overline{u^* p^*}, \overline{v^* p^*} \rangle\!\rangle, \quad (4)$$

a property that has been exploited by Cummins and Oey (1997), Niwa and Hibiya (2004), and Furuichi et al. (2008) for a model diagnosis. However, for mid-latitude RWs, the vector in (4) does not point in the direction of the group velocity of the waves (Longuet-Higgins 1964; Masuda 1978; Cai and Huang 2013). In order to retrieve the correct direction for the energy flux associated with mid-latitude RWs, Orlanski and Sheldon (1993, hereafter OS93) have suggested to modify (3), without affecting the horizontal divergence of the energy flux, as

$$\frac{\partial}{\partial t^*}\frac{1}{2}\overline{\left(u^{*2} + v^{*2} + \frac{p^{*2}}{c^{*2}}\right)} + \nabla^* \cdot \left\langle\!\!\left\langle \overline{u^* p^*} + \frac{\partial}{\partial y^*}\left(\frac{\overline{p^{*2}}}{2f^*}\right), \overline{v^* p^*} - \frac{\partial}{\partial x^*}\left(\frac{\overline{p^{*2}}}{2f^*}\right) \right\rangle\!\!\right\rangle = 0, \quad (5)$$

where each of u^* and v^* should be the sum of the geostrophic and ageostrophic components and $f^* = f_0^* + \beta^* y^*$ is understood. The energy flux vector in (5) consists of two terms,

$$\overline{\mathbf{V}^* p^*} - \nabla^* \times [\overline{p^{*2}}/(2f^*)]\, \mathbf{z}, \quad (6)$$

where $\overline{\mathbf{V}^* p^*}$ is as in the gravity wave literature (i.e., $\mathbf{V}^*$ is the sum of the geostrophic and ageostrophic components of velocity). The second term in (6) is the additional rotational component required to reproduce the direction of the group velocity of mid-latitude RWs ($\mathbf{z}$ is the upward vertical unit vector). In Longuet-Higgins (1964), the second term of (6) has been expressed as $-\nabla^* \times [f^* \overline{\psi^{*2}}/2]\, \mathbf{z}$ where ψ^* is a streamfunction based on the assumption of horizontally nondivergent velocity. This assumption is hardly used in modern oceanography owing to the smallness of the deformation radius. In quasi-geostrophic theory, $\psi^* = p^*/f^*$ from which the connection with (6) is clear.

The question naturally arises as to whether or not it is possible to find a general expression for the additional rotational flux, $\mathbf{R}^*$, that holds for waves at all latitudes and is such that the corresponding energy flux $\overline{\mathbf{V}^* p^*} + \mathbf{R}^*$ always points in the direction of the group velocity and thus constitutes a general expression for the energy flux associated with waves at all latitudes. This is the main subject of the present study. In this study, we focus on wave types for which the group velocity has been well formulated in the literature/textbook, as listed in Table 2. Of particular interest is the energy flux associated with equatorial RWs given that the expression in (6) is singular at the equator. The assumption of horizontally nondivergent velocity in Longuet-Higgins (1964) is also inappropriate for equatorial regions. In the next section, by investigating the analytical solution of equatorial waves, we derive an exact universal expression for the rotational flux which, after being added to $\overline{\mathbf{V}^* p^*}$, is able to indicate the direction of the group velocity for linear waves at all latitudes.

Table 2 Characteristics of different waves at various latitudes

	$\overline{\mathbf{V}^* p^*}$ parallel to group velocity	$q^* = 0$ ($\varphi^{\mathrm{app}*} = 0$)
Equatorial Rossby wave	No	No
Equatorial mixed Rossby-gravity wave	Depends on frequency	No
Equatorial inertia-gravity wave	Roughly yes	No
Equatorial Kelvin wave	Yes	Yes
Coastal Kelvin wave	Yes	Yes
Mid-latitude Rossby wave	No	No ($\varphi^{\mathrm{app}*} \simeq p^*/f^*$)
Mid-latitude inertia-gravity wave	Yes	Yes

Analytical investigation

We begin by revisiting analytical expressions for the profile of the energy flux associated with equatorial waves. This investigation allows us to derive an expression for the energy flux that points in the direction of the group velocity for waves at all latitudes.

Energy flux associated with equatorial waves

We assume linear waves in the absence of a mean flow on an equatorial β-plane. As in Matsuno (1966) and Gill (1982), we use a time scale $1/\sqrt{c^*\beta^*}$ and a length scale $\sqrt{c^*/\beta^*}$ to nondimensionalize the equation system (1a)–(1c) to give

$$u_t - yv + p_x = 0, \quad (7a)$$

$$v_t + yu + p_y = 0, \quad (7b)$$

$$p_t + u_x + v_y = 0, \quad (7c)$$

where symbols without an asterisk indicate nondimensionalized quantities and subscripts indicate partial differentiations. Manipulation of (7a)–(7c) yields prognostic equations for EPV and wave energy in a nondimensionalized form to read,

$$\partial_t \underbrace{(v_x - u_y - yp)}_{\equiv q} + v = 0, \quad (8)$$

$$\partial_t \overline{(u^2 + v^2 + p^2)}/2 + \nabla \cdot \langle\langle \overline{up}, \overline{vp} \rangle\rangle = 0, \quad (9)$$

where $\partial_t \equiv \frac{\partial}{\partial t}$, $\nabla \equiv \langle\langle \frac{\partial}{\partial x}, \frac{\partial}{\partial y} \rangle\rangle$, and for $A = u, v$, or $p, \overline{A} = 0$ for sinusoidally varying waves.

In what follows, we assume $v \not\equiv 0$ which is appropriate for equatorial RWs, RGWs, and IGWs (i.e., waves other than equatorial KWs). Then, we consider zonally propagating free waves with a relationship $v \propto \cos\theta$, $u \propto \sin\theta$, and $p \propto \sin\theta$ where $\theta \equiv kx - \omega t$ is wave phase with k and ω being wavenumber and wave frequency, respectively. Substitution of these relationships to (7a)–(7c), followed by some manipulation, yields a characteristic equation for the meridional structure of v to read,

$$v_{yy} + (\omega^2 - k^2 - k/\omega - y^2)v = 0. \quad (10)$$

Matsuno (1966) has derived a solution for (7a)–(7c) and (10) to yield,

$$v = \mathscr{A} \cos\theta \exp(-y^2/2)H^{(n)}, \quad (11a)$$

$$u = (\omega y v_\theta - k v_{y\theta})/(\omega^2 - k^2), \quad (11b)$$

$$p = (k y v_\theta - \omega v_{y\theta})/(\omega^2 - k^2), \quad (11c)$$

where $\mathscr{A}$ is wave amplitude and the symbol $H^{(n)}$ is the Hermite polynomial with n being the meridional mode number[3]. The subscript θ represents partial differentiation in terms of the wave phase [i.e., $v_\theta \equiv \partial v/\partial\theta = -\mathscr{A} \sin\theta \exp(-y^2/2)H^{(n)}$].

Substitution of (11a) to (10) yields,

$$\omega^3 - (k^2 + 2n + 1)\omega - k = 0, \quad (12)$$

which is a unified dispersion relation for equatorial RWs, RGWs, and IGWs. Partial differentiation of (12) with respect to wavenumber k yields a unified expression for the group velocity of equatorial waves,

$$\frac{\partial\omega}{\partial k} = \frac{2k\omega + 1}{3\omega^2 - (k^2 + 2n + 1)} = \frac{2\omega^2 + \omega/k}{2\omega^3/k + 1}, \quad (13)$$

where $2\omega^3/k$ in the denominator has often been ignored in previous studies when focusing on low-frequency equatorial waves (e.g., equatorial RWs; Gill 1982).

We now investigate the energy flux associated with (7a)–(7c). It is known that, for zonally propagating equatorial waves, the meridional integral of $\overline{up}$ is equal to the group velocity times the meridional integral of the wave energy (Philander 1989):

$$\int_{-\infty}^{+\infty} \overline{up}\, dy = (\partial\omega/\partial k) \int_{-\infty}^{+\infty} \overline{(u^2 + v^2 + p^2)}/2\, dy. \quad (14a)$$

It should be noted that the identity (14a) does not hold if it is evaluated without the meridional integral:

$$\overline{up} \neq (\partial\omega/\partial k)\overline{(u^2 + v^2 + p^2)}/2. \quad (14b)$$

For low-frequency equatorial waves (with $\omega < 1$—see Fig. 1—, i.e., all equatorial RWs and westward propagating RGWs), the meridional profiles of $\overline{up}$ and $(\partial\omega/\partial k)\overline{(u^2 + v^2 + p^2)}/2$ are shown by the dashed green and solid black lines, respectively, in Fig. 2. It is clear that, when compared at a given latitude, $\overline{up}$ is not equal to the group velocity times wave energy. In particular, the meridional profile of $\overline{up}$ is sign-indefinite for low-frequency equatorial waves (Fig. 2). On the other hand, as shown by the dashed green and solid black lines in Fig. 3 for high-frequency equatorial waves (with $\omega > 1$—see Fig. 1—, i.e., all equatorial IGWs and eastward propagating RGWs), the meridional profile of $\overline{up}$ provides a much better approximation for the group velocity times wave energy. The solid blue line, dashed orange line, and purple dots in Figs. 2 and 3 are explained later in the manuscript.

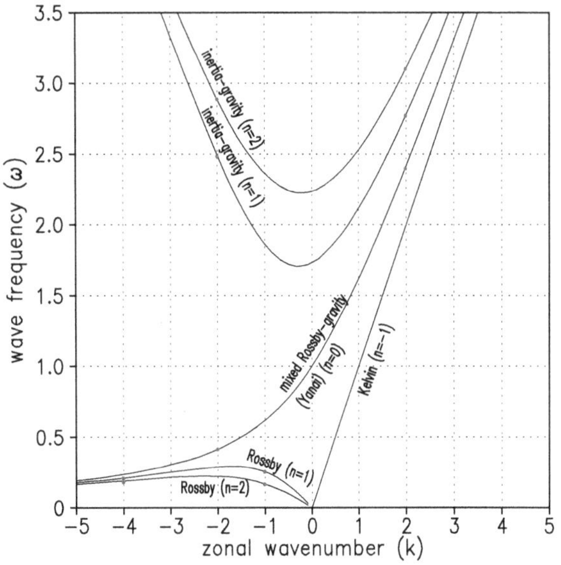

Fig. 1 Dispersion relation of free equatorial waves for a given meridional number n, as given by Eq. (12). Axes have been non-nondimensionalized by $\sqrt{\beta^* c^*}$ for wave frequency ω and by $\sqrt{\beta^*/c^*}$ for zonal wavenumber k. The *red dots* locate the parameters used to produce the 10 panels in Figs. 2 and 3

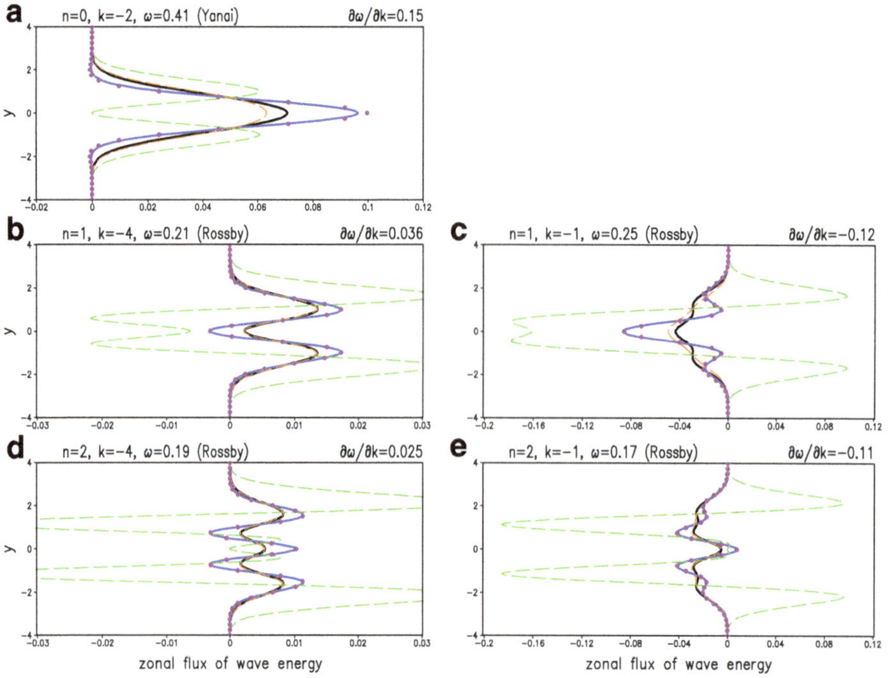

Fig. 2 Meridional profiles of the zonal component of phase-averaged energy flux associated with equatorial waves; the *solid black line* is $(\partial\omega/\partial k)\overline{(u^2+v^2+p^2)}/2$, the *dashed green line* is $\overline{up}$, the *dashed orange line* is $\overline{up} + (\overline{p\varphi^{app}}/2 + \overline{u_{tt}\varphi^{app}})_y$, the *solid blue line* is $\overline{up} + (\overline{p\varphi^{app}}/2)_y$, and the *purple dots* are $(\omega/k)(\overline{E - v^2}) = (\omega/k)\overline{(u^2 - v^2 + p^2)}/2$. All panels are for low-frequency equatorial waves with $\omega < 1$: **a** westward propagating RGWs, **b** short and **c** long RWs in the 1st meridional mode, and **d** short and **e** long RWs in the 2nd meridional mode. The associated values of meridional-mode number n, zonal wavenumber k, wave frequency ω, and group velocity $\partial\omega/\partial k$ are noted in each panel. For each of **a**–**e**, the wave amplitude $\mathscr{A}$ in (11a) has been set to normalize the meridional integral of wave energy: $\int_{-\infty}^{\infty}\overline{(u^2+v^2+p^2)}/2dy = 1$

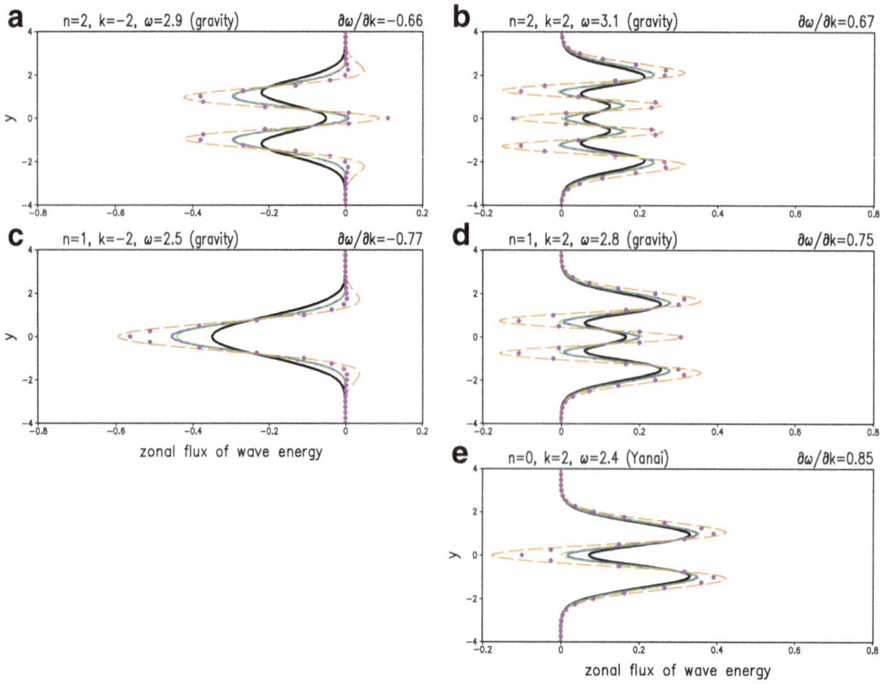

Fig. 3 Same as Fig. 2 except for high-frequency equatorial waves with $\omega > 1$: **a** westward and **b** eastward propagating IGWs in the 2nd meridional mode, **c** westward and **d** eastward propagating IGWs in the 1st meridional mode, and **e** eastward propagating RGWs

Identification of the additional rotational flux associated with equatorial waves

It is useful to derive the analytical expression for the difference between the left and right hand sides of (14b). A first step for identifying the difference is to decompose the zonal component of $\overline{up}$ into two parts, one that determines the meridional integral and one that does not affect it, as follows:

$$
\begin{aligned}
&\overline{up}\\
&= \left[y^2\overline{vv}(\omega k) - y\overline{v_y v}\left(\omega^2+k^2\right) + \overline{v_y v_y}(\omega k)\right] / \left(\omega^2-k^2\right)^2\\
&= \left\{\overline{v_{yy}v}(\omega k) + \overline{vv}\left(\omega^3 k - \omega k^3 - k^2\right)\right.\\
&\quad \left. - \left[\left(y\overline{vv}/2\right)_y - \overline{vv}/2\right]\left(\omega^2+k^2\right) + \overline{v_y v_y}(\omega k)\right\} / \left(\omega^2-k^2\right)^2\\
&= \left[\overline{vv}\left(2\omega^3 k - 2\omega k^3 - k^2 + \omega^2\right) + \left(\overline{v_y v}\right)_y(2\omega k)\right.\\
&\quad \left. - \left(y\overline{vv}\right)_y\left(\omega^2+k^2\right)\right] / \left[2\left(\omega^2-k^2\right)^2\right]\\
&= \overline{vv}\left(2\omega k+1\right) / \left[2\left(\omega^2-k^2\right)\right]\\
&\quad + \left[\overline{v_y v}(2\omega k) - y\overline{vv}\left(\omega^2+k^2\right)\right]_y / \left[2\left(\omega^2-k^2\right)^2\right],
\end{aligned}
\tag{14c}
$$

where the first equality has been derived using (11b)–(11c) and $\overline{\sin\theta\sin\theta} = \overline{\cos\theta\cos\theta}$ and the second equality has been derived using (10). Note that it is the second of the two terms whose meridional integral is zero (noting that v and yv go to zero at large distances from the equator).

We now decompose the wave energy[4] into two parts, one that determines the meridional integral and one does not. We then have

$$
\begin{aligned}
&\left(\overline{u^2+v^2+p^2}\right)/2\\
&= \overline{vv}/2 + \left[\left(y^2\overline{vv} + \overline{v_y v_y}\right)\left(\omega^2+k^2\right)\right.\\
&\quad \left. - \left(y\overline{v_y v}\right)(4k\omega)\right] / \left[2\left(\omega^2-k^2\right)^2\right]\\
&= \overline{vv}/2 + \left\{\left[y^2\overline{vv} - \overline{v_{yy}v} + \left(\overline{v_y v}\right)_y\right]\left(\omega^2+k^2\right)\right.\\
&\quad \left. - \left(y\overline{vv}\right)_y(2k\omega) + \left(\overline{vv}\right)(2k\omega)\right\} / \left[2\left(\omega^2-k^2\right)^2\right]\\
&= \left[\overline{vv}\left(k^4 - 2k^2\omega^2 + \omega^4 + \omega^4 - k^4 + k\omega - k^3/\omega\right)\right.\\
&\quad \left. + \left(\overline{v_y v}\right)_y\left(\omega^2+k^2\right) - \left(y\overline{vv}\right)_y(2k\omega)\right] / \left[2\left(\omega^2-k^2\right)^2\right]\\
&= \overline{vv}\left(2\omega^2 + k/\omega\right) / \left[2\left(\omega^2-k^2\right)\right]\\
&\quad + \left[\overline{v_y v}\left(\omega^2+k^2\right) - y\overline{vv}(2k\omega)\right]_y / \left[2\left(\omega^2-k^2\right)^2\right],
\end{aligned}
\tag{14d}
$$

where the first equality has been derived using (11b)–(11c) and $\overline{\sin\theta\sin\theta} = \overline{\cos\theta\cos\theta}$, and the third equality has been derived using (10). As before, it is the second of the two terms whose meridional integral is zero. Using (14c)–(14d), we now obtain an analytical expression for the difference between the right and left hand sides of (14b) to yield

$$
\begin{aligned}
&(\partial\omega/\partial k)\overline{(u^2+v^2+p^2)}/2 - \overline{up}\\
&= \frac{(\overline{v_y v})_y}{2(\omega^2-k^2)^2}\left[\frac{(2\omega^2+\omega/k)(\omega^2+k^2)}{2\omega^3/k+1} - 2\omega k\right]\\
&\quad - \frac{(y\overline{vv})_y}{2(\omega^2-k^2)^2}\left[\frac{(2\omega^2+\omega/k)2k\omega}{2\omega^3/k+1} - (\omega^2+k^2)\right]\\
&= \frac{(\overline{v_y v})_y(2\omega^4 + 2\omega^2k^2 + \omega^3/k + \omega k - 4\omega^4 - 2\omega k)}{2(\omega^2-k^2)^2(2\omega^3/k+1)}\\
&\quad - \frac{(y\overline{vv})_y(4\omega^3 k + 2\omega^2 - 2\omega^5/k - 2\omega^3 k - \omega^2 - k^2)}{2(\omega^2-k^2)^2(2\omega^3/k+1)}\\
&= \frac{(\overline{v_{y\theta}v_\theta})_y(\omega/k - 2\omega^2) - (y\overline{v_\theta v_\theta})_y(1 - 2\omega^3/k)}{2(\omega^2-k^2)(2\omega^3/k+1)}\\
&= \frac{[\overline{(\omega v_{y\theta} - kyv_\theta)v_\theta}]_y + [\overline{(-kv_{y\theta} + \omega yv_\theta)v_\theta}]_y(2\omega^2)}{2k(\omega^2-k^2)(2\omega^3/k+1)}\\
&= \frac{-(\overline{pv_\theta})_y - (\overline{2u_{tt}v_\theta})_y}{2k(1+2\omega^3/k)},
\end{aligned}
\tag{14e}
$$

where the first and second equalities have been derived using (13), the third equality has been derived using $\overline{\cos\theta\cos\theta} = \overline{\sin\theta\sin\theta}$, and the last equality has been derived using (11b)–(11c). The last line of (14e) has been written as the meridional gradient of scalar quantities. Thus, the meridional integral of (14e) vanishes for equatorial waves (with a meridionally decaying structure) and is consistent with (14a).

Using (14e), we can now rewrite the zonal component of the group velocity times wave energy as

$$
(\partial\omega/\partial k)\overline{(u^2+v^2+p^2)}/2 = \overline{up} + (\overline{p\varphi}/2 + \overline{u_{tt}\varphi})_y, \tag{15a}
$$

$$
\varphi \equiv -v_\theta/(k+2\omega^3), \tag{15b}
$$

where the scalar quantity φ has been introduced. We have confirmed that, as long as φ is set by (15b), the meridional profile of the zonal energy flux, $\overline{up} + (\overline{p\varphi}/2 + \overline{u_{tt}\varphi})_y$, in (15a) is precisely identical to $(\partial\omega/\partial k)\overline{(u^2+v^2+p^2)}/2$ for all types of equatorial waves in Figs. 2 and 3. Namely, all solid black lines in Figs. 2 and 3 may be drawn using either expression. As far as we know, (15a) and (15b) have not been mentioned in previous studies and therefore constitute a new result.

Inversion equations for Ertel's potential vorticity

The definition of φ, as given by (15b), is based on a Fourier expansion. However, we have found that (15b) may be rewritten into an expression which contains none of θ, k, and ω to read

$$
\begin{aligned}
\nabla^2\varphi - y^2\varphi - 3\varphi_{tt} &= -v_\theta/\omega\\
&= q,
\end{aligned}
\tag{16}
$$

where $\nabla^2 \equiv \partial_{xx} + \partial_{yy}$ is understood, the first line has been derived using (10), and the second line has been derived using (8) [i.e., $q_t = -\omega q_\theta = -v$ and thus $-\omega q_{\theta\theta} = \omega q = -v_\theta$]. The new Eq. (16) of EPV is the cornerstone of the present study, because it suggests a possibility for the scalar quantity φ to be estimated without using a Fourier analysis. This feature is important for identifying the direction of the energy flux of waves in the presence of coastal boundaries.

To summarize, in order to reproduce the profile of the group velocity times wave energy without relying on a Fourier analysis, we have obtained a new expression for the energy flux that has turned out to be associated with the streamfunction Eq. (16). Equation (16) may be rewritten into a dimensional form as

$$\nabla^{*2}\varphi^* - (f^*/c^*)^2\varphi^* - (3/c^{*2})\varphi^*_{t^*t^*} = q^*, \quad (17a)$$

where $\nabla^* \equiv \langle\langle \partial_{x^*}, \partial_{y^*} \rangle\rangle$ and $q^* = v^*_{x^*} - u^*_{y^*} - (f^*/c^{*2})p^*$. The exact profile of the group velocity times wave energy may be reproduced by the right hand side of (15a) and is here rewritten into a vector and dimensional form as

$$\overline{\mathbf{V}^* p^*} - \nabla^* \times [\,\overline{(p^*\varphi^*)}/2 + \overline{(u^*_{t^*t^*}\varphi^*)}/\beta^*]\,\mathbf{z}. \quad (17b)$$

The additional rotational flux in (17b) corrects the profile of the energy flux, without affecting the divergence of the energy flux. The quantity φ^* in (17b) is the solution of the accurate streamfunction Eq. (17a) associated with EPV in a dimensional form. We note in passing that for zonally propagating equatorial waves, as given by (11a)–(11c), $\overline{v^* p^*}$ vanishes owing to the phase relationship between v^* and p^* [see (11a) and (11c)] and the meridional component of the additional rotational flux, $-(\overline{p^*\varphi^*}/2 + \overline{u^*_{t^*t^*}\varphi^*}/\beta^*)_{x^*}$, also vanishes.

Equatorial KWs

So far, we have not investigated the energy flux of equatorial KWs. Since KWs are gravity waves, $\overline{\mathbf{V}^* p^*}$ becomes equal to the group velocity times wave energy. Namely, the additional rotational flux is absent. KWs are also characterized by $q^* = 0$; hence, the EPV equation (17a) yields $\varphi^* = 0$. The result is that, in the case of KWs, the expression for the energy flux, as given by (17b) reduces to $\overline{\mathbf{V}^* p^*}$, which is consistent with the nature of gravity waves.

Boundary conditions and the connection to mid-latitude regions

Consider a basin with closed zonal boundaries (i.e., the eastern and western coastlines of a basin of arbitrary shape). It is clear that the flux $\overline{\mathbf{V}^* p^*}$ in (17b) has no component normal to the zonal boundaries. Hence, the additional rotational flux in (17b) should also have no component crossing the closed boundaries. This requirement is satisfied in the present study by solving (17a) with a boundary condition of

$$\varphi^* = 0. \quad (17c)$$

In a general situation in the ocean, waves propagating eastward along the equatorial waveguide are partially redirected poleward as KWs along the eastern boundary where they can shed RWs that then propagate westward into the ocean interior (Cane and Moore 1981; Philander 1989; Chelton and Schlax 1996; Isachsen et al. 2007).

We now investigate whether or not the set of (17a) and (17b) is applicable to off-equatorial regions where small-amplitude perturbations are characterized by either mid-latitude RWs or IGWs. For perturbations associated with mid-latitude RWs, the solution φ^* of (17a) corresponds to the geostrophic streamfunction for which $\varphi^* \simeq p^*/f^*$ is a reasonable approximation in an interior region (i.e., far from coastal boundaries), noting that $\nabla^{*2}\varphi^*$ corresponds to $v^*_{x^*} - u^*_{y^*}$. The result is that the energy flux in (17b) automatically reduces to the expression of OS93 for mid-latitude RWs[5]. On the other hand, if perturbations associated with mid-latitude IGWs are given, the inversion Eq. (17a) of EPV, which equals zero, yields, with $\varphi^* = 0$ on the boundaries, $\varphi^* = 0$ everywhere. Thus, the energy flux in (17b) automatically reduces to $\overline{\mathbf{V}^* p^*}$ which represents the group velocity of mid-latitude IGWs times wave energy. We conclude that the set of (17a) and (17b) can represent the exact profile of the group velocity times wave energy associated with both mid-latitude IGWs and RWs, which may be reconfirmed using almost the same procedure as in the "Identification of the additional rotational flux associated with equatorial waves" section. See Appendix 1 for details.

Methods/Experimental

The rest of this manuscript presents an example illustrating the diagnosis of the energy flux from a model. To be useful for our discussion, the exact universal expression for both equatorial and mid-latitude waves, as given by the set of (17a) and (17b), is hereafter referred to as the level-0 energy flux. In practice, the level-0 expression of the energy flux is not straightforward to compute from model output, since the second-order time derivative term in (17a) makes it difficult to solve for φ^*.

For the present study, we investigate the consequence of artificially removing the second-order time derivative term from (17a) to give

$$\nabla^{*2}\varphi^{\mathrm{app}*} - (f^*/c^*)^2\varphi^{\mathrm{app}*} = q^*, \quad (18a)$$

which may be justified at least for low-frequency waves (e.g., both equatorial and mid-latitude RWs) based on scale analysis. The superscript of $\varphi^{\mathrm{app}*}$ indicates that the solution of (18a) may be regarded as an approximation for the solution φ^* of the accurate streamfunction Eq. (17a) associated with EPV. Then, we replace φ^* in (17b) with $\varphi^{\mathrm{app}*}$ to read

$$\overline{\mathbf{V}^* p^*} - \nabla^* \times [\,\overline{(p^*\varphi^{\mathrm{app}*})}/2 + \overline{(u^*_{t^*t^*}\varphi^{\mathrm{app}*})}/\beta^*]\,\mathbf{z}, \tag{18b}$$

which is diagnosable[6] from model output and is referred to as the level-1 expression of the energy flux in the present study. As shown by the dashed orange lines in Fig. 2, the level-1 expression provides a nice approximation for the group-velocity-based energy flux of low-frequency equatorial waves, but not for high-frequency equatorial waves in Fig. 3. Next, with the form of the additional rotational flux $-\nabla^* \times [\overline{p^{*2}}/(2f^*)]\,\mathbf{z}$ in (6) in mind, we investigate the consequence of simplifying (18b) as

$$\overline{\mathbf{V}^* p^*} - \nabla^* \times \overline{(p^*\varphi^{\mathrm{app}*}/2)}\mathbf{z}, \tag{18c}$$

which we refer to as the level-2 expression for the energy flux. As shown by the solid blue lines in Figs. 2 and 3, the level-2 expression provides an approximation for the group-velocity-based energy flux of both low- and high-frequency equatorial waves, although there can be some error. Further discussion of the level-2 approximation is given in Appendices 2 and 3 where it is noted that the level-2 approximation is comparable in accuracy to the pseudomomentum (or wave-activity) flux used in previous studies (Randel and Williamson 1990; Brunet and Haynes 1996; Fukutomi and Yasunari 2002; Wakata and Kitaya 2002; Kawatani et al. 2010).

We now contrast both the level-1 and level-2 energy fluxes with the expressions in previous studies, given by (6) and (4), using a solution from a linear shallow water model. This illustrates the potential of the expression given by (18b) and (18c) for use as a model diagnostic (see Table 3). Suitable for this purpose is an equatorial basin mode solution since it is associated with both equatorial and coastal waveguides as well as the radiation of mid-latitude RWs into the basin interior. Furthermore, as noted in the Introduction section, the equatorial basin mode, first studied by Cane and Moore (1981), has recently attracted attention because of its importance in the dynamics of the equatorial Atlantic Ocean. Indeed, the annual cycle, the semi-annual cycle, and the interannual variability associated with the Atlantic equatorial deep jets (Brandt et al. 2011) all appear to be resonant excitations of equatorial basin modes [see Brandt et al. (2016) and Claus et al. (2016) for more details].

Model set-up

To illustrate the importance of dissipation for explaining the observed cross-equatorial width of the equatorial deep jets, Greatbatch et al. (2012, hereafter G12) have simulated a forced/dissipative basin mode solution using a single-layer reduced-gravity linear model. The model is set up in spherical coordinates, with a rectangular domain in latitude/longitude space of roughly the same width as the Atlantic Ocean at the equator (that is 55° in longitude) and reaching to 10°N/S on either side of the equator[7]. All lateral boundaries are closed. In both G12 and Claus et al. (2014, hereafter C14), the model has been forced by an idealized oscillatory forcing with a period of 4.5 years in the zonal momentum equation to mimic the forcing of the jets, together with a lateral mixing of momentum that provides dissipation. [See Ascani et al. (2015) for a discussion on the forcing of the equatorial deep jets, the details of which are not important here]. It should be noted that 4.5 years is roughly the time taken for an equatorial KW and the reflected long gravest equatorial RW, to travel across the basin for the vertical mode that is closest to resonance. As noted in G12 and C14, the (westward) propagation speed of equatorial long RWs is three times less than the (eastward) propagation speed of equatorial KWs [see the dispersion relation (12)].

Our model has been set up as in G12 and C14. The gravity wave speed is set equal to $c^* = 0.17$ m/s [see the upper panel in Fig. 4 of C14]. The equatorial deformation radius becomes $\sqrt{c^*/\beta^*} = 87$ km, with a consequence that disturbances further than a few degrees from the equator in our model experiment may be regarded as mid-latitude

Table 3 List of energy flux vectors and EPV-based streamfunctions in dimensional form and their location in the text and figures

Approx.	Energy flux vector	Equation	Figs. 2 & 3	Figs. 6 & 7
Level-0	$\overline{\mathbf{V}^* p^*} - \nabla^* \times [\,\overline{(p^*\varphi^*)}/2 + \overline{(u^*_{t^*t^*}\varphi^*)}/\beta^*]\,\mathbf{z}$	(15a), (17b), (23a)	Solid black	–
Level-1	$\overline{\mathbf{V}^* p^*} - \nabla^* \times [\,\overline{(p^*\varphi^{\mathrm{app}*})}/2 + \overline{(u^*_{t^*t^*}\varphi^{\mathrm{app}*})}/\beta^*]\,\mathbf{z}$	(18b), (26b)	Dashed orange	–
Level-2	$\overline{\mathbf{V}^* p^*} - \nabla^* \times \overline{(p^*\varphi^{\mathrm{app}*})}/2\mathbf{z}$	(18c), (26c)	Solid blue	(c)
QG	$\overline{\mathbf{V}^* p^*} - \nabla^* \times [\overline{p^{*2}}/(2f^*)]\,\mathbf{z}$	(5), (6)	–	(b)
f-Plane	$\overline{\mathbf{V}^* p^*}$	(3), (4)	Dashed green	(a)
	Definition of EPV-based streamfunctions			
	$\nabla^{*2}\varphi^* - (f^*/c^*)^2\varphi^* - (3/c^{*2})\varphi^*_{t^*t^*} = q^*,$	(15b), (16), (17a), (23c)		
	$\nabla^{*2}\varphi^{\mathrm{app}*} - (f^*/c^*)^2\varphi^{\mathrm{app}*} = q^*,$	(18a), (26a)		

Table 4 Parameters in the model experiment of the present study

Long gravity wave speed	$c^* = 0.17$ m/s
Equatorial deformation radius	$\sqrt{c^*/\beta^*} = 87$ km
Equatorial inertial period	$2\pi/\sqrt{c^*\beta^*} = 37$ days
Forcing period	$T^* = 4.5$ years
Forcing amplitude	10^{-10} m/s^2
Forcing area	Full domain
Domain size	55° (zonal) × 20° (meridional)
Forcing Froude number	$(0.0023 \text{ m/s})/c^* = 0.014$
Horizontal resolution	0.1° (zonal) × 0.1° (meridional)
Lateral eddy viscosity	10 m^2/s

RWs, even though they are part of the equatorial basin mode resonance. As in G12, our model has been formulated in a spherical coordinate system with a grid spacing of 0.1° in both longitude and latitude. The coefficient[8] of eddy viscosity has been set to 10 m^2/s. From an initial condition of no motion and no pressure anomaly, the model has been integrated for 20 cycles (i.e., 90 years) using the oscillatory forcing which is sufficient for a steady oscillatory state to be reached. Since the model code is fully non-linear, we have set the amplitude of the forcing to a small value, 1.0×10^{-10} m/s^2 to ensure that linear dynamics prevails. Indeed, the magnitude of velocity associated with the gravest basin mode may be scaled as 10^{-10} m/s^2 $\times$ 4.5 years/(2π) = 0.0023 m/s, which results in a Froude number of $(0.0023 \text{ m/s})/c^* = 0.014$ (nondimensional). These parameters are summarized in Table 4. Below, we show results from an experiment which corresponds to the "full" case in G12. In particular, the oscillatory zonal forcing is spatially uniform and acts over the whole model domain. All the model results shown below are averages over the last model cycle.

Results and discussion

At each time step of the model output, we have calculated the EPV-based streamfunction φ^{app*} (contours in

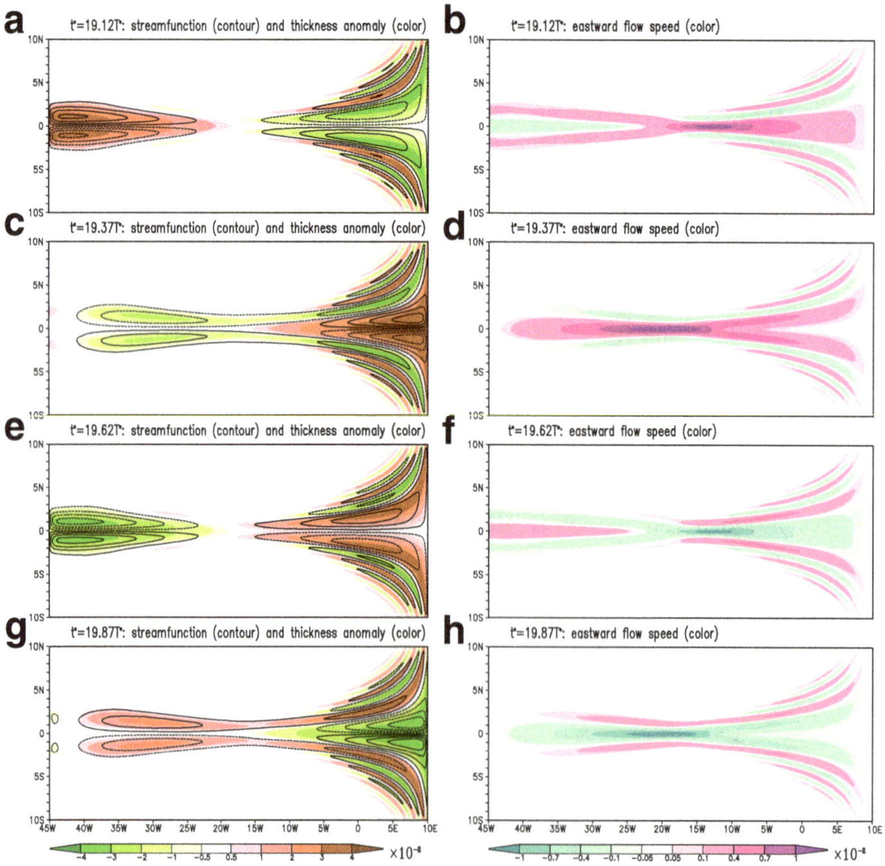

Fig. 4 *Left panels* are the snapshots of the EPV-based streamfunction φ^{app*} (*solid* and *dotted line* contours for positive and negative values, respectively, with an interval of 60 m^2/s) and thickness anomaly (*color shading*, normalized). *Right panels* are the snapshots of the zonal component of velocity u^* (*color shading*, m/s). These snapshots have been adapted from the 20th cycle of the model experiment, wherein the period of the oscillatory forcing is $T^* = 4.5$ years; the reference layer thickness is $H^* = 100$ m, and the equatorial deformation radius is $\sqrt{c^*/\beta^*} = 87$ km. See also movie (Additional file 1)

the left panels of Fig. 4) by solving the spherical coordinate version of (18a) with the boundary condition of $\varphi^{app*} = 0$. The color shading in Fig. 4 shows the snapshots of thickness anomaly (left panels) and the zonal component of velocity u^* (right panels). The movie of the model experiment is found in Additional file 1. RWs are identified by the correlation (anticorrelation) between the EPV-based streamfunction and thickness anomaly in the northern (southern) hemisphere. This follows from the correspondence between the EPV-based streamfunction and the geostrophic streamfunction for the case of mid-latitude RWs, as noted earlier. As noted in G12 and C14, the (westward) propagation speed of equatorial long RWs is three times smaller than the (eastward) propagation speed of equatorial KWs [see the dispersion relation (12)]. It takes a three-quarter cycle (i.e., $3T^*/4$) for equatorial long RWs to travel westward from the eastern boundary to the western boundary of the model domain (see red lines in Fig. 5a). After reflection at the western boundary, it takes only a quarter of a cycle (i.e., $T^*/4$) for equatorial KWs to travel eastward to the eastern boundary of the model domain (see blue lines in Fig. 5a), where some disturbances are deflected poleward along the eastern boundary to be the source of mid-latitude RWs which then propagate westward (Fig. 5b).

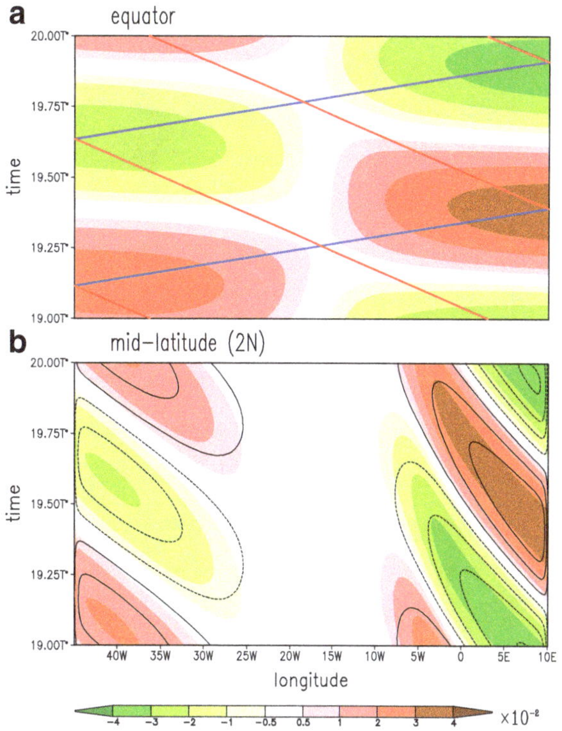

Fig. 5 a Hovmoller diagram at the equator for thickness anomaly (*color shading*, normalized) with the *blue* and *red lines* indicating the phase speeds of equatorial Kelvin waves and long Rossby waves, respectively. **b** Hovmoller diagram at 2°N for thickness anomaly (*color shading*, normalized) and the EPV-based streamfunction (*contours*, plotted as in the *left panels* of Fig. 4)

In Fig. 6, the divergence of the horizontal energy flux, given by $\nabla^* \cdot \overline{\mathbf{V}^*p^*}$, is shown for the whole model domain using color shading. Red indicates regions of a net energy input, and blue indicates regions of a net dissipation. It is clear that the main region of energy input is in the central part of the basin along the equator, where the strongest zonal velocities are found, and that the main regions of energy loss are associated with the RWs that radiate away from the eastern boundary. Arrows in Fig. 6a show the energy flux used in the gravity-wave literature, $\overline{\mathbf{V}^*p^*}$, which is mostly westward along the equator and eastward in the immediate off-equatorial region. This can be clearly seen in Fig. 7a which shows a blow-up of the eastern equatorial region. Figures 6b and 7b show the energy flux given by (6), which has been adapted from OS93, where only regions more than 1° latitude away from the equator are plotted to avoid the singularity in the Coriolis parameter f^* at the equator. From these figures (especially the blow-up of the eastern equatorial region in Fig. 7a, b), it is clear that the energy flux is strongly reversed when compared to $\overline{\mathbf{V}^*p^*}$ in the immediate off-equatorial region and is now strongly eastward in association with RWs that are radiated from the eastern boundary.

From Figs. 6c and 7c, it is clear that when the set of Eqs. (18a), (18c) and (17c) is used to estimate the energy flux, the westward flux associated with the off-equatorial RWs is part of a recirculation of energy in the eastern part of the basin (Fig. 7c) with eastward energy flux along the equator and westward energy flux off the equator. The eastward flux along the equator in Figs. 6c and 7c is in the opposite direction to the westward $\overline{\mathbf{V}^*p^*}$ flux in Figs. 6a and 7a along the equator in the same region. This indicates the role of the rotational flux contribution in (18c) which counters the westward $\overline{\mathbf{V}^*p^*}$ flux along the equator. This westward flux is associated with the equatorial RWs but represents an overestimation of the energy flux associated with these waves (see Fig. 2). When the rotational flux is added, what emerges is the eastward flux associated with the KW which, in turn, leads to a poleward flux arising from KWs propagating along the eastern boundary and, in turn, leads to the westward flux associated with the off-equatorial RWs that are excited at the eastern boundary. Here, in terms of the transfer of wave energy, the equatorial waveguide has been connected to the eastern coastal waveguide and, in turn, to the basin interior at off-equatorial latitudes, which is at the heart of the present study.

Finally, we note that the forcing period of $T^* = 4.5$ years is much longer than the equatorial inertial period of $2\pi/\sqrt{c^*\beta^*} = 37$ days. It can be said that the simulated equatorial basin mode consists of low-frequency equatorial waves, as in Fig. 2, and mid-latitude RWs. We recall the

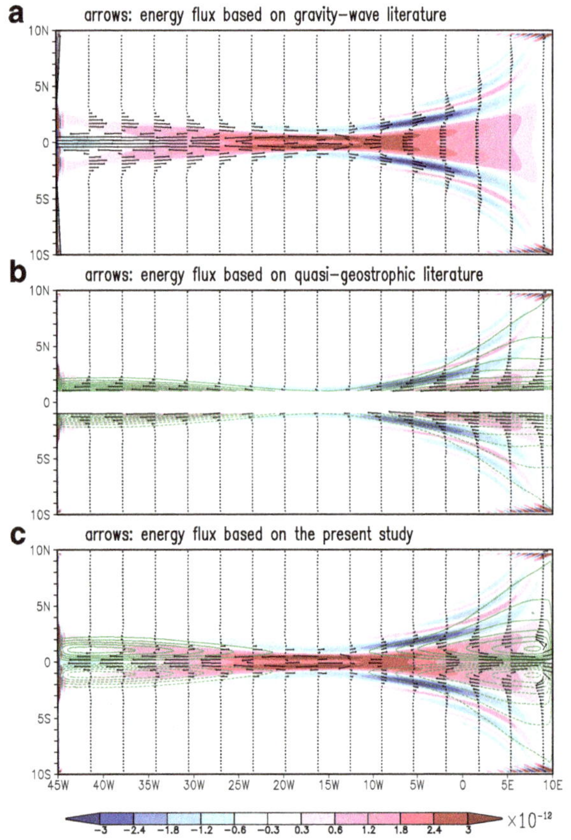

Fig. 6 Comparisons of three expressions for the horizontal flux of wave energy (*arrows*): **a** the pressure flux [see Eq. (3)], **b** the pressure flux plus the additional rotational flux of Orlanski and Sheldon (1993) [see Eq. (4)], and **c** the pressure flux plus the additional rotational flux of the present study [see Eq. (15)]. In both **b** and **c**, the additional fluxes have been calculated using the rotation operator in the spherical coordinate system to be consistent with the model formulation. *Green contours* in **b** and **c** show the distributions of $\overline{p^{*2}/(2f^*)}$ and $\overline{p^*\varphi^{\mathrm{app}*}}/2$, respectively (*solid* and *dotted lines* indicate positive and negative values, respectively, with an interval of 10^{-2} m^4/s^3). *Color shading* in all panels shows the *horizontal* divergence of the time-averaged energy flux. Note that the additional rotational flux in **b** and **c** has no influence on the divergence. All quantities have been calculated from the output of the same experiment with a time-average between $t^* = 19T^*$ and $20T^*$

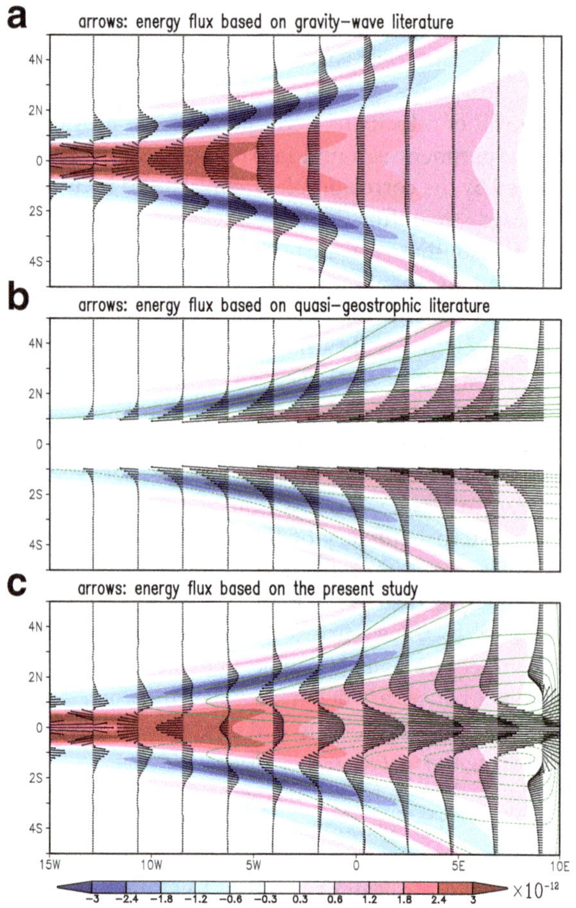

Fig. 7 Same as Fig. 6 except that this is an enlarged view of the eastern equatorial region of the model domain

small difference between the solid blue and dashed orange lines in Fig. 2, the former and the latter of which may be written as $\overline{u^*p^*} + (\overline{p^*\varphi^{\mathrm{app}*}}/2)_{y^*}$ and $\overline{u^*p^*} + (\overline{p^*\varphi^{\mathrm{app}*}}/2 + \overline{u^*_{t^*t^*}\varphi^{\mathrm{app}*}}/\beta^*)_{y^*}$, respectively, in a dimensional form (see level-2 and level-1, respectively, in Table 3). Since arrows in Figs. 6c and 7c have been plotted using the expression which corresponds to the solid blue line in Fig. 2, we have checked for any improvement by using the expression which corresponds to the dashed orange lines in the same figure. The checking has been done by comparing the distribution of $\overline{p^*\varphi^{\mathrm{app}*}}/2$ and $\overline{u^*_{t^*t^*}\varphi^{\mathrm{app}*}}/\beta^*$, from which we have learned that the latter quantity (not shown) is three orders of magnitude smaller than the former. Thus, we conclude that, in the diagnosis of the simulated basin mode, the expression of the energy flux, as given by (18c), has provided a nice approximation for the group velocity times wave energy.

Conclusions

In previous studies of the ocean, the energy flux of waves in model output has been diagnosed using $\overline{\mathbf{V}^*p^*}$, where $\mathbf{V}^*$ is the horizontal component of velocity perturbation and p^* corresponds to the pressure perturbation. This is appropriate for understanding the energy flux associated with mid-latitude inertia-gravity waves (IGWs). For mid-latitude Rossby waves (RWs), however, the direction of $\overline{\mathbf{V}^*p^*}$ differs from the group velocity and hence the energy flux, by a rotational vector flux with zero divergence. The rotational flux to be added to $\overline{\mathbf{V}^*p^*}$ for estimating the group velocity of mid-latitude RWs has previously been derived using quasi-geostrophic equations and is singular at the equator.

By investigating the analytical solution of both equatorial waves ("Analytical investigation" section) and mid-latitude waves (Appendix 1), we have derived an exact universal[9] expression for the rotational flux which, after being added to $\overline{\mathbf{V}^* p^*}$, is able to indicate the profile of the group velocity times wave energy for linear waves at all latitudes. This is what we call the level-0 expression of the energy flux. The level-0 energy flux is written using the solution φ^* of (17a), previously unmentioned in the literature, which we refer to as the accurate streamfunction associated with Ertel's potential vorticity (EPV). Equation (17a) is the cornerstone of the present study, because it suggests a possibility for the energy flux to be estimated (i) without using a Fourier analysis nor ray theory and (ii) in the presence of coastal boundaries, which will allow for tropical-extratropical interactions in model output to be diagnosed in terms of an energy cycle in a future study. Presently, the level-0 energy flux is not practical for use as a model diagnostic, since the second-order time derivative term in (17a) makes it difficult to solve for φ^*. Thus, we hope that a future study is able to develop a numerical algorithm to solve (17a) for φ^*. We also note the need to extend the theory to a continuously stratified ocean and also to test out the theory in the presence of a sheared mean flow, both of which topics await a future study. This is a new step from the recent understanding of energetics in the atmosphere and ocean that had been focused on, for example, the global mapping of energy conversion rates associated with various physical processes (e.g., baroclinic and barotropic instabilities) and external forcing (Iwasaki 2001; Aiki and Richards 2008; Zhai et al. 2012).

The potential of our analysis as a model diagnostic is illustrated in the present study for a forced/dissipative equatorial basin mode simulated by a single-layer model. The model result includes both mid-latitude RWs (maintained by coastal KWs propagating poleward along the eastern boundary) and equatorial RWs (maintained by the reflection of equatorial KWs at the eastern boundary). We have used approximate expressions for the energy flux (what we call the level-1 and level-2 energy fluxes) that is based on the inversion equation (18a) of EPV and which is shown to be good approximations to the level-0 expression in the case of the model run being considered. Since (18a) is seamlessly solvable at all latitudes with $\varphi^{\mathrm{app}*} = 0$ at coastlines, the source of the westward energy flux of mid-latitude RWs in the model output has been successfully illustrated in the present study. To our knowledge, this is the first attempt to diagnose the energy cycle of a tropical-extratropical interaction associated with the connection of the equatorial and coastal waveguides.

Endnotes

[1] While the energy flux of waves at all latitudes is considered in the present study, the pseudomomentum (or wave-activity) flux of waves at all latitudes is considered in Aiki et al. (2015, hereafter ATG15). Both the formulations of the present study and ATG15 may be reproduced even if a spherical coordinate system is used. The use of a Cartesian horizontal coordinate system in both the present study and ATG15 is for the purpose of simplicity, which will allow for the results of the two studies to be linked in a future study. A related discussion appears in Appendix 3.

[2] What we call pressure, energy, and momentum in the present study are actually dynamic pressure, energy density, and momentum density, respectively, following ATG15.

[3] $dH^{(n)}/dy = 2nH^{(n-1)}$, $H^{(n+1)} = 2yH^{(n)} - 2nH^{(n-1)}$, $H^{(0)} = 1$, $H^{(1)} = 2y$, $H^{(2)} = 4y^2 - 2$, $H^{(3)} = 8y^3 - 12y$, $H^{(4)} = 16y^4 - 48y^2 + 12$.

[4] The factor $\partial\omega/\partial k$ to calculate the energy flux is added in (14e).

[5] The second term in the square brackets of (17b) vanishes as $\overline{u^*_{t^*t^*}\varphi^*} \simeq \overline{(-p^*_{y^*t^*t^*}/f^*)(p^*/f^*)} = 0$ where the phase relationship of plane waves is understood.

[6] We use the term "diagnosable" to indicate that the quantity is readily estimated from quantities in model output without relying on a Fourier analysis.

[7] In a related paper, Claus et al. (2014) also used this solution to investigate the influence of the barotropic mean flow on the Atlantic equatorial deep jets. The Atlantic equatorial deep jets are resonant with the gravest basin mode for a high-order baroclinic mode (typically the 15th vertical normal mode) and consist of vertically stacked zonal jets that oscillate at a given depth with a period of around 4.5 years.

[8] This is lower than the value recommended by G12 for capturing the observed width of the deep jets but is chosen here since it is not so large as to prevent focusing of RWs on the equator. In the inviscid solution of Cane and Moore (1981), there is a singularity on the equator at the center of the basin due to RW focusing as described by Schopf et al. (1981).

[9] In the present manuscript, we have used the term "exact" to refer to the level-0 expression, in contrast to approximate expressions (i.e., level-1 and -2). Likewise, we have used the term "universal" to indicate the

ability to handle all wave types in Table 2, for which the group velocity has been well formulated in the literature/textbook.

[10]Although it is not in the list of wave types in Table 2, IGWs on a mid-latitude β-plane may be characterized as $\alpha \ll 1, \delta^2 \leq 1, \gamma^2 < 1$ where $\alpha \ll 1$ corresponds to (19b). Thus, the net content in the square brackets on the last line of (24c) becomes $O(1)$. Given α in front of $c^*\overline{v^*v^*}$ on the last line of (24c), we may justify (23d) for IGWs on a mid-latitude β-plane. It can be said that the right hand side of (24c) becomes significantly nonzero when the assumption of plane waves in the meridional direction becomes inconsistent (Anderson and Gill 1979).

[11]While the pseudomomentum flux itself $(\overline{E^* - v^*v^*})$ is diagnosable from model output, the pseudomomentum-flux-based expression of the energy flux $(\overline{E^* - v^*v^*})\omega^*/k^*$ is not easily diagnosable from model output because of multiplication by the phase speed (see Appendix 3 for details).

Appendix 1

Is the streamfunction Eq. (17a) associated with EPV applicable to mid-latitude waves?

Manipulation of the shallow water equation system (1a)–(1c) yields a characteristic equation associated with the meridional component of velocity to read

$$v^*_{t^*t^*t^*} - c^{*2}\left(v^*_{x^*x^*} + v^*_{y^*y^*}\right)_{t^*} + f^{*2}v^*_{t^*} - \beta^*c^{*2}v^*_{x^*} = 0, \tag{19a}$$

which is applicable to both mid-latitude and equatorial regions. In what follows, we consider plane waves on either an f-plane or a mid-latitude β-plane (i.e., $f^* = f_0^* + \beta^*y^*$ and $|f_0^*| \gg |\beta^*y^*|$) and thus assume

$$f^{*2} \simeq f_0^{*2}. \tag{19b}$$

Then, (19a) may be simplified as

$$v^*_{t^*t^*t^*} - c^{*2}\left(v^*_{x^*x^*} + v^*_{y^*y^*}\right)_{t^*} + f_0^{*2}v^*_{t^*} - \beta^*c^{*2}v^*_{x^*} = 0. \tag{19c}$$

The Coriolis parameter f_0^* in (19c) is constant that allows us to assume a horizontally monochromatic wave in a complex form

$$v^* = \mathscr{A}^*e^{i\theta}, \tag{20a}$$

where i is the unit imaginary number, $\mathscr{A}^*$ is wave amplitude, and $\theta = k^*x^* + l^*y^* - \omega^*t^*$ is wave phase (k^* and l^* are the zonal and meridional components of a wavenumber vector, respectively, and ω^* is wave phase). For simplicity, all $\mathscr{A}^*$, k^*, l^*, and ω^* are assumed to be constant. Substitution of (20a) to both (1a) and (1c) yields a solution for u^* and p^* to read

$$u^* = \left(f^*\omega^*v^*_\theta + c^{*2}k^*l^*v^*\right)/\left(\omega^{*2} - c^{*2}k^{*2}\right), \tag{20b}$$

$$p^* = \left(f^*k^*v^*_\theta + \omega^*l^*v^*\right)c^{*2}/\left(\omega^{*2} - c^{*2}k^{*2}\right), \tag{20c}$$

where $f^* = f_0^* + \beta^*y^*$. On the other hand, substitution of (20a) to (19c) yields

$$\omega^{*3} - c^{*2}\left(k^{*2} + l^{*2}\right)\omega^* - f_0^{*2}\omega^* - \beta^*c^{*2}k^* = 0, \tag{21}$$

which is a universal expression for the dispersion relation of the various types of waves in mid-latitude regions. For example, substitution of $\beta^* = 0$ to (21) yields a classical dispersion relation for mid-latitude IGWs (i.e., waves on an f-plane), and substitution of $\omega^{*2} \ll c^{*2}k^{*2}$ to (21) yields a classical dispersion relation for mid-latitude RWs.

An expression for the zonal component of group velocity may be derived using (21) to read

$$\frac{\partial\omega^*}{\partial k^*} = \frac{2c^{*2}k^*\omega^* + \beta^*c^{*2}}{3\omega^{*2} - c^{*2}\left(k^{*2} + l^{*2}\right) - f_0^{*2}} = \frac{2c^{*2}\omega^{*2}k^* + \beta^*c^{*2}\omega^*}{2\omega^{*3} + \beta^*c^{*2}k^*}. \tag{22a}$$

We now identify the content of $\left(\overline{A^*B^*}\right)_{y^*}$ in the following equation:

$$\overline{u^*p^*} + \left(\overline{A^*B^*}\right)_{y^*} = \frac{\partial\omega^*}{\partial k^*}\frac{1}{2}\left(\overline{u^{*2} + v^{*2} + \frac{p^{*2}}{c^{*2}}}\right), \tag{22b}$$

where each of A^* and B^* are quantities associated with the set of u^*, v^*, p^*, c^*, and f^*. A first step for investigating (22b) is to decompose $\overline{u^*p^*}$ into two parts: one that is associated with the numerator of (22a) and one that is written as the meridional derivative of a scalar quantity, as follows:

$$\begin{aligned}\overline{u^*p^*} &= \frac{\overline{v^*v^*}\left(f^{*2} + c^{*2}l^{*2}\right)c^{*2}\omega^*k^*}{\left(\omega^{*2} - c^{*2}k^{*2}\right)^2} \\ &\simeq \frac{\overline{v^*v^*}\left(\omega^{*3} - c^{*2}k^{*2}\omega^* - \beta^*c^{*2}k^*\right)c^{*2}k^*}{\left(\omega^{*2} - c^{*2}k^{*2}\right)^2} \\ &= \frac{\overline{v^*v^*}c^{*2}\omega^*k^*}{\left(\omega^{*2} - c^{*2}k^{*2}\right)} - \frac{\left(f^*\overline{v^*v^*}\right)_{y^*}c^{*4}k^{*2}}{\left(\omega^{*2} - c^{*2}k^{*2}\right)^2} \\ &= \frac{\overline{v^*v^*}\left(2c^{*2}\omega^*k^* + \beta^*c^{*2}\right)}{2\left(\omega^{*2} - c^{*2}k^{*2}\right)} - \frac{\left(f^*\overline{v^*v^*}\right)_{y^*}c^{*2}\left(\omega^{*2} + c^{*2}k^{*2}\right)}{2\left(\omega^{*2} - c^{*2}k^{*2}\right)^2},\end{aligned} \tag{22c}$$

where the first equality has been derived using both (20b)–(20c) and the set of $\overline{v^*v^*} = \overline{v^*_\theta v^*_\theta}$ and $\overline{v^*_\theta v^*} = 0$ and the approximate equality in the middle has been derived using both the dispersion relation (21) and (19b). Then, we decompose the wave energy in (22b) into two parts, one that is associated with the denominator of (22a) and one that is written as the meridional derivative of a scalar quantity. We then have

$$\begin{aligned}
\frac{1}{2}\overline{\left(u^{*2}+v^{*2}+\frac{p^{*2}}{c^{*2}}\right)}
&= \frac{\overline{v^*v^*}\left(\omega^{*2}-c^{*2}k^{*2}\right)^2}{2\left(\omega^{*2}-c^{*2}k^{*2}\right)^2} \\
&\quad + \frac{\overline{v^*v^*}\left(\omega^{*2}f^{*2}+c^{*4}k^{*2}l^{*2}+k^{*2}f^{*2}c^{*2}+\omega^{*2}l^{*2}c^{*2}\right)}{2\left(\omega^{*2}-c^{*2}k^{*2}\right)^2} \\
&= \frac{\overline{v^*v^*}\left[\left(\omega^{*2}-c^{*2}k^{*2}\right)^2+\left(f^{*2}+c^{*2}l^{*2}\right)\left(\omega^{*2}+c^{*2}k^{*2}\right)\right]}{2\left(\omega^{*2}-c^{*2}k^{*2}\right)^2} \\
&\simeq \frac{\overline{v^*v^*}\left(\omega^{*2}-c^{*2}k^{*2}\right)^2}{2\left(\omega^{*2}-c^{*2}k^{*2}\right)^2} \\
&\quad + \frac{\overline{v^*v^*}\left(\omega^{*2}-c^{*2}k^{*2}-\beta^*c^{*2}k^*/\omega^*\right)\left(\omega^{*2}+c^{*2}k^{*2}\right)}{2\left(\omega^{*2}-c^{*2}k^{*2}\right)^2} \\
&= \frac{\overline{v^*v^*}\omega^{*2}}{\left(\omega^{*2}-c^{*2}k^{*2}\right)} - \frac{\left(f^*\overline{v^*v^*}\right)_{y^*}c^{*2}k^*\left(\omega^{*2}+c^{*2}k^{*2}\right)}{2\omega^*\left(\omega^{*2}-c^{*2}k^{*2}\right)^2} \\
&= \frac{\overline{v^*v^*}\left(2\omega^{*3}+\beta^*c^{*2}k^*\right)}{2\omega^*\left(\omega^{*2}-c^{*2}k^{*2}\right)} - \frac{\left(f^*\overline{v^*v^*}\right)_{y^*}2c^{*2}\omega^*k^*}{2\left(\omega^{*2}-c^{*2}k^{*2}\right)^2},
\end{aligned} \tag{22d}$$

where the first equality has been derived using both (20b)–(20c) and the set of $\overline{v^*v^*}=\overline{v^*_\theta v^*_\theta}$ and $\overline{v^*_\theta v^*}=0$ and the approximated equality in the middle has been derived using both the dispersion relation (21) and (19b). The set of (22c) and (22d) allows us to identify the content of $(\overline{A^*B^*})_{y^*}$ in (22b) to read

$$\begin{aligned}
&\frac{\partial\omega^*}{\partial k^*}\frac{1}{2}\overline{\left(u^{*2}+v^{*2}+\frac{p^{*2}}{c^{*2}}\right)} - \overline{u^*p^*} \\
&\simeq \frac{-(f^*\overline{v^*v^*})_{y^*}c^{*2}}{2(\omega^{*2}-c^{*2}k^{*2})^2}\left\{\frac{(2c^{*2}\omega^{*2}k^*+\beta^*c^{*2}\omega)2k^*\omega^*}{2\omega^{*3}+\beta^*c^{*2}k^*}\right. \\
&\quad \left. -\left(\omega^{*2}+c^{*2}k^{*2}\right)\right\} \\
&= \frac{-(f^*\overline{v^*v^*})_{y^*}c^{*2}}{2(\omega^{*2}-c^{*2}k^{*2})^2}\left\{\frac{(4c^{*2}\omega^{*3}k^{*2}+2\beta^*c^{*2}\omega^{*2}k^*)}{2\omega^{*3}+\beta^*c^{*2}k^*}\right. \\
&\quad \left. + \frac{(-2\omega^{*5}-2c^{*2}\omega^{*3}k^{*2}-\beta^*c^{*2}\omega^{*2}k^*-\beta^*c^{*4}k^{*3})}{2\omega^{*3}+\beta^*c^{*2}k^*}\right\} \\
&= \frac{-(f^*\overline{v^*_\theta v^*_\theta})_{y^*}c^{*2}(\beta^*c^{*2}k-2\omega^{*3})}{2(\omega^{*2}-c^{*2}k^{*2})(2\omega^{*3}+\beta^*c^{*2}k^*)} \\
&= \frac{-(f^*\overline{v^*_\theta v^*_\theta})_{y^*}c^{*2}[1-2\omega^{*3}/(\beta^*c^{*2}k^*)]}{2(\omega^{*2}-c^{*2}k^{*2})[2\omega^{*3}/(\beta^*c^{*2}k^*)+1]} \\
&= \frac{-[\overline{(f^*k^*v^*_\theta+\omega^*l^*v^*)c^{*2}v^*_\theta}]_{y^*}}{2k^*(\omega^{*2}-c^{*2}k^{*2})[2\omega^{*3}/(\beta^*c^{*2}k^*)+1]} \\
&\quad + \frac{[\overline{(f^*\omega^*v^*_\theta+c^{*2}k^*l^*v^*)v^*_\theta}]_{y^*}2\omega^{*2}/\beta^*}{2k^*(\omega^{*2}-c^{*2}k^{*2})[2\omega^{*3}/(\beta^*c^{*2}k^*)+1]} \\
&= \frac{-(\overline{p^*v^*_\theta})_{y^*}-(\overline{2u^*_{t^*t^*}v^*_\theta})_{y^*}/\beta^*}{2k^*[1+2\omega^{*3}/(\beta^*c^{*2}k^*)]},
\end{aligned} \tag{22e}$$

where the last equality has been derived using (20a)–(20c). Equation (22e) may be rewritten as

$$\overline{u^*p^*}+(\overline{p^*\varphi^*}/2+\overline{u^*_{t^*t^*}\varphi^*}/\beta^*)_{y^*} = \frac{\partial\omega^*}{\partial k^*}\frac{1}{2}\overline{\left(u^{*2}+v^{*2}+\frac{p^{*2}}{c^{*2}}\right)}, \tag{23a}$$

where

$$\varphi^* \equiv \frac{-v^*_\theta}{k^*+2\omega^{*3}/(\beta^*c^{*2})}, \tag{23b}$$

has been introduced. The definition of φ^*, as given by (23b), is based on a Fourier expansion and may be rewritten into an expression which contains none of θ, k^*, l^*, and ω^* to read

$$\begin{aligned}
\nabla^{*2}\varphi^* - (f^*_0/c^*)^2\varphi^* - (3/c^{*2})\varphi^*_{t^*t^*} &= -\beta^*v^*_\theta/\omega^* \\
&= q^*,
\end{aligned} \tag{23c}$$

where the first equality has been derived using (20b) and the second equality has been derived using (2) [i.e., $q^*_{t^*} = -\omega^*q^*_\theta = -\beta^*v^*$ and thus $-\omega^*q^*_{\theta\theta} = \omega^*q^* = -\beta^*v^*_\theta$]. As far as we know, the set of (23a) and (23c) has not been mentioned in previous studies for mid-latitude waves and has turned out to be almost the same as the set of (17b) and (17a) that has been derived for equatorial waves.

We now consider the meridional flux of wave energy. We would like to show that

$$\overline{v^*p^*} - \underbrace{(\overline{p^*\varphi^*}/2+\overline{u^*_{t^*t^*}\varphi^*}/\beta^*)_{x^*}}_{0} = \frac{\partial\omega^*}{\partial l^*}\frac{1}{2}\overline{\left(u^{*2}+v^{*2}+\frac{p^{*2}}{c^{*2}}\right)}. \tag{23d}$$

It turns out that the second term on the left hand side, associated with the additional rotational flux, vanishes when evaluated using the analytical solution of waves [i.e., $(\overline{v^*v^*})_{x^*} = k\overline{v^*_\theta v^*} = 0$], which is as in Longuet-Higgins (1964). This is attributed to the assumption of all $\mathscr{A}^*$, k^*, l^*, and ω^* being constant in particular in the zonal direction. An expression for the meridional component of group velocity may be derived from (21) to read

$$\begin{aligned}
\frac{\partial\omega^*}{\partial l^*} &= \frac{2c^{*2}l^*\omega^*}{3\omega^{*2}-c^{*2}(k^{*2}+l^{*2})-f_0^{*2}} \\
&= \frac{2c^{*2}l^*\omega^{*2}}{2\omega^{*3}+\beta^*c^{*2}k^*}.
\end{aligned} \tag{24a}$$

Then, we calculate the left hand side of (23d) using (20a)–(20b) as

$$\overline{v^*p^*} = \frac{\overline{v^*v^*}c^{*2}\omega^*l^*}{\omega^{*2}-c^{*2}k^{*2}}, \tag{24b}$$

where $\overline{v^*_\theta v^*}=0$ has been used. We now calculate the difference of the meridional component of the group velocity times wave energy and $\overline{v^*p^*}$ using the set of (22d), (24a), and (24b) to yield

$$\begin{aligned}
&\frac{\partial\omega^*}{\partial l^*}\frac{1}{2}\left(\overline{u^{*2}+v^{*2}+\frac{p^{*2}}{c^{*2}}}\right)-\overline{v^*p^*} \\
&= -\frac{2c^{*4}k^*l^*\omega^{*3}}{(2\omega^{*3}+\beta^*c^{*2}k^*)}\frac{(f\overline{v^*v^*})_{y^*}}{(\omega^{*2}-c^{*2}k^{*2})^2} \\
&= -\frac{2\beta^*c^{*3}k^*l^*}{(2+\beta^*c^{*2}k^*/\omega^{*3})}\frac{c^*\overline{v^*v^*}}{\omega^{*4}(1-c^{*2}k^{*2}/\omega^{*2})^2} \\
&= -\left[\frac{2\delta^2}{(2+\alpha\delta^2\gamma)(1-\gamma^2)^2}\frac{c^{*2}k^*l^*}{\omega^{*2}}\right]\alpha c^*\overline{v^*v^*}, \qquad (24c)
\end{aligned}$$

where the last line has been written using the set of nondimensional parameters. These are defined as

$$\alpha \equiv \beta^*c^*/f_0^{*2}, \;\; \delta \equiv f_0^*/\omega^*, \;\; \gamma \equiv c^*k^*/\omega^*. \qquad (24d)$$

It can be said that the last line of (24c) represents the contribution of higher order terms in an asymptotic expansion based on α, δ, and γ. This contribution should not be confused with the universal expression of the additional rotational flux which has already been clarified at (23a) and (23d). It should be also noted that the net content within the square brackets on the last line of (24c) is nondimensional, for which we shall make scale analysis in the next paragraph.

The quantity $\alpha c^*\overline{v^*v^*}$ on the last line of (24c) may be interpreted as a reference for the magnitude of the energy flux of mid-latitude RWs. Mid-latitude RWs may be characterized as

$$|\alpha\gamma| = \frac{\beta^*c^{*2}/f_0^{*2}}{|\omega^*/k^*|} \geq 1, \;\; \delta^2 \gg 1, \;\; \gamma^2 \gg 1. \qquad (25a)$$

Thus, the net content within the square brackets on the last line of (24c) approximates to zero, which justifies (23d) for mid-latitude RWs. On the other hand, for mid-latitude IGWs, $c^*\overline{v^*v^*}$ on the last line of (24c) represents a reference for the magnitude of the energy flux. IGWs on an f-plane may be characterized as

$$\alpha = 0, \;\; \delta^2 \leq 1, \;\; \gamma^2 < 1. \qquad (25b)$$

Thus, the last line of (24c) vanishes, which justifies (23d) for IGWs on an f-plane[10].

To summarize, the streamfunction Eq. (17a) associated with EPV and the universal expression of the additional rotational flux in (17b) applies to both mid-latitude and equatorial waves, in particular for wave types considered in the present study, as listed in Table 2.

Appendix 2

Approximate expressions for the energy flux

The exact profile of the group velocity times wave energy is given by the set of (15a) and (16), which is what we call the level-0 energy flux. Owing to the last term on the left hand side of (16) that contains the second-order partial differentiation with respect to time, the procedure of inverting EPV, without using a Fourier analysis, is still complicated.

Hence, we investigate the consequence of artificially removing the second-order time derivative term from (16) as

$$\nabla^2\varphi^{\mathrm{app}} - y^2\varphi^{\mathrm{app}} = q, \qquad (26a)$$

where the superscript of φ^{app} indicates that the solution of (26a) may be regarded as an approximation for the solution φ of the accurate streamfunction Eq. (16) associated with EPV. We have calculated the meridional profiles of

$$\overline{up} + (\overline{p\varphi^{\mathrm{app}}}/2 + \overline{u_{tt}\varphi^{\mathrm{app}}})_y, \qquad (26b)$$

as shown by the dashed orange lines in Fig. 2 for low-frequency equatorial waves (e.g., equatorial RWs) and in Fig. 3 for high-frequency equatorial waves (e.g., equatorial IGWs). Since this is an analytical investigation, we have used $\varphi^{\mathrm{app}} = -v_\theta/(k-\omega^3)$ which has been derived from the EPV inversion Eq. (26a) with the use of the characteristic Eq. (10). All panels in Fig. 2 show a nice agreement between the dashed orange line given by (26b) and the solid black line, $(\partial\omega/\partial k)(\overline{u^2+v^2+p^2})$. By contrast, all panels in Fig. 3 show a finite disagreement between the dashed orange line given by (26b), $\overline{up} + (\overline{p\varphi^{\mathrm{app}}}/2 + \overline{u_{tt}\varphi^{\mathrm{app}}})_y$, and the solid-black line, $(\partial\omega/\partial k)(\overline{u^2+v^2+p^2})$.

It would be nice if there is a unified approximation for the energy flux that is able to represent the profile of the group velocity times the energy of both low- and high-frequency equatorial waves. We have found that this requirement is roughly satisfied if (26b) is simplified as

$$\overline{up} + (\overline{p\varphi^{\mathrm{app}}}/2)_y, \qquad (26c)$$

where $\varphi^{\mathrm{app}} = -v_\theta/(k-\omega^3)$ is the solution of (26a). The profile of (26c) is shown by the solid blue lines in Figs. 2 and 3 for low- and high-frequency equatorial waves, respectively. This expression provides what we think is a potentially useful approximation for the group velocity times wave energy (the solid black lines) for all types of equatorial waves, as we show in the "Methods/Experimental" section.

In the present study, (26b) and its vector and dimensional form (18b) are referred to as the level-1 energy flux. Likewise, (26c) and its vector and dimensional form (18c) are referred to as the level-2 energy flux.

Why do we appreciate the level-2 energy flux regardless of the error? An expression for pseudomomentum (or wave-activity) flux has long been used for the model diagnosis of the direction of the group velocity of waves in the atmosphere (and also the ocean), including in low-latitude regions (Ripa 1982; Hoskins et al. 1983; Plumb 1986; Haynes 1988; Randel and Williamson 1990; Brunet and Haynes 1996; Fukutomi and Yasunari 2002; Wakata and Kitaya 2002; Kawatani et al. 2010). Using the analytical solution of equatorial waves, we have calculated the profile of the traditional pseudomomentum flux[11] times the phase velocity of waves (see Appendix 3), as shown by the purple dots in Figs. 2 and 3. Interestingly, for low-frequency waves, the profile of the pseudomomentum-flux-based expression (the purple dots) is almost the same as that of the level-2 energy flux (the blue solid line). On the other hand, for high-frequency waves, the profile of the pseudomomentum-flux-based expression (the purple dots) is similar to that of the level-1 energy flux (the orange dashed line) and quite different from the exact, level-0 energy flux to which the level-2 energy flux is a better approximation. Thus, the level-2 energy flux is, in general, an improvement on the traditional model diagnosis of group velocity based on the pseudomomentum flux.

Concerning extension to mid-latitude waves, both the level-1 and level-2 energy fluxes satisfy all conditions noted in the last paragraph of the "Boundary conditions and the connection to mid-latitude regions" section. Note that the inversion Eq. (18a) of EPV is seamlessly solvable at all latitudes with the boundary condition of $\varphi^{\mathrm{app}*} = 0$. To summarize, the set of (18a) and (18c) [together with the boundary condition (17c)]—what we call the level-2 expression—originates from a trade-off between mathematical exactness and practical accessibility. The mathematical exactness for retrieving the group velocity of equatorial waves times wave energy has been achieved by the set of (17a) and (17b)—what we call the level-0 expression. However, its accessibility is harmed by the second-order time derivative term in the streamfunction equation (16) associated with EPV. On the other hand, concerning the practical accessibility, the set of (18a) and (18c)—the level-2 expression—has the advantages that (i) it is seamlessly solvable at all latitudes and (ii) it provides a unified expression for all types of waves with which to estimate the direction of the group velocity. We have noted, for equatorial waves, that the profile of the level-2 energy flux is somewhat better than that of the traditional pseudomomentum flux. It should be also noted that the energy flux given by (18c) satisfies the boundary condition of no flux through coastlines [using (17c)], an issue not considered in previous studies for the pseudomomentum flux. With these requirements in mind, we hope that future studies can lead to either an improved approximation or a numerical algorithm for the level-0 energy flux.

Appendix 3

Similarity between the level-2 energy flux of this study and the pseudomomentum flux in previous studies

Ripa (1982) has derived a conservation equation for pseudomomentum (or wave activity) associated with ageostrophic waves. His equation may be reproduced using (1a)–(1c) as

$$\frac{\partial}{\partial t^*}\underbrace{\left(\frac{p^*u^*}{c^{*2}} - \frac{q^{*2}}{2\beta^*}\right)}_{\text{IB pseudomomentum}} + \nabla^* \cdot \underbrace{\langle\!\langle E^* - v^*v^*,\ v^*u^* \rangle\!\rangle}_{\text{IB flux}} = 0, \tag{27a}$$

$$E^* \equiv \frac{1}{2}\left(u^{*2} + v^{*2} + \frac{p^{*2}}{c^{*2}}\right), \tag{27b}$$

where the prognostic quantity may be referred to as the impulse-bolus (IB) pseudomomentum (Aiki et al. 2015, hereafter ATG15) and E^* is the wave energy. Note that the IB pseudomomentum given here is the shallow water version of that given by Eq. (27a) in ATG15. It has been known that the expression of the flux in (27a) can indicate the direction of the group velocity of different types of waves, in particular, mid-latitude RWs and IGWs (Hoskins et al. 1983; Plumb 1986; Haynes 1988). Another nice feature of the IB pseudomomentum Eq. (27a) is that it does not contain a singularity at the equator. In order to investigate the origin of these features, ATG15 have shown in their Eq. (18a) an identity between the IB pseudomomentum and the classical energy-based (CE) pseudomomentum to read (again, written here for the shallow water equations)

$$\underbrace{\frac{E^*}{(\omega^*/k^*)}}_{\text{CE pseudomomentum}} = \underbrace{\frac{p^*u^*}{c^{*2}} - \frac{q^{*2}}{2\beta^*}}_{\text{IB pseudomomentum}} - \frac{\partial}{\partial y^*}\left(\frac{u^*q^*}{2\beta^*}\right) + \frac{\partial}{\partial x^*}\left(\frac{v^*q^*}{2\beta^*}\right), \tag{28a}$$

which may be derived from (1a)–(1c) of the present study. Application of a low-pass temporal filter to (27b), and then, understanding the phase relationship between $v^* = -q^*_{t^*}/\beta^*$ and q^* yields

$$\frac{\overline{E^*}}{(\omega^*/k^*)} = \overline{\frac{p^*u^*}{c^{*2}} - \frac{q^{*2}}{2\beta^*}} - \frac{\partial}{\partial y^*}\left(\overline{\frac{u^*q^*}{2\beta^*}}\right). \tag{28b}$$

Substitution of (28b) to a low-pass time-filtered version of (28a) yields

$$\frac{\partial}{\partial t^*}\overline{E^*} + \frac{\omega^*}{k^*}\nabla^* \cdot \left\langle\!\!\left\langle \overline{E^* - v^*v^*},\ \overline{v^*u^*} + \frac{\partial}{\partial t^*}\left(\overline{\frac{u^*q^*}{2\beta^*}}\right) \right\rangle\!\!\right\rangle = 0, \tag{29}$$

which is a prognostic equation for the wave energy wherein the zonal component of the flux is proportional to that in the IB pseudomomentum equation (27a).

It is easy to expect that the expression of the flux in (29) can indicate the direction of the group velocity of mid-latitude RWs and IGWs (Hoskins et al. 1983; Plumb 1986; Haynes 1988). For equatorial waves, here, we investigate the meridional profile of $(\overline{E^* - v^*v^*})\omega^*/k^*$ as shown by the purple dots in Figs. 2 and 3 for low- and high-frequency waves, respectively. For low-frequency waves (Fig. 2), the meridional profile of $(\overline{E^* - v^*v^*})\omega^*/k^*$ (the purple dots) is almost the same as that of the level-2 energy flux (the blue solid line), showing that the level-2 energy flux and the IB flux are closely related. For high-frequency waves (Fig. 3), the meridional profile of $(\overline{E^* - v^*v^*})\omega^*/k^*$ (the purple dots) is nearly the same as that of the level-1 energy flux (the orange dashed line), indicating that the level-2 energy flux is somewhat better than the IB flux.

In fact, without relying on the level-0 expression, we have arrived at the level-2 expression of the energy flux by extending the investigation of ATG15 concerning the algebraic structure of the IB flux (to be explained in a future study). ATG15 have addressed the importance of a wave-induced scalar quantity and symbolized it as Λ: it vanishes for mid-latitude IGWs (i.e., waves with no perturbation of EPV) and becomes nonzero for mid-latitude RWs (i.e., wave with a perturbation of EPV). Here, we suggest that $\overline{\Lambda} = (\overline{p^*\eta^*})_{y^*}/2$ is closely linked to $(\overline{p^*\varphi^{\text{app}*}})_{y^*}/2$ in the present study (η^* is meridional displacement). This is why the level-2 expression for the energy flux in the present study can indicate the direction of the group velocity of different types of waves, an issue we shall discuss in a future study.

Note that the IB flux in (27a) has already been used for the model diagnosis of waves in low-latitude regions (Randel and Williamson 1990; Brunet and Haynes 1996; Fukutomi and Yasunari 2002; Wakata and Kitaya 2002; Kawatani et al. 2010). We suggest that, despite the certain inaccuracy associated with equatorial waves as compared with the level-0 expression, the level-2 expression of the energy flux in the present study will be at least as useful as the IB flux which has long been used in the atmospheric (and oceanic) literature. For oceanic applications, the level-2 energy flux brings two new advantages over the IB flux: (i) the level-2 energy flux satisfies a no-normal-flux boundary condition at coastlines, and (ii) the wave energy is a sign-definite quantity while the IB pseudomomentum is not.

Overall, we address the balance of (i) model accessibility, (ii) unified treatment for different types of waves, (iii) mathematical accuracy, and (iv) boundary conditions at coastlines. With these requirements in mind, we hope future studies can lead to either an improved approximation or a numerical algorithm for the level-0 energy flux, wherein the profile of the IB flux will provide a reference for accuracy because the IB flux has long been used in previous studies.

Abbreviations

EPV: Ertel's potential vorticity; IGW: Inertia gravity wave; KW: Kelvin wave; RGW: Mixed Rossby-gravity wave; RW: Rossby wave

Acknowledgements

This manuscript has been improved by comments from two anonymous reviewers. HA thanks Paal Erik Isachsen for the helpful discussions and RJG is grateful to the GEOMAR for ongoing support.

Funding

This study was supported by JSPS KAKENHI Grant Numbers 26400474 and 15H02129 and also by the Deutsche Forschungsgemeinschaft as part of the Sonderforschungsbereich 754 "Climate - Biogeochemistry Interactions in the Tropical Ocean," by the German Federal Ministry of Education and Research as part of the cooperative project SACUS (03G0837A), and by the European Union 7th Framework Programme (FP7 2007-2013) under grant agreement 603521 PREFACE project.

Authors' contributions

HA proposed the topic and performed the analytical investigation. RJG helped write the manuscript. MC helped with the numerical investigation. All authors read and approved the final manuscript.

Competing interests

The authors declare that they have no competing interest.

Author details

[1]Institute for Space-Earth Environmental Research, Nagoya University, Nagoya City, 464-8601 Aichi, Japan. [2]Application Laboratory, Japan Agency for Marine-Earth Science and Technology, Yokohama, Japan. [3]GEOMAR Helmholtz-Zentrum für Ozeanforschung Kiel, Kiel, Germany. [4]Faculty of Mathematics and Natural Sciences, University of Kiel, Kiel, Germany.

References

Aiki H, Richards KJ (2008) Energetics of the global ocean: the role of layer-thickness form drag. J Phys Oceanogr 38:1845–1869

Aiki H, Takaya K, Greatbatch RJ (2015) A divergence-form wave-induced pressure inherent in the extension of the Eliassen-Palm theory to a three-dimensional framework for waves at all latitudes. J Atmos Sci 72:2822–2849

Anderson DLT, Gill AE (1979) Beta dispersion of inertial waves. J Geophys Res 84:1836–1842

Ascani F, Firing E, McCreary JP, Brandt P, Greatbatch RJ (2015) The deep equatorial ocean circulation in wind-forced numerical solutions. J Phys Oceanogr 45:1709–1734

Brandt P, Funk A, Hormann V, Dengler M, Greatbatch RJ (2011) Interannual atmospheric variability forced by the deep equatorial Atlantic Ocean. Nature 473:497–500

Brandt P, Claus M, Greatbatch RJ, Kopte R, Toole JM, Johns WE (2016) Annual and semi-annual cycle of equatorial Atlantic circulation associated with basin mode resonance. J Phys Oceanogr 46:3011–3029

Brunet G, Haynes PH (1996) Low-latitude reflection of Rossby wave trains. J Atmos Sci 53:482–496

Cai M, Huang B (2013) A new look at the physics of Rossby waves: a mechanical-Coriolis oscillation. J Atmos Sci 70:303–316

Cane MA, Moore DW (1981) A note on low-frequency equatorial basin modes. J Phys Oceanogr 11:1794–1806

Chelton DB, Schlax MG (1996) Global observations of oceanic Rossby waves. Science 272:234–238

Claus M, Greatbatch RJ, Brandt P (2014) Influence of the barotropic mean flow on the width and the structure of the Atlantic equatorial deep jets. J Phys Oceanogr 44:2485–2497

Claus M, Greatbatch RJ, Brandt P, Toole J (2016) Forcing of the Atlantic equatorial deep jets derived from observations. J Phys Oceanogr 46:3549–3562
Cummins PF, Oey LY (1997) Simulation of barotropic and baroclinic tides off northern British Columbia. J Phys Oceanogr 27:762–781
Fukutomi Y, Yasunari T (2002) Tropical-extratropical interaction associated with the 10–25-day oscillation over the western Pacific during the northern summer. J Meteo Soc Japan 80:311–331
Furuichi N, Hibiya T, Niwa Y (2008) Model-predicted distribution of wind-induced internal wave energy in the world's oceans. J Geophys Res 113:C09034
Gill AE (1982) Atmosphere–ocean dynamics. Academic Press, London
Greatbatch RJ, Brandt P, Claus M, Didwischus S-H, Fu Y (2012) On the width of the equatorial deep jets. J Phys Oceanogr 42:1729–1740
Haynes PH (1988) Forced, dissipative generalizations of finite-amplitude wave-activity conservation relations for zontal and nonzonal basic flows. J Atmos Sci 45:2352–2362
Hoskins BJ, James IN, White GH (1983) The shape, propagation and mean-flow interaction of large-scale weather systems. J Atmos Sci 40:1595–1612
Isachsen PE, LaCasce JJ, Pedlosky J (2007) Rossby wave instability and apparent phase speeds in large ocean basins. J Phys Oceanogr 37:1177–1191
Iwasaki T (2001) Atmospheric energy cycle viewed from wave-mean-flow interaction and Lagrangian mean circulation. J Atmos Sci 58:3036–3052
Johnson GC, Zhang D (2003) Structure of the Atlantic Ocean equatorial deep jets. J Phys Oceanogr 33:600–609
Kawatani Y, Sato K, Dunkerton TJ, Watanabe S, Miyahara S, Takahashi M (2010) The roles of equatorial trapped waves and internal inertia-gravity waves in driving the quasi-biennial oscillation. Part II: three-dimensional distribution of wave forcing. J Atmos Sci 67:981–997
Lübbecke JF, Böning CW, Keenlyside N, Xie S-P (2010) On the connection between Benguela and equatorial Atlantic Ninos and the role of the South Atlantic Anticyclone. J Geophys Res 115:C09015
Longuet-Higgins MS (1964) On group velocity and energy flux in planetary wave motion. Deep-Sea Res 11:35–42
Masuda A (1978) Group velocity and energy transport by Rossby waves. J Oceanogr Soc Jpn 34:1–7
Matsuno T (1966) Quasi-geostrophic motions in the equatorial area. J Meteo Soc Japan 44:25–43
Matthiessen J-D, Greatbatch RJ, Brandt P, Claus M, Didwischus S-H (2015) Influence of the equatorial deep jets on the north equatorial countercurrent. Ocean Dyn 65:1095–1102
McPhaden MJ, Ripa P (1990) Wave-mean flow interactions in the equatorial ocean. Annu Rev Fluid Mech 20:167–205
Merle J (1980) Annual and interannual variability of temperature in the eastern equatorial Atlantic—the hypothesis of an Atlantic El Nino. Oceanol Acta 3:209–220
Nakamura N, Solomon A (2011) Finite-amplitude wave activity and mean flow adjustments in the atmospheric general circulation. Part II: analysis in the isentropic coordinates. J Atmos Sci 68:2783–2799
Niwa Y, Hibiya T (2004) Three-dimensional numerical simulation of M2 internal tides in the East China Sea. J Geophys Res 109:C04027
Orlanski I, Sheldon J (1993) A case of downstream baroclinic development over western north America. Mon Wea Rev 121:2929–2950
Philander SGH (1989) El Nino, La Nina, and the Southern Oscillation. Academic Press, London
Plumb RA (1986) Three-dimensional propagation of transient quasi-geostrophic eddies and its relationship with the eddy forcing of the time mean flow. J Atmos Sci 43:1657–1678
Randel WJ, Williamson DL (1990) A comparison of the climate simulated by the NCAR community climate model (CCM1:R15) with ECMWF analysis. J Climate 3:608–633
Ripa P (1982) Nonlinear wave-wave interactions in a one-layer reduced-gravity model on the equatorial β plane. J Phys Oceanogr 12:97–111
Schopf PS, Anderson DLT, Smith R (1981) Beta-dispersion of low-frequency Rossby waves. Dyn Atmos Oceans 5:187–214
Takaya K, Nakamura H (1997) A formulation of a wave activity flux for stationary Rossby waves on a zonally varying basic flow. Geophys Res Lett 24:2985–2988
Thierry V, Treguier AM, Mercier H (2004) Numerical study of the annual and semi-annual fluctuations in the deep equatorial Atlantic Ocean. Ocean Model 6:1–30
Wakata Y, Kitaya S (2002) Annual variability of sea surface height and upper layer thickness in the Pacific Ocean. J Oceanogr 58:439–450
Yanai M, Maruyama T (1966) Stratospheric wave disturbances propagating over the equatorial pacific. J Meteo Soc Japan 44:291–294
Zhai X, Johnson HL, Marshall DP, Wunsch C (2012) On the wind power input to the ocean general circulation. J Phys Oceanogr 42:1357–1365

5

High-resolution simulations of turbidity currents

Edward Biegert, Bernhard Vowinckel, Raphael Ouillon and Eckart Meiburg*

Abstract

We employ direct numerical simulations of the three-dimensional Navier-Stokes equations, based on a continuum formulation for the sediment concentration, to investigate the physics of turbidity currents in complex situations, such as when they interact with seafloor topography, submarine engineering infrastructure and stratified ambients. In order to obtain a more accurate representation of the dynamics of erosion and resuspension, we have furthermore developed a grain-resolving simulation approach for representing the flow in the high-concentration region near and within the sediment bed. In these simulations, the Navier-Stokes flow around each particle and within the pore spaces of the sediment bed is resolved by means of an immersed boundary method, with the particle-particle interactions being taken into account via a detailed collision model.

Keywords: Turbidity currents, Navier-Stokes simulations, Continuum formulation, Grain-resolving simulations

Introduction

Turbidity currents are particle-laden flows in the ocean that are driven by gravity (Meiburg and Kneller 2010). Particle concentrations are usually sufficiently low far away from the sediment bed so that particle-particle interactions play a small or negligible role throughout most of the body of the current. In this region, the Boussinesq approximation of the Navier-Stokes equations, in conjunction with a continuum formulation for the sediment concentration, is well-suited to capture the dynamics of the flow. However, near the sediment bed particle concentrations can be very high, which can potentially result in complex non-Newtonian behavior, hindered settling, and other effects. Here, we describe the above two different simulation approaches, along with representative results, which open up a path towards multiscale flow simulations via the $\mu(I)$ rheology (Cassar et al. 2005; Boyer et al. 2011; Aussillous et al. 2013).

Methods

Continuum approach

Physical model and governing equations

In many situations of interest, compositional gravity currents and turbidity currents are driven by small density differences not exceeding $O(1\%)$. Under such conditions, the Boussinesq approximation can be employed, which treats the density as constant in the momentum equation with the exception of the body force terms. When dealing with turbidity currents, we account for the dispersed particle phase by means of a Eulerian-Eulerian formulation, which means that we employ a continuum equation for the particle concentration field, rather than tracking particles individually in a Lagrangian fashion.

In the following, it will be important to carefully distinguish between dimensional and dimensionless variables. Towards this end, we will employ the tilde symbol to indicate a dimensional variable, whereas variables without the tilde symbol are dimensionless. Under the Boussinesq approximation, the dimensional governing equations for compositional gravity currents driven by salinity and/or temperature gradients can be written as

$$\frac{\partial \tilde{u}_j}{\partial \tilde{x}_j} = 0, \tag{1}$$

$$\frac{\partial \tilde{u}_i}{\partial \tilde{t}} + \frac{\partial \left(\tilde{u}_i \tilde{u}_j\right)}{\partial \tilde{x}_j} = -\frac{1}{\tilde{\rho}_1}\frac{\partial \tilde{p}}{\partial \tilde{x}_i} + \tilde{\nu}\frac{\partial^2 \tilde{u}_i}{\partial \tilde{x}_j \partial \tilde{x}_j} + \frac{\tilde{\rho}\tilde{g}}{\tilde{\rho}_1} e_i^g, \tag{2}$$

$$\frac{\partial \tilde{\rho}}{\partial \tilde{t}} + \frac{\partial \left(\tilde{\rho}\tilde{u}_j\right)}{\partial \tilde{x}_j} = \tilde{\alpha}\frac{\partial^2 \tilde{\rho}}{\partial \tilde{x}_j \partial \tilde{x}_j}. \tag{3}$$

Here, $\tilde{u}_i$ denotes the velocity vector, $\tilde{p}$ the pressure, $\tilde{\rho}$ the density, $\tilde{g}$ the gravitational acceleration, e_i^g the unit vector

*Correspondence: meiburg@engineering.ucsb.edu
Department of Mechanical Engineering, University of California, Santa Barbara, Engineering II, Santa Barbara, CA 93106, USA

pointing in the direction of gravity, $\widetilde{\nu}$ the kinematic viscosity, and $\widetilde{\alpha}$ the molecular diffusivity of the density field. We nondimensionalize the above Eqs. (1)–(3) by a reference length scale, such as the domain half height $\widetilde{H}/2$ of a lock-exchange flow (Meiburg et al. 2015; Nasr-Azadani and Meiburg 2014; Necker et al. 2002, 2005), the current density $\widetilde{\rho}_1$, and the buoyancy velocity $\widetilde{u}_b$

$$\widetilde{u}_b = \sqrt{\widetilde{g}'\,\widetilde{H}/2}\,. \tag{4}$$

Here, $\widetilde{g}'$ indicates the reduced gravity

$$\widetilde{g}' = \widetilde{g}\,\frac{\widetilde{\rho}_1 - \widetilde{\rho}_2}{\widetilde{\rho}_1}\,. \tag{5}$$

where $\widetilde{\rho}_2$ represents the ambient density. After nondimensionalization, we obtain

$$\frac{\partial u_j}{\partial x_j} = 0, \tag{6}$$

$$\frac{\partial u_i}{\partial t} + \frac{\partial\left(u_i u_j\right)}{\partial x_j} = -\frac{\partial p}{\partial x_i} + \frac{1}{Re}\frac{\partial^2 u_i}{\partial x_j \partial x_j} + \rho e_i^g, \tag{7}$$

$$\frac{\partial \rho}{\partial t} + \frac{\partial\left(\rho u_j\right)}{\partial x_j} = \frac{1}{ReSc}\frac{\partial^2 \rho}{\partial x_j \partial x_j}\,. \tag{8}$$

Here, the nondimensional pressure p and density ρ are given by

$$p = \frac{\widetilde{p}}{\widetilde{\rho}_1 \widetilde{u}_b^2}\,, \quad \rho = \frac{\widetilde{\rho} - \widetilde{\rho}_2}{\widetilde{\rho}_1 - \widetilde{\rho}_2}\,. \tag{9}$$

The nondimensionlization of the governing equations gives rise to two dimensionless parameters in the form of the Reynolds number Re and the Schmidt number Sc

$$Re = \frac{\widetilde{u}_b \widetilde{H}}{2\widetilde{\nu}}\,, \quad Sc = \frac{\widetilde{\nu}}{\widetilde{\alpha}}\,. \tag{10}$$

While the Reynolds number indicates the ratio of inertial to viscous forces, the Schmidt number represents the ratio of kinematic fluid viscosity to molecular diffusivity of the density field.

When the driving density difference is due to gradients in particle loading, rather than salinity or temperature gradients, the above set of equations no longer provides a full description of the flow. Particles settle within the fluid, so that the scalar concentration field no longer moves with the fluid velocity. In addition, particle-particle interactions can result in such effects as hindered settling (Ham and Homsy 1988), increased effective viscosity, and non-Newtonian dynamics (Guazzelli and Morris 2011), thereby further complicating the picture. However, away from the sediment bed, turbidity currents are often quite dilute, with the volume fraction of the suspended sediment phase being well below $O(1\%)$. Under such conditions, particle-particle interactions can usually be neglected, so that the particle settling velocity remains the key difference (along with erosion) that distinguishes turbidity currents from compositional gravity currents.

Due to the small particle volume fraction of dilute turbidity currents, the volumetric displacement of fluid by the particulate phase can usually be neglected, allowing us to consider the fluid velocity field to be divergence-free. Rather, the particle-fluid interaction occurs primarily through the exchange of momentum, so that it suffices to account for the presence of the particles in the fluid momentum equation. In the following, we assume that the particle diameter $\widetilde{d}_p$ is smaller than the smallest length scale of the flow, such as the Kolmogorov scale in turbulent flow. In addition, we consider only particles whose aerodynamic response time $\widetilde{t}_p$ is significantly smaller than the smallest time scale of the flow $\widetilde{t}_f$, so that the particle Stokes number $St = \widetilde{t}_p/\widetilde{t}_f \ll O(1)$ (Raju and Meiburg 1995). Here, the aerodynamic response time is defined as

$$\widetilde{t}_p = \frac{\widetilde{\rho}_p \widetilde{d}_p^2}{18\widetilde{\mu}}\,, \tag{11}$$

with $\widetilde{\rho}_p$ indicating the particle material density and $\widetilde{\mu}$ denoting the dynamic viscosity of the fluid. Such particles can then be assumed to move with a velocity $\widetilde{u}_{p,i}$ that is obtained by superimposing the local fluid velocity $\widetilde{u}_i$ and the particle settling velocity $\widetilde{u}_s e_i^g$

$$\widetilde{u}_{p,i} = \widetilde{u}_i + \widetilde{u}_s e_i^g\,, \tag{12}$$

where $\widetilde{u}_s$ follows from balancing the gravitational force with the Stokes drag force

$$\widetilde{F}_i = 3\pi\widetilde{\mu}\widetilde{d}_p(\widetilde{u}_i - \widetilde{u}_{p,i}) \tag{13}$$

as

$$\widetilde{u}_s = \frac{\widetilde{d}_p^2(\widetilde{\rho}_p - \widetilde{\rho})\widetilde{g}}{18\widetilde{\mu}}\,. \tag{14}$$

Note that this implies that the particle velocity field is single-valued and divergence-free, so that monodisperse particles do not, for example, accumulate near stagnation points or get ejected from vortex centers. Hence, we can describe the spatio-temporal evolution of the particle number concentration field $\widetilde{c}$ in a Eulerian fashion by the transport equation

$$\frac{\partial\widetilde{c}}{\partial\widetilde{t}} + \frac{\partial\left(\widetilde{c}\left(\widetilde{u}_j + \widetilde{u}_s e_j^g\right)\right)}{\partial\widetilde{x}_j} = \widetilde{\alpha}\frac{\partial^2\widetilde{c}}{\partial\widetilde{x}_j\partial\widetilde{x}_j}\,. \tag{15}$$

The diffusion term in Eq. (15) represents a model for the decay of concentration gradients due to the hydrodynamic diffusion of particles and/or slight variations in particle size and shape (Davis and Hassen 1988; Ham and Homsy 1988).

The motion of the fluid phase is described by the incompressible continuity equation and the Navier-Stokes equation augmented by the force exerted on the fluid by the particles, which is equal and opposite to the Stokes drag force acting on the particles. In a dimensional form, these equations read

$$\frac{\partial \widetilde{u}_j}{\partial \widetilde{x}_j} = 0, \tag{16}$$

$$\frac{\partial \widetilde{u}_i}{\partial \widetilde{t}} + \frac{\partial \left(\widetilde{u}_i \widetilde{u}_j\right)}{\partial \widetilde{x}_j} = -\frac{1}{\widetilde{\rho}}\frac{\partial \widetilde{p}}{\partial \widetilde{x}_i} + \widetilde{\nu}\frac{\partial^2 \widetilde{u}_i}{\partial \widetilde{x}_j \partial \widetilde{x}_j} + \frac{\widetilde{c}}{\widetilde{\rho}}\widetilde{F}_i, \tag{17}$$

As we had done for compositional gravity currents, we use the domain half height $\widetilde{H}/2$ and buoyancy velocity $\widetilde{u}_b$ for nondimensionalization. The reduced gravity $\widetilde{g}'$ appearing in the calculation of $\widetilde{u}_b$ can now be calculated as

$$\widetilde{g}' = \frac{\pi(\widetilde{\rho}_p - \widetilde{\rho})\widetilde{c}_0 \widetilde{d}_p^3}{6\widetilde{\rho}}\widetilde{g}, \tag{18}$$

where $\widetilde{c}_0$ indicates a reference number concentration of particles in the suspension. After nondimensionalization, we obtain

$$\frac{\partial u_j}{\partial x_j} = 0, \tag{19}$$

$$\frac{\partial u_i}{\partial t} + \frac{\partial \left(u_i u_j\right)}{\partial x_j} = -\frac{\partial p}{\partial x_i} + \frac{1}{Re}\frac{\partial^2 u_i}{\partial x_j \partial x_j} + c e_i^g, \tag{20}$$

$$\frac{\partial c}{\partial t} + \frac{\partial \left(c\left(u_j + u_s e_j^g\right)\right)}{\partial x_j} = \frac{1}{ReSc}\frac{\partial^2 c}{\partial x_j \partial x_j}. \tag{21}$$

For polydisperse suspensions containing particles of different sizes, the above approach can easily be extended by solving one concentration equation for each particle size and corresponding settling velocity (Nasr-Azadani and Meiburg 2014). Note that the set of governing equations for turbidity currents (19)–(21) differs from the corresponding set for compositional gravity currents (6)–(8) only by the additional settling velocity term in the concentration equation. In the following, we employ Eqs. (19)–(21) for both types of currents, with the tacit assumption that the settling velocity vanishes for compositional gravity currents.

Direct numerical simulations (DNS) represent the most accurate computational approach for studying gravity currents. In DNS, all scales of motion, from the integral scales dictated by the boundary conditions down to the dissipative Kolmogorov scale determined by viscosity, are explicitly resolved. However, for the case of turbidity currents, when the particle diameter is smaller than the Kolmogorov scale, the fluid motion around each particle is usually not resolved, due to the prohibitive computational cost. Nevertheless, the drag law accurately captures the exchange of momentum between the two phases at scales smaller than the Kolmogorov scale, so that the approach described above is still referred to as DNS.

Consistent with the above arguments, the grid spacing required for DNS is of the order of the Kolmogorov scale, while the time step needs to be of the same order as the time scales of the smallest eddies. Due to the large disparity between integral and Kolmogorov scales at high Reynolds numbers, the computational cost of DNS scales as Re^3, so that the DNS approach is effectively limited to laboratory scale Reynolds numbers. The first DNS simulations of gravity currents in a lock-exchange configuration were reported by Härtel et al. (2000) for $Re = 1225$. Necker et al. (2002) extended this work to turbidity currents at $Re = 2240$. More recent simulations of lock-exchange gravity currents by Cantero et al. (2008) were able to reach $Re = 15,000$, which corresponds to a laboratory scale current of height 0.5 m with a front velocity of 3 cm/s.

DNS simulations can provide detailed information on the structure and statistics of the flow, on the various components of its energy budget, on the mixing behavior, and many additional aspects. As a case in point, the simulations by Härtel et al. (2000) explored the detailed flow topology near the current front and demonstrated that the stagnation point is located a significant distance behind the nose of the current. DNS results are furthermore very useful for testing the accuracy and identifying any deficiencies in larger-scale LES and RANS models (Yeh et al. 2013). Thus, while they are currently limited to laboratory scale currents, DNS simulations represent an excellent research tool for exploring the detailed physics of moderate Reynolds number gravity currents and for constructing larger-scale models for higher Reynolds number applications.

Results and discussion

Continuum approach results

We illustrate the ability of the continuum approach to reproduce lab-scale experiments by presenting the results of highly resolved simulations of a turbidity current moving down a slope into a stratified saline ambient. The numerical setup directly replicates the experiments conducted by Snow and Sutherland (2014) and is presented in Fig. 1. The density inside the ambient increases linearly from ρ_T at the top to ρ_B at the bottom such that

$$\rho_2(y) = \rho_B + (\rho_T - \rho_B) \cdot \frac{y}{H} \tag{22}$$

The channel has a constant width denoted as W. The lock region is initially at rest with density ρ_1, chosen such that $\rho_B > \rho_1 > \rho_T$. At $t = 0$, the lock is released and the particle-laden flow moves down the slope forming a turbidity current interacting with the ambient fluid. Here, both the particle concentration c and salinity s contribute to the Boussinesq term in Eq. (20) such that Eq. (21) has to be solved for each scalar field, with $u_s = 0$ in the case of the salinity field. When a settling velocity is used, the particles (concentration field) are allowed to settle through the lower boundary so that an erodible bed is not formed.

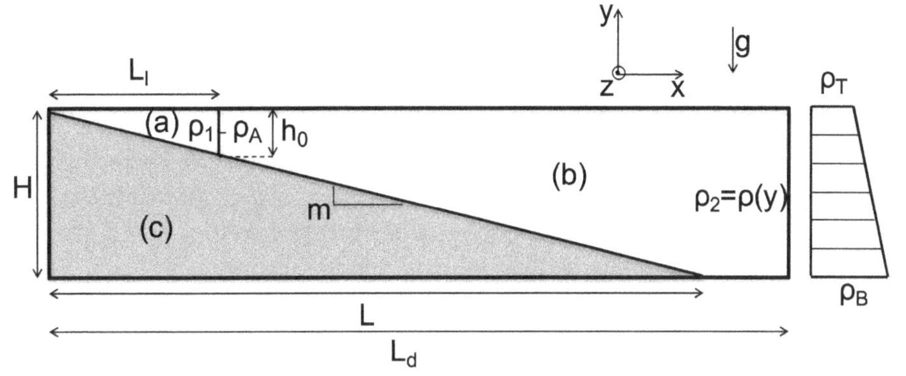

Fig. 1 Problem setup and configuration. **a** Particle-laden fluid. **b** Ambient stratified fluid. **c** Solid region. H and L denote the height and length of the ramp, L_d is the horizontal length of the domain, h_0 and L_l denote the height and horizontal length of the lock, and m is the slope. ρ_1 is the bulk density of the lock, ρ_T is the initial density at the top of the ambient, ρ_B is the initial density at the bottom of the ambient, and ρ_A is the initial density in the ambient at half-lock depth

We define a nondimensional buoyancy frequency N to quantify stratification such that

$$N = \sqrt{\frac{\tilde{\rho}_0(\tilde{\rho}_B - \tilde{\rho}_T)}{\tilde{\rho}_T(\tilde{\rho}_1 - \tilde{\rho}_A)}\frac{\tilde{H}}{\tilde{h}_0}}, \tag{23}$$

where $\tilde{\rho}_0$ is a reference density.

A finite difference code is used to solve the equations, and the MPI library is used for parallelization. A third-order Runge-Kutta scheme with a three substep method is used to discretize the equations in time. The wall-normal viscous and diffusive terms are solved implicitly while the convective terms and the remaining viscous and diffusive terms are treated explicitly. To impose incompressibility, a projection method is used (Spalart et al. 1991) and a direct solver is used for the resulting Poisson equation. Slip-wall boundary conditions are used at the top and right walls and z-periodic boundary conditions are used at the lateral walls. The domain is assumed to be sufficiently long to neglect boundary effects in x, and the width of the domain is chosen so that the periodic boundary condition does not impact the flow development. Finally, an immersed boundary method is used to impose the no-slip condition on the slope (Nasr-Azadani and Meiburg 2011).

In Fig. 2, we present a time series depicting the evolution of a typical turbidity current moving down a slope and intruding when its density matches that of the stratified ambient. The spanwise-averaged particle concentration is represented on a linear gray scale for various times. Upon release of the lock, the current starts moving down the slope and a trail of large Kelvin-Helmholtz rollers forms in the tail. These large instabilities then break into fully three-dimensional turbulence creating smaller dissipative vortices ($t > 10$). The absence of large distinct structures indicates the presence of fully developed turbulence in the tail of the current.

The direct impact of stratification is seen at later times ($t \approx 15$) when the current intrudes into the ambient, i.e. separates from the surface of the slope. The effects of stratification on intrusion depth are key in understanding the evolution of the suspended mass, deposition profiles and energy budgets of turbidity currents. Intrusion only occurs when the density of the current reaches the density of the ambient.

Using numerical simulations, we are able to investigate the fundamental mechanisms that control the propagation of the turbidity current and monitor all the relevant dynamic variables. For instance, we can investigate the initial perturbation that leads to the spanwise breakdown of the large Kelvin-Helmoltz structures that initially appear in the tail of the current. Figure 3 is a representation of the concentration isosurface $c = 0.25$ at $t_1 = 10$ for a typical turbidity current at $Re = 6000$. The flow at that instant is not yet fully turbulent but displays strong spanwise instabilities characterized at the head by the lobe-and-cleft instability. This instability is responsible for the breakdown of the large Kelvin-Helmholtz rollers that only high-resolution 3D simulations are able to capture.

Quantitative analysis of the velocity of the current as a function of the buoyancy frequency shows very good agreement between the numerical simulations and the experiments. The front velocity was measured by Snow and Sutherland (2014) in a series of experiments of which we report three. The margin of error for the experimental measurements is typically of the order of $\pm 20\%$ while the relative difference with the numerical simulations was found to be of 14, -11, and 16% for three widely different settling velocities, Reynolds numbers, and buoyancy frequencies. The simulation parameters and relative error for the front velocity are summarized in Table 1. Numerical results also agree almost perfectly with analytical results in the limit of no stratification, where the Froude number is expected to be $Fr = 0.5$. Numerical results consistently yielded a Froude number of $Fr = 0.496$ for $N = 0$, which

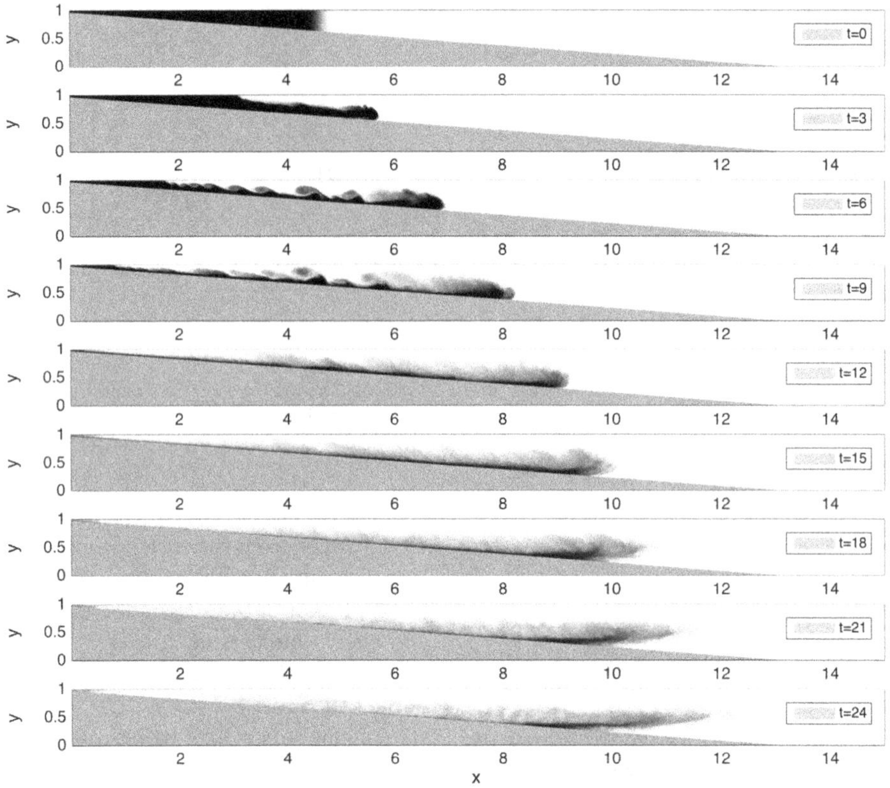

Fig. 2 Spanwise averaged particle concentration pseudo-color plot for various times ($Re = 15000, v_s = 0.001, N = 2.11, m = 0.0744$). This represents the case of a turbidity current with non-zero settling velocity moving down a uniformly stratified ambient. The density of the lock is initially larger than the ambient fluid outside of the lock such that the current starts moving downslope upon release. It then encounters region of denser and denser saline ambient, until its density reaches that of the ambient and it intrudes horizontally. The flow is 2D at very early stages but quickly grows unstable under the effect of a spanwise instability at the head. This leads to the breakdown of 2D vorticies into much smaller 3D mixing structures at the head and all along the body of the current at the interface with the ambient

corresponds to a relative error of 0.8% when compared to the analytical result.

The depth at which the current intrudes into the ambient also reveals a good agreement between the two approaches and validates the ability of numerical simulations to reproduce the dynamical features of turbidity currents moving into a stratified ambient. While it is extremely challenging to experimentally measure the velocity, particle concentration, and salinity fields of such 3D turbulent flows, direct numerical simulations give access to an entirely new set of data and opens the door to more accurate prediction tools and a deeper understanding of the underlying physics of gravity and turbidity currents in realistic environments at the scale of the lab.

Methods

Grain-resolving approach

Physical model and governing equations

When the concentration of particles grows large, particle-particle interactions become important and the aforementioned continuum approach is no longer applicable. For such cases, we have to account for the rheology of dense suspensions. A key element of progress with regard to the rheology of dense suspensions over the last decade has been the development of the so-called $\mu(I)$ approach, cf. (Guazzelli and Morris 2011; Boyer et al. 2011). Grain-resolving simulations of the type to be discussed in the following are expected to provide a tool for further investigating the validity of the assumptions underlying the derivation of the $\mu(I)$ rheology. One way to approach the simulation of dense suspensions is to fully resolve the particles interacting with the fluid by tracking each individual particle, evaluating the fluid no-slip condition at the particle surface and accounting for all the forces acting on the particle. Such simulations typically use grid resolutions of 10–25 grid cells per particle diameter to resolve the flow and are thus limited in scope to domains whose dimensions measure only tens to hundreds of diameters in length (Vowinckel et al. 2017; Kidanemariam and Uhlmann 2017). Thus, if simulating sand grains 100 μm in diameter, the domain dimensions would range in length from 1 mm to 1 cm. However, the idea is to

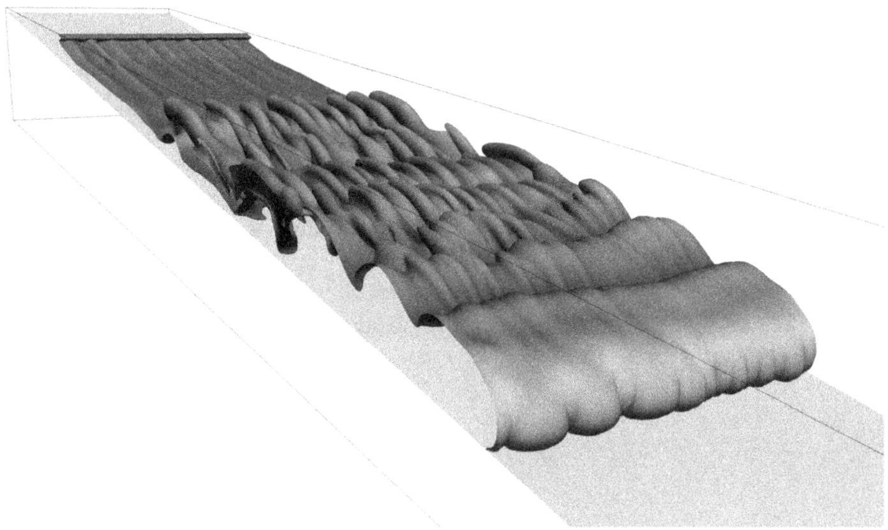

Fig. 3 Concentration isosurface $c = 0.25$ at $t = 10$ ($Re = 6000, v_s = 0.005, N = 2.09, m = 0.0744$). The iso-surface reveals the global structure of the envelope of a hyperpycnal current upon destibalization. The lobe-and-cleft instability is clearly visible at the head of the current as hills and crests form in the spanwise direction. This instability propagates in the body of the current and destabilises the large 2D vortical structures initially present. The largest wavelength corresponds to the initial dominant mode of the instability and dictates the width of the domain necessary to observe the instability at this given Reynolds number

use grain-resolved simulations to develop better models of sediment transport to be used in larger-scale simulations.

We employ an immersed boundary method (IBM) developed by Uhlmann (2005) and Kempe and Fröhlich (2012) with a modified collision model adjusted for the present context, as explained below. This method solves the Navier-Stokes equations everywhere in the domain, including nearby and within the particles:

$$\frac{\partial \mathbf{u}}{\partial t} + \nabla \cdot (\mathbf{uu}) = -\frac{1}{\rho_f}\nabla p + \nu_f \nabla^2 \mathbf{u} + \mathbf{f}_{IBM} \qquad (24)$$

where ρ_f is the fluid density and $\mathbf{f}_{IBM}$ is the IBM force, which acts as a source term to enforce the no-slip condition at the particle surfaces. This force effectively couples the particle and fluid momentum equations. Though there are many ways to carry out this coupling, the method we employ for the particles uses regularized Dirac delta functions, which interpolate fluid velocities onto the particle surface and spread $\mathbf{f}_{IBM}$ onto the fluid (Roma et al. 1999).

Table 1 Comparison of numerically and experimentally measured front velocity

	Exp 1	Exp 2	Exp 3
Re	16,850	15,000	35,000
N	3.66	2.39	2.77
v_s	0.001	0.0046	0.00046
$(U_e - U_s)/U_e$	14%	−11%	16%

Data from Snow and Sutherland (2014). The experimentally measured and numerically computed front velocities are denoted as U_e and U_s respectively

Note that this implementation is different from that used to create the sloped lower wall in the turbidity current simulations of the previous section.

The equations of motion for the particles are given by the momentum equations for the translational velocity $\mathbf{u}_p = (u_p, v_p, w_p)^T$

$$m_p \frac{\mathrm{d}\mathbf{u}_p}{\mathrm{d}t} = \underbrace{\oint_{\Gamma_p} \boldsymbol{\tau} \cdot \mathbf{n}\, \mathrm{d}A}_{=\mathbf{F}_h} + \underbrace{V_p\left(\rho_p - \rho_f\right)\mathbf{g}}_{=\mathbf{F}_g} + \mathbf{F}_l + \mathbf{F}_c, \quad (25)$$

the angular velocity $\boldsymbol{\omega}_p = (\omega_{p,x}, \omega_{p,y}, \omega_{p,z})^T$

$$I_p \frac{\mathrm{d}\boldsymbol{\omega}_p}{\mathrm{d}t} = \underbrace{\oint_{\Gamma_p} \mathbf{r} \times (\boldsymbol{\tau} \cdot \mathbf{n})\, \mathrm{d}A}_{=\mathbf{T}_h} + \mathbf{T}_l + \mathbf{T}_c, \qquad (26)$$

and the position $\mathbf{x}_p = (x_p, y_p, z_p)^T$

$$\frac{\mathrm{d}\mathbf{x}_p}{\mathrm{d}t} = \mathbf{u}_p. \qquad (27)$$

Here, m_p is the particle mass, Γ_p the fluid-particle interface, $\boldsymbol{\tau}$ the hydrodynamic stress tensor, ρ_p the particle density, V_p the particle volume, $\mathbf{g}$ the gravitational acceleration, $I_p = 8\pi\rho_p R_p^5/15$ the moment of inertia, and R_p the particle radius. Furthermore, the vector $\mathbf{n}$ is the outward-pointing normal on the interface Γ_p, $\mathbf{r} = \mathbf{x} - \mathbf{x}_p$ is the position vector of the surface point with respect to the center of mass $\mathbf{x_p}$ of a particle, $\mathbf{F}_l$ and $\mathbf{T}_l$ are the force and torque due to lubrication forces, and $\mathbf{F}_c$ and $\mathbf{T}_c$ are the force and torque due to particle collisions. We evaluate the IBM force $\mathbf{f}_{IBM}$ as well as the hydrodynamic force,

$\mathbf{F}_h$, and torque, $\mathbf{T}_h$, using the approach of Kempe and Fröhlich (2012), fully resolving the hydrodynamic effects of the fluid on the particles as well as the particles on the fluid. The lubrication force, $\mathbf{F}_l$, and contact force, $\mathbf{F}_c$, model close-range particle-particle interactions. With the exception of the tangential lubrication force, the methods used to evaluate these forces are described and validated in detail by Biegert et al. (2017), but here, we present them briefly.

The lubrication force

$$\mathbf{F}_l = -\ 6\pi\rho_f\nu_f R_{\text{eff}}\left(\frac{R_{\text{eff}}}{\max(\zeta_n,\zeta_{\min})}\mathbf{g}_n + F_t^*\mathbf{g}_t + F_r^*\left(R_p\boldsymbol{\omega}_p\times\mathbf{n} + R_q\boldsymbol{\omega}_q\times\mathbf{n}\right)\right) \tag{28}$$

and torque

$$\mathbf{T}_l = 8\pi\rho_f\nu_f R_{\text{eff}}^2\left[\mathbf{g}_t T_t^* + T_r^*(R_p\boldsymbol{\omega}_p\times\mathbf{n} + R_q\boldsymbol{\omega}_q\times\mathbf{n})\right]\times\mathbf{n} \tag{29}$$

are added to account for short-range hydrodynamic forces that are unresolved by the fluid grid. Here, $R_{eff} = R_pR_q/(R_p + R_q)$ is an effective radius accounting for size differences between particles p and q, $\mathbf{g}_n$ and $\mathbf{g}_t$ are the relative velocities in the normal and tangential directions, respectively, between the two particle surfaces at the point of contact, ζ_n is the surface distance between the two particles, and $\zeta_{n,min} = 3\times10^{-3}R_p$ is a limiter preventing the lubrication force from reaching its singularity at $\zeta_n\to0$. The terms F_t^*, F_r^*, T_t^*, and T_r^* were obtained via asymptotic expansions by Goldman et al. (1967):

$$F_t^* \sim \frac{8}{15}\ln\left(\frac{\max(\zeta_n,\zeta_{\min})}{R_{\text{eff}}}\right) - 0.9588 \tag{30}$$

$$F_r^* \sim -\frac{2}{15}\ln\left(\frac{\max(\zeta_n,\zeta_{\min})}{R_{\text{eff}}}\right) - 0.2526 \tag{31}$$

$$T_t^* \sim -\frac{1}{10}\ln\left(\frac{\max(\zeta_n,\zeta_{\min})}{R_{\text{eff}}}\right) - 0.1895 \tag{32}$$

$$T_r^* \sim \frac{2}{5}\ln\left(\frac{\max(\zeta_n,\zeta_{\min})}{R_{\text{eff}}}\right) - 0.3817. \tag{33}$$

As indicated in (25) and (26), we also account for particle-particle contacts through $\mathbf{F}_c$ and $\mathbf{T}_c$. These contact forces are composed of components normal and tangential to the particle surface, represented by $\mathbf{F}_n$ and $\mathbf{F}_t$, respectively, which act at the point of contact between the two particles so that the resulting force and torque on the particle are given by

$$\mathbf{F}_c = \mathbf{F}_n + \mathbf{F}_t \tag{34}$$

$$\mathbf{T}_c = R_p\,\mathbf{n}\times\mathbf{F}_t, \tag{35}$$

where $\mathbf{n}$ is the outward-pointing normal vector from the contact point. A nonlinear spring-dashpot model is used for the normal contact force

$$\mathbf{F}_n = -k_n|\zeta_n|^{3/2}\mathbf{n} - d_n\mathbf{g}_n, \tag{36}$$

where the stiffness and damping coefficients, k_n and d_n, respectively, are adaptively calibrated for every collision. A linear spring-dashpot model is used for the tangential contact force

$$\mathbf{F}_t = \min\left(-k_t\boldsymbol{\zeta}_t - d_t\mathbf{g}_t, ||\mu\mathbf{F}_n||\mathbf{t}\right), \tag{37}$$

where $\boldsymbol{\zeta}_t$ is the tangential displacement vector representing accumulated slip between the two surfaces, μ is the coefficient of friction between the surfaces, and $\mathbf{t}$ is the unit normal vector in the tangential direction. The Coulomb friction criterion, represented by $||\mu\mathbf{F}_n||$, allows the two surfaces to slip past one another when large stresses are present. Similar to the normal coefficients, the tangential coefficients of stiffness and damping, k_t and d_t, are also adaptively calibrated.

Results and discussion

Grain-resolving results

Pressure-driven flow over dense sediment

To address the bulk behavior of a dense granular bed sheared by a laminar Poiseuille flow, we carried out numerical simulations to reproduce the experimental results of Aussillous et al. (2013), who studied pressure-driven flows over glass spheres with a mean diameter $D_p = 1.1$ mm and a standard deviation of $\sigma(D_p) = 0.1$ mm as a sediment material. This experimental work provides investigations over a range of submergences h_f/D_p and Reynolds numbers in the laminar regime, where h_f is the height of the clear-water layer above the sediment bed illustrated in Fig. 4. We define h_f to be the height above which the average particle volume fraction $\phi < 0.05$, which is the threshold for negligible impact of particle-particle interaction on the flow (Capart and Fraccarollo 2011).

We executed several simulations in an attempt to match the experimental results of Aussillous et al. (2013) at different flow rates and fluid heights. To this end, we simulated a monodisperse granular sediment bed sheared by a pressure-driven Poiseuille flow with periodic conditions in both the streamwise (x) and spanwise (z) directions, respectively. A no-slip condition was applied at the top and bottom wall as well as at the particle surface. A detailed comparison validating the simulation results has been presented in great detail in Biegert et al. (2017). Here, we extend this work to show data from the same physical setup, but with an increased flow rate, such that the gross of the particles are set into motion and interact in a complex network. The physical and numerical parameters associated with this simulation are listed in Table 2.

Some qualitative inferences can already be drawn from Fig. 4, where we have nondimensionalized velocities by

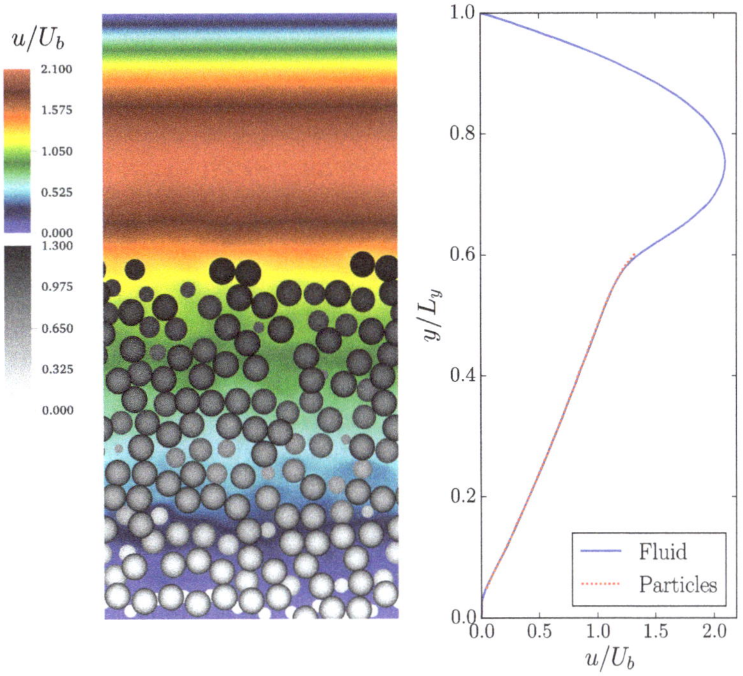

Fig. 4 Pressure-driven flow over a dense sediment bed. The left figure shows an instantaneous slice through the domain, the color scale illustrating streamwise fluid velocity, and the gray scale illustrating particle velocity. The right figure shows the time-and-space-averaged streamwise velocity of the fluid and particles

the bulk fluid velocity $U_b = \frac{1}{L_y}\int_0^{L_y} u\,dy$. In the clear-water layer, a parabolic profile obeying the analytical solution of the classical Poiseuille flow can be observed. The lower end of this parabolic region, however, is not a no-slip wall, but a moving granular bed, which causes the symmetry axis of the flow profile to shift from $h_f/2$ to a lower position. Inside the granular bed, a linear shear flow profile develops and since all particles are moving, this profile continues all the way to the bottom wall of the domain. This interesting behavior and the wealth of data obtained from the grain-resolving simulations opens up a wide range of analytical tools in terms of statistical description as well as physical modeling, which will be our focus in the future.

Table 2 Simulation parameters of the pressure-driven flow scenario, where $u_\tau = \sqrt{\tau_w/\rho_f}$ is the friction velocity at the fluid/particle interface, U_b is the bulk (average) velocity of the fluid, τ_w is the shear stress at the fluid/particle interface, ν_f the kinematic viscosity, L_x, L_y, and L_z are the spatial extents of the computational domain in the Cartesian space, and h is the grid cell size

$Re = U_b L_y/\nu_f$	9.9
$D^+ = u_\tau D_p/\nu_f$	0.39
$Sh = \tau_w/[(\rho_p - \rho_f)gD]$	0.97
ρ_p/ρ_f	2.1
$L_x \times L_y \times L_z$	$11.26D_p \times 22.52D_p \times 11.26D_p$
h_f/D_p	8.7
D_p/h	22.7

Shearing of dense suspensions

To further investigate the rheologic properties of dense suspensions, we simulated a shear flow of two parallel walls with a spacing of H moving in opposite directions. A no-slip condition is applied for the moving walls at the top and bottom of the domain and at the particle surface. The walls move with a relative velocity $\Delta U_w = U_t - U_b$, where U_t is the velocity of the top wall and U_b is the velocity of the bottom wall. Periodic conditions are applied in streamwise (x) and spanwise (z) directions. The fluid in between the two walls is a dense mixture of spherical particles, where "dense" indicates a volume fraction $\phi_v = N_p V_p/V_f > 0.4$. Here, $V_p = 4\pi R_p^3/3$ is the volume of a single particle, N_p is the number of particles in the computational domain V_f, and R_p is the particle radius. The dimension of the computational domain can be treated as part of the physical problem. We have considered two computational domains: scenario A with

Table 3 Physical and numerical simulation parameters for simulations of shear flows with dense suspensions

e_{dry}	μ_k	μ_s	ν	ζ_{min}	ρ_p/ρ_f	H/D_p	D_p/h	Re
0.97	0.15	0.8	0.22	$3 \cdot 10^{-3} R_p$	1.011	10	25.6	10

Table 4 Simulation scenarios

Scenario	$L_x \times L_y \times L_z$	Re	ϕ_v	$t_s\dot{I}$	$t_a\dot{I}$
Re10p42	$2H \times 1H \times 1H$	10	0.42	85	35
Re10p54	$2H \times 1H \times 1H$	10	0.54	150	30
Re40p54	$1H \times 2H \times 1H$	40	0.54	20	20

dimensions $L_x \times L_y \times L_z = 2H \times 1H \times 1H$ and scenario B with $L_x \times L_y \times L_z = 1H \times 2H \times 1H$. Thus, the relative submergence becomes $H/D_p = 10$ and $H/D_p = 20$, respectively, where D_p is the particle diameter. For both scenarios A and B, the shear rate $\dot{I} = \Delta U_w/H$ was kept constant so that the particle Reynolds number $Re_p = \dot{I}D_p^2/\nu_f = 0.1$ also stays constant but the channel Reynolds number for scenario B is increased by a factor of 4 with respect to scenario A. This yields channel Reynolds numbers of $Re = 10$ (scenario A) and $Re = 40$ (scenario B). The particles have a density of $\rho_p/\rho_f = 1.011$, which is close to neutrally buoyant conditions. We assume material parameters corresponding to glass or silicate materials as thoroughly validated in Biegert et al. (2017). In particular, we choose $e_{dry} = 0.97$, $\mu_k = 0.15$, $\mu_s = 0.8$, and $\nu = 0.22$, where e_{dry} is the wall-normal restitution coefficient for dry collisions, μ_k and μ_s are the kinetic and static friction coefficients for oblique collisions, and ν is Poisson's ratio. Every particle is discretized by 25.6 grid cells per diameter. The particle parameters are summarized in Table 3. Three simulations were conducted with varying volume fractions of the mixture and different relative gap sizes H/D_p to explore the effects of these two parameters. All simulations were initialized and run for a start-up time t_s until a true steady steady state had been established. Subsequently, data was collected for the averaging time t_a to reach converged statistics for the profiles presented in the following. The different scenarios are displayed in Table 4.

An instantaneous snapshot of the particle distribution colored by the particle velocity is given in Fig. 5. As desired, particles are dragged along the moving walls whenever they collide with them. These particles moving with the wall transfer kinetic energy through collisions towards the channel center. In addition, the moving walls establish a background profile for the fluid velocity, which should be close to the linear shear profile commonly observed in Couette-type flows. The two mechanisms from collision and hydrodynamic interactions establish a shear flow profile within the suspension. Looking at the wall-normal profiles of the porosity, we can see a distinct pattern of oscillations (Fig. 6). This crystal-like layering of the particles reflects the fact that all particles are the same size. While a strongly layered structure is visible for all three simulations close to the wall, less pronounced layers form in the channel center for the two scenarios *Re10p42* and *Re40p54*. For these two scenarios, particles have more space to rearrange due to the lower volume fraction and the larger relative gap size, respectively. Horizontal-and-time-averaged profiles of the streamwise component of the fluid and particle velocity show that particles move with almost the same velocity as the fluid flow (Fig. 7). Slight distortions can be seen, especially for case *Re10p54*, illustrating local effects of the particle clustering on the global velocity profile.

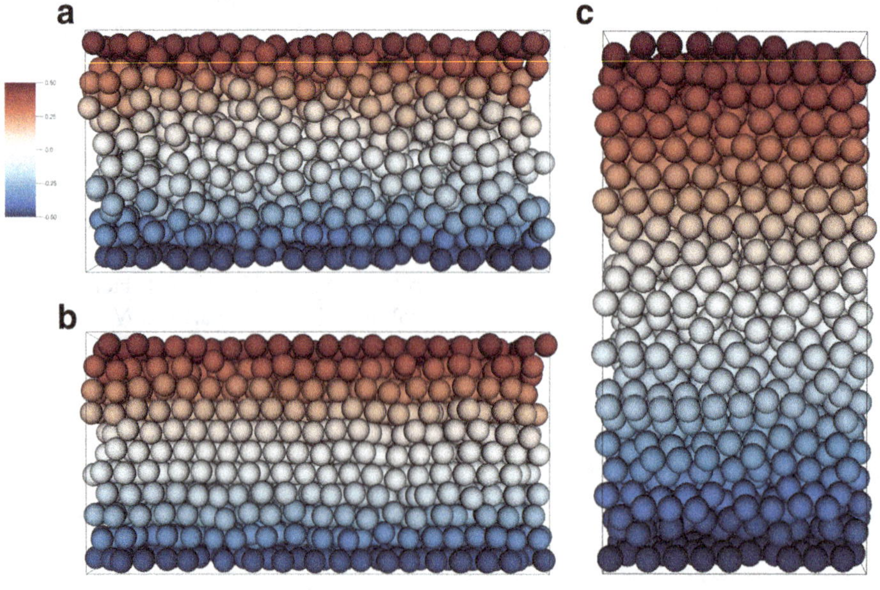

Fig. 5 Shearing of dense suspensions. Color bar indicates streamwise velocity of particles. Shown are instantaneous snapshots for **a** $Re = 10$ and $\phi_v = 0.42$, **b** $Re = 10$ and $\phi_v = 0.54$, and **c** $Re = 40$ and $\phi_v = 0.54$

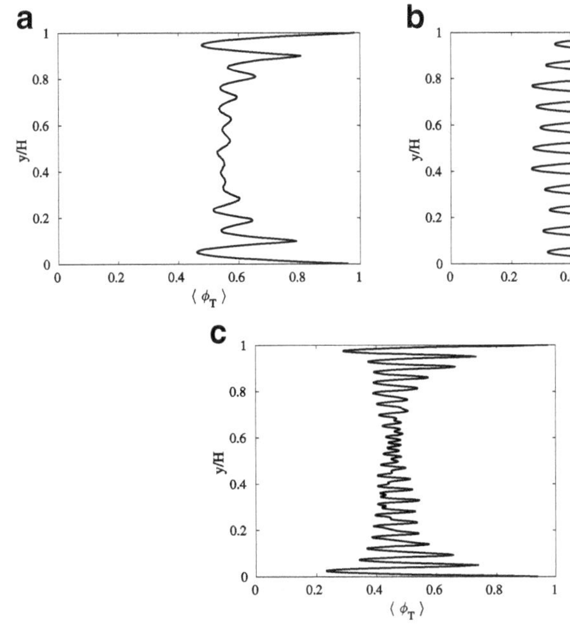

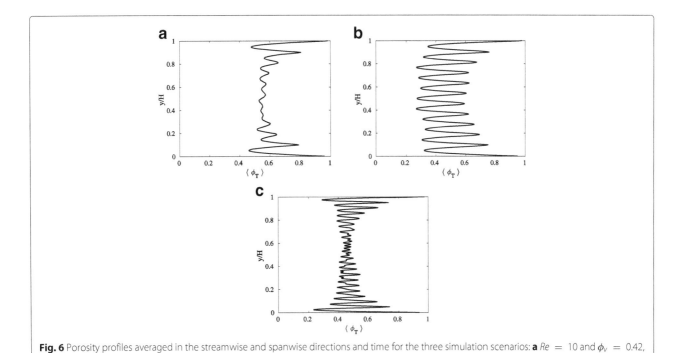

Fig. 6 Porosity profiles averaged in the streamwise and spanwise directions and time for the three simulation scenarios: **a** $Re = 10$ and $\phi_v = 0.42$, **b** $Re = 10$ and $\phi_v = 0.54$, and **c** $Re = 40$ and $\phi_v = 0.54$

The present study of a Couette-type flow supplements our simulations of pressure-driven flow described in the previous section to fully understand the rheologic behavior of dense suspensions of particles with different inertia in flows with different momentum supply.

Internal waves propagating over fully resolved sediment beds

We also studied the hydrodynamic forces acting on a fully resolved sediment bed induced by a gravity current. Here, the key issue is to explore how a jump in the hydrostatic pressure traveling along the surface of the sediment bed propagates within the bed. To this end,

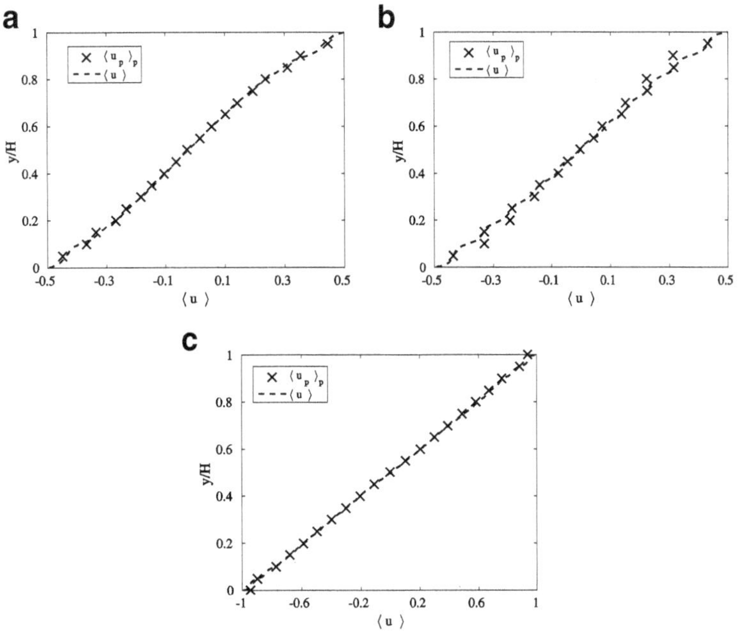

Fig. 7 Average streamwise particle velocity profiles for the three simulation scenarios: **a** $Re = 10$ and $\phi_v = 0.42$, **b** $Re = 10$ and $\phi_v = 0.54$, and **c** $Re = 40$ and $\phi_v = 0.54$

we solved the advection-diffusion Eq. (8) and supplemented (24) with the term stemming from the Boussinesq approximation. The Peclet number was chosen to be $Pe = ReSc = 10^4$. The initial configuration is similar to a lock-exchange, but the particles are submerged in a layer of high concentration as well. The geometry is $L_x/H \times L_y/H \times L_z/H = 6 \times 1 \times 1$, where H is the channel height from the no-slip wall on the bottom to the free-slip wall on the top, which is equivalent to L_y. The Reynolds number for the present scenario was chosen to be $Re = u_b h/\nu_f = 164$, where $u_b = \sqrt{g'h}$ is the buoyancy velocity, h is the lock height, $g' = \frac{\rho_p - \rho_f}{\rho_f} g$ is the specific gravity, and g is the gravitational acceleration. Free-slip walls were applied to the left, back, and front wall (Fig. 8a). The right boundary was set to be a convective outflow condition. The relative submergence of a particle is $H/D_p = 10$. Every particle is discretized by $D_p/h = 16$.

Immediately after being released, the block of heavy fluid starts to propagate along the rough wall, forming an internal wave at the interface between light and heavy fluid, which are indicated in Fig. 8 as blue and red fluid, respectively. Particles in Fig. 8b are colored by the lift forces acting in a vertical direction. We can see that particle in front of the wave start to experience a lift force even though the wave front has not yet reached it. The grain-resolving simulation approach now allows us to track the drag and lift on individual particles as a function of time to elucidate this effect in more detail. This has been done for the particles colored in red in Fig. 9a. The lift force normalized by the buoyant weight of the particles $F_g = \rho_f V_p g$ is shown in Fig. 9b. Every curve represents a particle shown in Fig. 9a, and it becomes obvious that the lift experienced by the particles along the transect appears to be similar with decaying intensity along the flow direction.

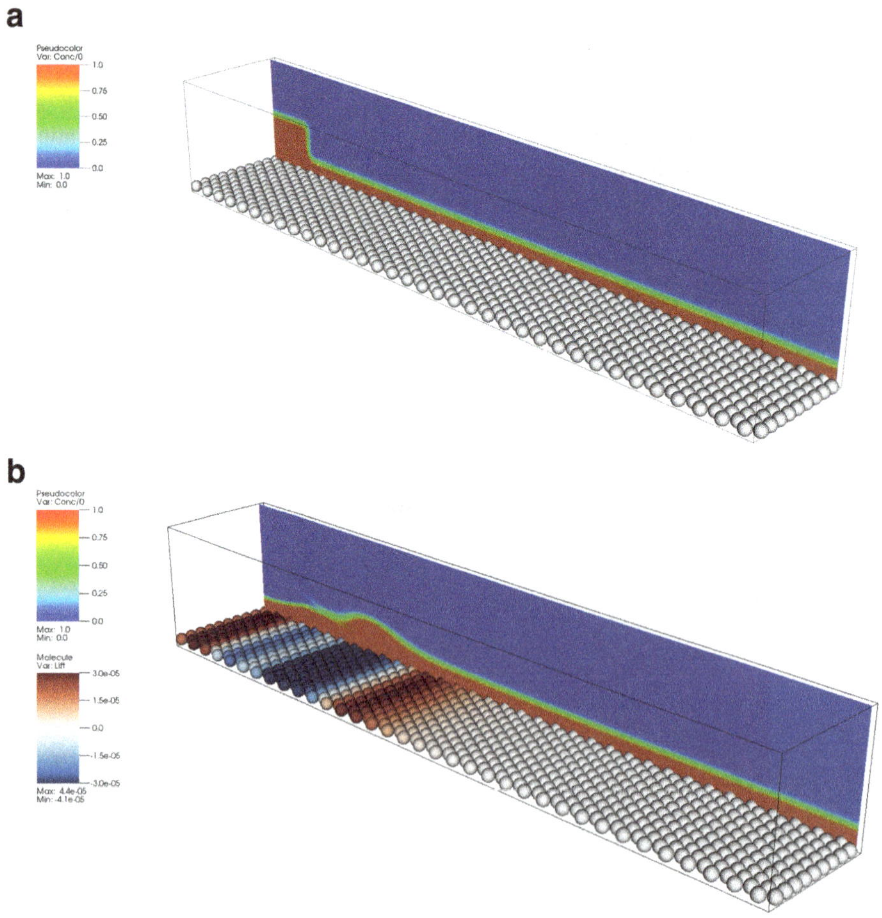

Fig. 8 Instantaneous snapshots for the internal wave study. Fixed particles are arranged in a hexagonal packing immersed in the heavy (red) fluid. **a** Initially, a lock of dense fluid is arranged at one end of the domain, which then **b** generates a bore propagating to the right at a later time. The vertical slice shows the concentration profile of the heavy (red) and light (blue) fluids. Particles are colored by the lift force acting on them; red and blue indicate positive and negative forces, respectively

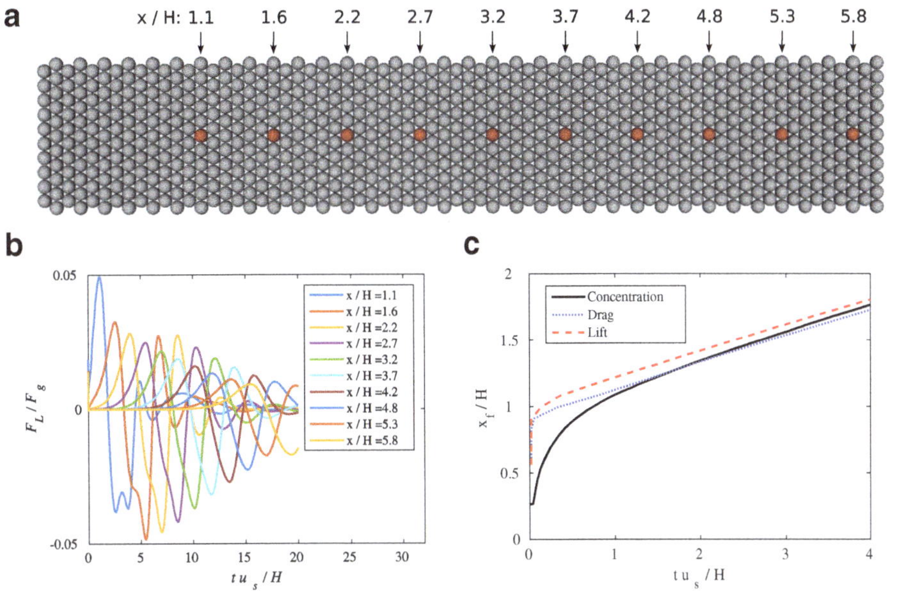

Fig. 9 Lift/drag of the internal propagating wave acting on the centerline particles. **a** shows a top view of the particle bed, where lift forces are measured at the red-colored centerline particles. **b** shows the lift force normalized by the gravitational force, F_L/F_G, versus time for the centerline particles, each particle represented by a separate curve. **c** shows the position of the internal wave front versus time, where the position is determined either from the concentration profile, the drag force acting on the particle bed, or the lift force acting on the particle bed

We can track the front of the current using the location in the horizontal profile where the interface between light and heavy fluid starts to increase in height. Alternatively, we can track the front using the location where particles start to experience an enhanced lift force. A comparison of these two methods is shown in Fig. 9c for the first few time units simulated. It can be seen that, indeed, during the initial stage of the simulation, where we can see a steep front of the internal wave, the force signal propagates quicker through the sediment bed than the actual propagation speed of the current would suggest. This effect, however, levels off over time as the wave continues to travel over the rough bed, constantly losing energy due to viscous dissipation.

Conclusions

The modeling of dilute, non-eroding turbidity currents has reached a mature level, as evidenced by the fact that high-resolution simulations have been able to reproduce many of the observations made in laboratory experiments (e.g., Nasr-Azadani et al. 2013). We are now able to account for some topographical complexity via the immersed boundary method. Some of the remaining challenges concern the extension to the very large Reynolds number values of field-scale flows and the frequent interaction with ambient phenomena in the ocean such as internal waves and tides, as well as the accurate modeling of erosion and resuspension in such high Reynolds number flows. However, a similar level of maturity has not yet been achieved with regard to the modeling of highly concentrated turbidity currents with significant erosion, resuspension, and bedload transport. Especially the dynamics of the near-bed region of such high-concentration currents in the form of dense suspensions is still poorly understood, as it is governed by intense particle-fluid and particle-particle interactions that give rise to strongly non-Newtonian dynamics and to mass and momentum exchanges between the current and the sediment bed. As a result, insight into the erosional and depositional behavior of such currents and the coupling between the motion of the current above the sediment bed and the fluid flow inside the bed is just beginning to emerge. Key progress has been accomplished with regard to understanding the rheology of dense suspensions over the last decade, through the development of the so-called $\mu(I)$ approach, cf. (Guazzelli and Morris 2011; Boyer et al. 2011). Grain-resolving simulations based on the approach outlined here will provide a tool for further investigating the validity of the assumptions underlying the derivation of the $\mu(I)$ rheology. The computational approach outlined and tested in the present paper holds great promise, as it is able to capture the grain-resolved dynamics of thick, mobile sediment beds and their coupled dynamics with the flow above. Simulations on this basis provide the opportunity to understand erosion and dense suspension rheology from a fundamental perspective, which can lead to better models for use at larger scales. This multiscale approach would thus further enrich our understanding of turbidity currents.

Acknowledgements
Computational resources for this work were provided by the Extreme Science and Engineering Discovery Environment (XSEDE), supported by the National Science Foundation, USA, Grant No. TG-CTS150053.

Funding
This research is supported in part by the Department of Energy Office of Science Graduate Fellowship Program (DOE SCGF), made possible in part by the American Recovery and Reinvestment Act of 2009, administered by ORISE-ORAU under the contract no. DE-AC05-06OR23100. BV gratefully acknowledges the Feodor-Lynen scholarship provided by the Alexander von Humboldt Foundation, Germany.

Authors' contributions
EB and BV developed the grain-resolving simulation approach. RO carried out the continuum formulation simulations. EM proposed the topic and conceived the study. All authors read and approved the final manuscript.

Competing interests
The authors declare that they have no competing interests.

References
Aussillous P, Chauchat J, Pailha M, Médale M, Guazzelli É (2013) Investigation of the mobile granular layer in bedload transport by laminar shearing flows. J Fluid Mech 736:594–615
Biegert E, Vowinckel B, Meiburg E (2017) A collision model for grain-resolving simulations of flows over dense, mobile, polydisperse granular sediment beds. J Comput Phys 340:105–127
Boyer F, Guazzelli É, Pouliquen O (2011) Unifying suspension and granular rheology. Phys Rev Lett 107(18):1–5
Cantero MI, Balachandar S, García MH, Bock D (2008) Turbulent structures in planar gravity currents and their influence on the flow dynamics. J Geophys Res Oceans 113(C8):1–22
Capart H, Fraccarollo L (2011) Transport layer structure in intense bed-load. Geophys Res Lett 38(20):2–7
Cassar C, Nicolas M, Pouliquen O (2005) Submarine granular flows down inclined planes. Phys Fluids 17(10):103301
Davis RH, Hassen MA (1988) Spreading of the interface at the top of a slightly polydisperse sedimenting suspension. J Fluid Mech 196:107–134
Goldman AJ, Cox RG, Brenner H (1967) Slow viscous motion of a sphere parallel to a plane wall-I Motion through a quiescent fluid. Chem Eng Sci 22(4):637–651
Guazzelli É, Morris JF (2011) A physical introduction to suspension dynamics. Cambridge Texts in Applied Mathematics. Cambridge University Press, Cambridge
Ham JM, Homsy GM (1988) Hindered settling and hydrodynamic dispersion in quiescent sedimenting suspensions. Int J Multiphase Flow 14(5):533–546
Härtel C, Meiburg E, Necker F (2000) Analysis and direct numerical simulation of the flow at a gravity-current head. Part 1. Flow topology and front speed for slip and no-slip boundaries. J Fluid Mech 418:189–212
Kempe T, Fröhlich J (2012) An improved immersed boundary method with direct forcing for the simulation of particle laden flows. J Comput Phys 231(9):3663–3684
Kidanemariam AG, Uhlmann M (2017) Formation of sediment patterns in channel flow: minimal unstable systems and their temporal evolution. J Fluid Mech 818:716–743
Meiburg E, Kneller B (2010) Turbidity currents and their deposits. Annu Rev Fluid Mech 42:135–156
Meiburg E, Radhakrishnan S, Nasr-Azadani M (2015) Modeling gravity and turbidity currents: computational approaches and challenges. Appl Mech Rev 67(4):040802
Nasr-Azadani MM, Meiburg E (2011) TURBINS: an immersed boundary, Navier-Stokes code for the simulation of gravity and turbidity currents interacting with complex topographies. Comput Fluids 45(1, SI):14–28
Nasr-Azadani MM, Hall B, Meiburg E (2013) Polydisperse turbidity currents propagating over complex topography: comparison of experimental and depth-resolved simulation results. Comput Geosci 53:141–153
Nasr-Azadani MM, Meiburg E (2014) Turbidity currents intracting with three-dimensional seafloor topography. J Fluid Mech 745:409–443
Necker F, Härtel C, Kleiser L, Meiburg E (2002) High-resolution simulations of particle-driven gravity currents. Int J Multiphase Flow 28(2):279–300
Necker F, Härtel C, Kleiser L, Meiburg E (2005) Mixing and dissipation in particle-driven gravity currents. J Fluid Mech 545:339
Raju N, Meiburg E (1995) The accumulation and dispersion of heavy-particles in forced 2-dimensional mixing layers .2. The effect of gravity. Phys Fluids 7(6):1241–1264
Roma AM, Peskin CS, Berger MJ (1999) An adaptive version of the immersed boundary method. J Comput Phys 153(2):509–534
Snow K, Sutherland BR (2014) Particle-laden flow down a slope in uniform stratification. J Fluid Mech 755:251–273
Spalart PR, Moser RD, Rogers MM (1991) Spectral methods for the Navier-Stokes equations with one infinite and two periodic directions. J Comput Phys 96(2):297–324
Uhlmann M (2005) An immersed boundary method with direct forcing for the simulation of particulate flows. J Comput Phys 209(2):448–476
Vowinckel B, Nikora V, Kempe T, Fröhlich J (2017) Spatially-averaged momentum fluxes and stresses in flows over mobile granular beds: a DNS-based study. J Hydraulic Res 55(2):208–223
Yeh T-H, Cantero M, Cantelli A, Pirmez C, Parker G (2013) Turbidity current with a roof: success and failure of RANS modeling for turbidity currents under strongly stratified conditions. J Geophys Res Earth Surf 118(3):1975–1998

6

Heterogeneous interplate coupling along the Nankai Trough, Japan, detected by GPS-acoustic seafloor geodetic observation

Yusuke Yokota[1*], Tadashi Ishikawa[1], Mariko Sato[1], Shun-ichi Watanabe[1], Hiroaki Saito[1], Naoto Ujihara[1], Yoshihiro Matsumoto[1], Shin-ichi Toyama[1], Masayuki Fujita[1], Tetsuichiro Yabuki[1], Masashi Mochizuki[2] and Akira Asada[3]

Abstract

The recurring devastating earthquake that occurs in the Nankai Trough subduction zone between the Philippine Sea plate and the Eurasian plate has the potential to cause an extremely dangerous natural disaster in the foreseeable future. Many previous studies have assumed interplate-coupling ratios for this region along the trench axis using onshore geodetic data in order to understand this recursive event. However, the offshore region that has the potential to drive a devastating tsunami cannot be resolved sufficiently because the observation network is biased to the land area. Therefore, the Hydrographic and Oceanographic Department of Japan constructed a geodetic observation network on the seafloor along the Nankai Trough using a GPS-acoustic combination technique and has used it to observe seafloor crustal movements directly above the Nankai Trough subduction zone. We have set six seafloor sites and cumulated enough data to determine the displacement rate from 2006 to January 2011. Our seafloor geodetic observations at these sites revealed a heterogeneous interplate coupling that has three particular features. The fast displacement rates observed in the easternmost area indicate strong interplate coupling (>75%) around not only the future Tokai earthquake source region but also the Paleo-Zenisu ridge. The slow displacement rates near the trench axis in the Kumano-nada Sea, a shallow part of the 1944 Tonankai earthquake source region, show a lower coupling ratio (50% to 75%). The slow displacement rate observed in the area shallower than the 1946 Nankaido earthquake source region off Cape Muroto-zaki reflects weakening interplate coupling (about 50%) probably due to a subducting seamount. Our observations above the subducting ridge and seamount indicate that the effect of a subducting seamount on an interplate-coupling region depends on various conditions such as the geometry of the seamount and the friction parameters on the plate boundary.

Keywords: Seafloor geodetic observation; Nankai Trough; Interplate coupling; Seamount subduction

Background

Nankai trough subduction zone

In southwestern Japan, the oceanic Philippine Sea plate is subducting beneath the continental Eurasian plate northwestward along the Nankai Trough. In this region, interplate magnitude-8-class earthquakes have been documented with an average recurrence time of about 100 years (e.g., Ando 1975) due to this plate motion. The most recent events are the 1944 Tonankai (M7.9) and the 1946 Nankaido (M8.0) earthquakes. Thus, the next magnitude-8-class earthquake is predicted in the near future.

We analyzed the coseismic slip distributions of the 1944 and 1946 events using triangulation and leveling survey records (e.g., Sagiya and Thacher 1999 and Ito and Hashimoto 2004) and tsunami records (e.g., Tanioka and Satake 2001 and Baba and Cummins 2005). To understand the next disastrous event in advance, several research groups also analyzed the interseismic slip deficit for recent years using onshore global positioning systems (GPS) data (e.g., Nishimura and Hashimoto 2006 and Loveless and Meade 2010). The onshore geodetic

* Correspondence: eisei@jodc.go.jp
[1]Hydrographic and Oceanographic Department, Japan Coast Guard, 2-5-18 Aomi, Koto-ku, Tokyo 135-0064, Japan
Full list of author information is available at the end of the article

data do not have the power to resolve the offshore region as indicated in Yoshioka and Matsuoka (2013) (Figure 1). They pointed out the importance of seafloor geodetic observations for revealing the offshore region along the Nankai Trough.

Seafloor geodetic observation using a GPS-acoustic combination technique

In order to monitor seafloor crustal movements around the offshore plate boundary, the Hydrographic and Oceanographic Department of Japan (JHOD) has been carrying out seafloor geodetic observations using a GPS-acoustic combination technique on the landward side of the major trenches around Japan. Our past observations near the Japan Trench provided great advantage in understanding the 2005 Off-Miyagi Prefecture earthquake (M7.2) (e.g., Matsumoto et al. 2006 and Sato et al. 2011b), the 2011 Tohoku-oki earthquake (M9.0) (e.g., Sato et al. 2011a and Sato et al. 2013b), and others. Important observations have also been made by other research groups in Japan such as Tohoku University (e.g., Kido et al. 2006 and Kido et al. 2011) and Nagoya University (e.g., Tadokoro et al. 2012 and Yasuda et al. 2014).

In the Nankai Trough region, we were operating six seafloor sites along the trench axis at about 50 to 100 km intervals (Figure 1 and Table 1) before the 2011 Tohoku-oki earthquake. In this paper, we report displacement rates on the seafloor surface and discuss interplate coupling along the Nankai Trough during the period before the 2011 Tohoku-oki earthquake and, thus, unaffected by postseismic effects from this enormous event. Our observations directly revealed an

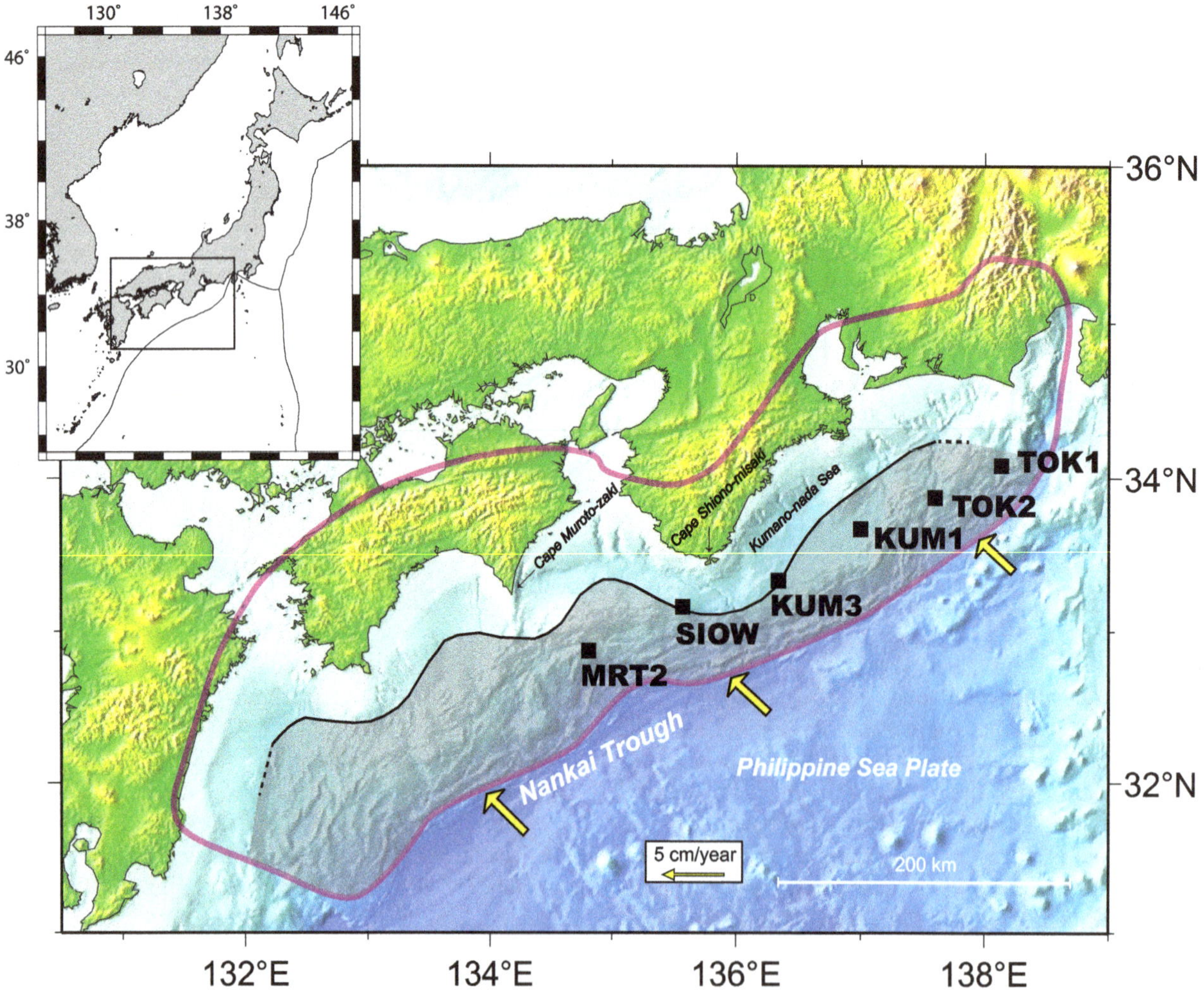

Figure 1 Locations of seafloor geodetic observation network sites along the Nankai Trough. Red squares show the sites set before the 2011 Tohoku-oki earthquake and used in this study. Yellow arrows indicate the convergence rate of the Philippine Sea plate under the Eurasian plate calculated using the NUVEL-1A model (DeMets et al. 1994). The purple region shows the maximum source model of the great Nankai Trough earthquake assumed by the Central Disaster Management Council of the Japanese government (CDMC) (2013). The gray shading near the trench axis indicates an area with small resolution values calculated using the onshore geodetic observations by Yoshioka and Matsuoka (2013).

Table 1 Positions of seafloor sites

Site name	Latitude (degree)	Longitude (degree)	Height (m)
TOK1	34.08231 N	138.13395 E	−2,374
TOK2	33.87698 N	137.59508 E	−1,524
KUM1	33.67026 N	136.99558 E	−1,957
KUM3	33.33287 N	136.34232 E	−1,957
SIOW	33.16056 N	135.57229 E	−1,524
MRT2	32.87264 N	134.81463 E	−1,402

uneven accumulation of interplate coupling along the Nankai Trough.

Methods

Measurement system

The GPS-acoustic combination technique was developed to detect seafloor displacements with an accuracy of a few (2 to 3) centimeters (e.g., Spiess et al. 1998 and Asada and Yabuki 2001). In Japan, our group first succeeded in detecting seafloor movements caused by plate convergence at the Japan Trench (Fujita et al. 2006) and has been deploying a seafloor observation network since 2000.

A schematic picture of our seafloor geodetic observation system is shown in Figure 2. This system consists of a seafloor unit with four acoustic mirror-type transponders and an on-board unit with a GPS antenna/receiver, an undersea acoustic transducer, and a dynamic motion sensor. The onboard acoustic transducers were mounted at the stern of survey vessels using an 8 m-long pole before 2007. We used to carry out surveys using a drifting observation method with the pole system before 2007, as shown in Figure 2a. We improved the observations by

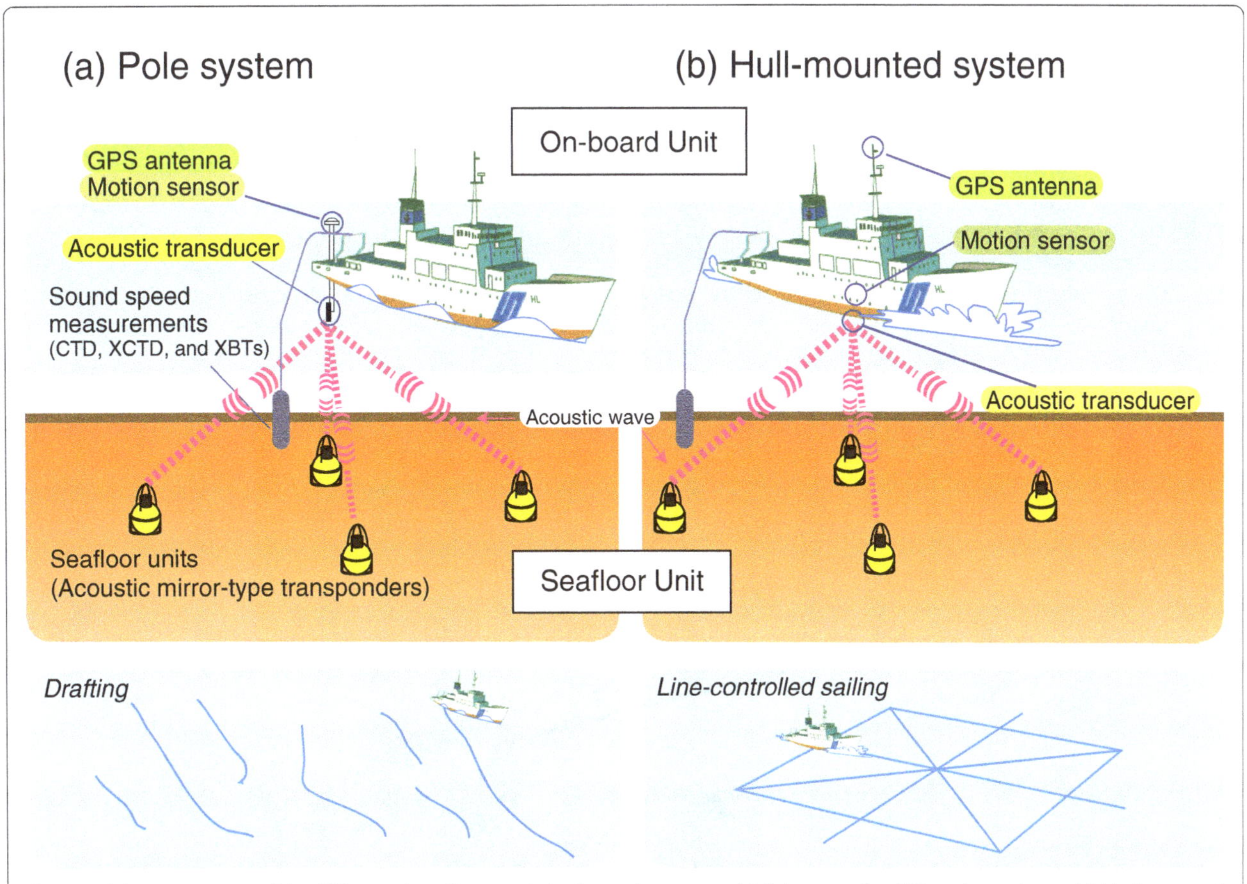

Figure 2 Schematic pictures of the GPS-acoustic seafloor geodetic observation systems. (**a**) Pole system for drifting observation and (**b**) hull-mounted system for line-controlled sailing observation.

switching to a line-controlled sailing observation method with a hull-mounted system after 2008 as shown in Figure 2b. Detail methodologies are introduced in Sato et al. (2013a).

The system measures ranges from the on-board transducer to the seafloor acoustic transponders through roundtrip acoustic travel times. Acoustic velocity profiles in the seawater are necessary to transform travel time into range. These are obtained using temperature and salinity profilers (conductivity temperature depth profiler (CTD), expendable conductivity temperature depth profiler (XCTD), and expendable bathy thermographs (XBTs)) every several hours. Kinematic GPS data are simultaneously gathered to determine the absolute position of the survey vessel. Attitude data on the survey vessel are also acquired on board by a dynamic motion sensor to determine the coordinates of the on-board transducer relative to those of the GPS antenna.

Analytical approach

The data analysis was done in three parts. First, we obtained the range between the on-board transducer and the seafloor transponder using a process that correlates measured acoustic wave data and velocity profiles. Second, the consecutive absolute positions of the GPS antenna on the vessel were determined by kinematic GPS analysis. Third, the position of the seafloor transponder was determined using a linearized inversion method based on a least squares formulation combining the results from prior analyses.

In the early stages of our observations, we performed this inversion analysis constrained by the height of the grouped transponders over all epochs (e.g., Fujita et al. 2006). The present analysis was constrained by the positional relationship of the grouped transponders for all epochs (e.g., Watanabe et al. 2014). The newer method was developed by Matsumoto et al. (2008) (Figure 3). To stabilize the estimates, over 5,000 shots of acoustic wave data per site were required for each observation epoch. We spent approximately 24 h performing an observation. More details on the analytical methodology are presented in Fujita et al. (2006), Matsumoto et al. (2008), Sato et al. (2013a), and Watanabe et al. (2014).

Results and discussion

Observation results

Our seafloor sites along the Nankai Trough are located on the offshore region that is predicted by the Central Disaster Management Council of the Japanese government (CDMC) to undergo the next great Nankai Trough earthquake (CDMC 2013) (available at http://www.bousai.go.jp/jishin/nankai/taisaku/pdf/1_1.pdf, in Japanese) (Figure 1). From the eastern side, sites TOK1, TOK2, KUM1, KUM3, SIOW, and MRT2 were set along the trench axis at 50 to 100 km intervals. TOK1 is located closest to the trench axis.

The observations began with site KUM1 in November 2000, with the other five sites being set between 2001 and 2004. They have been operating steadily since 2006. Here, we show observations before the 2011 Tohoku-oki earthquake. We carried out eight to ten observation campaigns in the period between 2006 and January 2011.

Figure 4 shows the time series of the estimated horizontal coordinates of these six sites. Each circle represents the coordinates of seafloor transponders for each epoch after 2006 relative to their locations in the first campaign. The positions are presented with respect to the stable part of the Eurasian plate, based on the

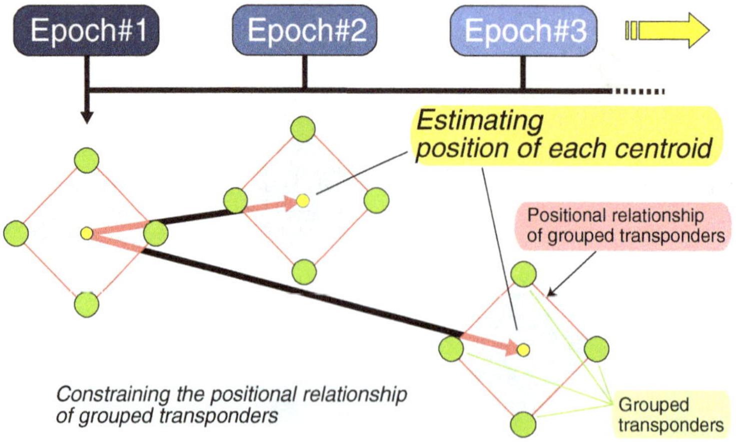

Figure 3 Schematic pictures of the estimation of seafloor positions. The positions were analyzed using the method developed by Matsumoto et al. (2008). In this analysis, we estimated the position (yellow circle) of the centroid of the grouped transponders (green circles) for each epoch, constrained by the positional relationship of the grouped transponders (red diamonds) in all epochs.

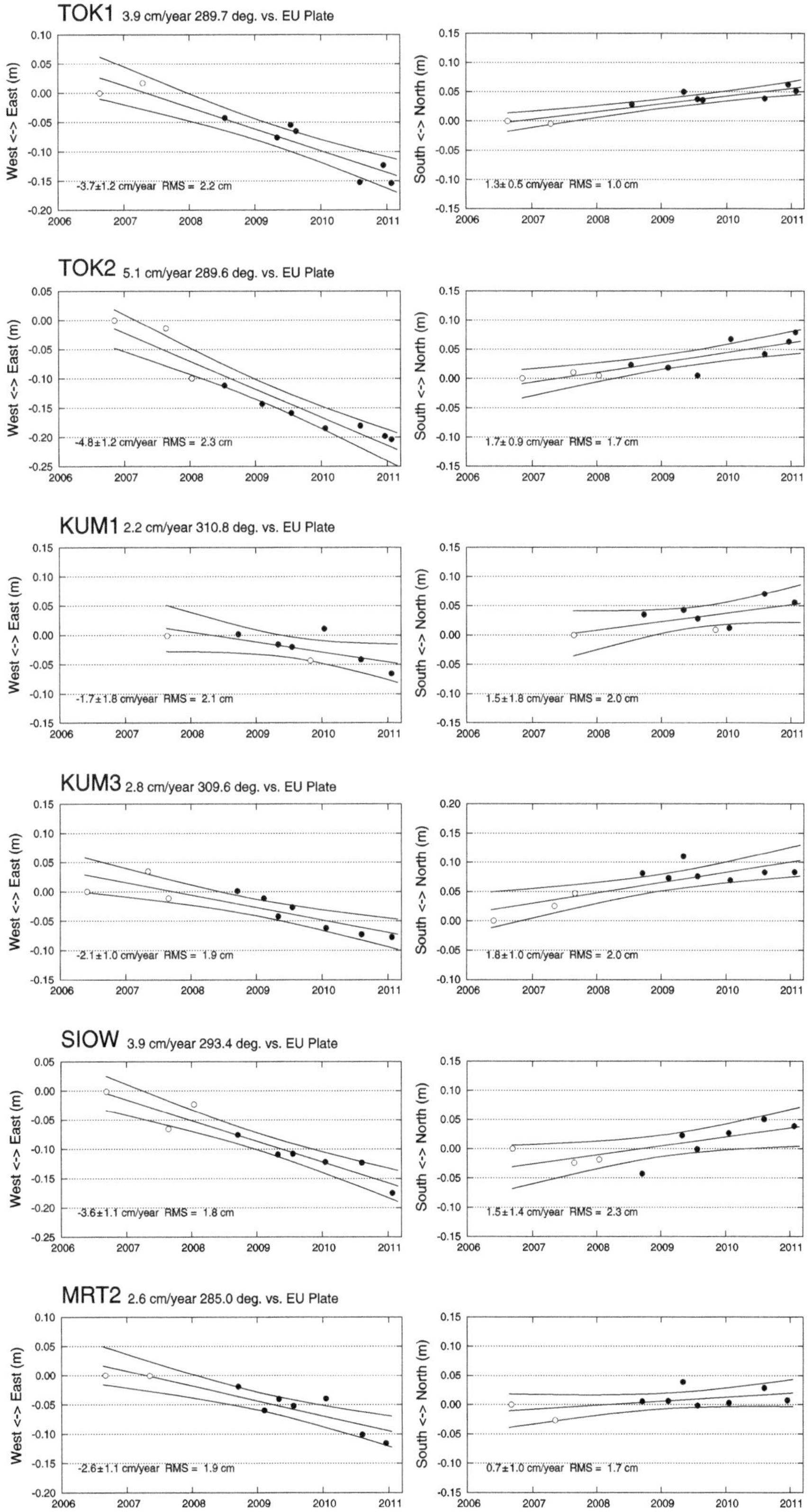

Figure 4 (See legend on next page.)

(See figure on previous page.)
Figure 4 Time series in east-west (left column) and north-south (right column) displacements. These data were obtained at the six seafloor sites TOK1, TOK2, KUM1, KUM3, SIOW, and MRT2 for the period from 2006 to January 2011. The position reference is the Eurasian plate. The open and solid red circles indicate results from drafting observations and results from sailing observations, respectively. The linear trends and the 95% two-sided confidence intervals are shown with red and blue lines, respectively.

NUVEL-1A velocity (DeMets et al. 1994). The data are summarized in Tables 2 and 3. Reference frame in Table 2 is ITRF2005 (Altamimi et al. 2007).

Seafloor displacement rates

Figure 4 also presents linear trends fitted to these time series, which gave us the offshore velocity fields with the 95% two-sided confidence intervals indicated by the blue hyperbolic lines. All the sites moved at stable displacement rates in the northwestward direction. Because no large earthquake occurred in this offshore region after the 2004 off-the-Kii-Peninsula earthquakes (M ~ 7), these movements reflect strain accumulation processes due to the Philippine Sea plate subduction under the upper Eurasian plate.

The estimated horizontal seafloor displacement rates are exhibited by arrows in Figure 5. The ellipses indicate the 2-sigma errors. The slowest and fastest velocities, 2.2 and 5.1 cm/year, were observed at KUM1 and TOK2, respectively, though the error ellipse of KUM1 is larger than that of the other sites because of its shorter observation period.

Figure 5 also presents the onshore velocity fields observed at GEONET stations operated by the Geospatial Information Authority of Japan (GSI) (Sagiya et al. 2000) for comparison. These onshore displacement rates were derived using the averages between March 2006 and February 2011 so as to match to our observation period.

Our seafloor observation results present uneven offshore velocity fields along the Nankai Trough. In particular, the 2 to 3 cm/year displacement rates at KUM1, KUM3, and MRT2 are slower than those of adjacent seafloor sites. In addition, the rates at KUM3 and MRT2 are slower than those of nearby onshore GPS stations.

The displacement rate at KUM1 is smaller than that at TOK2 (95% confidence level (CL)). Similarly, the displacement rates at KUM3 and MRT2 are smaller than those at SIOW (95% CL and 90% CL) and the onshore 'Kushimoto' (99.99% CL) and 'Muroto-4' (99.99% CL) GPS stations as shown in Figure 6. Because the relative velocity between the Eurasian plate and the Philippine Sea plate is approximately 4 to 5 cm/year (e.g., Seno et al. 1993 and DeMets et al. 1994), the plate boundary beneath these three offshore sites does not have high interplate-coupling ratios (<75%). Each observation is discussed in more detail from the eastern side below.

Tokai region

The direction and magnitude of the displacement rate at site TOK1 are approximately consistent with those of nearby onshore stations. In this easternmost plate boundary region beneath TOK1, no major earthquake has occurred in the past 160 years since the last historical earthquake in 1854 (Ando 1975). The CDMC (2013) therefore hypothesized that this will likely be the area of the next large event, the future Tokai earthquake (orange region in Figure 5). Our seafloor observations also confirm a high-coupling ratio (>75%) in this region near the trench axis.

The displacement rate of the next site (TOK2) is larger than not only those of nearby onshore and adjacent offshore sites but also the subducting plate velocity calculated on the basis of the NUVEL-1A model. This is possibly due to the Izu microplate (e.g., Sagiya 1999) and the spray fault along the Nankai Trough (e.g., Park et al. 2002). This observation provides evidence for almost full coupling in this interplate region, though TOK2 was not located in the source region of the future Tokai earthquake. This region also has the potential to produce a large earthquake, similar to the region around TOK1.

Tonankai region

The western side of the future Tokai earthquake source region is the location of the 1944 Tonankai earthquake. Baba and Cummins (2005) estimated the slip distribution (blue contour in Figure 5) using tsunami data. Since the tsunami data used have resolving power in the area near the trench axis, along with the seafloor geodetic data, this offshore slip distribution is reliable and can be compared with the seafloor displacement rates.

The seafloor geodetic sites operated by Nagoya University (Tadokoro et al. 2012) are located on the north side of our seafloor sites in the Kumano-nada Sea (KUM1 and KUM3) (Figure 5). Those seafloor sites moved slower than TOK2, like our result at KUM1, and faster than nearby onshore GPS stations.

Our result at KUM3 differs from Nagoya University's results. The displacement rate at KUM3 is smaller than the nearby 'Kushimoto' GPS station (99.99% CL), and its direction is comparatively clockwise because of a smaller westward component and a larger northward component than at 'Kushimoto' (99.99% CL) as shown in Figure 6. If the coupling ratio of this interplate region beneath KUM3 was the same as in the north deep region beneath Nagoya University's seafloor sites, KUM3 should move faster since it is closer to the plate boundary. Therefore, the coupling ratio is believed to be lower (50% to 75%) there than in the north region beneath the Kumano-nada Sea. This is consistent

Table 2 Estimated relative site positions

Epoch (year)	Eastward (m)	Northward (m)
TOK1		
2006.630[a]	0.000	0.000
2007.285[a]	0.030	−0.015
2008.537	−0.004	−0.001
2009.340	−0.022	0.008
2009.545	0.004	−0.008
2009.627	−0.005	−0.011
2010.595	−0.073	−0.023
2010.948	−0.037	−0.005
2011.068[b]	−0.066	−0.018
TOK2		
2006.847[a]	0.000	0.000
2007.633[a]	0.002	−0.002
2008.030[a]	−0.076	−0.014
2008.526	−0.078	−0.003
2009.104	−0.098	−0.017
2009.548	−0.105	−0.037
2010.063	−0.120	0.017
2010.595	−0.106	−0.017
2010.964	−0.116	−0.001
2011.066	−0.122	0.015
KUM1		
2007.638[a]	0.000	0.000
2008.726	0.025	0.018
2009.337	0.019	0.017
2009.551	0.019	−0.001
2009.827[ab]	0.002	−0.025
2010.041	0.060	−0.025
2010.597	0.019	0.025
2011.055	0.004	0.003
KUM3		
2006.405[a]	0.000	0.000
2007.337[a]	0.053	0.011
2007.655[a]	0.014	0.027
2008.704	0.047	0.046
2009.112	0.043	0.032
2009.334	0.017	0.065
2009.551	0.036	0.028
2010.060	0.011	0.013
2010.600	0.011	0.018
2011.058	−0.037	0.001
SIOW		
2006.690[ab]	0.000	0.000
2007.644[a]	−0.045	−0.039
2008.036[a]	0.005	−0.039
2008.707	−0.034	−0.074
2009.323	−0.055	−0.018
2009.553	−0.049	−0.045
2010.047	−0.053	−0.025
2010.603	−0.043	−0.010
2011.063	−0.085	−0.029
MRT2		
2006.666[a]	0.000	0.000
2007.348[a]	0.013	−0.037
2008.707	0.023	−0.026
2009.110	−0.010	−0.031
2009.332	0.014	−0.002
2009.556	0.007	−0.046
2010.049	0.030	−0.049
2010.603	−0.021	−0.031
2010.956	−0.028	−0.058

Reference frame is ITRF2005 (Altamimi et al. 2007). [a]Epoch with drafting observations. [b]Relatively unreliable due to less data.

with the 1944 Tonankai earthquake source region, which has a peak slip region in the deep side. These features are different from the easternmost offshore region around TOK1 and TOK2.

Nankaido region

The observation for the two westernmost sites SIOW and MRT2 contrast with each other. These sites are both located near the 1946 Nankaido earthquake source region estimated by Baba and Cummins (2005) (green contour in Figure 5). The displacement rate of SIOW is similar to nearby onshore data from Cape Shiono-misaki and suggests that the interplate region around SIOW has a high-coupling ratio (>75%). On the other hand, the displacement rate at MRT2 is clearly smaller than the onshore data from Cape Muroto-zaki, e.g., 'Muroto-4' (99.99% CL), as shown in Figure 6. Although the

Table 3 Velocities with respect to the Eurasian plate with component variances and covariances

Site name	Velocity (cm/year)		Variance and Covariance ((cm/year)^2)		
	East (V_e)	North (V_n)	V (V_e)	V (V_n)	Cov (V_e, V_n)
TOK1	−3.7	1.3	0.25	0.05	0.02
TOK2	−4.8	1.7	0.27	0.14	0.06
KUM1	−1.7	1.5	0.53	0.52	−0.11
KUM3	−2.1	1.8	0.19	0.20	0.06
SIOW	−3.6	1.5	0.21	0.33	0.03
MRT2	−2.6	0.7	0.23	0.18	−0.01

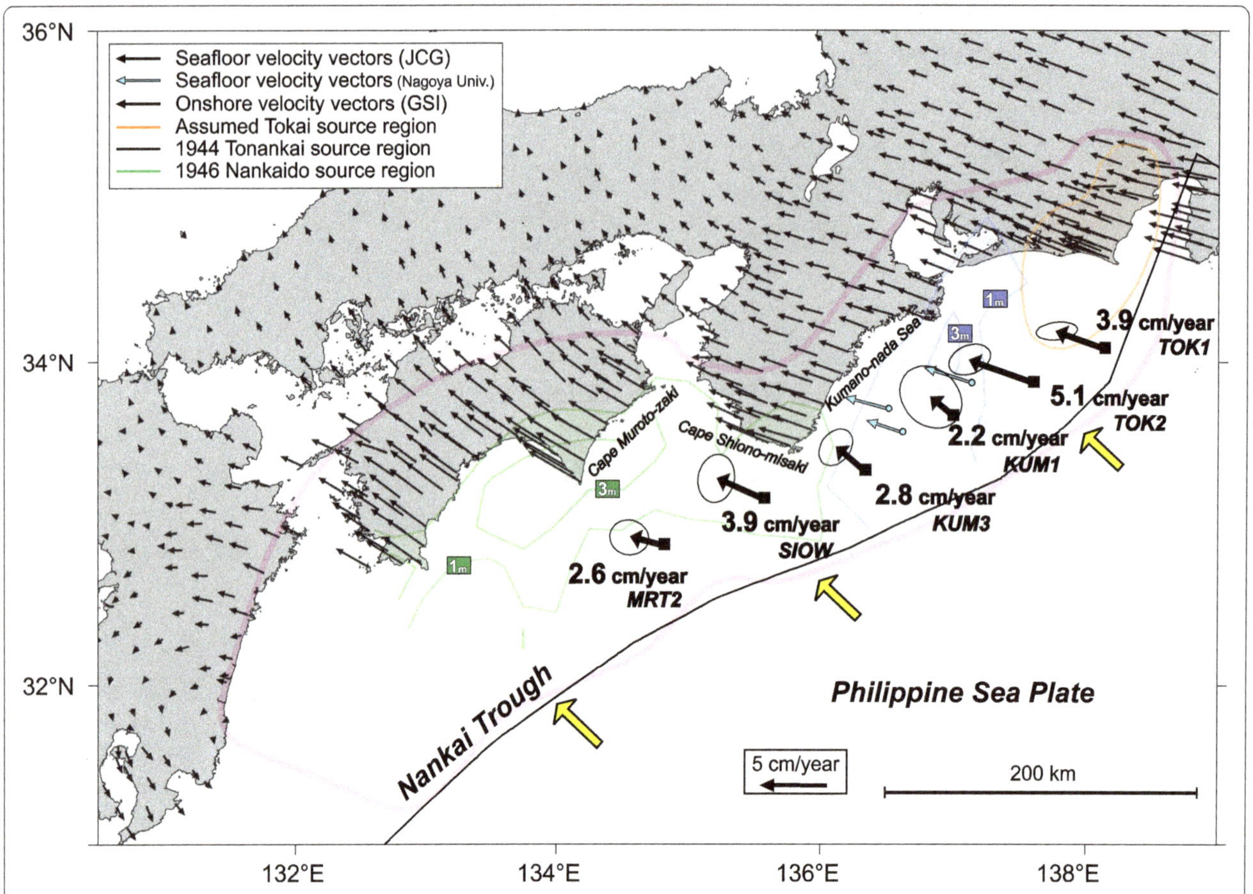

Figure 5 Horizontal velocity vectors detected by seafloor geodetic observations at six seafloor sites. Seafloor velocity vectors are shown by red arrows. Red ellipses indicate the 2-sigma errors. Light blue arrows indicate the seafloor geodetic observations by Nagoya University (Tadokoro et al. 2012). The onshore velocity vectors are calculated for the period from March 2006 to February 2011 using GEONET stations. They are shown as black arrows. Yellow arrows indicate the convergence rate of the Philippine Sea plate under the Eurasian plate (DeMets et al. 1994). The purple region shows the maximum source model of the Nankai Trough great earthquake assumed by the CDMC (2013). The orange region is the source model of the future Tokai earthquake assumed by the CDMC (2013). The blue and green contour lines are coseismic slip distributions associated with the 1944 Tonankai and 1946 Nankaido earthquakes, respectively, obtained by tsunami inversion analyses (Baba and Cummins 2005).

direction of SIOW's velocity vector is similar to those of onshore stations nearby, MRT2 differs in a counterclockwise fashion due to a smaller northward component than at 'Muroto-4' (99.99% CL). The movement at MRT2 is inconsistent with the onshore geodetic observations, suggesting the presence of a weak interplate-coupling region (about 50%) located off Cape Muroto-zaki. This is consistent with the 1946 Nankaido earthquake source region, which has a peak slip region in the deep side.

Relationship with subducting seamount and ridge

In the interplate region beneath TOK1 and TOK2, the Paleo-Zenisu ridge was detected by seismic refraction studies (Kodaira et al. 2003 and Park et al. 2003) (light blue polygon in Figure 7). These and several other studies (e.g., Kodaira et al. 2000 and Duan 2012) interpreted that a subducting seamount increases the normal stress and makes a locally strong coupling region. On the other hand, other seismic surveys (Ranero and von Huene 2000, Bangs et al. 2006, and Mochizuki et al. 2008) and a laboratory study (Dominguez et al. 2000) suggested that interplate coupling is weak on a subducting seamount due to elevated pore pressure. Therefore, the effect of a subducting seamount on an interplate-coupling state is now being debated. Our observations at TOK1 and TOK2 suggest the presence of strong interplate coupling around the Paleo-Zenisu ridge, consistent with the strengthening effect proposed by the former studies.

On the interplate boundary beneath MRT2, the seismic refraction study detected a subducting seamount (Kodaira et al. 2000) (light blue ellipse in Figure 7). The positional relationship between this seamount and the weak interplate-coupling region suggested by our seafloor observations at MRT2 suggests the possibility that the weakening effect proposed by the latter studies is at work.

Therefore, the Paleo-Zenisu ridge and the subducting seamount located off Cape Muroto-zaki have different

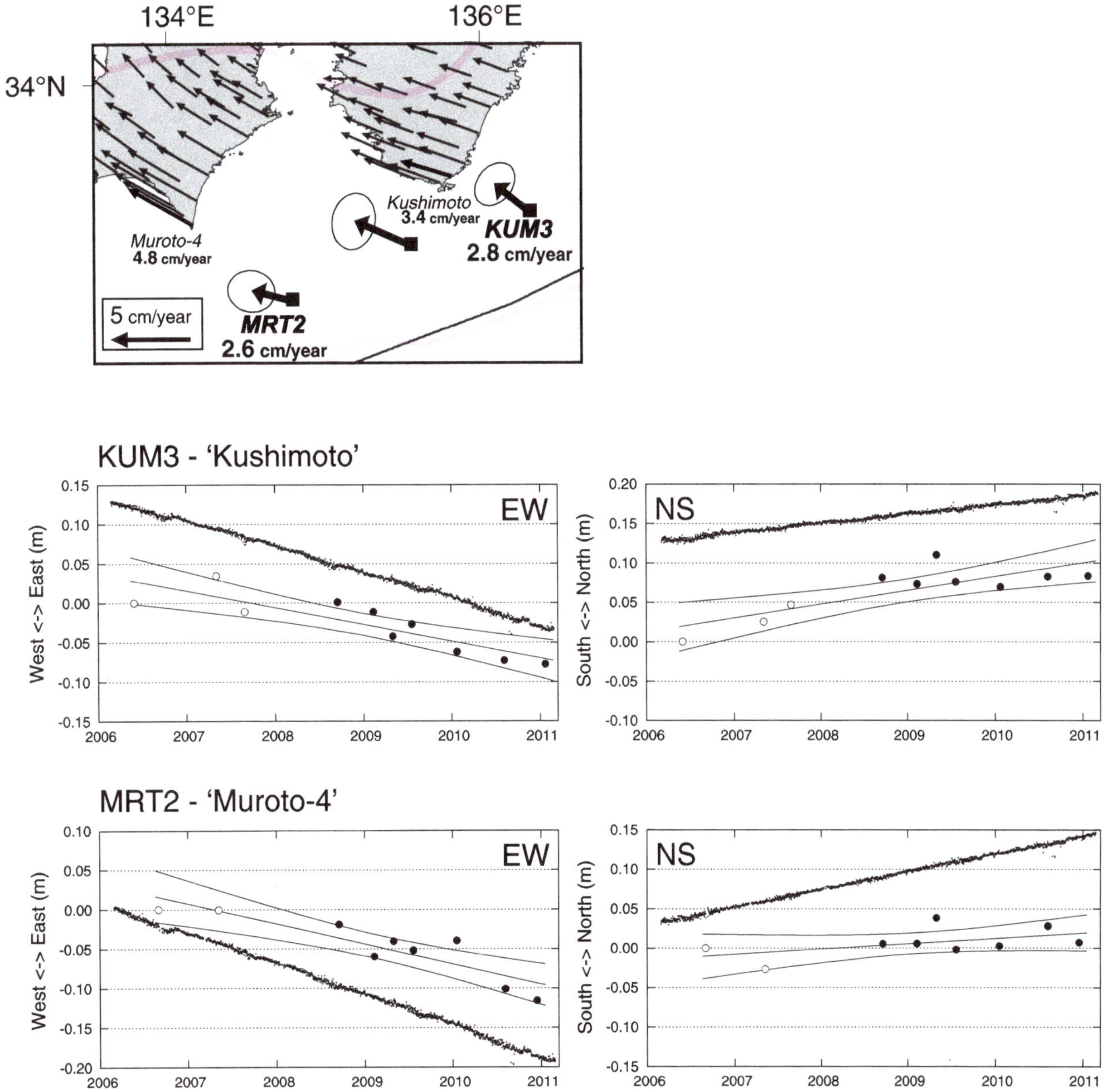

Figure 6 Comparisons of seafloor and onshore time series in east-west and north-south displacements. Time series of sites KUM3 and MRT2 compared with nearby onshore GPS data. Gray dots indicate the time series in the horizontal coordinates of the onshore 'Kushimoto' and 'Muroto-4' GPS station. The position reference is the Eurasian plate.

effects on interplate coupling. This indicates that the effect of a subducting seamount for a plate boundary differs according to various conditions, e.g., the height and width of the seamount, temperature and pressure on the plate boundary, physical properties, and convergence rate. Although the clearest difference between the Paleo-Zenisu ridge and the Muroto-zaki subducting seamount is their breadth in the east-west direction, it is difficult to identify that as the cause of the differing effects on interplate coupling because there are other unknown conditions. Recent numerical simulation studies also offer a similar suggestion (Yang et al. 2012, 2013). They point out that the effect of a subducting seamount varies complicatedly depending on the friction parameters, the depth, and the shape.

Conclusions

We have presented the results of seafloor geodetic observations conducted over about 5 years along the Nankai Trough and found the heterogeneous interplate coupling near the trough axis to have the following features. A strong coupling region is located in the easternmost interplate region around TOK1 above the Paleo-Zenisu ridge that is hypothesized to

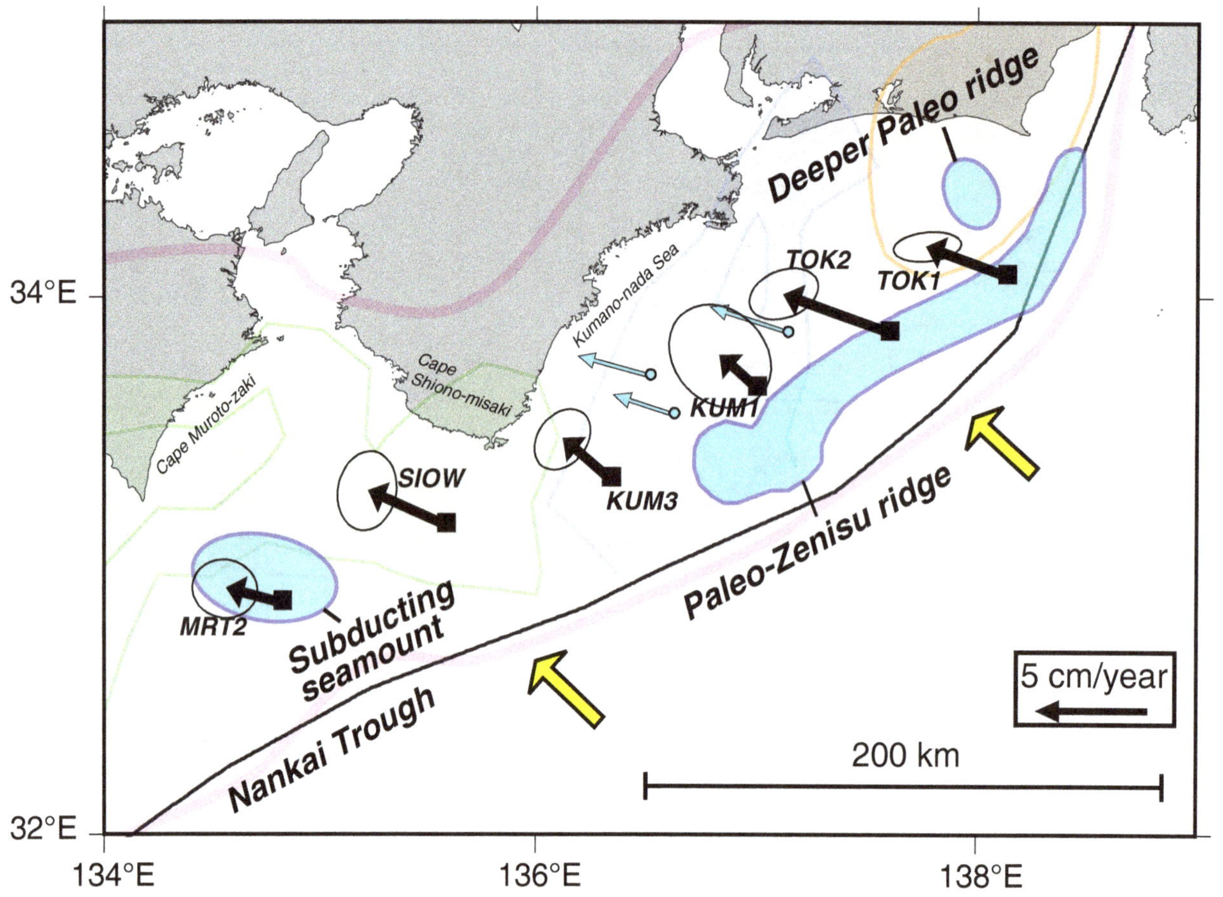

Figure 7 Seafloor velocity vectors and locations of the subducting seamount and ridges. The subducting seamount (Kodaira et al. 2000) and the Paleo-Zenisu and deeper Paleo ridges (Kodaira et al. 2003 and Park et al. 2003) are shown by light blue regions. The light blue and red arrows indicate the seafloor velocity vectors observed by Nagoya University (Tadokoro et al. 2012) and the authors, respectively.

be the source area of the future Tokai earthquake. This strong coupling region is believed extend to the southwest around TOK2. A shallow interplate region around KUM1 and KUM3 does not show a full coupling ratio, which is consistent with the 1944 Tonankai earthquake source region estimated using tsunami data. In the westernmost region, the slow displacement rate observed at MRT2 reflects weak interplate coupling, probably due to the subducting seamount off Cape Muroto-zaki. Our observations above the subducting ridge and seamount indicate that the effect of a subducting seamount on interplate-coupling region differs depending on various conditions.

As above, our seafloor geodetic observations suggested a complication of the interplate coupling on the assumed maximum source model of the Nankai Trough great earthquake. However, the blank offshore region on the west side of MRT2 has still not been revealed. In order to expand the seafloor geodetic observation network to this blank region and interpolate other blank regions, nine new seafloor sites were deployed in addition to the six existing sites after November 2011. These new sites will deliver useful information for discussing the earthquake model and disaster prevention in more detail.

Abbreviations

CDMC: the Central Disaster Management Council of the Japanese government; CL: confidence level; CTD: conductivity temperature depth profiler; GPS: global positioning system; GSI: Geospatial Information Authority of Japan; JHOD: Hydrographic and Oceanographic Department of Japan; XBT: expendable bathy thermograph; XCTD: expendable conductivity temperature depth profiler.

Competing interests

The authors declare that they have no competing interests.

Authors' contributions

YY, TI, MS, and SW proposed the topic and conceived and designed the study. YY, TI, MS, SW, HS, NU, YM, ST, MF, and TY carried out the observations and analyzed the data. YY, TI, MS, SW, HS, NU, YM, ST, MF, TY, MM, and AA helped in interpreting the data and approved the final manuscript.

Acknowledgements

We thank the Geospatial Information Authority of Japan for providing us with the GPS data from the terrestrial transponders for the kinematic GPS analysis and for releasing the daily coordinates of the sites on the GSI's website. We greatly thank

Professor Keiichi Tadokoro for providing us with Nagoya University's seafloor geodetic observations. We also thank Oscar L. Colombo of the NASA Goddard Space Flight Center for providing us with the kinematic GPS software IT (Interferometric Translocation) (Colombo et al. 2000). Many among the staff of the Hydrographic and Oceanographic Department of the Japan Coast Guard, including the crew of the S/Vs Meiyo and Kaiyo, have supported our observations and data analyses. Comments from anonymous reviewers have improved our manuscript. Some figures were produced using the GMT software (Wessel and Smith 1991). Our article is in the science section: 4) solid earth sciences.

Author details

[1]Hydrographic and Oceanographic Department, Japan Coast Guard, 2-5-18 Aomi, Koto-ku, Tokyo 135-0064, Japan. [2]National Research Institute for Earth Science and Disaster Prevention, 3-1 Tennodai, Tsukuba, Ibaraki 305-0006, Japan. [3]Institute of Industrial Science, University of Tokyo, 4-6-1 Komaba, Meguro-ku, Tokyo 153-8505, Japan.

References

Altamimi Z, Collilieux X, Legrand J, Garayt B, Boucher C (2007) ITRF2005: a new release of the International Terrestrial Reference Frame based on time series of station positions and Earth Orientation Parameters. J Geophys Res 112:B09401, doi:10.1029/2007JB004949

Ando M (1975) Source mechanism and tectonic significance of historical earthquakes along the Nankai Trough, Japan. Tectonophysics 27:119–140, doi:10.1016/0040-1951(75)90102-X

Asada A, Yabuki T (2001) Centimeter-level positioning on the seafloor. Proc Jpn Acad Ser B 77:7–12

Baba T, Cummins PR (2005) Contiguous rupture areas of two Nankai Trough earthquakes revealed by high-resolution tsunami waveform inversion. Geophys Res Lett 32:L083005, doi:10.1029/2004GL022320

Bangs NLB, Gulick SPS, Shipley TH (2006) Seamount subduction erosion in the Nankai Trough and its potential impact on the seismogenic zone. Geology 34:701–704, doi:10.1130/G22451.1

Colombo OL, Evans AG, Vigo-Aguiar MI, Ferrandiz JM, Benjamin JJ (2000) Long-baseline (>1000 km), sub-decimeter kinematic positioning of buoys at sea, with potential application to deep sea studies. Proc ION GPS 2000:1476–1484

Central Disaster Management Council of the Japanese Government (2013) http://www.bousai.go.jp/jishin/nankai/taisaku/pdf/1_1.pdf (in Japanese).

DeMets C, Gordon RG, Argus DF, Stein S (1994) Effect of recent revisions to the geomagnetic reversal time scale on estimates of current plate motions. Geophys Res Lett 21:2191–2194, doi:10.1029/94GL02118

Dominguez S, Malavieille J, Lallemand SE (2000) Deformation of accretionary wedges in response to seamount subduction: insights from sandbox experiments. Tectonics 19:182–196, doi:10.1029/1999TC900055

Duan B (2012) Dynamic rupture of the 2011 Mw 9.0 Tohoku-Oki earthquake: roles of a possible subducting seamount. J Geophys Res 117:B05311, doi:10.1029/2011JB009124

Fujita M, Ishikawa T, Mochizuki M, Sato M, Toyama S, Katayama M, Matsumoto Y, Yabuki T, Asada A, Colombo OL (2006) GPS/acoustic seafloor geodetic observation: method of data analysis and its application. Earth Planets Space 58:265_275, doi:10.1186/BF03351923

Ito T, Hashimoto M (2004) Spatiotemporal distribution of interplate coupling in southwest Japan from inversion of geodetic data. J Geophys Res 109:B02315, doi:10.1029/2002JB002358

Kido M, Fujimoto H, Miura S, Osada Y, Tsuka K, Tabei T (2006) Seafloor displacement at Kumano-nada caused by the 2004 off Kii Peninsula earthquakes, detected through repeated GPS/acoustic surveys. Earth Planets Space 58:911–915

Kido M, Osada Y, Fujimoto H, Hino R, Ito Y (2011) Trench-normal variation in observed seafloor displacements associated with the 2011 Tohoku-Oki earthquake. Geophys Res Lett 38:L24303, doi:10.1029/2011GL050057

Kodaira S, Takahashi N, Nakanishi A, Miura S, Kaneda Y (2000) Subducted seamount imaged in the rupture zone of the 1946 Nankaido earthquake. Science 289:104–106, doi:10.1126/science.289.5476.104

Kodaira S, Nakanishi A, Park J-O, Ito A, Tsuru T, Kaneda Y (2003) Cyclic ridge subduction at an inter-plate locked zone off central Japan. Geophys Res Lett 30:6, doi:10.1029/2002GL016595

Loveless JP, Meade BP (2010) Geodetic imaging of plate motions, slip rates, and partitioning of deformation in Japan. J Geophys Res 115:B02410, doi:10.1029/2008JB006248

Matsumoto Y, Fujita M, Ishikawa T, Mochizuki M, Yabuki T, Asada A (2006) Undersea co-seismic crustal movements associated with the 2005 Off Miyagi Prefecture Earthquake detected by GPS/acoustic seafloor geodetic observation. Earth Planets Space 58:1573–1576, doi:10.1186/BF03352663

Matsumoto Y, Fujita M, Ishikawa T (2008) Development of multi-epoch method for determining seafloor station position (in Japanese). Report of Hydrographic and Oceanographic Researches 26:16–22

Mochizuki K, Yamada T, Shinohara M, Yamanaka Y, Kanazawa T (2008) Weak interplate coupling by seamounts and repeating M ~ 7 earthquakes. Science 321:1194–1197, doi:10.1126/science.1160250

Nishimura S, Hashimoto M (2006) A model with rigid rotations and slip deficits for the GPS-derived velocity field in southwest Japan. Tectonophysics 421:187–207, doi:10.1016/j.tecto.2006.04.017

Park J-O, Tsuru GT, Kodaira S, Cummins PR, Kaneda Y (2002) Splay fault branching along the Nankai subduction zone. Science 297:1157–1160, doi:10.1126/science.1074111

Park J-O, Moore GF, Tsuru T, Kodaira S, Kaneda Y (2003) A subducted oceanic ridge influencing the Nankai megathrust earthquake rupture. Earth and Planet Sci Let 217:77–84, doi:10.1016/S0012-821X(03)00553-3

Ranero CR, von Huene R (2000) Subduction erosion along the Middle America convergent margin. Nature 404:748–752, doi:10.1038/35008046

Sagiya T (1999) Interplate coupling in the Tokai district, central Japan, deduced from continuous GPS data. Geophys Res Lett 26:2315–2318, doi:10.1029/1999GL900511

Sagiya T, Thacher W (1999) Coseismic slip resolution along a plate boundary megathrust: the Nankai Trough, southwest Japan. J Geophys Res 104(B1):1111–1129, doi:10.1029/98JB02644

Sagiya T, Miyazaki S, Tada T (2000) Continuous GPS array and present-day crustal deformation of Japan. Pure Appl Geophys 157:2303–2322, doi:10.1007/PL00022507

Sato M, Ishikawa T, Ujihara N, Yoshida S, Fujita M, Mochizuki M, Asada A (2011a) Displacement above the hypocenter of the 2011 Tohoku-oki earthquake. Science 332:1395, doi:10.1126/science.1207401

Sato M, Saito H, Ishikawa T, Matsumoto Y, Fujita M, Mochizuki M, Asada A (2011b) Restoration of interplate locking after the 2005 Off-Miyagi prefecture earthquake, detected by GPS/acoustic seafloor geodetic observation. Geophys Res Lett 38:L01312, doi:10.1029/2010GL045689

Sato M, Fujita M, Matsumoto Y, Saito H, Ishikawa T, Asakura T (2013a) Improvement of GPS/acoustic seafloor positioning precision through controlling the ship's track line. J Geod 87:825–842, doi:10.1007/s00190-013-0649-9

Sato M, Fujita M, Matsumoto Y, Ishikawa T, Saito H, Mochizuki M, Asada A (2013b) Interplate coupling off northeastern Japan before the 2011 Tohoku-oki earthquake, inferred from seafloor geodetic data. J Geophys Res 118:1–10, doi:10.1002/jgrb.50275

Seno T, Stein S, Gripp AE (1993) A model for the motion of the Philippine Sea plate consistent with NUVEL-1 and geological data. J Geophys Res 98(B10):17941–17948

Spiess FN, Chadwell CD, Hildebrand JA, Young LE, Purcell GH Jr, Dragert H (1998) Precise GPS/acoustic positioning of seafloor reference points for tectonic studies. Phys Earth and Planet Inter 108:101–112, doi:10.1016/S0031-9201(98)00089-2

Tadokoro K, Ikuta R, Watanabe T, Ando M, Okuda T, Nagai S, Yasuda K, Sakata T (2012) Interseismic seafloor crustal deformation immediately above the source region of anticipated megathrust earthquake along the Nankai Trough, Japan. Geophys Res Lett 39:L10306, doi:10.1029/2012GL051696

Tanioka Y, Satake K (2001) Detailed coseismic slip distribution of the 1944 Tonankai Earthquake estimated from tsunami waveforms. Geophys Res Lett 28:1075–1078, doi:10.1029/2000GL012284

Watanabe S, Sato M, Fujita M, Ishikawa T, Yokota Y, Ujihara N, Asada A (2014) Evidence of viscoelastic deformation following the 2011 Tohoku-oki earthquake revealed from seafloor geodetic observation. Geophys Res Lett 41:5789–5796, doi:10.1002/2014GL061134

Wessel P, Smith WHF (1991) Free software helps map and display data. Eos Trans AGU 72:441, doi:10.1029/90EO00319

Yang H, Liu Y, Lin J (2012) Effects of subducted seamounts on megathrust earthquake nucleation and rupture propagation. Geophys Res Lett 39:L24302, doi:10.1002/2012GL053892

Yang H, Liu Y, Lin J (2013) Geometrical effects of a subducted seamount on stopping megathrust ruptures. Geophys Res Lett 40:1–6, doi:10.1002/grl.50509

Yasuda K, Tadokoro K, Ikuta R, Watanabe T, Nagai S, Okuda T, Fujii C, Sayanagi K (2014) Interplate locking condition derived from seafloor geodetic data at the northernmost part of the Suruga Trough, Japan. Geophys Res Lett 41:5806–5812, doi:10.1002/2014GL060945

Yoshioka S, Matsuoka Y (2013) Interplate coupling along the Nankai Trough, southwest Japan, inferred from inversion analyses of GPS data: effects of subducting plate geometry and spacing of hypothetical ocean-bottom GPS stations. Tectonophysics 600:165–174, doi:10.1016/j.tecto.2013.01.023

7

Development of a database and visualization system integrating various models of seismic velocity structure and subducting plate geometry around Japan

Yasuko Yamagishi[1*], Ayako Nakanishi[2], Seiichi Miura[2], Shuichi Kodaira[2] and Hide Sakaguchi[1]

Abstract

To estimate strong ground motions caused by future earthquakes in Japan and to more accurately predict seismic hazards and tsunamis, it is necessary to accurately model the geometry of the subducting plate and the seismic velocity structure around Japan, particularly in offshore areas. Although various seismic velocity structure and plate boundary models have been proposed around Japan, they are all managed individually and differ in extent, data type, and format. Ensuring consistency among those models requires knowledge of their spatial distribution around the subduction zones of Japan. Here, we describe a database system to store and serve various velocity structure and plate geometry datasets from around Japan. Seismic structure models in this database include 3D seismic velocity models obtained by seismic tomography, 3D plate geometry models, 2D seismic velocity structure models, 2D plate geometry models obtained by offshore seismic surveys, and hypocenter distributions determined by offshore observations and the Japan Meteorological Agency. Using this database (currently available only in Japanese), users can obtain data from several structural models at once in the form of the original model data, equal-interval gridded data in a text file, and Keyhole Markup Language (KML) data. Users can grasp the distributions of all available seismic models and hypocenters using a web-based interface, simultaneously view various models and hypocenters as KML output files in Google Earth, and easily and freely handle the structural models in a selected area of interest using the gridded text-file output data. This system will be useful in creating more accurate models of the geometries of the subducting plate and the seismic velocity structure around Japan.

Keywords: Seismic velocity structure, Subducting plate, Database, Visualization, Google Earth, KML

Introduction

The Japanese Islands are in a complex tectonic setting, with the Pacific and Philippine Sea plates subducting beneath Japan from the east and southeast, respectively. Strong ground motions and tsunamis generated by large thrust earthquakes occurring on the subducting plate interfaces cause considerable damage to coastal areas, as illustrated by the 2011 Tohoku earthquake (e.g., Fujiwara et al. 2011).

To estimate strong ground motions and predict seismic hazards for such large thrust events, it is necessary to create accurate models of the geometries of the subducting plates and seismic velocity structure. In preparing the National Seismic Hazard Maps (2017), the Earthquake Research Committee, Headquarters for Earthquake Research Promotion in the Ministry of Education, Culture, Sports, Science and Technology (MEXT) constructed a standard velocity structure model around Japan based on data from Koketsu et al. (2008) and Fujiwara et al. (2009). This model, the Japan Integrated Velocity Structure Model (Koketsu et al. 2008), was created to simulate long-period ground motions and their associated seismic hazards. However, large thrust earthquakes often occur in deep offshore locations in subduction zones, where the velocity structure remains poorly understood because standard velocity structure models are based only on onshore

* Correspondence: yamagisi@jamstec.go.jp
[1]Department of Mathematical Science and Advanced Technology, Japan Agency for Marine-Earth Science and Technology, 3173-25, Showa-machi, Kanazawa-ku, Yokohama 236-0001, Japan
Full list of author information is available at the end of the article

geophysical and geological data. Nakamura et al. (2015) observed from ocean-bottom data that long-period (10–20 s) ground motions developed in ocean areas during a moderate (M_w 5.8) earthquake. It is thus necessary to consider the submarine velocity structure, including thick, low-velocity sedimentary layers, in simulations of long-period ground motions and associated seismic hazards because long-period ground motions in ocean areas could affect source analyses such as magnitude estimates and finite fault slips.

Various seismic velocity structure models have been constructed for the plate boundaries around Japan, including crustal structure and tomographic models based on offshore seismic surveys and observations. However, each model is different and managed individually by its constructor organization. It is therefore an opportune time to construct a new and more realistic model of the subducting plate geometry and seismic velocity structure. Integrated plate configuration data based on many seismic studies (Hirose 2013) and a visualization system for subsurface structures of Japan (Active Fault Database 2016) are available, although the structural information is of insufficient resolution for offshore application. To produce a fully integrated submarine velocity structure and subducting plate geometry model, we need to know the spatial distribution of the various subordinate models (i.e., 2D and 3D seismic velocity models and plate geometries) and visually compare their hypocentral distributions.

Here, we describe a newly developed database to store the various types of velocity structure and plate geometry data around Japan, available at http://www.kozo.jishin.go.jp. We introduce the concept, development, and maintenance of this database and provide some examples of its use.

Methods/Experimental

To better understand and conveniently compare submarine seismic velocity structure and subducting plate geometry models, we developed a database to store and visualize the various kinds of seismic velocity structure models around Japan. In this section, we first explain the storage of the seismic velocity structure models as data in the database and then describe the development and use of the database.

Data used for database development

We collected published and officially announced 2D velocity structure models and plate geometries along offshore seismic profile lines, 3D tomographic velocity models, and 3D plate geometries. We prepared four permission levels for the data and sources used to construct the database, depending on data availability: (1) only the areal extent of the data and the source reference can be indicated, (2) the data and source reference can be displayed but not distributed, (3) the data and reference are approved for both display and download, or (4) nothing is provided. For example, option 2 is chosen if the model was only reported in documents issued by governmental institutions.

2D velocity structure models and plate geometries along offshore seismic profile lines were provided by the University of Tokyo, Hokkaido University, Tokyo University of Marine Science and Technology, the Japan Coast Guard, and the Japan Agency for Marine-Earth Science and Technology (JAMSTEC). 3D velocity models estimated by seismic tomography and 3D subducting plate geometries were provided by the University of Tokyo, Tohoku University, Tokyo Institute of Technology, the Meteorological Research Institute, the National Research Institute for Earth and Disaster Resilience, and JAMSTEC. Hypocentral parameters were provided by the Japan Meteorological Agency (JMA) and JAMSTEC. We standardized the format of all models and parameters because their contents, notation systems, and sampling intervals were different depending on the data provider.

Development of the database

In addition to the original data files, our database provides reconstructed 250-m equal-interval gridded data files. We included a function to convert the data into Keyhole Markup Language (KML), an XML-based language schema for visualizing geographic data in Google Earth. Users can acquire various data files from several models at once, regardless of model type. We constructed a web-based graphical user interface (GUI) to search the database, display the extent of the data in each model, and retrieve the desired model data.

Database design

We constructed the database as a Java SE 8 application and developed the web GUI so that anyone can access the database via the Internet. The system uses Apache HTTP Server 2.2 and Apache Tomcat 8 as middleware. For searching and processing data, we developed a Java Servlet without a database management system (e.g., MySQL, PostgreSQL). Because the data archiving system uses only a file system (see Table 1), no knowledge of database language (e.g., SQL) is needed for management of the data. Instead, an information file must be created for each kind of model data (including the source reference, file name, and permission level of the data) and uploaded to the parent directory (Table 1).

File formats accepted by the database

In constructing the database, each model type presented various data formats. To standardize the data files, data should be written as geographic data comprising

Table 1 Structure of the file system used for data archiving

Directory structure	Contents
database/	Top-level directory of the database
database/data/	Information file for each model
database/data/2D_plate/	2D plate geometry model data files
database/data/2D_structure/	2D seismic velocity model data files
database/data/3D_plate/	3D plate geometry model data files
database/data/3D_structure/	3D seismic velocity model data files
database/data/Hypo/	Hypocenter distribution data files

latitude, longitude, and depth, plus either a seismic velocity (2D and 3D velocity models), seismic velocity perturbation (if available for 3D velocity models), or magnitude and origin time (hypocenter distributions). Except for original data files provided by various institutes, universities, and researchers, the database requires space-delimited plain text files organized for each model type as shown in Table 2. To provide 250-m equal-interval gridded data, the database requires horizontal equal-interval grid point data (except for hypocenter distributions); 2D and 3D seismic velocity models further require vertical equal-interval grid point data. For 3D seismic velocity and plate geometry models, the extent of the data is displayed in the web GUI as a rectangle on the map, in which the existence of blank or dummy data is permitted; thus, the database can accept "incomplete" data within the model extent. For 2D seismic velocity and plate geometry models, the data extent is displayed as a profile on the map, and data must exist across the entire profile. In addition to the formatted text file, the database can store original, unprocessed data files for download.

Searching the database

The database can be searched using user-specified parameters (Fig. 1). Users begin by setting the geographic region of interest as follows: (1) the user sets the geographic coordinates of the start and end points of a profile by drawing the profile on the Earth's surface, (2) the user sets the distance from the profile within which they wish to capture data, and (3) a rectangular region is created from the user-specified parameters (Fig. 2). All models containing data within the rectangular region of interest are searched. The user can then select the desired model(s) and output file type; the original data file (depending on permissions), the 250-m equal-interval gridded data file, and the KML file are available for download.

Users can select and acquire data files for 2D seismic velocity models, 2D plate geometry models, and hypocenter distributions simultaneously. However, 3D seismic velocity and plate geometry models must be selected individually, and only one data type (seismic velocity, Vp or Vs, and seismic velocity perturbation, dVp or dVs) can be selected at a time for 3D seismic velocity models. Because of their large data volume, when JMA hypocenter distributions are selected as output data, users must further specify a period of occurrence and maximum and minimum magnitudes.

A "ReadMe" file included with the downloaded data files provides the contents of each downloaded data file, the format of the equal-interval gridded data, the reference for the selected model, the user-selected search area, etc. The reference information provides users the method and observational data used to create the original model.

Generating 250-m equal-interval gridded data

The equal-interval gridded data files provided by the database are interpolated from the original data at equal horizontal and vertical intervals of 250 m using the bicubic method (ignoring the curvature of the Earth) and then written into a text file for output. The 250-m equal-interval data are produced automatically when

Table 2 Data types, contents, and structures required by the database

Data type	2D plate geometry	2D velocity structure	3D plate geometry	3D velocity structure	Hypocenter distribution
Contents	Depth of the plate boundary along a seismic line	2D velocity distribution along a seismic line	3D depth profile of the plate boundary	3D velocity distribution	Hypocentral parameter
Structure					Origin time
	Latitude (decimal degrees)	Latitude (decimal degrees)	Latitude (decimal degrees)	Latitude (decimal degrees)	Latitude (decimal degrees)
	Longitude (decimal degrees)	Longitude (decimal degrees)	Longitude (decimal degrees)	Longitude (decimal degrees)	Longitude (decimal degrees)
	Depth (km)	Depth (km)	Depth (km)	Depth (km)	Depth (km)
		Vp or Vs		Vp or/and Vs; dVp or/and dVs if available	Magnitude: Mj for JMA data M (Watanabe 1971) for other data.

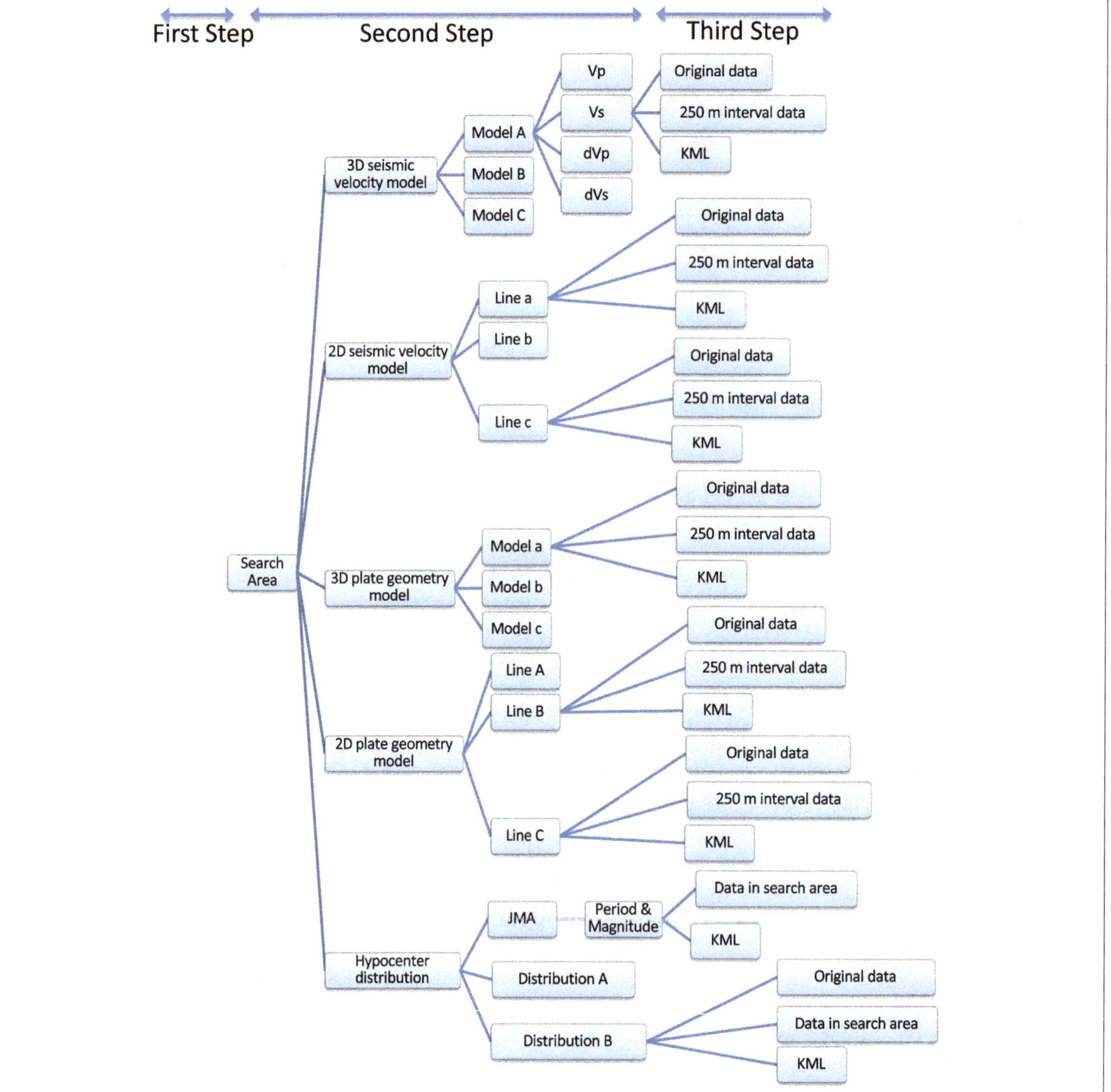

Fig. 1 Flowchart of the database search parameters. This search structure consists of three steps: definition of the search area, model selection, and selection of data for download. Once users set a search area, all models containing data within that search area are selected, regardless of model type. Each selection step is described in detail in the text

2D seismic velocity and plate geometry models are stored in the database and are regarded as existing in a vertical rectangle. For 3D seismic velocity and plate geometry models, the equal-interval data are produced each time the database is searched; 3D plate geometry models are interpolated from the original data only within the user-specified search region, whereas only the data in the vertical cross section just beneath the user-specified profile are newly interpolated for 3D seismic velocity models. The equal-interval data are regarded as existing within a rectangular parallelepiped in 3D seismic velocity models and are projected onto the Earth's surface as a horizontal rectangle in 3D plate geometry models. Grid points containing no data in the projected 3D models are given dummy values of − 9999, and new grid points interpolated from dummy values will be passed dummy values of − 9999. Dummy data are ignored upon conversion into KML.

Converting equal-interval data to KML

For simultaneous visualization of 2D and 3D seismic velocity structure models in Google Earth, the database converts the 250-m equal-interval gridded data

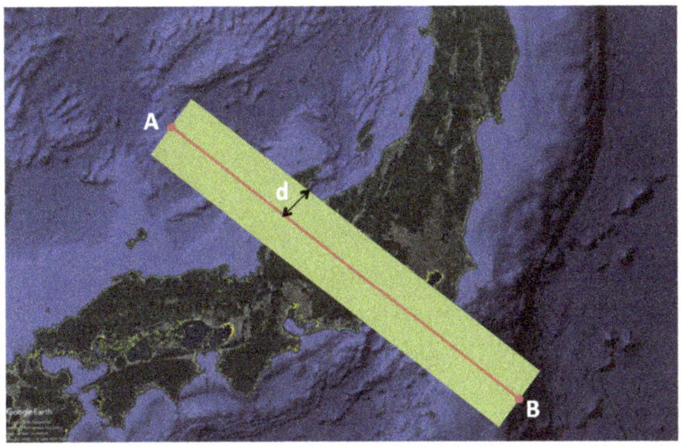

Fig. 2 Selection of the search region in the database system. Users input the latitudes and longitudes of points (A) and (B) to draw a profile (red line) and then set the distance from the profile (d) within which to capture model data (green rectangular region). Alternatively, the search profile can be drawn directly on the map

to KML when KML is selected as the output file. For 2D and 3D seismic velocity models, the KML files show vertical cross sections of the seismic velocity (or velocity perturbation in 3D) distributions just beneath the selected profile as color contour maps (Fig. 3a, b). Users can set the minimum and maximum seismic velocity/seismic velocity perturbation values shown by the color scale for 3D velocity models, whereas the maximum and minimum 2D seismic velocities shown by the color scale are 8.5 and 1.5 km/s, respectively. The 2D and 3D seismic velocity color contour maps are composed of polygons colored according to the value of the seismic velocity at each grid point. To minimize the KML file size, polygons are merged if the values of neighboring grids are very close. In Google Earth, the polygons are visible from the front and back; the KML files are thus created such that the south-facing side of the cross section is regarded as the "front" of the color contour map. For 2D plate

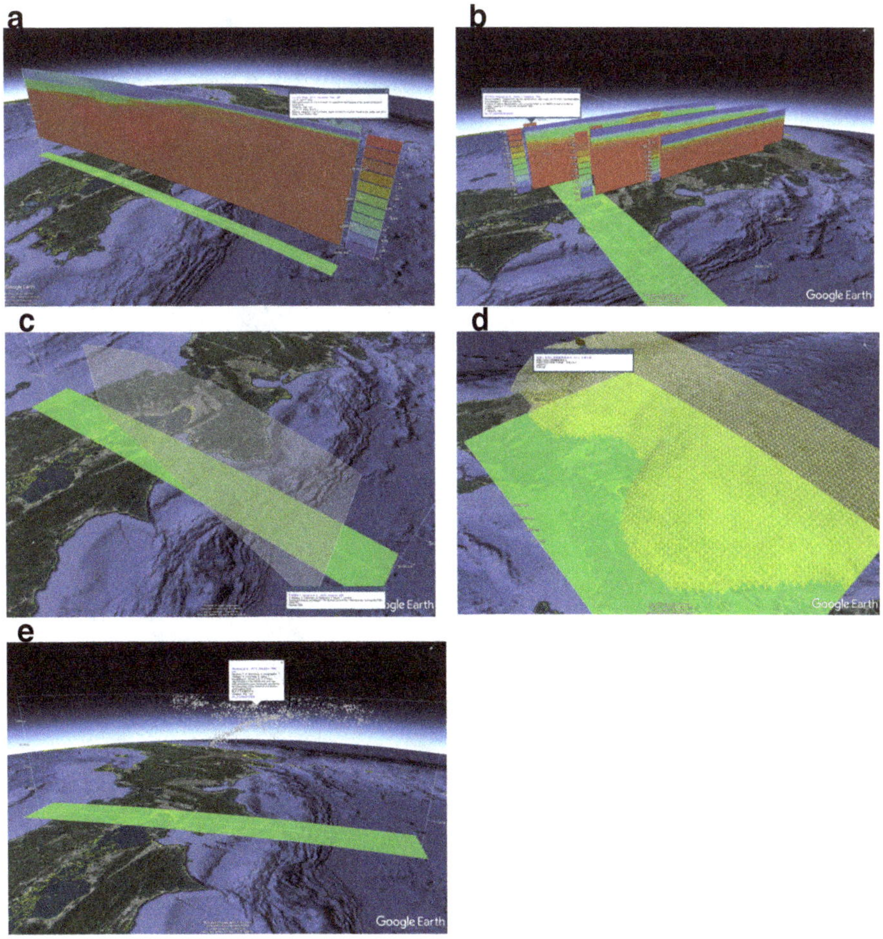

Fig. 3 Examples of seismic structural model visualizations using Google Earth. The green rectangles on the Earth's surface show the user-defined search area, and the source references of the models are shown in the white balloons. **a** 3D seismic velocity model (Liu and Zhao 2014; Liu et al. 2013). **b** 2D seismic velocity model (Kodaira et al. 2006). **c** 2D plate geometry model (Kodaira et al. 2000). **d** 3D plate geometry model (Nakanishi et al. 2018). **e** Hypocenter distributions (Akuhara et al. 2013)

geometry models, the depth to the top of the plate is represented as lines connecting the plate depths (Fig. 3c). For 3D plate geometry models, the plate geometry data are converted into KML as an aggregate surface of triangular polygons (Fig. 3d). Hypocenters are displayed as spheres (Fig. 3e). KML placemarks show the source data used in the models; upon clicking the placemark in Google Earth, the source reference appears in a white balloon (Fig. 3). As the Earth's interior is not visible in Google Earth, the data must be displayed outside of the Earth; the depths of the data are thus converted into altitudes relative to an "imaginary Earth's surface" altitude defined by the user.

Web GUI

A web-based GUI enables users to easily access the database via the Internet (presently only available in Japanese). A map embedded in the GUI shows the extent of all models in the database (Fig. 4a). In this view, source references can be viewed by clicking the profile line (2D plate geometry and seismic velocity models) or the margin of the bounding box (3D plate geometry and seismic velocity models) of individual models. In the GUI, users can easily adjust the geographic coordinates of the rectangular search region by clicking on the map, and the map is redrawn showing only the models that include data within the search region (Fig. 4b). For 3D seismic velocity and plate geometry models, the model area is shown by a rectangle, although some blank points may exist within that area. In these cases, the rectangle represents the maximum extent of the selected data. We developed this web GUI assuming use of the Google Chrome browser.

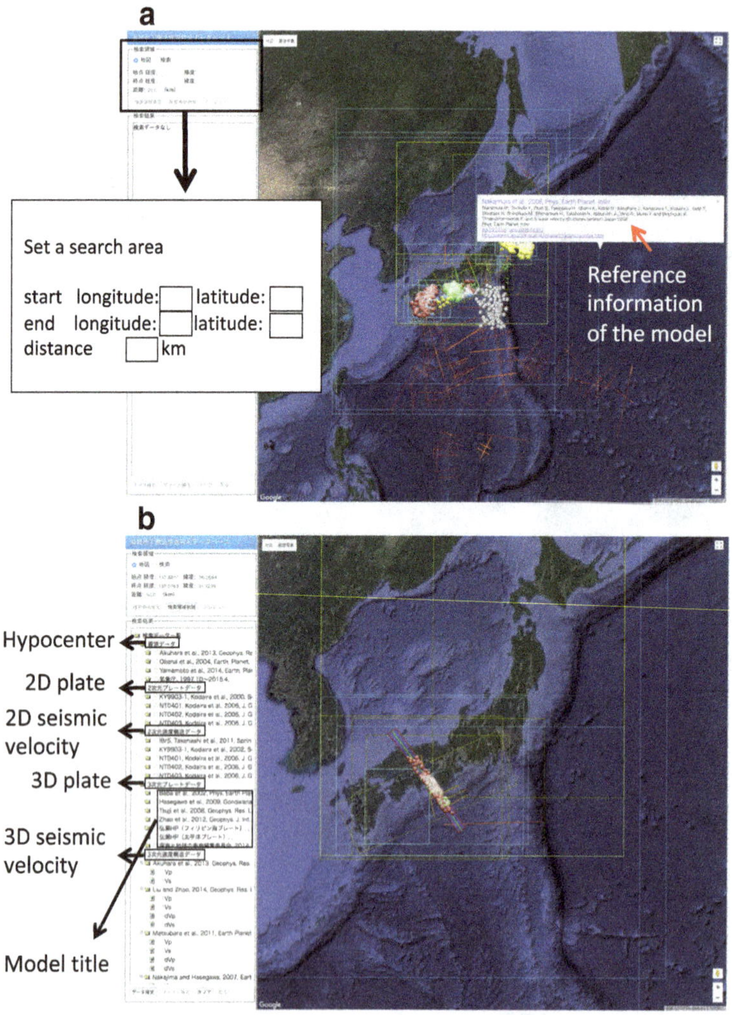

Fig. 4 Web GUI for searching the database. **a** The first page of the search tool. Green and red lines show the positions of 2D plate geometry and seismic velocity models, respectively. Yellow and green boxes show the extents of 3D plate geometry and seismic velocity models, respectively. Spheres represent hypocenters; hypocenters from the same model are shown in the same color. **b** The second page of the search tool. After search area selection, the map is refreshed to show only the models containing data within the search area. The left panel lists the titles of the selected models. The green line and red rectangle indicate the search area defined by the user in the first step (**a**)

Results

Towards the goal of constructing an integrated seismic structural model beneath the Pacific Ocean around Japan, we have developed a database that stores various seismic structural models (obtained by seismic surveys and tomographic analyses) and hypocentral parameters. Using this database, users can (1) acquire various data files from several models at once, regardless of the model type, (2) view the extent of the existing seismic structural models in the web GUI (Fig. 4a), and (3) easily visualize the structural data distribution, particularly in the offshore region, by interfacing with Google Earth (Fig. 3). Moreover, by using the original dataset or the gridded data provided by this database, users can analyze individual structural models. Therefore, this database will be useful in creating a better model of the subducting plate and seismic velocity structure around Japan.

Discussion

In this section, we introduce several example applications of the database. We first explain the use of the equal-interval gridded data and then show an example of simultaneous visualization of different seismic velocity models.

At present, the GUI is only available in Japanese. However, we believe that this database will be useful to all researchers interested in the seismic velocity structure and subducting plate geometry around Japan. We recognize that this system should be improved in the future to respond to users' requests, including an English GUI.

Use of 250-m equal-interval gridded data

The database provides three data output options: (1) the original data files, (2) 250 m equal-interval gridded data files, and (3) KML files. The unprocessed original data files are the intact datasets provided by various institutes or researchers; these are different from the formatted input data (text) files accepted by the database. JMA hypocenter data are the exception; the original data files are not provided because of their large volume. The equal-interval data files provide interpolated data gridded horizontally and vertically at 250-m intervals. The 250-m equal-interval data are provided in a vertical section beneath the user-specified profile for 3D seismic velocity models or in a rectangular parallelepiped under the user-specified search region for 3D plate geometry models. For 2D seismic velocity and

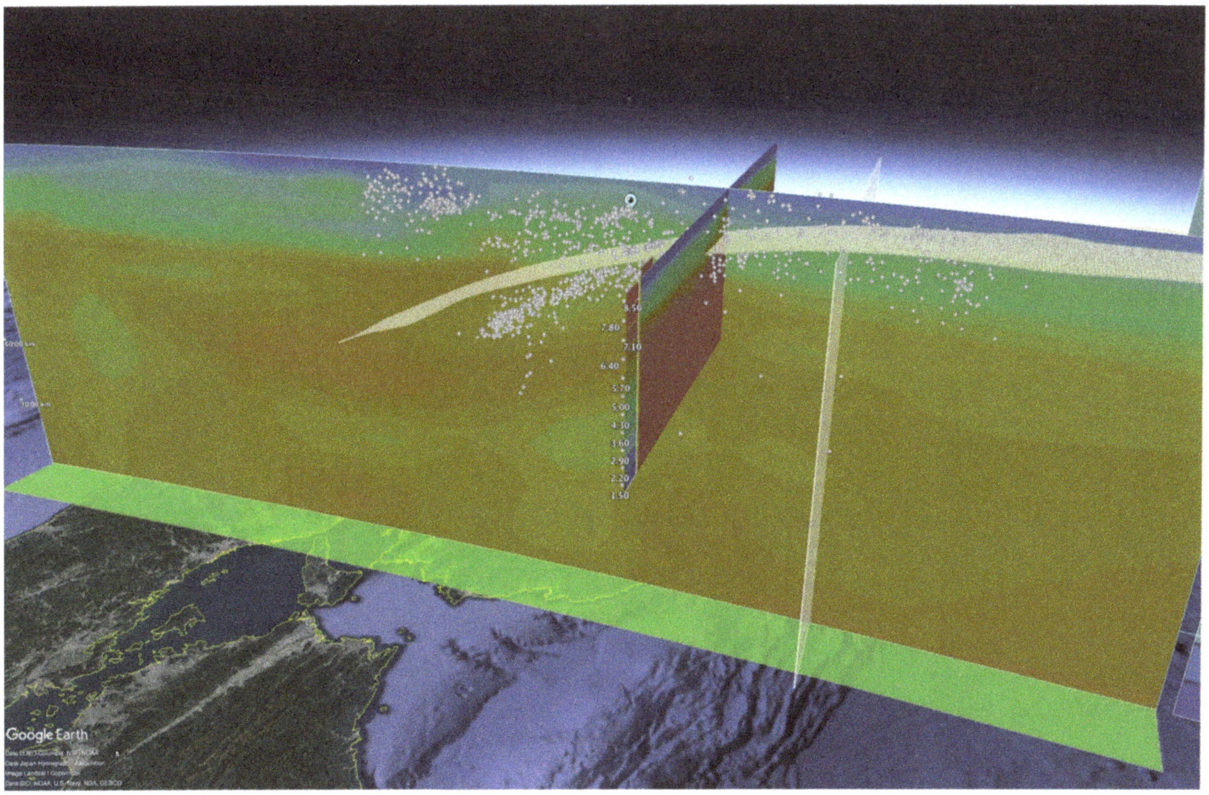

Fig. 5 Visualization of various seismic structural models under the Nankai Trough. The cross section of the 3D seismic velocity model is from Nakajima and Hasegawa (2007), the 3D plate geometry model from Nakanishi et al. (2018), the 2D plate geometry and seismic velocity models from Kodaira et al. (2006), and the hypocentral distribution from Akuhara et al. (2013). The green rectangle on the Earth's surface shows the user-defined search area. Differences between the individual models are notable; for example, hypocenters located at the surface of the subducting slab are consistently deeper than the top of the slab in the 3D plate geometry model

plate geometry models, the equal-interval data are provided beneath the entire survey line across the search area. Because the original data are often provided as a program or in a unique file format, complicated procedures may be necessary to process different types of original model data. The equal-interval gridded data are space-separated and provided as a plain text file in a simple format summarized by the "ReadMe" file so that users can easily analyze or visualize the models in their software of choice. If needed, users can easily convert the equal-interval gridded data to other data formats such as GeoCSV (Stults et al. 2015) or JSON (JavaScript Object Notation) for seismic tomographic data (Postpischl et al. 2011).

Visualization of seismic structure models

The database provides KML files produced from the 250-m equal-interval data for display in Google Earth. The layer system of Google Earth allows various geographic data to be displayed simultaneously on a virtual globe. Therefore, the database can simultaneously visualize several seismic structure models, regardless of the model types. Figure 5 shows a visual comparison of 3D plate geometry (Nakanishi et al. 2018) and seismic velocity models (Nakajima and Hasegawa 2007), 2D plate geometry and seismic velocity models (Kodaira et al. 2006), and hypocentral distributions (Akuhara et al. 2013). Differences between the 3D plate geometry model, 3D seismic velocity model, and hypocenters are notable at greater depths; for example, hypocenters located at the subducting slab interface are consistently deeper than the top of the slab in the 3D plate geometry model.

Conclusion

We developed a database system with a web GUI, available at http://www.kozo.jishin.go.jp, to provide seismic structure model data via the Internet. Using this database, users can easily obtain various kinds of seismic structure model data and visualize them simultaneously in Google Earth. By interfacing with Google Earth, this database allows users to easily understand the structural data distribution in offshore regions of Japan. This system also provides 250-m equal-interval gridded data from the desired structural models as a text file, which users can employ to analyze individual models without any complicated data preparation. As a result, we believe that the database will be useful to produce a more accurate seismic structure model beneath the Pacific Ocean around Japan.

Abbreviations

GUI: Graphical user interface; JAMSTEC: Japan Agency for Marine-Earth Science and Technology; JMA: Japan Meteorological Agency; KML: Keyhole Markup Language; MEXT: Ministry of Education, Culture, Sports, Science and Technology

Acknowledgements

This study was performed as a "research project for development of seismological information database for modeling seismic velocity structure offshore around Japan" funded by the Ministry of Education, Culture, Sports, Science and Technology (MEXT), Japan. We are grateful to all who have kindly agreed to the incorporation of their models and data into our database. The earthquake catalog used in this study is produced by the JMA in cooperation with MEXT. The catalog is based on seismic data provided by the National Research Institute for Earth Science and Disaster Prevention, JMA, Hokkaido University, Hirosaki University, Tohoku University, the University of Tokyo, Nagoya University, Kyoto University, Kochi University, Kyushu University, Kagoshima University, the National Institute of Advanced Industrial Science and Technology, the Geographical Survey Institute, Tokyo Metropolis, Shizuoka Prefecture, Hot Springs Research Institute of Kanagawa Prefecture, Yokohama City, and JAMSTEC. We are grateful to Riming Zhu (VisCore Co., Ltd.) and Kenji Morimoto (VINAS Co., Ltd.) for their skill in developing this database and the KML data conversion system.

Funding

This study was funded by the Ministry of Education, Culture, Sports, Science and Technology.

Authors' contributions

YY and AN proposed the topic. YY developed the database system. AN investigated existing structural models and coordinated the permissions levels of the data files with the providers. SM and SK helped collect the data files and develop the system. HS contributed to the discussion. All authors read and approved the final manuscript.

Competing interests

The authors declare that they have no competing interests.

Author details

[1]Department of Mathematical Science and Advanced Technology, Japan Agency for Marine-Earth Science and Technology, 3173-25, Showa-machi, Kanazawa-ku, Yokohama 236-0001, Japan. [2]Research and Development Center for Earthquake and Tsunami, Japan Agency for Marine-Earth Science and Technology, 3173-25, Showa-machi, Kanazawa-ku, Yokohama 236-0001, Japan.

References

Active Fault Database (2016) National Institute of Advanced Industrial Science and Technology, https://gbank.gsj.jp/subsurface/english/index.html. Accessed 21 Mar 2018

Akuhara T, Mochizuki K, Nakahigashi K, Yamada T, Shinohara M, Sakai S, Kanazawa T, Uehira K, Shimizu H (2013) Segmentation of the Vp/Vs ratio and low-frequency earthquake distribution around the fault boundary of the Tonankai and Nankai earthquakes. Geophys Res Lett 40:1306–1310. https://doi.org/10.1002/grl.50223

Fujiwara H, Kawai S, Aoi S, Morikawa N, Sennna S, Kudo N, Ooi M, Hao Kx-S, Hayakawa Y, Toyama N, Matsuyama N, Iwamoto L|K, Suzuki H, Ei R (2009) A study on subsurface structure model for deep sedimentary layers of Japan for strong-motion evaluation. Technical note of the National Research Institute for Earth Science and Disaster Prevention, No 337 (in Japanese)

Fujiwara T, Kodaira S, No T, Kaiho Y, Takahashi N, Kaneda Y (2011) The 2011 Tohoku-Oki earthquake: displacement reaching the trench axis. Science 334: 1240. https://doi.org/10.1126/science.1211554

Hirose F (2013) Fuyuki Hirose's HP, plate configuration data, http://www.mri-jma.go.jp/Dep/st/member/fhirose/en/en.PlateConfiguration.html. Accessed 21 Mar 2018

Kodaira S, Hori T, Ito A, Miura S, Fujie G, Park J-O, Baba T, Sakaguchi H, Kaneda Y (2006) A cause of rupture segmentation and synchronization in the Nankai trough revealed by seismic imaging and numerical simulation. J Geophys Res 111:B09301. https://doi.org/10.1029/2005JB004030

Kodaira S, Takahashi N, Nakanishi A, Miura S, Kaneda Y (2000) Subducted seamount imaged in the rupture zone of the 1946 Nankaido earthquake. Science 289:104–106. https://doi.org/10.1126/science.289.5476.104.

Koketsu K, Miyake H, Fujiwara H, Hashimoto T (2008) Progress towards a Japan integrated velocity structure model and long-period ground motion hazard map, in Proceedings of the 14th World Conference on Earthquake Engineering: Beijing, China, paper no. S10–038.

Liu X, Zhao D (2014) Structural control on the nucleation of megathrust earthquakes in the Nankai subduction zone. Geophys Res Lett 41:8288–8293. https://doi.org/10.1002/2014GL062002

Liu X, Zhao D, Li S (2013) Seismic imaging of the Southwest Japan arc from the Nankai trough to the Japan Sea. Phys Earth Planet Inter 216:59–73. https://doi.org/10.1016/j.pepi.2013.01.003

Nakajima J, Hasegawa A (2007) Subduction of the Philippine Sea slab beneath southwestern Japan: slab geometry and its relationship to arc magmatism. J Geophys Res 112:B08306. https://doi.org/10.1029/2006JB004770

Nakamura T, Takenaka H, Okamoto T, Ohori M, Tsuboi S (2015) Long-period ocean -bottom motions in the source areas of large subduction earthquakes, Scientific Reports 5, Article number: 16648. Doi:https://doi.org/10.1038/srep16648

Nakanishi A, Takahashi T, Yamamoto Y, Takahashi T, Citak SO, Nakamura T, Obana K, Kodaira S, Kaneda Y (2018) Three-dimensional plate geometry and P-wave velocity models of the subduction zone in SW Japan: Implication for seismogenesis. In: Byrne T, Fisher D, McNeil L, Saffer D, Ujiie K, Underwood M, Yamaguchi A (eds) Geology and tectonics of subduction zones: a tribute to Gaku Kimura. Geological Society of America Special Paper 534 (in press)

National Seismic Hazard Maps (2017) Report of Earthquake Research Committee, Headquarters for Earthquake Research Promotion, https://www.jishin.go.jp/evaluation/seismic_hazard_map/shm_report/shm_report_2017. Accessed 9 Apr 2018

Postpischl L, Danecek P, Morelli A, Pondrelli S (2011) Standardization of seismic tomographic models and earthquake focal mechanisms data sets based on web technologies, visualization with keyhole markup language. Comp Secsci 37:47–56. https://doi.org/10.1016/j.cageo.2010.05.006

Stults M, Arko R A, Davis E, Ertz D J, Turner M, Trabant C M, Valentine Jr D W, Ahem T K, Carbotte S M, Gurnis M, Meertens C, Ramamurthy M K, Zaslavsky L (2015) GeoCSV: tabular text formatting for geoscience data. Abstract IN11F-1809 presented at AGU Fall Meeting 2015, San Francisco, California, 14–18 December 2015. https://agu.confex.com/agu/fm15/webprogram/Paper84142.html. Accessed 8 June 2018

Watanabe H (1971) Determination of earthquake magnitude at regional distance in and near Japan. Zishin 24: 189–200. (in Japanese with English abstract)

8

Eigenvector of gravity gradient tensor for estimating fault dips considering fault type

Shigekazu Kusumoto

Abstract

The dips of boundaries in faults and caldera walls play an important role in understanding their formation mechanisms. The fault dip is a particularly important parameter in numerical simulations for hazard map creation as the fault dip affects estimations of the area of disaster occurrence. In this study, I introduce a technique for estimating the fault dip using the eigenvector of the observed or calculated gravity gradient tensor on a profile and investigating its properties through numerical simulations. From numerical simulations, it was found that the maximum eigenvector of the tensor points to the high-density causative body, and the dip of the maximum eigenvector closely follows the dip of the normal fault. It was also found that the minimum eigenvector of the tensor points to the low-density causative body and that the dip of the minimum eigenvector closely follows the dip of the reverse fault. It was shown that the eigenvector of the gravity gradient tensor for estimating fault dips is determined by fault type. As an application of this technique, I estimated the dip of the Kurehayama Fault located in Toyama, Japan, and obtained a result that corresponded to conventional fault dip estimations by geology and geomorphology. Because the gravity gradient tensor is required for this analysis, I present a technique that estimates the gravity gradient tensor from the gravity anomaly on a profile.

Keywords: Fault dip, Gravity gradient tensor, Eigenvector, Normal fault, Reverse fault, Kurehayama Fault

Introduction

In recent years, gravity gradiometry surveys have been widely conducted to obtain detailed subsurface structure data (e.g., Jekeli 1988; Dransfield 2010; Chowdhury and Cevallos 2013; Braga et al. 2014). Data collected by these surveys is the gravity gradient tensor defined by second derivatives of the gravity potential, and its response to subsurface structures is more sensitive than the gravity anomaly. At present, gravity gradiometry surveys have mainly been performed using a helicopter. Consequently, their observation interval is about 3 m on the flight profile, and the observation density is very high. The gravity gradiometry surveys allowed for high observation density, high resolution, and high sensitivity to the subsurface structures; therefore, these surveys contribute greatly to the earth science and resource engineering fields in terms of being useful and powerful tools for the estimation of subsurface structures.

Correspondence: kusu@sci.u-toyama.ac.jp
Graduate School of Science and Engineering for Research (Science), University of Toyama, 3910 Gofuku, Toyama 930-8555, Japan

Various analysis techniques using gravity gradient tensors have been suggested and discussed (e.g., Zhang et al. 2000; Beiki 2010; Martinez et al. 2013; Cevallos 2014; Li 2015). These are considered to be so-called inversion techniques. A semi-automatic interpretation method that can extract subsurface structure characteristics without geological and geophysical data input has also been developed and applied to field data (e.g., Cooper 2012; Ma 2013; Ferreira et al. 2013).

A typical semi-automatic interpretation method is an edge emphasis technique that uses extraction techniques to find locations (namely, edge) where the potential field changes abruptly due to density variations. The horizontal gravity gradient method and vertical gravity gradient method (e.g., Evjen 1936; Elkins 1951; Tsuboi and Kato 1952; Blakely and Simpson 1986) are classic edge emphasis techniques. In recent years, higher and keener extraction techniques have been suggested (e.g., Miller and Singh 1994; Cooper and Cowan 2006; Sertcelik and Kafadar 2012; Zhang et al. 2014). In addition, attention has been paid to techniques that evaluate the shape of the potential field (e.g., Koenderink and van Doorn 1992; Robert 2001; Zhou et al. 2013; Cevallos 2014).

Among these methodologies, a technique for estimating the dip of geological boundary using the gradient tensor of the potential fields has been developed (e.g., Beiki 2013). Beiki and Pedersen (2010) showed that the maximum eigenvector of the gravity gradient tensor points to the causative body (Fig. 1a). Since this property is common in the potential fields, Beiki (2013) applied it to a magnetic anomaly in the Åsele area (Sweden) and obtained useful information on the dip of the dike swarms. Kusumoto (2015), considering that the basement consists of an aggregate of high-density prisms (Fig. 1b), applied Beiki's technique (Beiki and Pedersen 2010; Beiki 2013) to the estimation of fault dips. This method provided results wherein the fault dip estimated by the gravity gradient tensor harmonized with the dip observed from seismic surveys (Kusumoto 2015, 2016a). In addition, the dip of an earthquake source fault of the Kumamoto Earthquake that occurred in April 2016 estimated from the gravity gradient tensor also corresponded with the dip of the fault model (normal fault of 60°), thus explaining the crustal movement observed by GNSS (Global Navigation Satellite System) (Kusumoto 2016b). The range for which this method is applicable is wide from low dip to high dip (e.g., Beiki 2013; Kusumoto 2015, 2016a, 2016b), although it has some numerical instability to the vertical fault (e.g., Kusumoto 2015).

Although analyses using the gravity gradient tensor have yielded excellent results in subsurface structure estimations and edge detections, gravity gradiometry surveys have been conducted in only a few areas, limiting the tensor data available. If we were to carry out these analyses in areas where gravity gradiometry surveys have not been conducted yet, we would have to use the tensor estimated from existing gravity anomaly data.

The procedure for estimating the gravity gradient tensor from gravity anomaly data has already been suggested by Mickus and Hinojosa (2001). This technique estimates the gravity gradient tensor from spatial distribution of gravity anomalies by the Fourier transform. Since the database of gravity anomalies has been prepared, studies using the gravity gradient tensor estimated by Mickus and Hinojosa's method will progress in the future. On the other hand, it is difficult to apply this method directly to gravity anomalies obtained by gravity surveys conducted on a profile employed frequently in active fault research.

In dense gravity surveys researching fault structures in detail, profiles were set perpendicular to the fault and short-spaced gravity observations were taken along the profiles (e.g., Iwano et al. 2001; Inoue et al. 2004). It is important to find the fault shape, especially its dip, in these studies because the fault dip affects the area of disaster occurrence (e.g., Abrahamson and Somerville 1996; Takemura et al. 1998) and is an important parameter in numerical simulations for hazard map creation (e.g., Irikura and Miyake 2011). Consequently, in two-dimensional gravity surveys for faults, a fault dip estimated from the eigenvectors of the gravity gradient tensor calculated from the gravity anomaly would be of additional value. In addition, since this analysis technique does not require vast calculation times, I expect it will be an effective new technique for analyzing high-resolution data obtained densely, i.e., through dense gravity surveys for fault research and also airborne gravity gradiometry surveys.

In this study, I first introduce the technique for the estimation of the gravity gradient tensor from a gravity anomaly on the profile. After that, I discuss the relationship between fault dips and eigenvectors of the gravity gradient tensor and apply its result to gravity anomaly data obtained on the profile crossing the Kurehayama Fault in Toyama, Japan.

Methods/Experimental

Gravity gradient tensor on the profile

Gravity gradient tensor Γ on the profile is defined as follows (e.g., Beiki and Pedersen 2011)

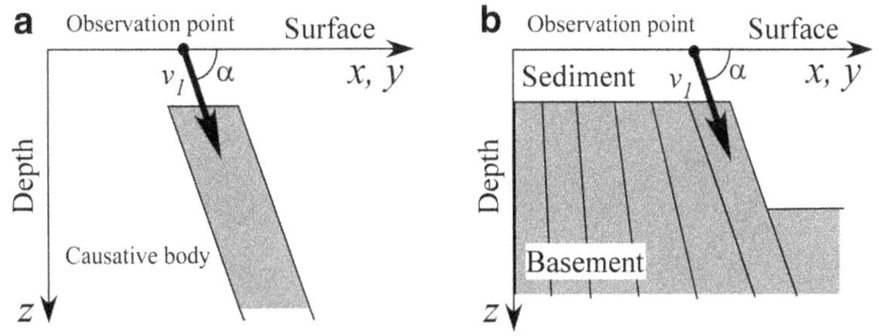

Fig. 1 Schematic illustration of the maximum eigenvectors for two-dimensional (2D) structures such as dykes and faults. **a** Basic model. In this figure, v_1 is the maximum eigenvector of the gravity gradient tensor and points to the causative body. The angle α between the surface and the maximum eigenvector is the dip of the causative body. **b** Fault model. A basement consists of an aggregate of high-density prisms, and the angle, α, indicates the fault dip

$$\Gamma = \begin{bmatrix} g_{xx} g_{xz} \\ g_{zx} g_{zz} \end{bmatrix} \tag{1}$$

Here, g_{xx}, g_{xz}, g_{zx}, and g_{zz} are each component of the tensor and are defined as the first derivative of gravity vector components g_x and g_z for each direction. In addition, gravity vectors g_x and g_z are given by the first derivative of gravity potential, W, namely, $g_x = \partial W/\partial x$ and $g_z = \partial W/\partial z$. As the gravity potential satisfies the Laplace equation, $\partial^2 W/\partial x^2 + \partial^2 W/\partial z^2 = g_{xx} + g_{zz} = 0$, we find the relationship $g_{zz} = -g_{xx}$. Also, the relationship is known to be $g_{xz} = g_{zx}$ because the gravity gradient tensor is a symmetric tensor (e.g., Torge 1989).

Relationship between subsurface structure and gravity anomaly

In the two-dimensional analyses, a structure in one direction is assumed to be infinite. Although this assumption is not realistic, it is a good approximation in fault structure analyses and gives us some practical analysis techniques. In calculations of the gravity gradient tensor from the gravity anomaly, we need gravity anomaly values at different heights. Consequently, I will show the relationship between two-dimensional subsurface structures and gravity anomalies in this subsection before estimating the gravity gradient tensor from the gravity anomaly.

As the simplest subsurface model, I set a two-dimensional double layer model consisting of a sedimentary layer and a basement (Fig. 2). Horizontal positions are given by x, and vertical positions are given by z. Depth is zero ($z = 0$) on the surface, and z increases with depth. As shown in Fig. 2, an average boundary depth between the sedimentary layer and basement is defined as $z = D$ (>0). When the boundary surface at point $Q(x')$ deviates by $h(x')$ from the average boundary depth (Fig. 2), gravity anomaly $g_z(x)$ at the point $P(x)$ on the surface caused by this deviation is given by the following equation (e.g., Blakely 1996).

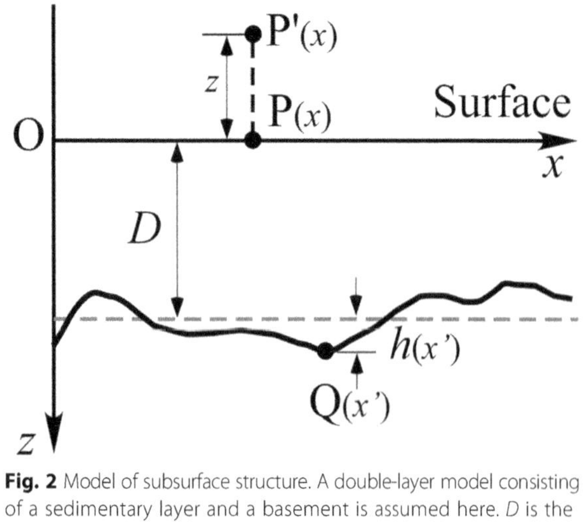

Fig. 2 Model of subsurface structure. A double-layer model consisting of a sedimentary layer and a basement is assumed here. D is the average depth of the stratum boundary, and $h(x')$ is the deviation of the boundary from the average. Here, the deviation is assumed to be very small, i.e., $h(x') << D$

$$g_z(x) = 2\gamma\Delta\rho \int_{-\infty}^{\infty} \int_{D}^{D+h(x')} \frac{z'}{(x-x')^2 + z'^2} dx' dz' \tag{2}$$

where γ is the gravitational constant and $\Delta\rho$ is the density contrast between the sedimentary layer and basement. The integration on z in Eq. (2) is solved as:

$$\int_{D}^{D+h(x')} \frac{z'}{(x-x')^2 + z'^2} dz' = \frac{1}{2} \log\left[\frac{(x-x')^2 + (D+h(x'))^2}{(x-x')^2 + D^2}\right] \tag{3}$$

here, if $h(x')$ is much smaller than D, namely, $h(x') << D$, $(D + h)^2$ is $\{D[1 + (h/D)]\}^2 \approx D^2(1 + 2\,h/D) = D^2 + 2Dh$, Eq. (3) would be rewritten as follows:

$$\int_{D}^{D+h(x')} \frac{z'}{(x-x')^2 + z'^2} dz' \approx \frac{1}{2} \log\left[\frac{(x-x')^2 + D^2 + 2Dh(x')}{(x-x')^2 + D^2}\right] = \frac{1}{2} \log\left[1 + \frac{2Dh(x')}{(x-x')^2 + D^2}\right] \tag{4}$$

In general, if $-1 < \xi \leq 1$ in $\log(1 + \xi)$, we have the following approximation (e.g., Gradshteyn and Ryzhik 2007)

$$\log(1+\xi) = \xi - \frac{1}{2}\xi^2 + \frac{1}{3}\xi^3 - \frac{1}{4}\xi^4 + \cdots = \sum_{p=1}^{\infty} (-1)^{p+1} \frac{\xi^p}{p} \tag{5}$$

The second term, $2Dh/[(x - x')^2 + D^2]$, in Eq. (4) is small because $D >> h$. We can use Eq. (5) to derive a linear approximate equation of Eq. (4). By neglecting higher terms of ξ, Eq. (3) or (4) is rewritten as follows:

$$\int_{D}^{D+h(x')} \frac{z'}{(x-x')^2 + z'^2} dz' \approx \frac{Dh(x')}{(x-x')^2 + D^2} \tag{6}$$

Consequently, we obtained the following equation.

$$g_z(x) \approx 2\gamma D\Delta\rho \int_{-\infty}^{\infty} \frac{h(x')}{(x-x')^2 + D^2} dx' \tag{7}$$

Here, I introduce a new function, ϕ, defined by:

$$\phi(x) = \frac{1}{x^2 + D^2} \tag{8}$$

and Eq. (7) is rewritten as follows:

$$g_z(x) = 2\gamma D\Delta\rho \int_{-\infty}^{\infty} \phi(x-x')h(x')dx' \quad (9)$$

This form is convoluted, and we obtain Eq. (10) by applying the Fourier transformation to Eq. (9)

$$G_z = 2\gamma D\Delta\rho\Phi H \quad (10)$$

where, G_z, Φ, and H are Fourier transforms of $g_z(x)$, $\phi(x)$, and $h(x)$, respectively. As is well known, the Fourier transform of Eq. (8) is (e.g., Blakely 1996; Gradshteyn and Ryzhik 2007)

$$\Phi = \frac{\pi}{D}e^{-D|k|} \quad (11)$$

Here, $|k| = ik_z = |k_x|$ (e.g., Blakely 1996) and k_x is the wave number in the x direction. Here, I employed the Fourier transform, F, of a function $f(x)$ defined as follows (e.g., Blakely 1996):

$$F = \int_{-\infty}^{\infty} f(x)e^{-ikx}dx \quad (12)$$

By Eq. (11), Eq. (10) is rewritten as:

$$G_z = 2\pi\gamma\Delta\rho He^{-D|k|} \quad (13)$$

This is the relationship between gravity anomaly on the profile and two-dimensional subsurface structure.

Relationship between gravity anomaly and gravity gradient tensor

As shown in the previous section, the gravity gradient tensor is given by the second derivative of the gravity potential. The relationship between gravity anomaly g_z and gravity potential W is

$$W = -\int g_z dz \quad (14)$$

From Eq. (13), the equation giving the gravity anomaly at point $P'(x)$ of an arbitrary height z from the surface (Fig. 2) is obtained in the Fourier domain as follows:

$$G_z = 2\pi\gamma\Delta\rho He^{-(D+z)|k|} \quad (15)$$

By integrating this equation to z and substituting $z = 0$, we obtain the gravity potential at the surface. If the Fourier transform of the gravitational potential is represented by U, from these calculations, the U would be given by G_z as follows:

$$U = \frac{1}{|k|}G_z \quad (16)$$

As the x direction component of gravity anomaly is given by the first derivative in the x direction of the gravity potential W, the g_x in the Fourier domain, G_x, would be given by a differential formula in the Fourier domain (e.g., Blakely 1996) as follows:

$$G_x = ik_xU \quad (17)$$

From Eq. (16), we obtained

$$G_x = \frac{ik_x}{|k|}G_z \quad (18)$$

g_{xx} in the Fourier domain is given by

$$G_{xx} = \frac{-k_x^2}{|k|}G_z \quad (19)$$

We can obtain g_{xx} by applying the inverse Fourier transform to G_{xx}, and g_{zz} would be obtained from the relationship of $g_{zz} = -g_{xx}$. The other component g_{zx} (=g_{xz}) would be given by:

$$G_{zx} = G_{xz} = ik_xG_z \quad (20)$$

where G_{zx} and G_{xz} are the Fourier transform of g_{zx} and g_{xz}.

Here, although I showed a technique to calculate the gravity gradient tensor in the Fourier domain, there is another technique to calculate the tensor by a simple finite-difference method (e.g., Blakely 1996) of gravity vectors g_x and g_z in the space domain.

Relationship between subsurface structures and eigenvectors

As indicated by Beiki and Pedersen (2010), the maximum eigenvector of the gravity gradient tensor points to the causative body of the gravity anomaly (Fig. 1a). They also pointed out that the minimum eigenvector of the tensor indicates the strike direction of structures such as dikes in three-dimensional analyses. Since there are two perpendicular eigenvectors of the gravity gradient tensor in the two-dimensional analyses, it is expected that the minimum eigenvector of the tensor will point to the low-density causative body or medium if the maximum eigenvector of the tensor points out high-density causative bodies such as a dike in a low-density layer such as a sedimentary layer.

To clear this inference, I calculated the gravity gradient tensor on the profile caused by the model shown in Fig. 3 and investigated the dips of the maximum and minimum eigenvectors of the tensor. The model shown in Fig. 3 has a width and height of 0.25 and 2.0 km, respectively.

Each component of the gravity gradient tensor caused by two-dimensional structures such as the dike shown in Fig. 3 is given by Telford et al. (1990). The relationship between eigenvectors and structural boundaries will be discussed widely in this study; I therefore employed calculation formulas given by Talwani et al. (1959). Talwani et al. (1959) show well-known calculation formulas giving g_x and g_z for two-dimensional arbitrary structures

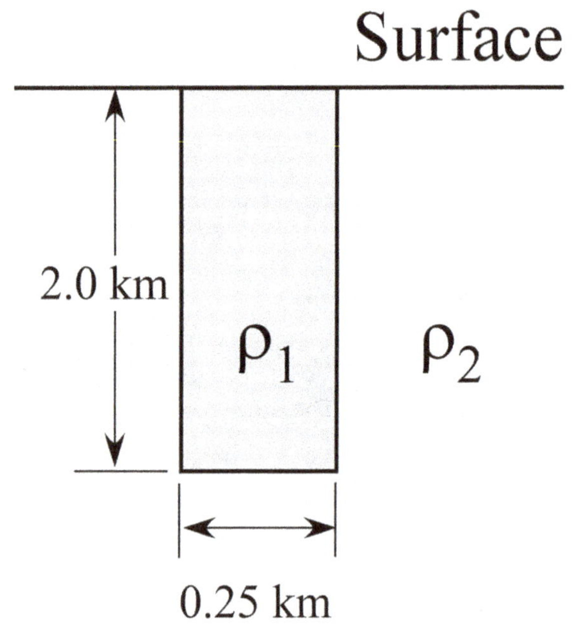

Fig. 3 Model of subsurface structures. Here, rectangular causative body of width and height of 0.25 and 2.0 km, respectively, is assumed. Densities of a medium and a causative body are ρ_2 and ρ_1, respectively

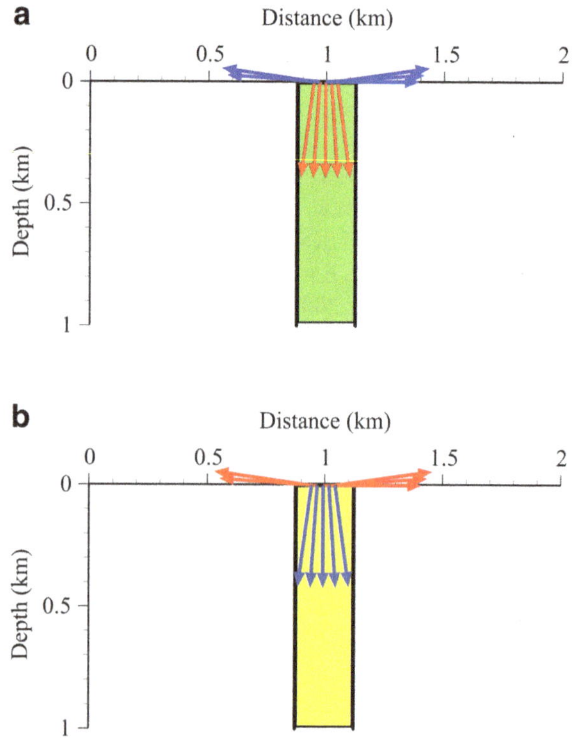

Fig. 4 Eigenvectors of the gravity gradient tensor caused by the subsurface model shown in Fig. 3. The maximum eigenvector and minimum eigenvector are shown by *red* and *blue*, respectively. **a** Eigenvectors on the high-density causative body (*light green*). When the high-density causative body was given in the low-density medium, the maximum eigenvector of the gravity gradient tensor points to the causative body and the minimum eigenvector points to the low-density medium. **b** Eigenvectors on the low-density causative body (*light yellow*). When the low-density causative body was given in the high-density medium, the minimum eigenvector of the tensor points to the causative body and the maximum eigenvector points to the high-density medium

closed by a polygon. In this study, I obtained g_{zx} ($=g_{xz}$) and g_{xx} components by the numerical differentiation of g_z and g_x, and the g_{zz} component was given by $g_{zz} = -g_{xx}$. A simple finite-difference method (e.g., Blakely 1996) was employed for these numerical differentiations. In addition, the dip of each eigenvector (α) was calculated by

$$\alpha = \arctan\left(\frac{v_z}{v_x}\right) \tag{21}$$

where v_x and v_z are x and z components of each eigenvector.

Results and discussion

Density structures and eigenvectors

Figure 4a shows distributions of the maximum (red) and minimum (blue) eigenvectors of the gravity gradient tensor caused by the model structure (Fig. 3) whose density contrast ($\Delta\rho = \rho_1 - \rho_2$) is 200 kg/m^3. Figure 4b shows distributions of the maximum (red) and minimum (blue) eigenvectors of the tensor caused by the model structure (Fig. 3) whose density contrast ($\Delta\rho$) is –200 kg/m^3. In each figure, the lengths of all the eigenvectors are the same.

From Fig. 4a, it is found that the maximum eigenvector of the gravity gradient tensor points to a high-density causative body if the body is embedded in the low-density medium. In this case, the minimum eigenvector of the tensor points to the low-density medium around the high-density body. On the other hand, the minimum eigenvector of the gravity gradient tensor points to a low-density causative body if the body is embedded in the high-density medium. In this case, the maximum eigenvector of the tensor points to the high-density medium around the low-density body. From these results, in the two-dimensional analyses, it was shown that the maximum eigenvector points to a high-density causative body and the minimum eigenvectors points to a low-density causative body.

In Fig. 4, there are vectors pointing to the area $z < 0$. This indicates that α is negative. Structures exist underground, and the negative α is not realistic. Consequently, I will add π to α if α is negative.

Fault types and eigenvectors

In calderas and/or sedimentary basins, high-density and low-density materials are in contact with each other via normal faults and/or reverse faults. In gravity anomalies and gravity gradient tensors, differences in fault type are defined as differences in density structure. As it was shown that the behavior of each eigenvector is dependent on the density structure in the previous subsection, I investigated the relationship between eigenvectors and fault type by the simplified sedimentary basin models.

Figure 5a is a simplified sedimentary basin model in which the sedimentary layer is in contact with the basement by normal faults, and Fig. 5b is a simplified sedimentary basin model in which the sedimentary layer is in contact with the basement by reverse faults. Density contrast between sedimentary layer and basement is assumed to be –200 kg/m^3.

In Fig. 6, I showed distributions of the maximum (red) and minimum (blue) eigenvectors of the gravity gradient tensor caused by these models. In each figure, the lengths of all the eigenvectors are the same, because we are interested in the fault dip and only angle information is necessary for this study.

From Fig. 6a, it is found that the dip of the maximum eigenvector of the gravity gradient tensor closely follows the dip of the normal fault. When the basement distributes near the surface, the maximum eigenvector points in the vertical direction to the high-density basement. The effect of the high-density basement is weak in the sedimentary layer area, while the effect of the low-density sedimentary layer is strong; therefore, the minimum eigenvector points in the vertical direction to the low-density sediment and the maximum eigenvector points in the horizontal direction.

When the boundary is a reverse fault, from Fig. 6b, it is found that the dip of the minimum eigenvector of the gravity gradient tensor indicates the dip of the fault well.

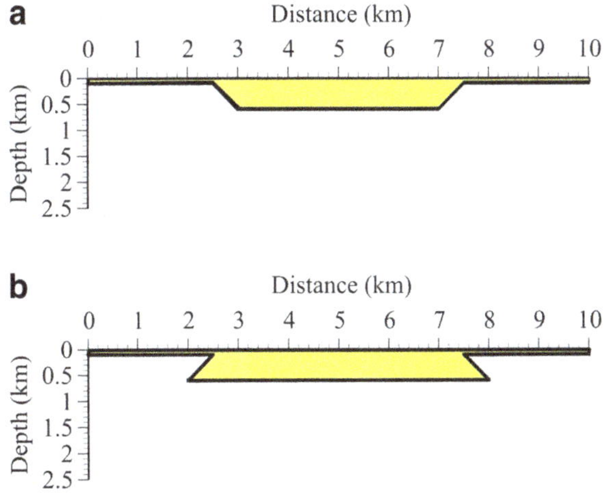

Fig. 5 Simplified sedimentary basin model. *Light yellow* and *white* areas indicate sedimentary layer and basement, respectively. **a** Sedimentary basin model where sedimentary layer is in contact with the basement by normal faults of 45° dip. **b** Sedimentary basin model where the sedimentary layer is in contact with the basement by reverse faults of 45° dip

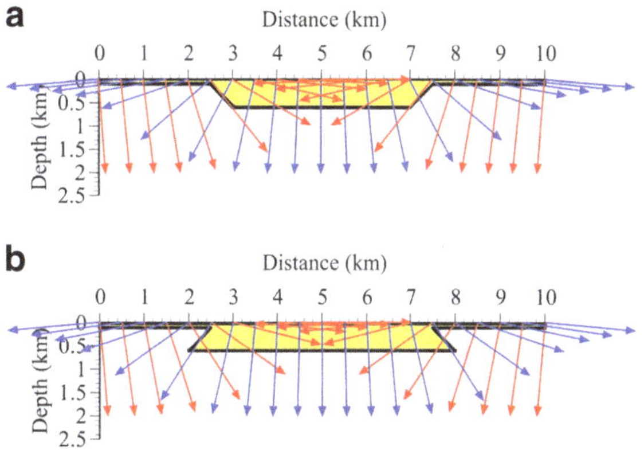

Fig. 6 Eigenvectors of the gravity gradient tensor caused by the simplified sedimentary basin models shown in Fig. 5. The maximum eigenvector and minimum eigenvector are indicated by *red* and *blue*, respectively. **a** Eigenvectors on the sedimentary basin formed by normal faults. When the sedimentary layer is in contact with the basement by normal fault, the dip of the maximum eigenvector follows the dip of the normal fault. **b** Eigenvectors on the sedimentary basin are formed by reverse faults. When the sedimentary layer is in contact with the basement by reverse fault, the dip of the minimum eigenvector follows the dip of the reverse fault

The maximum eigenvector on the basement points in the vertical direction to the high-density basement, and the minimum eigenvector points in the horizontal direction. Since the low-density sediment distributes near the surface in the sedimentary layer area, the minimum eigenvector points vertically.

From these results, it was concluded that if the structural boundary is a normal fault, its dip can be estimated from the dip of the maximum eigenvector of the gravity gradient tensor, and if the boundary is a reverse fault, its dip can be estimated from the dip of the minimum eigenvector of the tensor. In addition, in the area away from the boundary, it was found that the maximum eigenvector on the basement and the minimum eigenvector on the sediment point in the vertical direction, and the maximum eigenvector on the sediment and the minimum eigenvector on the basement point in the horizontal direction, regardless of whether the boundary is a normal fault or reverse fault.

Subsurface structures and eigenvectors

By simple numerical simulations, it was found that the maximum eigenvector of the gravity gradient tensor points to a high-density causative body and that the minimum eigenvector points to a low-density causative body. In addition, it was found that the dip of the maximum eigenvector of the tensor closely follows the dip of the normal fault and that the dip of the minimum eigenvector closely follows the dip of the reverse fault.

As mentioned above, Beiki and Pedersen (2010) have already pointed out that the maximum eigenvector of the gravity gradient tensor points to the high-density causative body. The result in Fig. 4a confirms that their results are true for the two-dimensional analyses as well. When the basement distributes near the surface, the maximum eigenvector points in the vertical direction. This property also shows that Beiki and Pedersen (2010) are correct, and the idea of the basement as an aggregate of high-density prisms (Fig. 1b), suggested by Kusumoto (2015, 2016b), would not be incorrect.

As to why the dip of normal fault was given by the dip of the maximum eigenvector of the gravity gradient tensor, I considered that the lower part of the boundary structure (fault) exists inside the low-density area more than its upper part. Therefore, because the gravity gradient tensor is most sensitive to the subsurface structures near the surface, the structure shown in Fig. 5a was considered a high-density body that intruded into the low-density layer, and the dip of the normal fault was given by the dip of the maximum eigenvector. I believe Kusumoto (2015, 2016a, 2016b) was able to obtain results that coincided with seismic surveys since he estimated the fault dip in normal fault regions by the maximum eigenvector of the tensor.

On the other hand, when the maximum eigenvector points to high-density causative bodies embedded in low-density medium or low-density causative bodies embedded in a high-density medium, the minimum eigenvector points to the low-density mediums or to the causative bodies. Beiki and Pedersen (2010) have not explicitly referred to analyses of low-density causative bodies using eigenvectors. Since it is necessary to analyze anomalies caused by low-density bodies in the field, it seems that the result, in which the minimum eigenvector points to the low-density bodies, would play an important role in subsurface structure estimation, although this is the result of two-dimensional analysis.

In addition, it was found that the dip of the minimum eigenvector of the gravity gradient tensor gave the dip of the reverse fault. As to the reason why the dip of reverse fault was given by the minimum eigenvector of the gravity gradient tensor, I considered that the lower part of the boundary structure (fault) exists inside the high-density area more than its upper part. Namely, because this structure was considered a low-density body that intruded into the high-density layer, the dip of the reverse fault was given by the dip of the minimum eigenvector of the gravity gradient tensor.

As is understood from the results and discussions obtained in this study, selecting a suitable eigenvector for estimating the fault dip is important. If the study area is not too wide and prior geological information is available, the eigenvector that should be employed for estimating the fault dip correctly would be selected based on the information. If the study area was a fault area where normal faults were mainly distributed, the maximum eigenvector of the gravity gradient tensor would be employed for estimating the fault dip. If the study area was a fault area where reverse faults were mainly distributed, the minimum eigenvector would be employed.

In the three-dimensional study for high-density causative bodies, it is pointed out that the minimum eigenvector is parallel to the strike direction of the structure (Beiki and Pedersen 2010; Beiki 2013). However, in the two-dimensional analyses, the strike direction of the structure is perpendicular to x- and z-axes and does not appear in the analyses. As it is difficult to directly compare the properties of the minimum eigenvector obtained in different dimensions, in the future, it would be necessary to discuss detailed properties of the minimum eigenvector.

Application to field data

As an application of the techniques, I estimated the dip of the Kurehayama Fault located in Toyama, Japan. The Kurehayama Fault is a reverse fault located at the center of the Toyama basin, and it strikes in the NNE-SSW direction (Fig. 7). The length of the fault is about 22 km, and the fault dip is about 45° (e.g., The Headquarters for Earthquake Research Promotion 2008; Toyama City 2013). The Toyama City has carried out seismic surveys and dense gravity surveys crossing this fault (Toyama City 2013). Toyama City (2013) set three profiles crossing the Kurehayama Fault, and the dense gravity surveys of 50 m spaced measurements have been conducted on these profiles, although spacing of several hundred meters has been usually employed for these surveys. Here, I used gravity anomaly data on the profile located at the shoreline. Figure 8 shows the Bouguer anomaly in which the Bouguer density of 2260 kg/m^3 was assumed (Toyama City 2013). The indication "Kurehayama Fault" shown in this figure indicates a rough fault location.

I applied the techniques to the Bouguer anomaly and obtained the gravity gradient tensor shown in Fig. 9. Figure 10 shows distributions of the maximum eigenvector (red) and the minimum eigenvector (blue) of the gravity gradient tensor. Since the Kurehayama Fault is a reverse fault, I focus on the dip of the minimum eigenvector. From Fig. 10, it is found that the dip (α) of the Kurehayama Fault was about 138°. Since the angle α is measured clockwise from the surface (x-axis), it seems that the obtained dip indicates the dip of the reverse fault of 42°. This fault dip is consistent with conventional data.

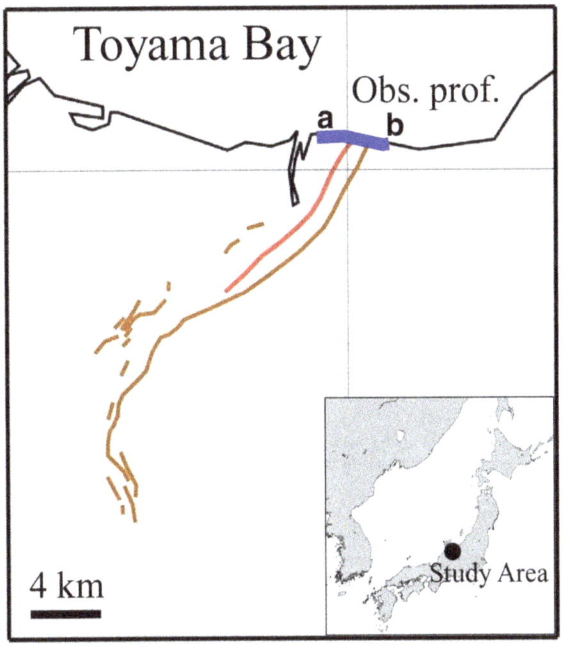

Fig. 7 Location map of the study area. Kurehayama Fault is a reverse fault located in the center of the Toyama Basin, Toyama Prefecture, Japan. Its location has been estimated by topographic, geological, and geophysical data. The *red line* and *brown lines* denote the estimated location of the Kurehayama Fault, Toyama City (Toyama City 2013), and The Headquarters for Earthquake Research Promotion (The Headquarters for Earthquake Research Promotion 2008), respectively. *Blue line a - b* indicates the dense gravity survey profile, which has gravity observation points at about 50 m intervals

The estimated fault dip would be the dip near the surface because the method employs the gravity gradient tensor, which is sensitive to subsurface structures near the surface. Since it is important to know quantitatively which depth the estimated fault dip is, in the future, it would be necessary to develop a technique

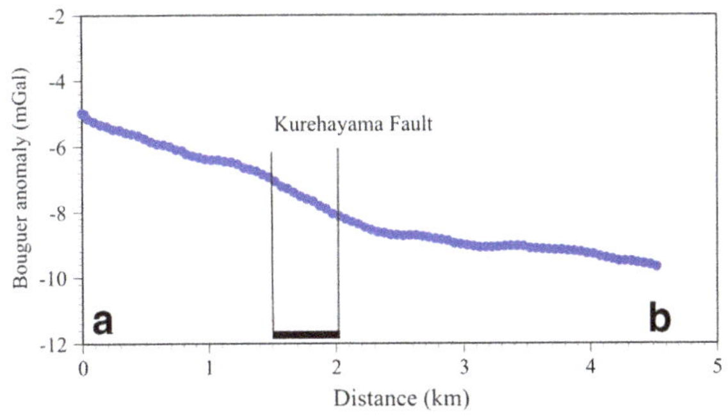

Fig. 8 Bouguer anomalies on the profiles (after Toyama City 2013). The Bouguer density of 2260 kg/m^3 is assumed. The "Kurehayama Fault" shown in this figure indicates a rough fault location by Toyama City (Toyama City 2013). The unit of the gravity anomaly is given in milligal, and mGal = 10^{-5} m/s^2

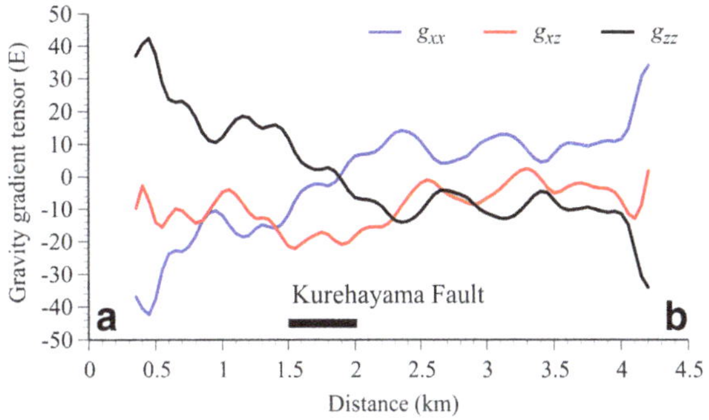

Fig. 9 Gravity gradient tensor (g_{xx}, g_{xz} (=g_{zx}), g_{zz}) on the profile. These are estimated from the Bouguer anomalies on the profile shown in Fig. 8. The component of g_{xx} and g_{xz} is calculated by a finite-difference method of gravity vectors g_x and g_z in the space domain. The "Kurehayama Fault" shown in this figure indicates a rough fault location by Toyama City (Toyama City 2013). The unit of the gravity gradient tensor is given in *E* (Eötvös), and 1 *E* = 0.1 mGal/km

estimating the depth of the estimated dip or the dip in the arbitrary depth.

Conclusions

In this study, I showed techniques for estimating the gravity gradient tensor from gravity anomalies on the profile and for estimating the fault dip by eigenvector of

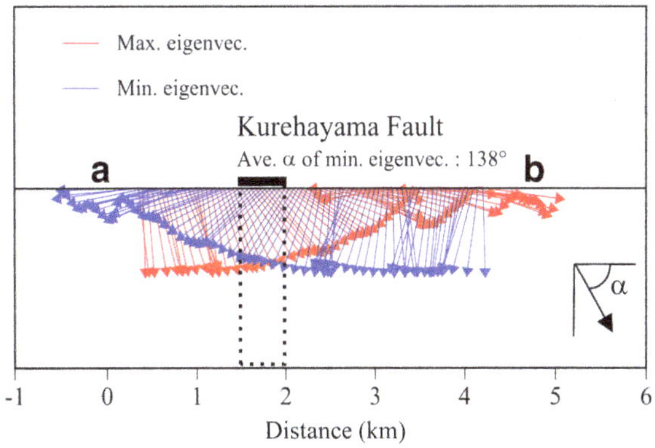

Fig. 10 Eigenvectors of the gravity gradient tensor on the profile shown in Fig. 7. The maximum eigenvector and minimum eigenvector are indicated by *red* and *blue*, respectively. The dips of eigenvectors are given clockwise from *x*-axis to *z*-axis. Since it is known that the Kurehayama Fault is a reverse fault, we focus on the minimum eigenvector of the tensor. The "Kurehayama Fault" shown in this figure indicates a tentative fault location in Toyama City (Toyama City 2013). The average dip of the minimum eigenvector in the Kurehayama Fault zone shown by a rectangle with dashed lines is about 138°, and this angle indicates that the Kurehayama Fault would be a reverse fault of 42°. In addition, the maximum eigenvectors on the *right side* of the "Kurehayama Fault" shown in this figure point to the vertical direction, and the minimum eigenvectors in the *left side* of the "Kurehayama Fault" point to the vertical direction

the observed or calculated gravity gradient tensor on the profile. I also investigated its properties by numerical simulations.

From numerical simulations, it was found that the maximum eigenvector of the tensor points to a high-density causative body and that the dip of the maximum eigenvector closely follows the dip of the normal fault. In addition, if the basement distributes near the surface, the maximum eigenvector points to the vertical direction. They have been pointed out already in previous studies, and the results shown in here confirmed that their results are true for the two-dimensional analyses as well. On the other hand, it was found that the minimum eigenvector of the tensor points to a low-density causative body and that the dip of the minimum eigenvector closely follows the dip of the reverse fault. Since eigenvector analyses of the anomalies caused by the low-density causative body have not been discussed explicitly in previous studies, these results would play an important role in estimations of subsurface structures in the future. From these results, it was found that the eigenvector of the gravity gradient tensor for estimating fault dips is determined by fault type, and we would estimate the fault dip correctly if we were to employ suitable eigenvectors based on prior information.

As an application of suggestions, I estimated the dip of the Kurehayama Fault located in Toyama, Japan, and obtained the fault dip of about 42° as the dip of the minimum eigenvector of the gravity gradient tensor because the fault is the reverse fault. This dip harmonized with conventional geological information.

Since the analysis technique shown in this study does not require complex calculations and vast calculation times, it will be an effective technique for analyzing high-resolution data obtained densely by not only dense gravity surveys for fault research but also airborne gravity or gravity gradiometry surveys.

Acknowledgements
The author is most grateful to the two anonymous reviewers for their constructive reviews and comments on the manuscript. In addition, the author is most grateful to Yuichi Hayakawa for his editorial advices and cooperation. The manuscript was improved by these reviewers' comments and suggestions. This work was supported partially by JSPS (Japan Society for the Promotion of Science) KAKENHI Grant Numbers 15K14274, 16H05651, 17K01325. The author is grateful to JSPS.

Funding
This work was supported partially by JSPS KAKENHI Grant Numbers 15K14274, 16H05651, and 17K01325.

Author's contributions
SK planned this study and conducted all the calculations and discussion. He also drafted this manuscript.

Authors' information
SK is an associate professor at the University of Toyama.

Competing interests
The author declares no competing interests.

References
Abrahamson NA, Somerville P (1996) Effects of the hanging wall and footwall on ground motions recorded during the Northridge earthquake. Bull Seism Soc Am 86:S93–S99

Beiki M (2010) Analytic signals of gravity gradient tensor and their application to estimate source location. Geophysics 75:I59–I74

Beiki M (2013) TSVD analysis of Euler deconvolution to improve estimating magnetic source parameters: an example from the Asele area, Sweden. J Appl Geophys 90:82–91

Beiki M, Pedersen LB (2010) Eigenvector analysis of gravity gradient tensor to locate geologic bodies. Geophysics 75:I37–I49

Beiki M, Pedersen LB (2011) Window constrained inversion of gravity gradient tensor data using dike and contact models. Geophysics 76:I59–I70

Blakely RJ (1996) Potential theory in gravity and magnetic applications. Cambridge University Press, Cambridge

Blakely R, Simpson RW (1986) Approximating edges of source bodies from magnetic or gravity anomalies. Geophysics 51:1494–1498

Braga MA, Endo I, Galbiatti HF, Carlos DU (2014) 3D full tensor gradiometry and Falcon systems data analysis for iron ore exploration: Bau Mine, Quadrilatero Ferrifero, Minas Gerais, Brazil. Geophysics 79:B213–B220

Cevallos C (2014) Automatic generation of 3D geophysical models using curvatures derived from airborne gravity gradient data. Geophysics 79:G49–G58

Chowdhury PR, Cevallos C (2013) Geometric shapes derived from airborne gravity gradiometry data: new tools for the explorationist. Lead Edge 32:1468–1474

Cooper GRJ (2012) The removal of unwanted edge contours from gravity datasets. Expl Geophys 44:42–47

Cooper GRJ, Cowan DR (2006) Enhancing potential field data using filters based on the local phase. Comp Geosci 32:1585–1591

Dransfield M (2010) Conforing Falcon gravity and the global gravity anomaly. Geophys Prospect 58:469–483

Elkins TA (1951) The second derivative method of gravity interpretation. Geophysics 16:29–50

Evjen HM (1936) The place of the vertical gradient in gravitational interpretations. Geophysics 1:127–137

Ferreira FJF, de Souza J, de B e S Bongiolo A, de Castro LG (2013) Enhancement of the total horizontal gradient of magnetic anomalies using the tilt angle. Geophysics 78:J33–J41

Gradshteyn IS, Ryzhik IM (2007) Table of integrals, series, and products, 7th edn. Academic press Elsevier, Oxford

Inoue N, Tanaka Y, Itoh H, Iwano S, Kitada N, Fukuda Y, Takemura K (2004) Density of sediment in Kyoto basin inferred from 2D gravity analysis along Horikawa-Oguraike and Kuzebashi seismic survey lines. Zisin 2(57):45–54 (in Japanese with English abstract)

Irikura K, Miyake H (2011) Recipe for predicting strong ground motion from crustal earthquake scenarios. Pure Appl Geophys 168:85–104. doi:10.1007/s00024-010-0150-9

Iwano S, Fukuda Y, Ishiyama T (2001) An estimation of fault related structures by means of one-dimensional gravity surveys - case studies at the Katagihara Fault and the Fumotomura Fault. J Geog 110:44–57 (in Japanese with English abstract)

Jekeli C (1988) The gravity gradiometer survey system (GGSS). EOS Trans AGU 69:105

Koenderink JJ, van Doorn AJ (1992) Surface shape and curvature scales. Im Vis Comp 10:557–564

Kusumoto S (2015) Estimation of dip angle of fault or structural boundary by eigenvectors of gravity gradient tensors. Butsuri-Tansa 68:277–287 (in Japanese with English abstract)

Kusumoto S (2016a) Structural analysis of caldera and buried caldera by semi-automatic interpretation techniques using gravity gradient tensor: a case study in central Kyushu Japan. In: Nemeth K (ed) Updates in Volcanology - From volcano modelling to volcano geology. InTech, Rijeka

Kusumoto S (2016b) Dip distribution of Oita-Kumamoto Tectonic Line located in central Kyushu, Japan, estimated by eigenvectors of gravity gradient tensor. Earth Plan Space 68:153. doi:10.1186/s40623-016-0529-7

Li X (2015) Curvature of a geometric surface and curvature of gravity and magnetic anomalies. Geophysics 80:G15–G26

Ma G (2013) Edge detection of potential field data using improved local phase filter. Expl Geophys 44:36–41

Martinez C, Li Y, Krahenbuhl R, Braga MA (2013) 3D inversion of airborne gravity gradiometry data in mineral exploration: a case study in the Quadrilatero Ferrifero, Brazil. Geophysics 78:B1–B11
Mickus KL, Hinojosa JH (2001) The complete gravity gradient tensor derived from the vertical component of gravity: a Fourier transform technique. J Appl Geophys 46:159–174
Miller HG, Singh V (1994) Potential field tilt—a new concept for location of potential field sources. J Appl Geophys 32:213–217
Robert A (2001) Curvature attributes and their application to 3D interpreted horizons. First Break 19:85–99
Sertcelik I, Kafadar O (2012) Application of edge detection to potential field data using eigenvalue analysis of structure tensor. J Appl Geophys 84:86–94
Takemura M, Moroi T, Yashiro K (1998) Characteristics of strong ground motions as deduced from spatial distributions of damages due to the destructive inland earthquakes from 1891 to 1995 in Japan. Zisin 2(50):485–505 (in Japanese with English abstract)
Talwani M, Lamar WJ, Landisman M (1959) Rapid gravity computations for two-dimensional bodies with application to the Mendocino submarine fracture zone. J Geophys Res 64:49–59
Telford WM, Geldart LP, Sheriff RE (1990) Applied geophysics. Cambridge University Press, Cambridge
The Headquarters for Earthquake Research Promotion (2008) Evaluations of Tonami Fault zone and Kurehayama Fault zone. In: The Headquarters for Earthquake Research Promotion web site. http://www.jishin.go.jp/main/chousa/katsudansou_pdf/56_tonami_kureha_2.pdf (in Japanese). Accessed 21 Nov 2016
Torge W (1989) Gravimetry. Walter de Gruyter, Berlin
Toyama City (2013) Research report on Kurehayama Fault (2). Toyama-shi, Toyama (in Japanese)
Tsuboi C, Kato M (1952) The first and second vertical derivatives of gravity. J Phys Earth 1:95–96
Zhang C, Mushayandebvu MF, Reid AB, Fairhead JD, Odegrad ME (2000) Euler deconvolution of gravity tensor gradient data. Geophysics 65:512–520
Zhang X, Yu P, Tang R, Xiang Y, Zhao C-J (2014) Edge enhancement of potential field data using an enhanced tilt angle. Expl Geophys 46:276–283. doi:10.1071/EG13104
Zhou W, Du X, Li J (2013) The limitation of curvature gravity gradient tensor for edge detection and a method for overcoming it. J Appl Geophys 98:237–242

9

Temporal change in seismic velocity associated with an offshore M_W 5.9 Off-Mie earthquake in the Nankai subduction zone from ambient noise cross-correlation

Tatsunori Ikeda[1*] and Takeshi Tsuji[1,2]

Abstract

The Nankai subduction zone off the Kii Peninsula, Japan, has a large potential to generate megathrust earthquakes in the near future. To investigate the temporal variation of stress or strain in the Nankai subduction zone, we estimated the temporal variation of seismic velocity by using cross-correlations of ambient noise in the frequency range 0.7–2.0 Hz, which was dominated by ACR waves, recorded by the DONET offshore seismic network from 1 October 2014 to 30 November 2017. The 1 April 2016 Off-Mie earthquake (M_W 5.9) and its aftershocks occurred beneath the seismic network. Our results document a clear decrease in seismic velocity at the time of the earthquake. These coseismic velocity drops were correlated with peak ground velocities at each station, suggesting that dynamic stress changes due to strong ground motions are a primary factor in coseismic velocity variations. Differences in the sensitivity of seismic velocity changes to peak ground velocity may reflect subsurface conditions at each station, such as geological structures and effective pressure conditions. We also observed a long-term increase in seismic velocities, independent of the 2016 earthquake, that may reflect tectonic strain accumulation around the Nankai subduction zone. After removing the long-term trend, we found that the coseismic velocity drops had not completely recovered by the end of the observation period, possibly indicating nonlinear effects of the 2016 earthquake. Our results suggest that ambient noise cross-correlation might be used to monitor the stress state in the Nankai accretionary prism in offshore environments, which would contribute to a better understanding of earthquake processes.

Keywords: Nankai trough, 2016 Off-Mie earthquake, Monitoring, Ambient noise, Seismic interferometry

Introduction

The Nankai subduction zone, where the Philippine Sea plate is subducting beneath the Japanese Islands at approximately 4.1–6.5 cm/year (Fig. 1; Seno et al. 1993; Miyazaki and Heki 2001), is a well-studied plate convergent margin (e.g., Tobin and Saffer 2009; Moore et al. 2009; Bangs et al. 2009; Park et al. 2010). Damaging great earthquakes in excess of M_W 8 have occurred here every ~ 200 years (Ando 1975) and pose a severe threat to the large cities in this part of Japan. In light of the high seismic risk, geophysical and drilling data have been intensively acquired in the Nankai accretionary prism. Because high pore pressure near a fault acts to reduce the effective stress (which presumably lowers the fault's strength), several studies have used seismic velocity to estimate the pore pressure around seismogenic faults (e.g., Tsuji et al. 2008, 2014; Tobin and Saffer 2009). Shear stress, which is also important to evaluate fault stability, has been evaluated using borehole breakouts (Lin et al. 2010) as well as seismic anisotropy (Tsuji et al. 2011). These studies have found that the direction of maximum horizontal stress changes from perpendicular to the trench seaward of the outer ridge (Fig. 1), to parallel to the trench landward of the outer ridge, to nearly parallel to the direction of plate convergence still further landward (Lin et al. 2010; Chang et al. 2010). These variations in stress orientations have been explained in terms of variations in static stress during the earthquake cycle (Wang and Hu 2006). Therefore,

* Correspondence: ikeda@i2cner.kyushu-u.ac.jp
[1]International Institute for Carbon-Neutral Energy Research (WPI-I2CNER), Kyushu University, Fukuoka 819-0395, Japan
Full list of author information is available at the end of the article

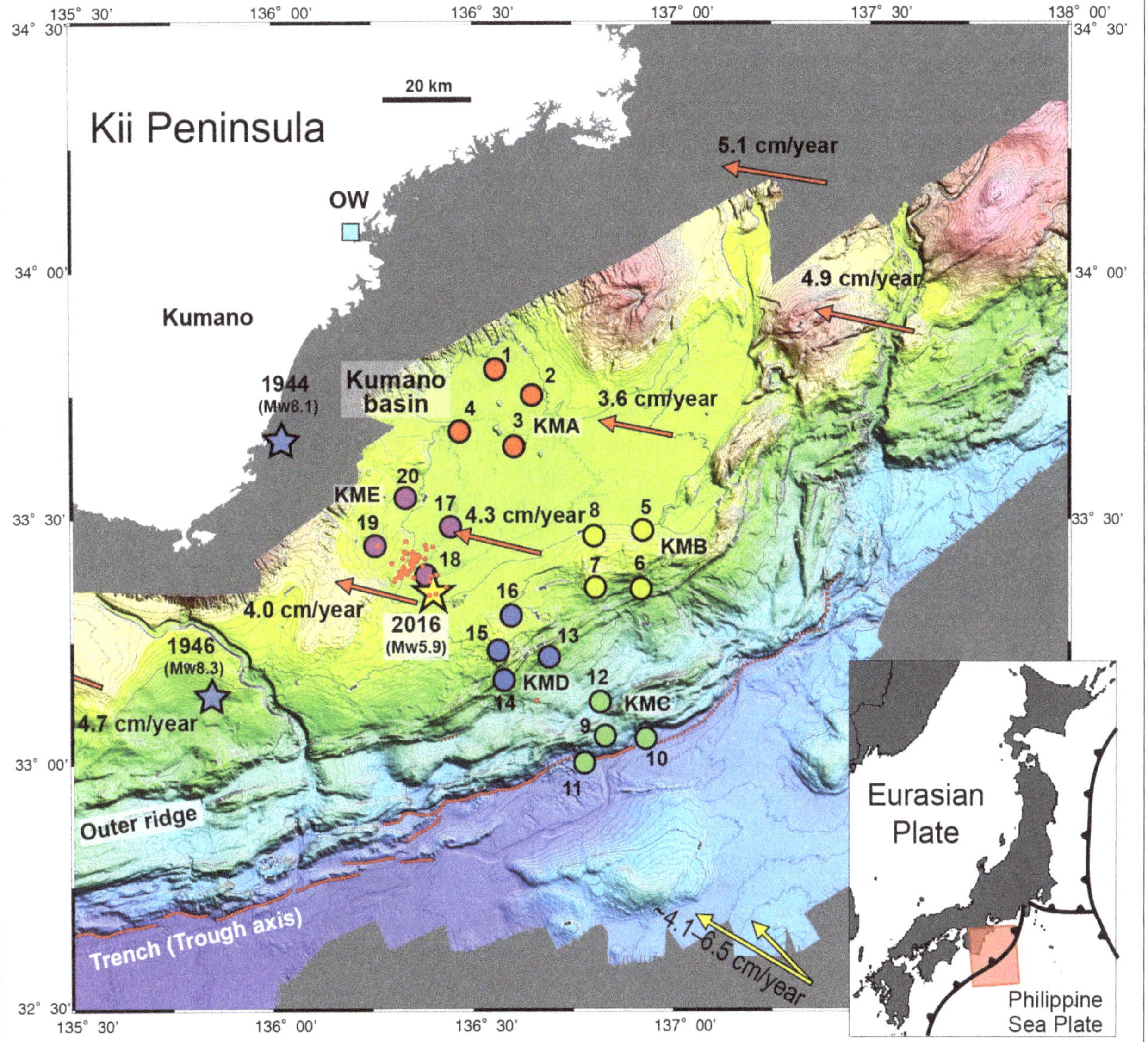

Fig. 1 Tectonic, bathymetric, and geologic features of the Nankai trough region (after Tsuji et al. 2017). Circles represent DONET stations with different colors for each node (KMA, KMB, KMC, KMD, and KME), and the square represents the OW tidal gauge station. The yellow star is the epicenter of the 2016 Off-Mie earthquake, and the red dots represent epicenters of foreshocks and aftershocks from 3 h before to 48 h after the mainshock (Wallace et al. 2016). Blue stars represent the epicenters of the 1944 Tonankai and the 1946 Nankaido earthquakes. Red arrows represent seafloor velocity vectors from Yokota et al. (2016)

determinations of the stress orientation and magnitude should provide useful information for evaluating and monitoring seismogenic faults (e.g., Crampin et al. 2008). As part of that effort, we need to monitor the temporal variation of stress or strain within the accretionary prism.

To monitor seismic activity near the Nankai subduction zone, the Japan Agency for Marine-Earth Science and Technology (JAMSTEC) deployed a network of real-time seafloor observatories off the Kii Peninsula in southwestern Japan called the Dense Oceanfloor Network System for Earthquakes and Tsunamis (DONET) (Kaneda et al. 2015; Kawaguchi et al. 2015). DONET has been used effectively to identify and monitor seismic activity including slow earthquakes such as low-frequency tremors, impulsive low-frequency earthquakes, very low frequency events, and slow slip events, which are important for understanding slip behavior along plate boundary faults (e.g., Nakano et al. 2013, 2014, 2018; Suzuki et al. 2016; Araki et al. 2017; Toh et al. 2018).

On 1 April 2016, the Off-Mie earthquake (M_W 5.9) occurred ~ 50 km off the Kii Peninsula at depth of 11.4 km directly beneath DONET (Wallance et al. 2016; Fig. 1). It was classified as a plate interface event from the analysis of ocean bottom seismometer data, seafloor and subseafloor geodetic data from DONET, and tsunami modeling (Wallace et al. 2016). This earthquake appears to have ruptured the same plate boundary fault responsible for great interplate earthquakes such as the 1944 Tonankai earthquake (M_W 8.1), although its rupture area was much smaller than that of the 1944 earthquake. Tsuji et al. (2017) showed that the fault planes of the 2016 Off-Mie earthquake and its aftershocks were influenced by the geometry of the plate boundary décollement and the older landward part of the accretionary prism along the coast of the Kii Peninsula. The aftershocks of the 2016 event occurred where the décollement soles onto the top of the oceanic crust beneath the old prism.

Seismic interferometry using ambient noise has been widely used to estimate seismic velocity structures (e.g., Shapiro et al. 2005; Lin et al. 2009; Nakata et al. 2015) and to monitor temporal changes in seismic velocity associated with earthquakes, volcanic activity, and environmental influences (e.g., Brenguier et al. 2008a, 2008b; Obermann et al. 2013, 2014; Nimiya et al. 2017; Wang et al. 2017). For the case of large earthquakes, coseismic and postseismic velocity variations detected by cross-correlation of ambient noise may be related to changes in stress or strain in the subsurface (e.g., Brenguier et al. 2008a; Taira et al. 2015; Hobiger et al. 2016; Nimiya et al. 2017). Tonegawa et al. (2015) used ambient noise cross-correlation to extract virtual seismograms propagating between pairs of stations in the Nankai subduction zone. On the basis of these cross-correlation functions, they inferred the presence of acoustic-coupled Rayleigh (ACR) waves traveling in the ocean and marine sediments, excited by small earthquakes in the Nankai subduction zone.

To investigate temporal changes in stress or strain in the Nankai subduction zone, particularly those associated with the 2016 Off-Mie earthquake, we estimated temporal changes in seismic velocity from cross-correlations of ambient noise. Our results demonstrated that seismic velocity decreased during the 2016 earthquake and increased during the postseismic period. We also observed long-term and short-term variations in seismic velocity that were independent of the 2016 earthquake. In this paper, we discuss possible mechanisms of these seismic velocity variations and implications of our results for understanding future earthquakes.

Data

We used the vertical component of continuous ambient seismic noise recorded by a subset of 20 DONET broadband seismometers (DONET1 hereafter), grouped in five nodes of four seismometers (Fig. 1), from 1 October 2014 to 30 November 2017. Seismic data in the National Research Institute for Earth Science and Disaster Resilience (NIED) data server were not available from node KMD for the period 12 to 31 March 2016, and from node KME for the period 1 October 2014 to 24 March 2015 and after 1 June 2016.

Methods

To extract virtual seismograms propagating between pairs of stations, we computed cross-correlations between two seismic records. We first divided each day of ambient noise data into 30-min segments offset by 15 min (i.e., with 50% overlap). We then applied a band-pass filter in the frequency range 0.7–2.0 Hz, which was dominated by ACR waves in the study area (Tonegawa et al. 2015). To reduce the effects of earthquakes, instrumental irregularities, and non-stationary noise sources on the cross-correlations, we removed segments with root-mean-square (RMS) amplitudes greater than three times the median RMS amplitude of the station or greater than 1.5 times that of adjacent segments. The cross-correlations between stations within each node were computed by power-normalized cross-correlation (cross-coherence) in the frequency domain (e.g., Nakata et al. 2011, 2015).

To estimate temporal changes in seismic velocity from the cross-correlations, we used the stretching interpolation technique (e.g., Sens-Schönfelder and Wegler 2006; Minato et al. 2012; Nimiya et al. 2017). The method elongates the time axis and finds the trace most similar to the reference trace:

$$f_\varepsilon^{\mathrm{cur}}(t) = f^{\mathrm{cur}}(t(1+\varepsilon)), \tag{1}$$

$$CC(\varepsilon) = \frac{\int f_\varepsilon^{\mathrm{cur}}(t) f^{\mathrm{ref}}(t)dt}{\left(\int \left(f_\varepsilon^{\mathrm{cur}}(t)\right)^2 dt \int \left(f^{\mathrm{ref}}(t)\right)^2 dt\right)^{\frac{1}{2}}}, \tag{2}$$

where ε is a stretching parameter, f^{ref} is the reference trace, f^{cur} is the current trace, t is time, and $CC(\varepsilon)$ is the correlation coefficient between the reference and current traces. We applied a grid search algorithm to find the value of ε that maximizes $CC(\varepsilon)$. The parameter ε corresponds to a relative time shift ($\frac{\Delta t}{t}$) and relates to a velocity change ($\frac{\Delta v}{v}$) as follows:

$$\varepsilon = \frac{\Delta t}{t} = -\frac{\Delta v}{v}. \tag{3}$$

The stretching interpolation technique was applied to the coda of cross-correlations, because the coda is more sensitive to velocity changes than the direct wave (e.g., Meier et al. 2010) and less sensitive to variations in noise sources (e.g., Colombi et al. 2014; Chaves and Schwartz 2016). The reference trace was obtained by stacking all available cross-correlations, and the current trace was defined as a 30-day stack of cross-correlations to stabilize the monitoring results. We selected a window of 30 s for the coda starting 10 s after the arrival of waves with 0.9 km/s apparent velocity. The measured velocity change was considered to represent the velocity change in the middle of the 30-day window. By moving the 30-day window used for the current trace, we estimated the daily variation of the velocity change between pairs of stations in each node. To emphasize seismic velocity changes associated with the 2016 Off-Mie earthquake, we subtracted average values of ε for a 60-day period (18 January to 17 March 2016) before the earthquake (ε_0) from the estimated values of ε as follows:

$$\varepsilon' = \varepsilon - \varepsilon_0. \tag{4}$$

The relative velocity changes $\frac{dv}{v} = -\varepsilon'$ between two stations were estimated by applying the stretching interpolation technique separately for the causal and acausal parts of the cross-correlation codas. To evaluate the standard deviation of $\frac{dv}{v}$, we used the theoretical formula proposed by Weaver et al. (2011) as follows:

$$\sigma_d = \frac{\sqrt{1 - CC_{\max}^2}}{2CC_{\max}} \sqrt{\frac{6R}{\omega_c^2 (t_2^3 - t_1^3)}}, \tag{5}$$

where t_1 and t_2 are the beginning and end of the time window for the coda, ω_c is the central angular frequency, and $CC_{\max}$ is the maximum correlation coefficient corresponding to the estimated velocity change using Eq. (2). R is the parameter related to the frequency bandwidth (Weaver et al. 2011), and it was calculated by integrating square of auto-correlation of the coda corresponding to the reference trace.

To detect the trend of seismic velocity change at each node, we averaged velocity variations obtained from both causal and acausal parts using all possible station pairs within each node. In averaging, we used the inverse of the square of the standard deviation using Eq. (5) as a weighting factor and calculated the errors by error propagation.

Furthermore, to determine the spatial variation of seismic velocity changes, we applied a simple tomography algorithm in which, assuming that the seismic velocity change between a pair of stations represents the average of the actual velocity changes at each station, we obtained seismic velocity changes at each station by seeking the least-squares solution (Hobiger et al. 2012). We defined velocity changes between a pair of stations from estimates of the both causal and acausal parts of the codas weighted by the inverse of the square of the standard deviation using Eq. (5), which was also used as the weighting factor in the tomography.

Results

Figure 2 shows reference cross-correlations obtained by stacking all available segments in the analyzed period after removing segments based on RMS amplitudes. In the cross-correlations in the northern part of the study area (nodes KMA and KME; Fig. 2a, e), we observed clear seismic waves with group velocities of ~ 1.5 km/s. The signals were stronger when they propagated northward (e.g., from KMA-2 to KMA-1, KMA-3 to KMA-1, and KME-17 to KME-20). Similar features were observed by Tonegawa et al. (2015), who concluded that they were ACR waves traveling in the ocean and marine sediments, persistently excited by small earthquakes near

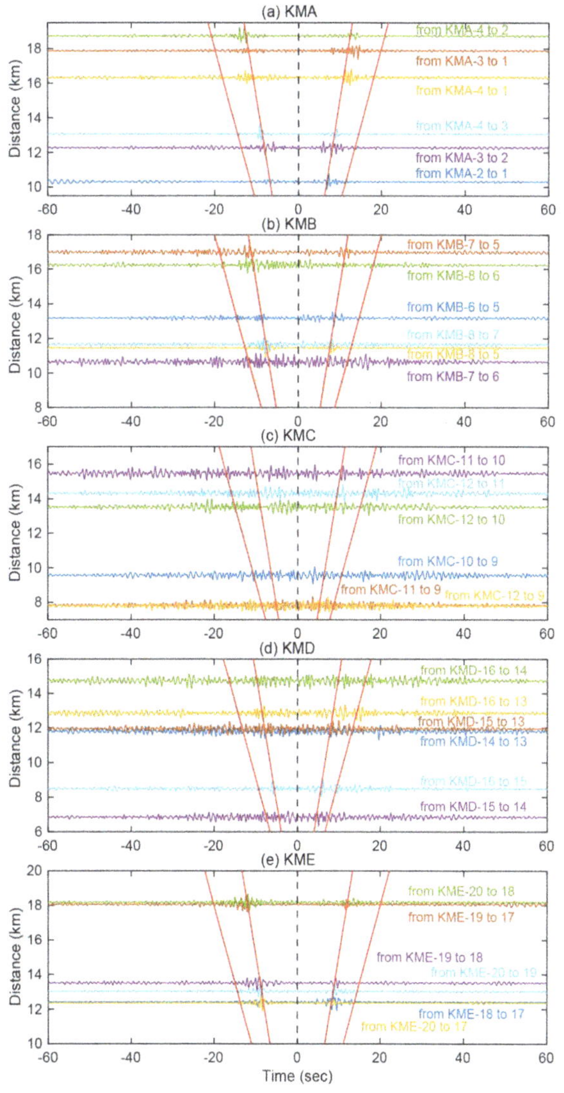

Fig. 2 Reference cross-correlations between station pairs within the five DONET1 nodes. Each node is labeled with the station pair and the propagation direction in positive time. The red lines represent velocities of 0.9 km/s (outer pair) and 1.5 km/s (inner pair). **a** KMA. **b** KMB. **c** KMC. **d** KMD. **e** KME

the Nankai subduction zone. These direct ACR waves were weak in the cross-correlations in the southern part of the study area (nodes KMB, KMC, and KMD; Fig. 2b–d).

By applying the stretching interpolation technique for each set of cross-correlation data, we estimated seismic velocity changes between all possible pairs of stations within each node (Additional file 1). Figure 3 shows examples of temporal variation of cross-correlation data and seismic velocity. We then obtained the average seismic velocity variations at each node (Fig. 4) and velocity variation at each station by using the tomography approach (Fig. 5). These velocity changes clearly decreased at the time of the 2016 Off-Mie earthquake by ~ 0.05 to 0.35% in the averaged velocity changes at each node (Fig. 4). The largest velocity decrease occurred in node KME, the closest node to the epicentral region (Fig. 6; Additional file 2). We observed gradual velocity increases at most of the stations after the earthquake. Before the earthquake, we also observed a long-term variation (gradual increase) and a short-term (several months) fluctuation in the velocity changes (e.g., Fig. 4) that were apparently independent of the 2016 earthquake.

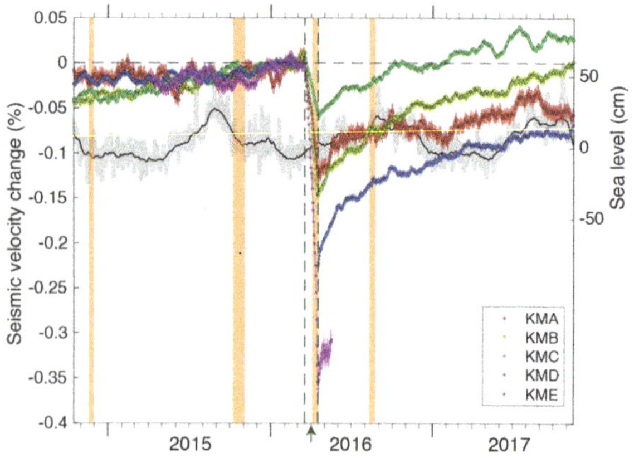

Fig. 4 Temporal variation of the average seismic velocity in each node (colored dots). Error bars correspond to the errors calculated from the standard deviations of seismic velocity change using Eq. (5) by error propagation. The arrow at the bottom indicates the day of the 2016 Off-Mie earthquake, and the vertical dashed lines show the 30-day time window influenced by the mainshock. Gray and black curves represent sea levels at station OW computed by daily and 30-day stacking of hourly data, respectively. Orange areas represent the periods of slow slip events reported by Araki et al. (2017)

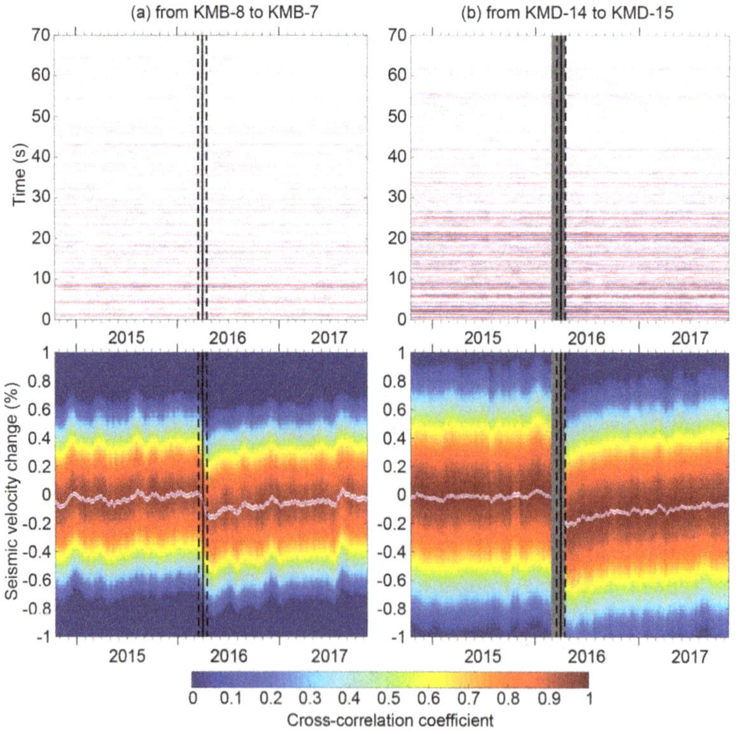

Fig. 3 Examples of temporal variations of cross-correlations and seismic velocities for station pairs. **a** KMB-7 and KMB-8. **b** KMD-14 and KMD-15. Upper plots are daily cross-correlations constructed from the 30-day stacked moving average. Lower plots are seismic velocity variations between two stations obtained from the cross-correlations in the upper panel (thin central magenta line, with standard deviation calculated using Eq. (5), shown in white) with respect to changes before the 2016 Off-Mie earthquake. The background colors indicate the cross-correlation coefficient. Vertical solid lines show the day of the 2016 Off-Mie earthquake and the dashed lines show the 30-day time window influenced by the mainshock. The gray shaded area in **b** represents the set of 30-day periods affected by missing data from node KMD from 12 to 31 March 2016

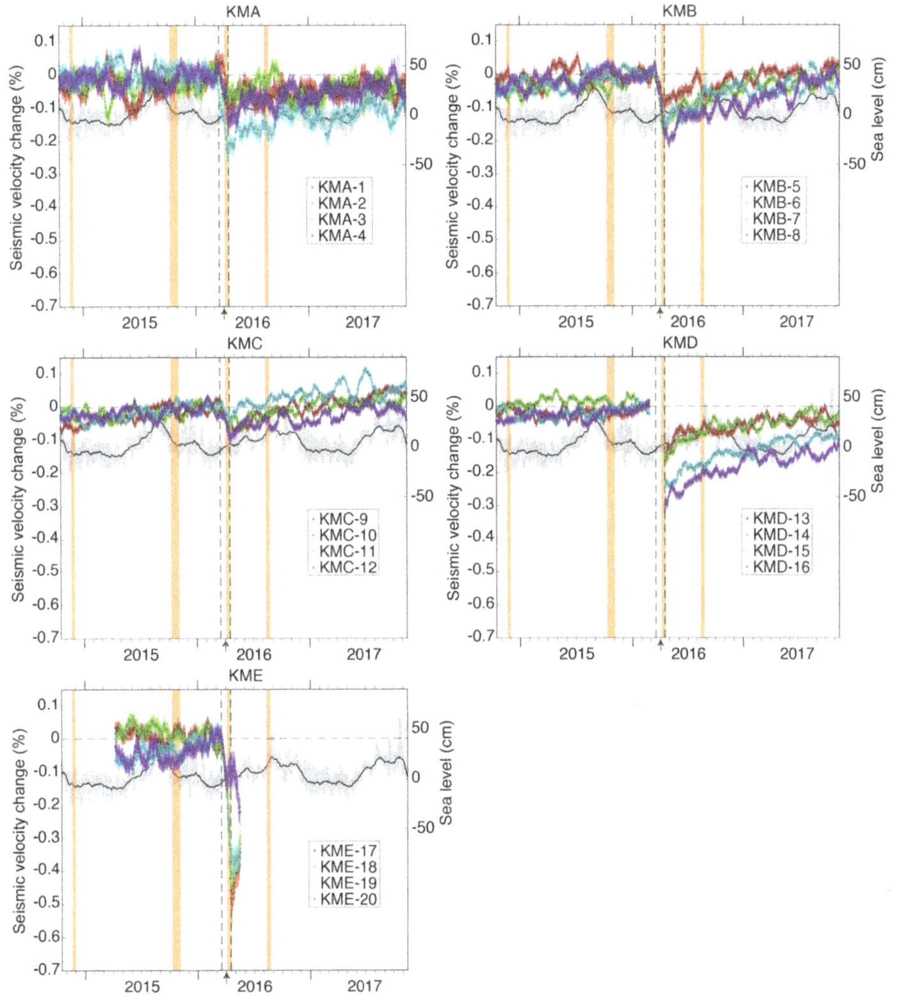

Fig. 5 Temporal variation of seismic velocity at each station (colored dots). Symbols are the same as in Fig. 4

Discussion

Temporal stability and depth sensitivity of coda

In the cross-correlations between pairs of stations, direct ACR waves were not clearly detected in the southern part of the study area (nodes KMB, KMC, and KMD; Fig. 2b–d). This is probably because of the proximity of those stations to the source region of the persistent seismic noises as some of the stations are located in the source region. Nevertheless, we obtained larger values of correlation coefficients through the stretching interpolation in the southern part, indicating that our record of the coda was robust, particularly in the southern part (Additional file 1). The scattering of ACR waves due to the complicated bathymetry and sediment structure in the southern part of study area in Tonegawa et al. (2015) might enhance the directivity of ACR waves and be responsible for the high temporal stability of the cross-correlations.

According to the normal mode calculation of the eigenfunctions of stress (τ_{zz}) by Tonegawa et al. (2015), the stress component (τ_{zz}) of ACR waves propagating in the study area has amplitude (sensitivity) within ocean and marine sediments at 1 Hz. If we assume that the codas of cross-correlations we analyzed were dominated by ACR waves, the estimated velocity changes would be sensitive to a depth of several kilometers below the seafloor as well as in the water layer, although it is difficult to clarify the depth sensitivity because the predicted mode transition is complicated in the studied frequency range (Tonegawa et al. 2015).

Possible mechanism of coseismic velocity change

The opening and closing of cracks due to stress changes are often considered as a mechanism of seismic velocity changes associated with earthquakes. Previous studies have demonstrated that coseismic velocity decreases are correlated with dynamic rather than static changes in stress or strain (e.g., Rubinstein and Beroza 2004; Hobiger et al. 2012, 2016; Brenguier et al. 2014; Taira and

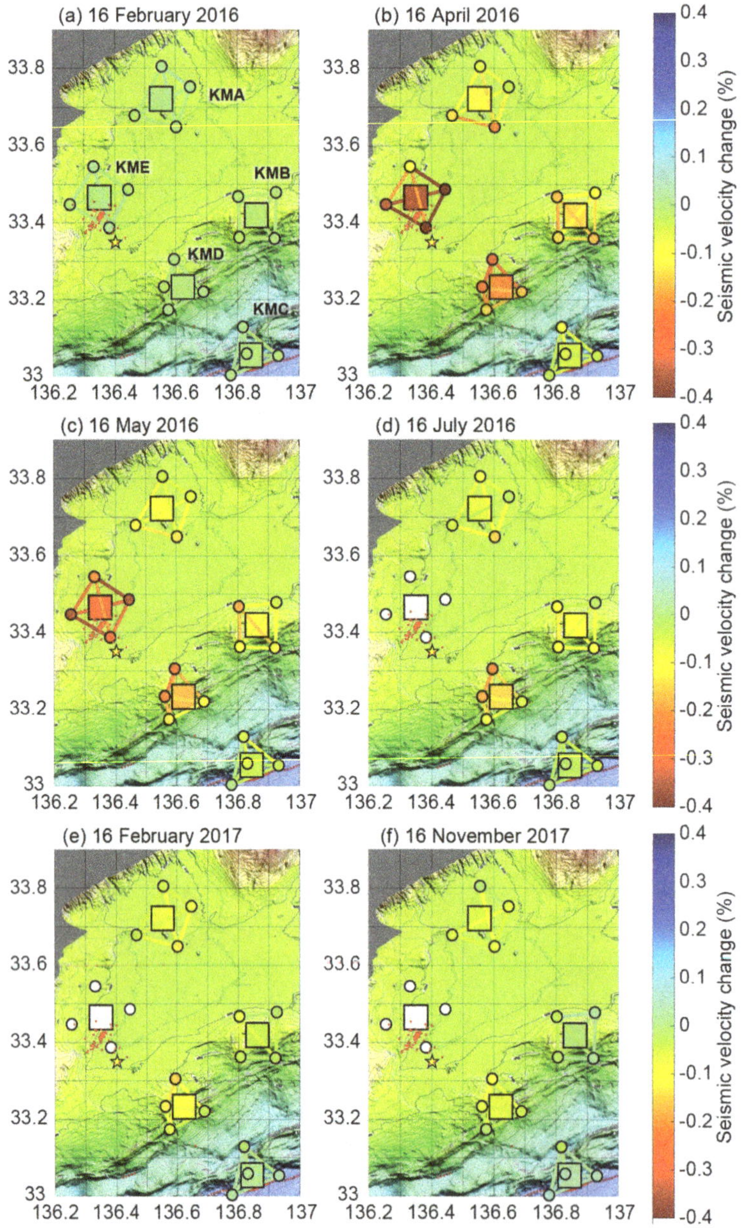

Fig. 6 Snapshots of spatial and temporal seismic velocity variations. Plots show 30-day windows centered on **a** 16 February 2016, **b** 16 April 2016 (including the day of the mainshock), **c** 16 May 2016, **d** 16 July 2016, **e** 16 February 2017, and **f** 16 November 2017. Squares represent the locations of nodes, and their colors represent averaged velocity changes at each node. Circles represent the locations of DONET stations, and their colors represent velocity changes at each station obtained by the tomography approach. Colored lines represent velocity changes between each station pair. The star marks the epicenter of the 2016 Off-Mie earthquake, and red dots are the epicenters of earthquakes in the period from 3 h before to 48 h after the mainshock (Wallace et al. 2016). Additional file 2 presents the complete time series as an animation

Brenguier 2016) and can be explained by the opening and growth of cracks in response to strong ground shaking (e.g., Rubinstein and Beroza 2004; Wu et al. 2009). We therefore compared our estimated coseismic velocity changes at each station with strong ground motion, using peak ground velocity (PGV) as an indicator (Fig. 7). We defined PGV as larger values of PGV of two horizontal components recorded at each station, after bandpass filtering from 0.1 to 10 Hz, by using strong motion data record by DONET1. Seismic velocity changes 15 days after the 2016 earthquake (including the day of the mainshock) at each station were defined as coseismic velocity changes. We observed that the coseismic velocity drops tended to increase with increasing PGV (Fig. 7), which suggests that coseismic velocity changes were influenced by dynamic stress changes due to strong ground shaking.

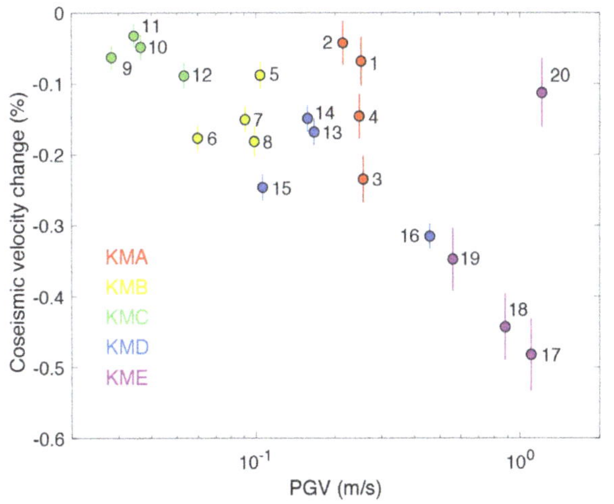

Fig. 7 Comparison between coseismic velocity changes and PGV at each station, denoted by its number. The colors of the station symbols indicate their nodes

The coseismic velocity drops at stations near land (KMA-1, KMA-2, and KME-20) were less sensitive to PGVs than those at other stations. DONET1 stations near land (in nodes KMA and KME) are distributed on the thick sediment deposits of the Kumano basin, whereas other nodes are distributed at the seaward edge of the thin sediments of the Kumano basin or on the accretionary wedge (e.g., Tsuji et al. 2014; Kubo et al. 2018). Such spatial variations in geological structures would reflect scattering in the relationship between coseismic velocity changes and PGVs (Fig. 7). Furthermore, the coseismic velocity changes could be influenced by differences in effective pressure conditions below each station because the sensitivity of seismic velocity to stress change increases as effective pressure decreases (e.g., Toksöz et al. 1976; Zinszner et al. 1997; Shapiro 2003; Brenguier et al. 2014). The high sensitivity of seismic velocity changes to ground shaking might reflect high pore pressures, because fractures are developed around the trough axis (e.g., Kamei et al. 2012; Tsuji et al. 2014, 2015). Because it is difficult to relate our estimated velocity changes to a particular depth, however, comparison of our results with the seismic structure is of limited value.

Curve fitting for observed velocity change

We observed the velocity recovery during the postseismic period as a gradual increase in seismic velocity, except at station KME-20 (Figs. 4 and 5). However, at node KMC (e.g., green line in Fig. 4), the velocity came to exceed its pre-earthquake value approximately 8 months after the earthquake and continued to increase until the end of the study period. Indeed, the seismic velocity variations exhibited a long-term increase before the 2016 earthquake.

Based on the approach by Hobiger et al. (2014, 2016) and Taira et al. (2018), to evaluate the long-term increase and the postseismic velocity change, we fitted the averaged seismic velocity changes at each node $f(T)$ to the following curves.

$$f(T) = A + BT + \left[C + D\exp\left(-\frac{T - T_{eq}}{E}\right)\right] H(T - T_{eq}), \quad (6)$$

where T is the day, A, B, C, D, and E are the five fit parameters, $H(T)$ is the Heaviside function, and T_{eq} is the time of occurrence of the Off-Mie earthquake. A is a constant offset, and B is a linear trend of velocity change. C is a non-recovering coseismic velocity change during the observation time, and D is a coseismic velocity change recovering on an exponential scale with time constant E. Linear regressions were applied for the velocity variations before the Off-Mie earthquake to obtain the parameters A and B (Table 1). The parameters C, D, and E were then obtained from the velocity changes after the earthquake through non-linear regressions (Table 1). We did not apply the curve fitting for velocity changes for nodes KMA and KME because the linear trend and the exponential decay in Eq. (6) were difficult to estimate due to large fluctuations, and the data after the earthquake were insufficient for node KME. In both regressions, we used the inverse of the square of the standard deviation as the weight factor and calculated the errors by error propagation.

Long-term increase in seismic velocity

Linear regressions for the average velocity variations at each node yielded velocity increases of 0.013–0.033%/year (Table 1). The 2011 M_W 9.0 Tohoku-Oki earthquake may be partly responsible for the long-term increase in seismic velocity, although the survey area is located ~ 800 km from the epicenter. Coseismic velocity drops and postseismic velocity recovery associated with the 2011 Tohoku-Oki earthquake were widely observed in Japan from cross-correlation of ambient noise using land stations (Hobiger et al. 2016; Wang et al. 2017). If the long-term velocity increase is attributed to postseismic recovery of the Tohoku-Oki earthquake assuming the exponential recovery model, however, the coseismic velocity drops would be much larger than those caused by the Off-Mie earthquake, which we do not expect. We therefore suggest additional factors are required to explain the long-term velocity increase.

Another possible explanation for the long-term increase in seismic velocity is the influence of subduction of the Philippine Sea plate beneath the Eurasian plate (Fig. 1). At land stations in the Tokai region of Japan, a long-term increase in S-wave velocity and a change in

Table 1 Parameters of the best-fitting curves to observed seismic velocity changes. Initial values used in non-linear regression are also listed

Node	B (%/year)	C (%)	D (%)	E (year)
KMB	0.0315 ± 0.0005	− 0.0623 ± 0.0004	− 0.0946 ± 0.0012	0.3249 ± 0.0079
KMC	0.0329 ± 0.0004	− 0.0243 ± 0.0003	− 0.0437 ± 0.0011	0.2770 ± 0.0114
KMD	0.0132 ± 0.0004	− 0.0916 ± 0.0005	− 0.1102 ± 0.0008	0.4559 ± 0.0081
Initial value		− 0.0500	− 0.0500	0.5000

anisotropy, documented by Tsuji et al. (2016) using a continuous controlled seismic source system (e.g., Yamaoka et al. 2008) is consistent with the tectonic strain field of NW-SE compression estimated geodetically by the GNSS Earth Observation Network System. Geodetic observations show that the seafloor around DONET1 is moving 3.6–4.3 cm/year toward the WNW (Fig. 1; Yokota et al. 2016). If this movement continuously generates strain, the increases in seismic velocity that we observed might be used as an indicator of strain accumulation in the Nankai accretionary prism.

The different characteristics of long-term velocity increase at each node could be influenced by local geological structures and stress state. The previous studies using borehole breakouts and seismic anisotropy revealed that the direction of maximum horizontal stress is almost parallel to the direction of plate convergence around DONET1 (Lin et al. 2010; Tsuji et al. 2011). However, the direction of maximum horizontal stress around node KMD is locally perpendicular to the direction of the plate convergence probably reflecting the development of the trough-parallel strike-slip faults in the outer ridge (e.g., Tsuji et al. 2014). The smaller value of the velocity increase rate at node KMD may reflect such spatial variation of stress state and fracture intensity. The long-term velocity increase observed at node KMC directly reflects strain accumulation within the accretionary prism, because node KMC is located on the toe of the accretionary prism where strain accumulation is dominant. On the other hand, nodes KMA and KME are located on the forearc basin above the accretionary prism, thus the long-term velocity increase could be slower, although we had difficulty in estimating a linear trend of velocity increase for nodes KMA and KME due to large fluctuation.

Postseismic velocity recovery

When we removed the long-term increasing trend from the estimated velocity changes by subtracting $A + BT$ from the data, it appeared that the recovery process was not complete at the end of the study period (Fig. 8). The exponential model of Eq. (6) allows us to separate the non-recovering from the recovering coseismic velocity variations (Hobiger et al. 2014, 2016). When we calculated the ratio between the non-recovering coseismic velocity changes C and the total coseismic velocity changes C + D, the non-recovering rates ranged from ~ 35 to 45%, indicating nonlinear effects of the main shock. We note that the non-recovering velocity change means that the velocity changes are not significantly recovering over the observation time, but they are supposed to recover over much longer timescales (Hobiger et al. 2014). The recovery time constants ranged from 0.28 to 0.46 year (Table 1). The non-recovering velocity ratios and the recovery time constants are roughly consistent with the results associated with large earthquakes at land stations in Japan (Hobiger et al. 2014, 2016). However, a detailed comparison with the results using land stations is not straightforward because ACR waves probably dominant in the coda we analyzed would have different frequency-dependent depth sensitivity, compared to those derived from coda of cross-correlation using land stations.

Comparison with sea levels and slow slip events

In addition to the seismic velocity variations mentioned above, we observed a short-term (several months) fluctuation, particularly in DONET1 stations near land (in nodes KMA and KME). The patterns are complicated because they are not correlated with other stations even in the same nodes. Seismic velocity changes measured at land stations

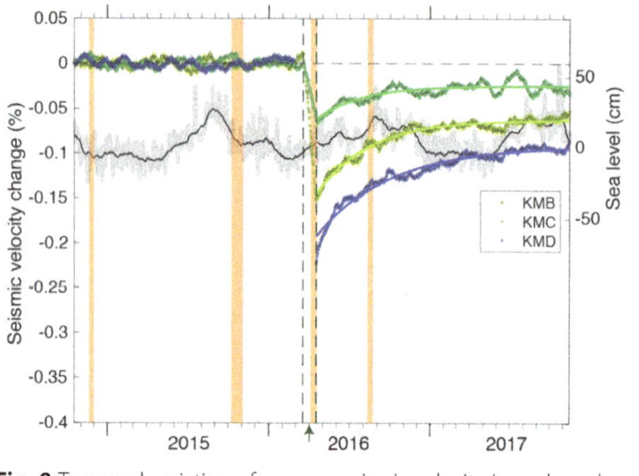

Fig. 8 Temporal variation of average seismic velocity in each node (colored dots) after removing long-term trend. Symbols are the same as in Fig. 4. The solid lines represent the exponential models of Eq. (6) obtained by curve fitting

near the Sea of Japan are strongly correlated with changes in sea level without any remarkable phase shift, probably reflecting an instantaneous elastic response of the lithosphere to ocean loading (Wang et al. 2017). However, the correlation between seismic velocity changes in the DONET1 area and sea level is not clear (Figs. 4, 5, and 8; Additional file 1). Another possible influence on the short-term variation is slow slip events. In their study of slow slip events observed using subseafloor borehole observatories installed in the study area, Araki et al. (2017) reported that eight events accommodated ~ 30–55% (~ 11–22 cm) of the total plate convergence budget (~ 39 cm) from 2011 to 2016. In our results, the estimated seismic velocity changes do not show clear temporal change at the periods of the slow slip events reported by Araki et al. (2017) (orange in Figs. 4, 5, and 8; Additional file 1). We infer that the short-term variation was largely influenced by other factors, such as variations in the sources of ambient noise.

Conclusions

In this study, we estimated temporal changes in the velocity of ACR waves in the Nankai subduction zone by using cross-correlations of ambient noise. Our results showed that drops in seismic velocity were associated with the 2016 Off-Mie earthquake and were followed by postseismic recovery. The correlation between the coseismic velocity drops and PGVs from the earthquake suggests that the coseismic velocity changes were mainly caused by dynamic stress changes due to strong ground shaking. We also observed a long-term increase in seismic velocity. After removing this long-term trend, our record of velocity changes showed that ~ 35 to 45% of the coseismic velocity drop had not recovered by the end of the study period, suggesting that the mainshock had non-linear effects on velocities. If the long-term increasing trend is related to strain accumulation from plate subduction, our approach may be useful in monitoring variations of the regional stress state, which is difficult to observe in offshore environments, and may provide vital information for understanding future earthquakes.

Abbreviations

ACR waves: Acoustic-coupled Rayleigh waves; DONET: Dense Oceanfloor Network System for Earthquakes and Tsunamis; JAMSTEC: Japan Agency for Marine-Earth Science and Technology; PGV: Peak ground velocity

Acknowledgements

We thank two anonymous reviewers for their constructive comments. We also thank T. Taira for valuable discussions. We used the DONET data obtained from the NIED server (http://www.hinet.bosai.go.jp/?LANG=en). The sea level data at station OW were obtained from the Japan Meteorological Agency (http://www.jma.go.jp/jma/index.html). We used Generic Mapping Tools (Wessel and Smith 1998) for generating Fig. 1. We gratefully acknowledge the support of I2CNER, sponsored by the World Premier International Research Center Initiative, MEXT, Japan.

Funding

This work was supported by the Japan Society for the Promotion of Science through KAKENHI Grant Number 16K18332 and 17H05318, and by the Ministry of Education, Culture, Sports, Science and Technology of Japan (MEXT) through the World Premier International Research Center Initiative.

Authors' contributions

TT conceived the study. TI analyzed the data and drafted the manuscript. TT joined in the interpretation and helped to draft the manuscript. Both authors read and approved the final manuscript.

Competing interests

The authors declare that they have no competing interests.

Author details

[1]International Institute for Carbon-Neutral Energy Research (WPI-I2CNER), Kyushu University, Fukuoka 819-0395, Japan. [2]Department of Earth Resources Engineering, Kyushu University, Fukuoka 819-0395, Japan.

References

Ando M (1975) Source mechanisms and tectonic significance of historical earthquakes along the Nankai Trough, Japan. Tectonophysics 27:119–140. https://doi.org/10.1016/0040-1951(75)90102-X

Araki E, Safer DM, Kopf AJ, Wallace LM, Kimura T, Machida Y, Ide S, Davis E, IODP Expedition 365 (2017) Recurring and triggered slow-slip events near the trench at the Nankai Trough subduction megathrust. Science 356:1157–1160. https://doi.org/10.1126/science.aan3120

Bangs NLB, Moore GF, Gulick SPS, Pangborn EM, Tobin HJ, Kurumoto S, Taira A (2009) Broad, weak regions of the Nankai Megathrust and implications for shallow coseismic slip. Earth Planet Sci Lett 284:44–49. https://doi.org/10.1016/j.epsl.2009.04.026

Brenguier F, Campillo M, Hadziioannou C, Shapiro NM, Nadeau RM, Larose E (2008a) Postseismic relaxation along the San Andreas fault at Parkfield from continuous seismological observations. Science 321:1478–1481. https://doi.org/10.1126/science.1160943

Brenguier F, Campillo M, Takeda T, Aoki Y, Shapiro NM, Briand X, Emoto K, Miyake H (2014) Mapping pressurized volcanic fluids from induced crustal seismic velocity drops. Science 345:80–82. https://doi.org/10.1126/science.1254073

Brenguier F, Shapiro NM, Campillo M, Ferrazzini V, Duputel Z, Coutant O, Nercessian A (2008b) Towards forecasting volcanic eruptions using seismic noise. Nat Geosci 1:126–130. https://doi.org/10.1038/ngeo104

Chang C, McNeill LC, Moore JC, Lin W, Conin M, Yamada Y (2010) In situ stress state in the Nankai accretionary wedge estimated from borehole wall failures. Geochem Geophys Geosyst 11. https://doi.org/10.1029/2010GC003261

Chaves EJ, Schwartz SY (2016) Monitoring transient changes within overpressured regions of subduction zones using ambient seismic noise. Sci Adv 2:e1501289–e1501289. https://doi.org/10.1126/sciadv.1501289

Colombi A, Chaput J, Brenguier F, Hillers G, Roux P, Campillo M (2014) On the temporal stability of the coda of ambient noise correlations. Comptes Rendus Geosci 346:307–316. https://doi.org/10.1016/j.crte.2014.10.002

Crampin S, Gao Y, Peacock S (2008) Stress-forecasting (not predicting) earthquakes: a paradigm shift. Geology 36:427. https://doi.org/10.1130/G24643A.1

Hobiger M, Wegler U, Shiomi K, Nakahara H (2012) Coseismic and postseismic elastic wave velocity variations caused by the 2008 Iwate-Miyagi Nairiku earthquake, Japan. J Geophys Res Solid Earth 117:1–19. https://doi.org/10.1029/2012JB009402

Hobiger M, Wegler U, Shiomi K, Nakahara H (2014) Signle-station cross-correlation analysis of ambient seismic noise: application to stations in the surroundings of the 2008 Iwate-Miyagi Nairiku earthquake. Geophys J Int 198:90–109. https://doi.org/10.1093/gji/ggu115

Hobiger M, Wegler U, Shiomi K, Nakahara H (2016) Coseismic and post-seismic velocity changes detected by passive image interferometry: comparison of one great and five strong earthquakes in Japan. Geophys J Int 205:1053–1073. https://doi.org/10.1093/gji/ggw066

Kamei R, Pratt RG, Tsuji T (2012) Waveform tomography imaging of a megasplay fault system in the seismogenic Nankai subduction zone. Earth Planet Sci Lett 317–318:343–353. https://doi.org/10.1016/j.epsl.2011.10.042

Kaneda Y, Kawaguchi K, Araki E, Matsumoto H, Nakamura T, Kamiya S, Ariyoshi K, Hori T, Baba T, Takahashi N (2015) Development and application of an advanced ocean floor network system for megathrust earthquakes and tsunamis. In: Seafloor Observatories. Springer Berlin Heidelberg, Berlin, Heidelberg, pp 643–662

Kawaguchi K, Kaneko S, Nishida T, Komine T (2015) Construction of the DONET real-time seafloor observatory for earthquakes and tsunami monitoring. In: Seafloor Observatories. Springer Berlin Heidelberg, Berlin, Heidelberg, pp 211–228

Kubo H, Nakamura T, Suzuki W, Kimura T, Kunugi T, Takahashi N, Aoi S (2018) Site amplification characteristics at Nankai seafloor observation network, DONET1, Japan, Evaluated Using Spectral Inversion. Bull Seismol Soc Am. https://doi.org/10.1785/0120170254

Lin F-CC, Ritzwoller MH, Snieder R (2009) Eikonal tomography: surface wave tomography by phase front tracking across a regional broad-band seismic array. Geophys J Int 177:1091–1110. https://doi.org/10.1111/j.1365-246X.2009.04105.x

Lin W, Doan ML, Moore JC, Mcneill L, Byrne T, Ito T, Saffer D, Conin M, Kinoshita M, Sanada Y, Moe KT, Araki E, Tobin H, Boutt D, Kano Y, Hayman NW, Flemings P, Huftile G, Cukur D, Buret C, Schleicher A, Efimenko N, Kawabata K, Buchs D, Jiang S, Kameo K, Horiguchi K, Wiersberg T, Kopf A, Kitada K et al (2010) Present-day principal horizontal stress orientations in the Kumano forearc basin of the southwest Japan subduction zone determined from IODP NanTroSEIZE drilling site C0009. Geophys Res Lett 37. https://doi.org/10.1029/2010GL043158

Meier U, Shapiro NM, Brenguier F (2010) Detecting seasonal variations in seismic velocities within Los Angeles basin from correlations of ambient seismic noise. Geophys J Int 181:985–996. https://doi.org/10.1111/j.1365-246X.2010.04550.x

Minato S, Tsuji T, Ohmi S, Matsuoka T (2012) Monitoring seismic velocity change caused by the 2011 Tohoku-oki earthquake using ambient noise records. Geophys Res Lett 39. https://doi.org/10.1029/2012GL051405

Miyazaki S, Heki K (2001) Crustal velocity field of southwest Japan: subduction and arc-arc collision. J Geophys Res Solid Earth 106:4305–4326. https://doi.org/10.1029/2000JB900312

Moore GF, Park JO, Bangs NL, Gulick SP, Tobin HJ, Nakamura Y, Sato S, Tsuji T, Yoro T, Tanaka H, Uraki S, Kido Y, Sanada Y, Kuramoto S, Taira A (2009) Structural and seismic stratigraphic framework of the NanTroSEIZE stage 1 transect. Proc IODP 314:315–316. https://doi.org/10.2204/iodp.proc.314315316.102

Nakano M, Hori T, Araki E, Kodaira S, Ide S (2018) Shallow very-low-frequency earthquakes accompany slow slip events in the Nankai subduction zone. Nat Commun 9:984. https://doi.org/10.1038/s41467-018-03431-5

Nakano M, Nakamura T, Kamiya S, Kaneda Y (2014) Seismic activity beneath the Nankai trough revealed by DONET ocean-bottom observations. Mar Geophys Res 35:271–284. https://doi.org/10.1007/s11001-013-9195-3

Nakano M, Nakamura T, Kamiya S, Ohori M, Kaneda Y (2013) Intensive seismic activity around the Nankai trough revealed by DONET ocean-floor seismic observations. Earth, Planets Sp 65:5–15. https://doi.org/10.5047/eps.2012.05.013

Nakata N, Chang JP, Lawrence JF, Boué P (2015) Body wave extraction and tomography at Long Beach, California, with ambient-noise interferometry. J Geophys Res Solid Earth 120:1159–1173. https://doi.org/10.1002/2015JB011870

Nakata N, Snieder R, Tsuji T, Larner K, Matsuoka T (2011) Shear wave imaging from traffic noise using seismic interferometry by cross-coherence. Geophysics 76:SA97–SA106. https://doi.org/10.1190/geo2010-0188.1

Nimiya H, Ikeda T, Tsuji T (2017) Spatial and temporal seismic velocity changes on Kyushu Island during the 2016 Kumamoto earthquake. Sci Adv 3: e1700813. https://doi.org/10.1126/sciadv.1700813

Obermann A, Froment B, Campillo M, Larose E, Planès T, Valette B, Chen JH, Liu QY (2014) Seismic noise correlations to image structural and mechanical changes associated with the Mw 7.9 2008 Wenchuan earthquake. J Geophys Res Solid Earth 119:3155–3168. https://doi.org/10.1002/2013JB010932

Obermann A, Planès T, Larose E, Campillo M (2013) Imaging preeruptive and coeruptive structural and mechanical changes of a volcano with ambient seismic noise. J Geophys Res Solid Earth 118:6285–6294. https://doi.org/10.1002/2013JB010399

Park J-O, Fujie G, Wijerathne L, Hori T, Kodaira S, Fukao Y, Moore GF, Bangs NL, Kurumoto S, Taira A (2010) A low-velocity zone with weak reflectivity along the Nankai subduction zone. Geology 38:283–286. https://doi.org/10.1130/G30205.1

Rubinstein JL, Beroza GC (2004) Evidence for widespread nonlinear strong ground motion in the M_W 6.9 Loma Prieta earthquake. Bull Seismol Soc Am 94:1595–1608. https://doi.org/10.1785/012004009

Seno T, Stein S, Gripp AE (1993) A model for the motion of the Philippine Sea plate consistent with NUVEL-1 and geological data. J Geophys Res Solid Earth 98:17941–17948. https://doi.org/10.1029/93JB00782

Sens-Schönfelder C, Wegler U (2006) Passive image interferometry and seasonal variations of seismic velocities at Merapi Volcano, Indonesia. Geophys Res Lett 33:L21302. https://doi.org/10.1029/2006GL027797

Shapiro NM, Campillo M, Stehly L, Ritzwoller MH (2005) High-resolution surface-wave tomography from ambient seismic noise. Science 307:1615–1618. https://doi.org/10.1126/science.1108339

Shapiro SA (2003) Elastic piezosensitivity of porous and fractured rocks. Geophysics 68:482–486. https://doi.org/10.1190/1.1567215

Suzuki K, Nakano M, Takahashi N, Hori T, Kamiya S, Araki S, Nakata R, Kaneda Y (2016) Synchronous changes in the seismicity rate and ocean-bottom hydrostatic pressures along the Nankai trough: a possible slow slip event detected by the Dense Oceanfloor Network system for Earthquakes and Tsunamis (DONET). Tectonophysics 680:90–98. https://doi.org/10.1016/j.tecto.2016.05.012

Taira T, Brenguier F (2016) Response of hydrothermal system to stress transients at Lassen Volcanic Center, California, inferred from seismic interferometry with ambient noise. Earth Planets Sp 68:162. https://doi.org/10.1186/s40623-016-0538-6

Taira T, Brenguier F, Kong Q (2015) Ambient noise-based monitoring of seismic velocity changes associated with the 2014 Mw 6.0 South Napa earthquake. Geophys Res Lett 42:6997–7004. https://doi.org/10.1002/2015GL065308

Taira T, Nayak A, Brenguier F, Manga M (2018) Monitoring reservoir response to earthquakes and fluid extraction, Salton Sea geothermal field, California. Sci Adv 4:e1701536. https://doi.org/10.1126/sciadv.1701536

Tobin HJ, Saffer DM (2009) Elevated fluid pressure and extreme mechanical weakness of a plate boundary thrust, Nankai Trough subduction zone. Geology 37:679–682. https://doi.org/10.1130/G25752A.1

Toh A, Obana K, Araki E (2018) Distribution of very low frequency earthquakes in the Nankai accretionary prism influenced by a subducting-ridge. Earth Planet Sci Lett 482:342–356. https://doi.org/10.1016/j.epsl.2017.10.062

Toksöz MN, Cheng CH, Timur A (1976) Velocities of seismic waves in porous rocks. GEOPHYSICS 41:621–645. https://doi.org/10.1190/1.1440639

Tonegawa T, Fukao Y, Takahashi T, Obana K, Kodaira S, Kaneda Y (2015) Ambient seafloor noise excited by earthquakes in the Nankai subduction zone. Nat Commun 6:6132. https://doi.org/10.1038/ncomms7132

Tsuji S, Yamaoka K, Ikuta R, Watanabe T, Katsumata A, Kunitomo T (2016) Seismic velocity change in Tokai region detected by Morimachi ACROSS. In: Japan Geoscience Union Meeting

Tsuji T, Ashi J, Strasser M, Kimura G (2015) Identification of the static backstop and its influence on the evolution of the accretionary prism in the Nankai Trough. Earth Planet Sci Lett 431:15–25. https://doi.org/10.1016/j.epsl.2015.09.011

Tsuji T, Dvorkin J, Mavko G, Nakata N, Matsuoka T, Nakanishi A, Kodaira S, Nishizawa O (2011) V P / V S ratio and shear-wave splitting in the Nankai Trough seismogenic zone: insights into effective stress, pore pressure, and sediment consolidation. Geophysics 76:WA71–WA82. https://doi.org/10.1190/1.3560018

Tsuji T, Kamei R, Pratt RG (2014) Pore pressure distribution of a mega-splay fault system in the Nankai Trough subduction zone: insight into up-dip extent of the seismogenic zone. Earth Planet Sci Lett 396:165–178. https://doi.org/10.1016/j.epsl.2014.04.011

Tsuji T, Minato S, Kamei R, Tsuru T, Kimura G (2017) 3D geometry of a plate boundary fault related to the 2016 Off-Mie earthquake in the Nankai subduction zone, Japan. Earth Planet Sci Lett 478:234–244. https://doi.org/10.1016/j.epsl.2017.08.041

Tsuji T, Tokuyama H, Costa Pisani P, Moore G (2008) Effective stress and pore pressure in the Nankai accretionary prism off the Muroto Peninsula, southwestern Japan. J Geophys Res Solid Earth 113:1–19. https://doi.org/10.1029/2007JB005002

Wallace LM, Araki E, Saffer D, Wang X, Roesner A, Kopf A, Nakanishi A, Power W, Kobayashi R, Kinoshita C, Toczko S, Kimura T, Machida Y, Carr S (2016) Near-field observations of an offshore M w 6.0 earthquake from an integrated seafloor and subseafloor monitoring network at the Nankai Trough, Southwest Japan. J Geophys Res Solid Earth 121:8338–8351. https://doi.org/10.1002/2016JB013417

Wang K, Hu Y (2006) Accretionary prisms in subduction earthquake cycles: the theory of dynamic Coulomb wedge. J Geophys Res Solid Earth 111:B06410. https://doi.org/10.1029/2005JB004094

Wang Q-Y, Brenguier F, Campillo M, Lecointre A, Takeda T, Aoki Y (2017) Seasonal crustal seismic velocity changes throughout Japan. J Geophys Res Solid Earth 122:7987–8002. https://doi.org/10.1002/2017JB014307

Weaver RL, Hadziioannou C, Larose E, Campillo M (2011) On the precision of noise correlation interferometry. Geophys J Int 185:1384–1392. https://doi.org/10.1111/j.1365-246X.2011.05015.x

Wessel P, Smith WHF (1998) New, improved version of generic mapping tools released. EOS Trans Am Geophys Union 79:579–579. https://doi.org/10.1029/98EO00426

Wu C, Peng Z, Ben-Zion Y (2009) Non-linearity and temporal changes of fault zone site response associated with strong ground motion. Geophys J Int 176:265–278. https://doi.org/10.1111/j.1365-246X.2008.04005.x

Yamaoka K, Kunitomo T, Miyakawa K, Kobayashi K, Kumazawa M (2008) A trial for monitoring temporal variation of seismic velocity using an ACROSS system. Island Arc 10:336–347. https://doi.org/10.1111/j.1440-1738.2001.00332.x

Yokota Y, Ishikawa T, Watanabe S, Tashiro T, Asada A (2016) Seafloor geodetic constraints on interplate coupling of the Nankai Trough megathrust zone. Nature 534:374–377. https://doi.org/10.1038/nature17632

Zinszner B, Johnson PA, Rasolofosaon PNJ (1997) Influence of change in physical state on elastic nonlinear response in rock: significance of effective pressure and water saturation. J Geophys Res Solid Earth 102:8105–8120. https://doi.org/10.1029/96JB03225

10

A record of the upper Olduvai geomagnetic polarity transition from a sediment core in southern Yokohama City, Pacific side of central Japan

Chie Kusu[1*], Makoto Okada[2], Atsushi Nozaki[3], Ryuichi Majima[4] and Hideki Wada[5]

Abstract

A detailed paleomagnetic record of the upper Olduvai polarity transition was obtained from a 106.72 m-long sediment core drilled in southern Yokohama City, located on the northern Miura Peninsula, on the Pacific side of central Japan. The core spans the upper part of the Nojima Formation and the lowermost part of the Ofuna Formation, both of which correspond to the middle Kazusa Group (Lower Pleistocene forearc basin fill). The record was reconstructed using discrete specimens taken throughout mudstone and/or sandy mudstone sequences in the Nojima Formation. In this record, the virtual geomagnetic pole (VGP) fluctuation accompanying the polarity transition was determined to occur between depths of 66.99 and 63.60 m. These depths have been dated at 1784.4 and 1779.9 ka, respectively, and the duration of the polarity transition is estimated to be 4.5 kyr using an age model based on a $\delta^{18}O$ record from that core. The VGP paths during the transition do not appear to show any preferred longitudinal bands. However, the VGP positions cluster in five areas: (A) eastern Asia near Japan, (B) the Middle East, (C) eastern North America (North Atlantic), (D) off southern Australasia, and (E) the southern South Atlantic off South Africa. The primary locations of the observed VGP clusters coincide with the areas on the Earth's surface that possess a strong downward flux of the vertical component of the present geomagnetic non-axial dipole field. The relative paleointensity rapidly decreased approximately 1 kyr before the beginning of the polarity transition and gradually recovered to its initial level in 12 kyr.

Keywords: Geomagnetism, Paleomagnetism, Geomagnetic polarity transition, Olduvai subchron, VGP path, VGP cluster, Reversal, Relative paleointensity, Sediment core

Introduction

Studies of geomagnetic reversals, which are among the most conspicuous phenomena associated with the Earth's magnetic field, provide valuable information about geodynamo processes. Previous paleomagnetic investigations have documented the details of geomagnetic field behavior during geomagnetic polarity transitions. The virtual geomagnetic poles (VGPs) during polarity transitions, as inferred from sediments, tend to pass within two preferred longitudinal bands over the Americas and antipodally over eastern Asia and through Australia (Tric et al. 1991; Clement 1991; Laj et al. 1991). Conversely, late Cenozoic volcanic records from intervals spanning polarity transitions suggest that VGPs were clustered mainly within two regions near the southern portions of South America and Western Australia (Hoffman 1991, 1992). These regions lie within the two preferred longitudinal bands identified in the sedimentary records. However, the VGPs inferred from the volcanic records appear not to have moved continuously but to have "jumped" between the regions. The two preferred bands coincide with the centers of the radial magnetic flux of the present geomagnetic field on the Earth's surface when the axial dipole component is removed (Hoffman 1992). Laj et al. (1991) reported that the two preferred

* Correspondence: kusu-chie-sf@ynu.jp
[1]Graduate School of Environment and Information Sciences, Yokohama National University, 79-7 Tokiwadai, Hodogaya-ku, Yokohama 240-8501, Japan
Full list of author information is available at the end of the article

bands coincide well with regions of fast seismic wave propagation (regions of cooler temperature) observed in the lowest mantle. Therefore, they suggested that the transitional field was constrained by temperature patterns at the core-mantle boundary, and its configuration might have persisted over the past 10 Myr. Recently, several high-resolution sedimentary records from deep-sea cores have shown that transitional VGPs still passed within the preferred longitudinal bands but also stayed within regions similar to the VGP clusters observed in volcanic records (e.g., Mazaud and Channell 1999; Ohno et al. 2008; Mazaud et al. 2009).

The detailed processes relating to geomagnetic polarity reversals remain unclear, although it is generally accepted that field intensities during polarity transitions decline to ~10 % of their usual values, which is close to the magnitude of the non-dipole component of the current geomagnetic field (Merrill and McFadden 1999). Absolute paleomagnetic intensities are typically only recovered from the thermal remanent magnetization of volcanic rocks, but this does not yield a continuous record. Tauxe (1993) proposed a method to deduce relative paleointensities using sedimentary records, and continuous relative paleointensity records have thus been obtained from deep-sea sediments. In recent years, studies of relative paleointensity have progressed (e.g., Tauxe and Yamazaki 2007), and global stacks of relative paleointensity variations spanning the last 1 to 3 Myr have been generated (e.g., Valet et al. 2005; Yamazaki and Oda 2005; Channell et al. 2009). Those records, which have focused on reconstructing long-term variations in geomagnetic dipole moment, have been successful in revealing the broad features of geomagnetic dipole fluctuations. However, the details of paleointensity variations during polarity transitions are poorly documented because the accumulation rates of deep-sea sediments used to generate the stacked records are quite low (approximately a few cm/kyr). Because the average duration of polarity transitions has been estimated at 7 kyr, based on available sediment records spanning the four most recent transitions (Clement 2004), records with resolutions higher than 1 kyr are needed to observe detailed variations in paleointensity as well as direction during polarity transitions.

It is generally considered that the magnetization of sediments is stabilized through post-depositional remanent magnetization (PDRM) processes (Irving and Major 1964; Kent 1973). The fixing of magnetic particles in a sedimentary column is thought to occur gradually within a zone due to sediment compaction and dewatering, which generates an offset between the depth where the magnetization is fixed and the sediment surface (e.g., Verosub 1977; Hyodo 1984). This offset is referred to as the "lock-in depth" of PDRM and has been the focus of much debate for a long time (e.g., Okada and Niitsuma 1989; deMenocal et al. 1990; Tauxe et al. 1996, 2006; Channell and Guyodo 2004). Suganuma et al. (2010, 2011) convincingly showed that magnetization in a sedimentary column takes place within a zone just below the base of the bioturbation zone. Based on direct comparisons of ^{10}Be flux and paleomagnetic records through the Matuyama-Brunhes transition in deep-sea cores, they deduced by means of a Gaussian lock-in function that the thickness of the zone of magnetization is approximately 17 cm, regardless of the sedimentation rate. This indicates that sedimentary sequences with accumulation rates that exceed 17 cm/kyr are enough to reconstruct records of past geomagnetic field fluctuations with a time resolution of 1 kyr. Although the PDRM process is uncertain and still under debate, it is possible to minimize the length of delay and smoothing width in remanence acquisition due to the PDRM process by using a sediment sequence that has as high a sedimentation rate as possible.

In this paper, we present an ultra-high-resolution paleomagnetic record of the upper Olduvai polarity transition from a sediment core comprising a homogeneous siltstone whose sedimentation rate exceeded 70 cm/kyr. We also present the VGP paths and relative paleointensity variations during the transition.

Methods/Experimental

Geology of the core site

The core used in this study (Core M; 35.354 6° N, 139.606 3° E) was recovered from the middle part of the lower Pleistocene Kazusa Group in September 2010 using a rotary drilling method and an 86 mm diameter bit. The core is 106.72 m long and composed of sediments with almost 100 % recovery (Fig. 1). The Kazusa Group, a forearc basin fill, is exposed in the northern Miura Peninsula, central Japan. Core M is composed of the upper part of the Nojima Formation (alternations of mudstones and sandy mudstones below a core depth of 20.5 m) and overlies the lowermost part of the Ofuna Formation (massive mudstones from the top of the core to 20.5 m) in the middle part of the Kazusa Group.

The depths at which the sediments of the Nojima and Ofuna Formations were deposited are estimated to 400 to 500 m in the lower horizon of the Nojima Formation (Utsunomiya and Majima 2012) and 200 to 300 m in the upper horizon of the Ofuna Formation (Tate and Majima 1998; Kitazaki and Majima 2003). The strike and dip of the bedding around the Core M site are N63° W and 12° NE, respectively, based on outcrop observations of the NOT-1 tuff bed (Nozaki et al. 2014; the YH02 tuff bed of Takahashi et al. 2005 and Kusu et al. 2014), which is intercalated with sediments between 14.46 and 14.41 m depth in Core M. The detailed lithologies of Core M will be described in a separate paper.

In May 2004, Core I (105.00 m in total length; 35.357 5° N, 139.589 5° E) was drilled and recovered approximately 1.6 km west of the Core M drill site (Fig. 1c), and Kusu et al. (2014) published a paleomagnetic study of the upper Olduvai polarity transition and the vicinity in this core. In Core I, the upper Olduvai polarity transition was determined to lie between 86.77 and 84.64 m depth in the Nojima Formation on the basis of paleomagnetic inclination data at a horizon below the NOT-1 tuff bed that is intercalated between 27.03 and 26.83 m core depth. Core I was shown to provide a detailed record of the paleomagnetic behavior during the upper Olduvai transition; however, the record consists only of *in situ* paleomagnetic inclinations. Because Core I was stored as horizontally unoriented pieces of sections, information on the core's orientation and thus the declinations were lost. Conversely, when the newly recovered Core M was drilled, its orientation was recorded and described, thus enabling the reconstruction of the paleomagnetic vectors through the upper Olduvai polarity transition.

Paleomagnetic and rock-magnetic measurements

Core M was not initially oriented in a horizontal plane. Thus, the absolute declination values could not be directly obtained; however, the relative declination has been preserved within a continuously recovered interval without any core break. Because the core barrel used for the drilling was 2 m long, a 2 m-long continuous core segment was recovered at each stroke in the drilling operation. To determine the relative declinations for as long a continuous interval as possible, for each core segment, we confirmed whether its end surface could be connected to the next core segment and carefully checked the connection between the two segments without any horizontal rotation to ensure the continuity of the core. Additionally, when the dip of a bedding plane was identified from the sedimentary structures in the core, we drew a vertical orientation line at the position of the maximum dip direction on the core surface. Even when the dip could not be identified, we drew a vertical line on the core surface to ensure the directional continuity of the core segments.

For paleomagnetic and rock-magnetic measurements, 25.4 mm diameter mini-cores were collected horizontally from the side surface of the core along the vertical line marked on the core surface. The mini-cores were taken using a core-picker every 1 m over the entire core interval, every 10 cm between 85.00 and 75.20 m and

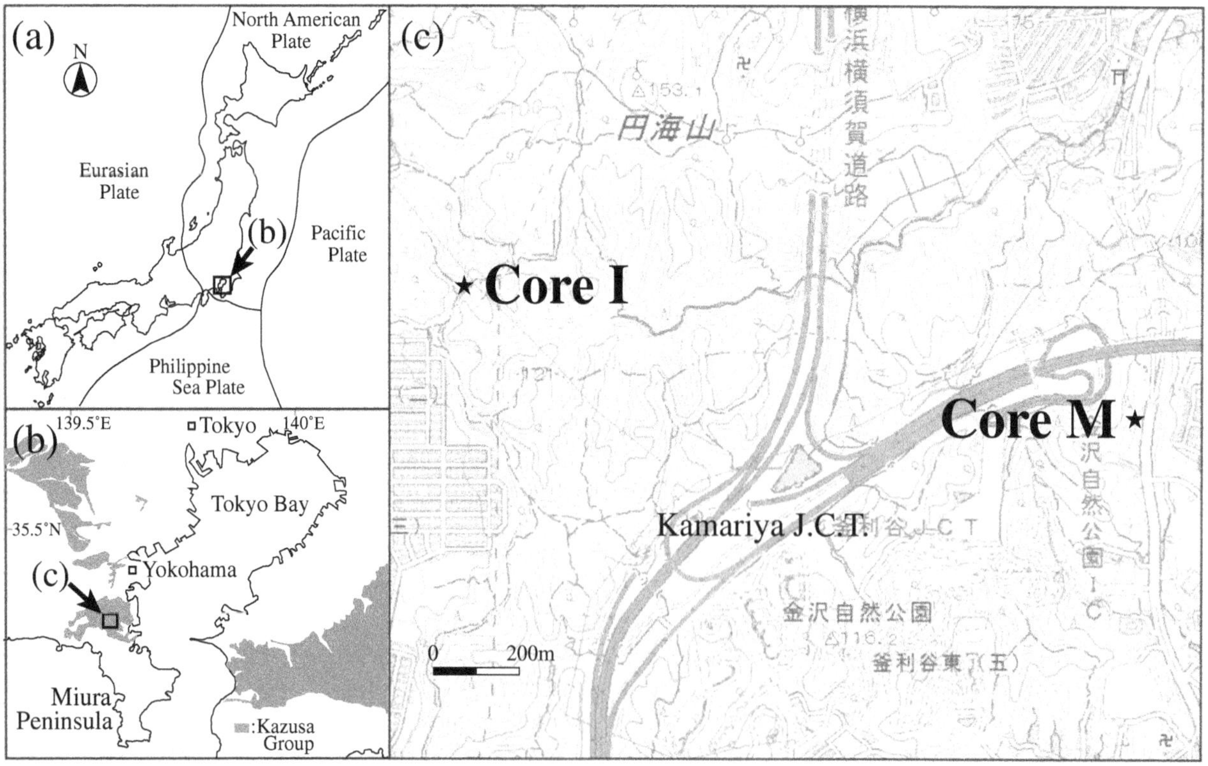

Fig. 1 Location maps of the study site. **a** Tectonic plates under the Japanese Islands (modified from Fig. 1 of Seno and Takano 1989). **b** Distribution of the Kazusa Group in the southern Kanto region (compiled from Mitsunashi and Suda 1980; Mitsunashi et al. 1980; Unozawa et al. 1983; Eto 1986; and Sakamoto et al. 1987). **c** Locality map showing the core sites (*solid stars*) plotted on a 1:25,000 scale topographic map published by the Geospatial Information Authority of Japan

between 60.00 and 50.00 m and every 5 cm between 75.20 and 69.84 m. Each mini-core was cut into approximately 2 cm-long sections. Between depths of 69.84 and 60.00 m, the core was quartered vertically using a rock cutter, and one quarter of the core was cut continuously into approximately 2 cm-long cubic blocks from the center portion of the core. In total, specimens were obtained from 766 horizons in Core M.

Before demagnetization experiments were carried out, low-field magnetic susceptibility (volumetric) and anisotropy of magnetic susceptibility (AMS) measurements were performed on all specimens using a KLY-3 Kappabridge susceptibility meter (AGICO, Czech Republic). The natural remanent magnetization (NRM) was measured using an RF-SQUID Cryogenic magnetometer model 750R (2G Enterprises, USA). Progressive alternating field demagnetization (AFD) with a static 3-axis was performed in 5 mT increments up to 60 mT using a separate AF demagnetizer (DEM-8601C, Natsuhara-Giken, Japan), and progressive thermal demagnetization (THD) was performed in 50 °C increments up to 600 °C in air using a thermal demagnetizer (TD-48, ASC Scientific, USA). The measurements were performed at Ibaraki University, Japan. To deduce relative paleointensity, progressive AFD must be employed as the main demagnetization technique because after the demagnetization of the NRMs, the same specimen is repeatedly used for the acquisition of artificial remanent magnetizations to estimate relative paleointensities. The results of the progressive AFD and THD were displayed using an orthogonal vector diagram (Zijderveld 1967). Characteristic remanent magnetization (ChRM) directions were determined by principal component analysis (Kirschvink 1980) using more than five vector endpoints.

Relative paleointensity was estimated by normalizing for the magnetic grain content using anhysteretic remanent magnetization (ARM). Subsequent to the AFD of the NRM, the ARM was imparted to each specimen using a 30 μT DC field in an alternating field that decreased from a peak of 60 mT. After the ARM acquisition, AFD was then performed at 20, 30, 40, 50, and 60 mT to calculate the paleointensity proxy (NRM/ARM). Thermomagnetic analyses were performed on two samples (from depths of 70.01 and 58.58 m, near the polarity transition) using a Curie Balance NMB-89 (Natsuhara-Giken, Japan). The samples were heated to 700 °C and cooled in air and in a vacuum with an applied field of 0.3 T. Magnetic hysteresis was measured on eight samples at 2–3 m intervals between 70 and 50 m in Core M using a MicroMag 2900 alternating-force gradient magnetometer (Princeton Measurements, USA) to peak fields of ±1 T. The thermomagnetic experiments and hysteresis measurements were performed at the Kochi Core Center, Japan.

Oxygen isotope measurements

Stable oxygen and carbon isotopic ratios were measured to establish an age model based on correlation with the LR04 curve, which is a stacked $\delta^{18}O$ curve that uses oxygen isotope ratios measured in benthic foraminifers from 57 globally distributed deep-sea cores (Lisiecki and Raymo 2005). The isotopic measurements were performed on tests of the planktonic foraminifer *Globorotalia inflata*. Because *G. inflata* precipitates its test at approximately 300–500 m below the sea surface where water temperatures are relatively constant, the *G. inflata* $\delta^{18}O$ profile tends to be parallel to that of benthic foraminifers (Oba et al. 2006). To extract the *G. inflata* tests, a total of 104 sediment samples were collected from the core at approximately 1 m intervals. Each dried sediment sample (approximately 80 g) was crushed and disaggregated using a sodium sulfate solution. From each sediment sample, we picked tests of *G. inflata* that were unbroken, untarnished, "glassy," and larger than 250 μm in size. Each test was immersed in ethanol, broken using a needle, and then ultrasonically cleaned to remove filling impurities.

The isotopic measurements were performed at Shizuoka University, Japan, using the procedure of Wada et al. (1984). The 20–30 individual tests used for each measurement were reacted in saturated phosphoric acid at 60 °C, and the evolved CO_2 gas was analyzed with a MAT-250 mass spectrometer. The obtained value was converted into a value against a PDB standard employing the US National Bureau of Standards NBS-20 (Vienna PDB). The standard deviations of the in-house standard (total of 19 measurements) were 0.04 ‰ for $\delta^{18}O$ and 0.02 ‰ for $\delta^{13}C$, which represented the precisions of the isotopic measurements.

Results

Rock magnetism

The magnetic fabrics were analyzed via AMS. The directions of the maximum susceptibility (K_{max}), intermediate susceptibility (K_{int}), and minimum susceptibility axes (K_{min}) of the AMS ellipsoids were plotted on lower hemisphere equal area projections. Figure 2 shows the AMS data for depths of 67.85–66.11 m within the same core segment for which the directional continuity was confirmed. The data are not corrected for tilt. The directions of K_{max} and K_{int} indicate a girdle distribution perpendicular to K_{min}, which indicates a flattening fabric, and all results of the AMS analysis showed similar flattening fabrics (foliations). The mean inclination of K_{min} between depths of 67.85 and 66.11 m was 74°, and the dip of the foliation plane formed by K_{max} and K_{int} is therefore estimated to be 16°. The estimated dip angle of the bedding plane at the core site is approximately 12°, which suggests that the foliation plane reasonably

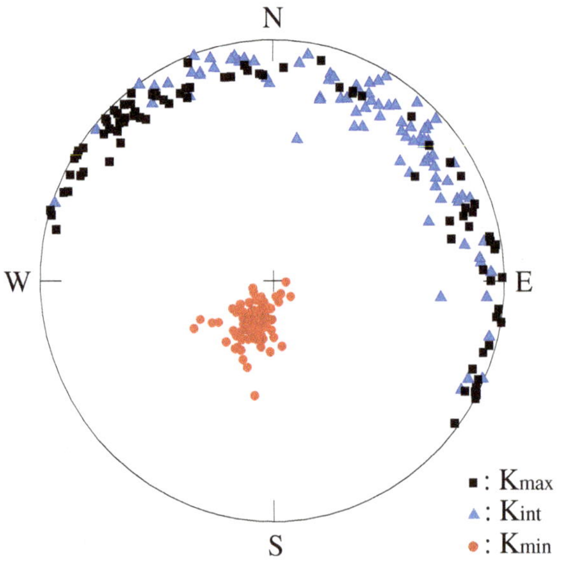

Fig. 2 Lower hemisphere equal area projections of the anisotropy of magnetic susceptibility (AMS). AMS data from the 67.85–66.11 m depth interval in Core M. The *squares*, *triangles*, and *circles* represent the directions of the maximum, intermediate, and minimum susceptibilities, respectively

coincides with the bedding plane. In the four sections of the core, bedding planes comprised of intercalated thin sandstone beds were directly examined. In each segment, the difference between the observed maximum dip direction and the direction of the foliation plane was less than 10°. Accordingly, we treat the foliation plane, as deduced from the AMS data, as a bedding plane at the site.

Thermomagnetic curves from samples at depths of 70.01 and 58.58 m are shown in Fig. 3. Both curves show an identical Curie temperature of approximately 570 °C, which indicates that magnetite (or Ti-poor titanomagnetite) is the predominant magnetic carrier. The heating curves in air as well as in vacuum show a slight increase above 450 °C, but the cooling curves do not show such inflections. When iron sulfide minerals such as pyrite are present in sediments, they are changed into magnetite and maghemite and, finally, into hematite through heating by oxidation that occurs primarily above 450 °C, which produces a heating curve that at first shows an increase in the magnetization and then a decrease (Passier et al. 2001). Therefore, although the thermomagnetic curves suggest a minor amount of iron sulfide minerals, (titano)magnetite is considered to be the main magnetic mineral in the samples.

A plot of anhysteretic remanent magnetization susceptibility (κ_{ARM}) versus magnetic susceptibility (κ) for Core M and Core I is shown in Fig. 4a. According to the grain size estimates of King et al. (1983), the mean grain sizes of the magnetic grains in the specimens from Core M and Core I should primarily range from 1 to 5 μm. Figure 4b shows a Day plot (Day et al. 1977) using

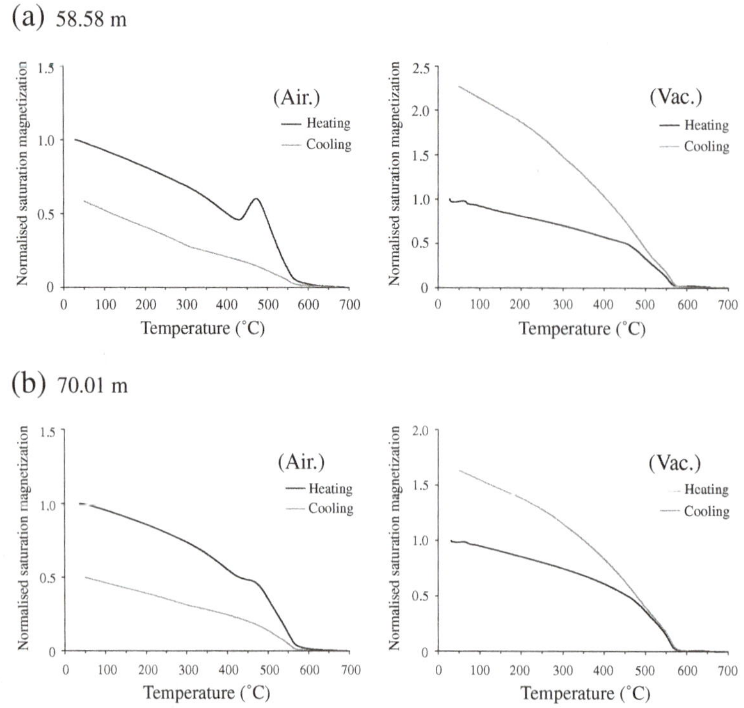

Fig. 3 Strong-field thermomagnetic behaviors. **a** sample from the depth of 58.58 m, **b** sample from the depth of 70.01 m in Core M. Air.: Under an air condition. Vac.: Under a vacuum condition

selected specimens from Core M and Core I. Both plots indicate that pseudo-single domain (PSD) or mixture of single domain (SD) and multidomain (MD) magnetite grains are the predominant magnetic carriers. The Core I data are from Kusu et al. (2014).

A plot of the magnetic susceptibility (κ) and anhysteretic remanent magnetization susceptibility ($κ_{ARM}$) on a logarithmic horizontal scale is shown in Fig. 6g. The κ and $κ_{ARM}$ show similar gradual fluctuations, which range within an order of magnitude. Additionally, the fluctuation patterns of κ and $κ_{ARM}$ strongly resemble each other. Therefore, it is possible that the changes in the values of κ and $κ_{ARM}$ are related to changes in the magnetic mineral content in the specimen, and the changes in the magnetic mineral content with depth are relatively small.

Remanent magnetization

Typical orthogonal vector diagrams for NRMs during progressive AFD and THD are shown in Fig. 5. Most of the results show that the secondary magnetization components were removed by AF demagnetization at a peak field strength of 15 mT. Large fluctuations in the inclinations can be recognized between depths of 67.03 and 64.38 m (Fig. 5e), and steep upward inclinations are observed between depths of 64.36 and 63.20 m (Fig. 5d). In the lower part, below a depth of 67.05 m, the inclinations are positive (Fig. 5b, f), and the maximum angular dispersion (MAD) values for the ChRMs are generally less than 5°. However, values slightly over 10° are seen in a few specimens just below a depth of 67.05 m (Fig. 6d). In the upper part, above a depth of 63.18 m, the inclinations are negative (Figs. 5a, c), and the MAD values for the ChRMs are generally less than 10°, approximately 80 % of which are less than 5° (Fig. 6d). ChRMs were not obtained from five specimens (from depths of 66.61, 66.59, 66.55, 66.49, and 65.81 m), which had very weak remanences (Fig. 5g) and/or MAD values exceeding 25°. From these inclinations, we infer that the Olduvai normal polarity subchronozone occupies the lower part below a depth of 67.05 m, and the post-Olduvai Matuyama reversed polarity chronozone occupies the upper part above a depth of 63.18 m.

Orientation of the core

The core was not oriented, and absolute paleomagnetic declinations therefore could not be derived directly from the measured ChRMs. To obtain the declinations of the ChRMs, we reconstructed the core orientation via the following process. First, we calculated the dip azimuth of the foliation plane, which was deduced from the AMS in each core segment where the continuity of the cores was confirmed. We then adjusted the dip azimuth to the north direction. Using this process, we connected core segments between depths of 80 and 50 m. However, significantly different declinations of ChRMs were observed at three depth intervals: 79.78 to 76.59 m, 73.71 to 72.05 m, and 51.48 to 51.17 m (intervals are indicated by the gray bars in Fig. 6b). We then readjusted the azimuth directions at the three intervals to minimize the differences in the declinations between the three intervals and other depths. Next, the dip azimuth observed in the core was adjusted to the dip azimuth of the bedding plane observed around the core site (strike and dip: N63° W and 12° NE), and the tilt was then corrected. Finally, the mean declination of the normal polarity (below 67.05 m) was adjusted to 0°.

Paleomagnetic directional changes and relative paleointensity

The resultant declinations and inclinations of the ChRMs are shown in Fig. 6a, c. The reversed polarity declinations were generally aligned to 180°, and the

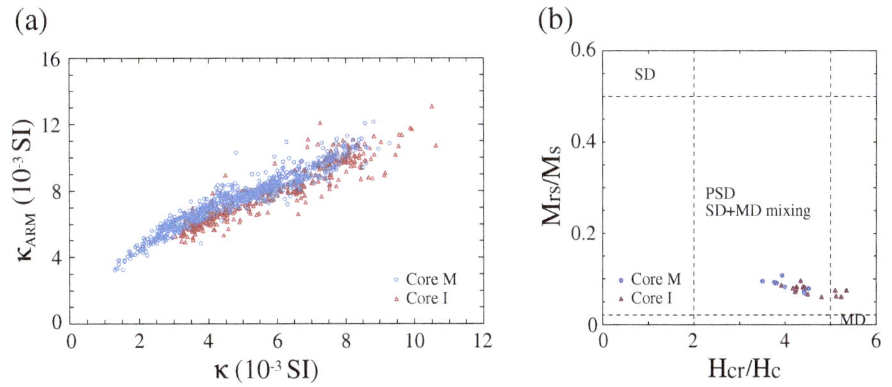

Fig. 4 Rock-magnetic results. **a** Anhysteretic remanent magnetization susceptibility ($κ_{ARM}$) plotted against volume magnetic susceptibility (κ). **b** The Day plot used selected specimens from Core M and Core I (Day et al. 1977). *Mrs* saturation remanent magnetization, *Ms* saturation magnetization, *Hcr* remanent coercivity, *Hc* coercivity. *SD*, *PSD* and *MD* indicate single domain, pseudo-single domain, and multidomain fields, respectively, with the limits given by Dunlop (2002)

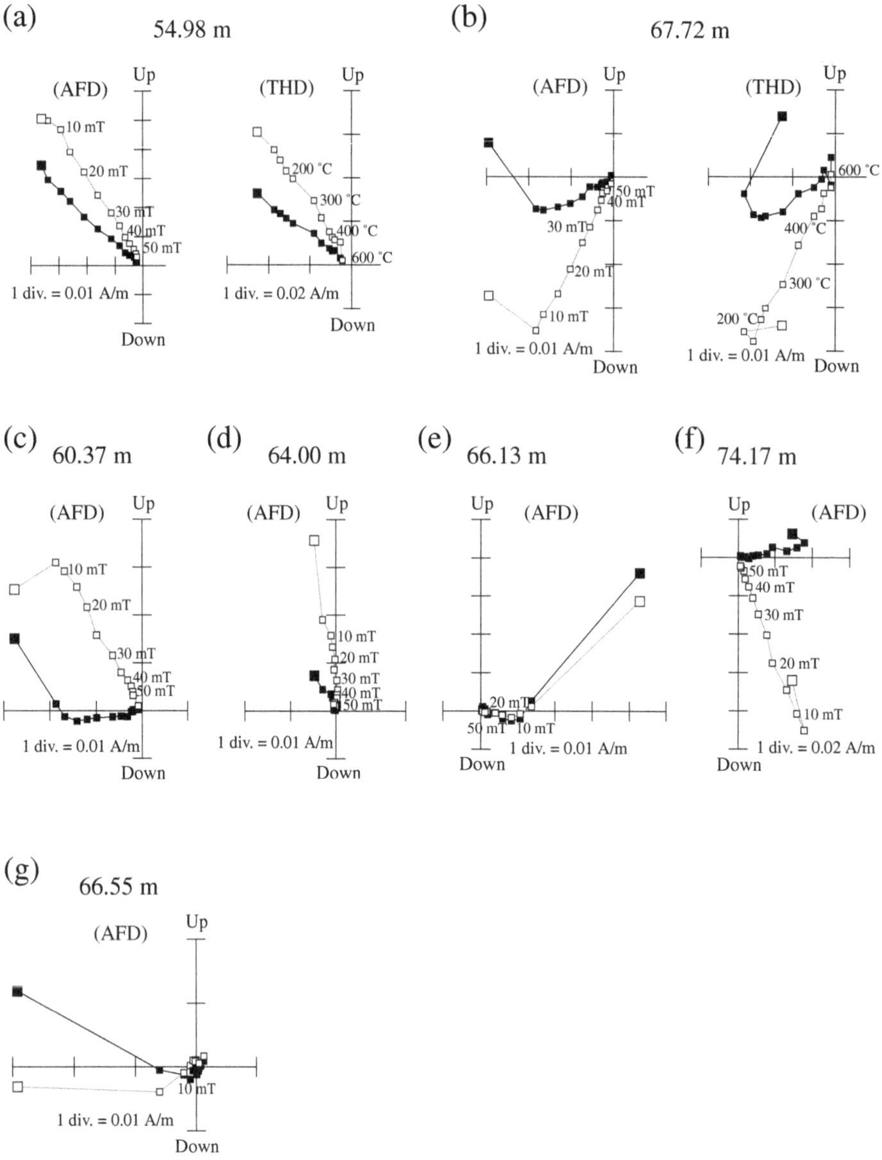

Fig. 5 Examples of demagnetization diagrams. Typical results from the orthogonal vector diagrams for data obtained during progressive thermal demagnetizations (*THD*) and alternating-field demagnetizations (*AFD*) for samples from Core M. **a**, **c** Samples from the post-reversal zone. **d**, **e** Samples from the transitional zone. **b**, **f** Samples from the pre-reversal zone. **g** Sample from a section yielding weak remanence intensity. The *solid* and *open symbols* indicate the projection of the vector endpoints onto the horizontal and vertical planes, respectively

inclinations after the tilting correction were aligned to ±53°, which would be expected given a geocentric axial dipole (GAD) at the core site. Large fluctuations in the declinations and inclinations were observed between depths of 67.03 and 64.38 m, and steep upward inclinations were observed at approximately 63.5 m depth. The VGP latitudes calculated from the ChRMs are shown in Fig. 6e. The VGP moves across the equator from the Northern Hemisphere to the Southern Hemisphere between 65.69 and 65.67 m depth. The VGP begins to move rapidly at a depth of 67.03 m, and the VGP settles at approximately 63.44 m after moving to the Southern Hemisphere.

The rock-magnetic measurement results indicate that the predominant carrier of magnetic remanences in the sediments is PSD, or a mixture of SD and MD-sized (titano)magnetite (Figs. 3 and 4), and the fluctuations in κ and κ_{ARM} range within an order of magnitude (Fig. 6g). From these results, the sediment satisfies the criteria for relative paleointensity reconstruction proposed by Tauxe (1993). To obtain an appropriate proxy for relative paleointensity, we compared the NRM/ARM ratios from three different coercivity fractions, which were between 20 and 40 mT, 30 and 50 mT, and 30 and 40 mT. The variations of the three coercivity fractions were almost

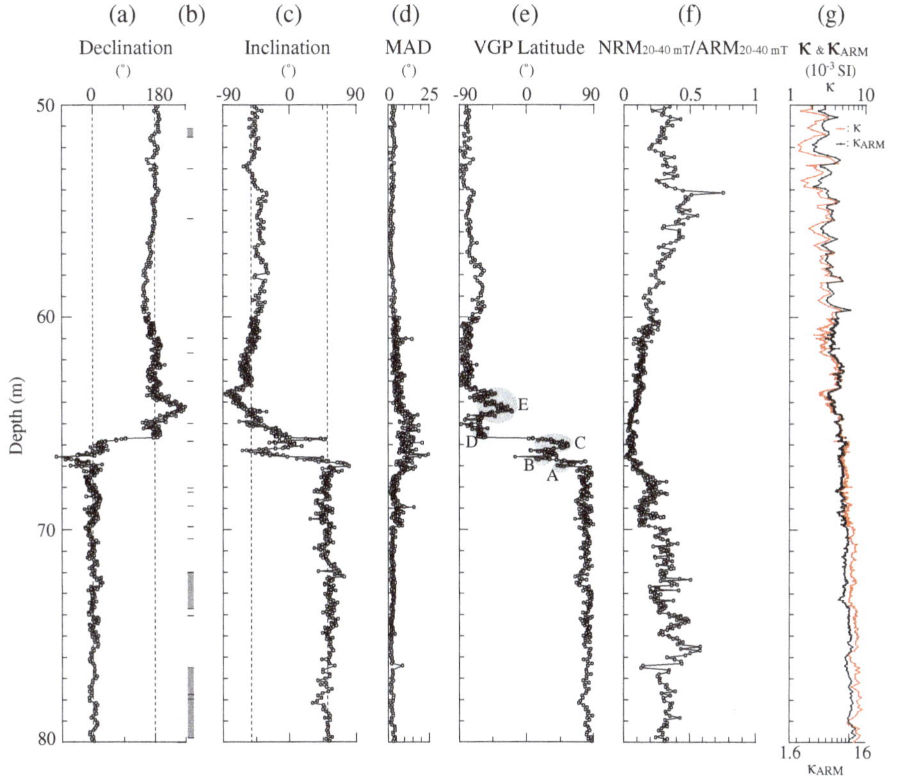

Fig. 6 Rock-magnetic and paleomagnetic results from a depth interval of 80–50 m in Core M. **a** Paleomagnetic declinations. **b** Section breaks. **c** Paleomagnetic inclinations. **d** Maximum angular dispersions (*MAD*) of principal component analysis. **e** VGP latitudes, in which the letters *A–E* are the VGP cluster locations see in Fig. 7a. **f** Relative paleointensities ($NRM_{20-40\ mT}/ARM_{20-40\ mT}$). **g** Volume magnetic susceptibilities (κ) and anhysteretic remanent magnetization susceptibilities (κ_{ARM}). The *vertical dotted lines* are the expected declinations and inclinations for the two polarities at the core site from the geocentric axial dipole field

identical, and we therefore used the NRM/ARM ratio between 20 and 40 mT as a proxy for relative paleointensity in this study (Fig. 6f).

VGP and paleointensity

The VGP paths between 70.0 and 60.0 m are shown in Fig. 7a. In the upper Olduvai polarity transition, the VGP apparently did not move within a preferred longitudinal band but settled in several VGP cluster areas. The VGP can be observed in five areas (Fig. 7a): (A) eastern Asia near Japan, (B) the Middle East, (C) eastern North America (North Atlantic), (D) off southern Australasia, and (E) the southern South Atlantic off South Africa. The VGP apparently moved rapidly between the clusters.

In the relative paleointensity of the normal polarity, a maximum value is observed at 75.50 m depth (Fig. 6f) and it then decreases with the short-term oscillations. Before the onset of the paleomagnetic directional change, the relative paleointensity has a peak at 67.75 m depth (with a value approximately 56 % of the maximum value), and values then decrease at 67.15 m to approximately 11 % of the maximum value. Subsequently, the relative paleointensity slightly recovers up to 66.89 m. In this interval, the directional change begins at a depth of 67.03 m, and the VGP near the North Pole rapidly moves to cluster A (Fig. 6e). The relative paleointensity subsequently drops at 66.5 m depth. Between 66.53 and 65.81 m, the relative paleointensity is approximately 9 % of the maximum value. In the same interval, the VGP settles in cluster C. The relative paleointensity then remains low up to 64.56 m. In this interval, the VGP in the Northern Hemisphere moves into the Southern Hemisphere between depths of 65.69 and 65.67 m, and the VGP moves to cluster D at ~65.17 m. Afterwards, the relative paleointensity gradually increases. From 64.65 to 63.46 m, the VGP settles in cluster E, and then, the VGP moves to the area near the Antarctic.

During the polarity reversal, especially between depths of 67.03 and 63.46 m, the relative paleointensity drops to approximately 12 % of the maximum value. After the large movements in the VGP, the relative paleointensity remains low for a time and then gradually recovers. At 58.38 m depth, the relative paleointensity returns to the peak value observed at 67.75 m, just before the paleomagnetic directional change began, and at about

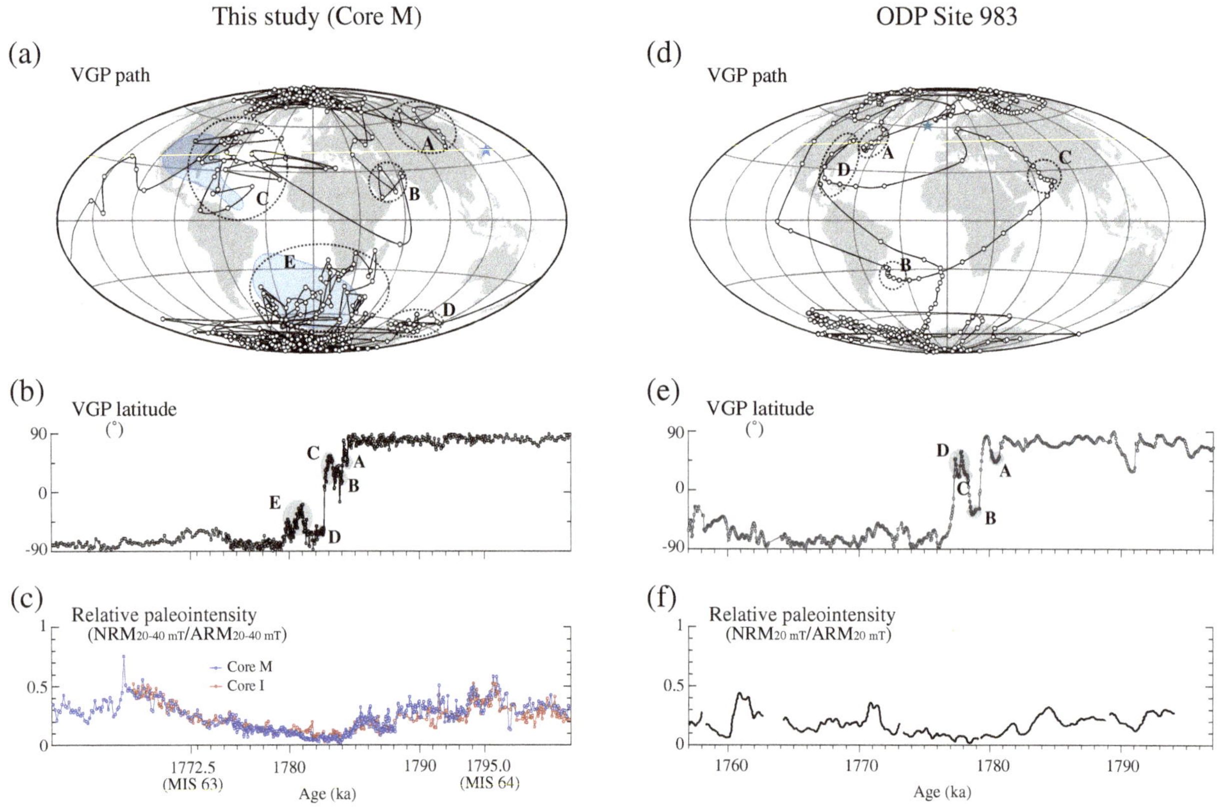

Fig. 7 Paleomagnetic results in the vicinity of the upper Olduvai polarity reversal. Comparison between Core M and ODP Site 983. **a** VGP path. **b** VGP latitude. **c** Relative paleointensity ($NRM_{20-40\ mT}/ARM_{20-40\ mT}$) from Core M and Core I. **d** VGP path. **e** VGP latitude. **f** Relative paleointensity ($NRM_{20\ mT}/ARM_{20\ mT}$) from ODP Site 983. For Site 983, the VGP positions and age model are based on Channell et al. (2002), and the relative paleointensity is based on Mazaud and Channell (1999). The declination and inclination data from ODP Site 983 are plotted using the PANGEA database (Channell et al. 2013). The *dotted circles* in **a** and **d** and the *shaded circles* in **b** and **e** indicate VGP clusters, and the *bold letters* represent the temporal order. The *blue shaded areas* in **a** indicate that the vertical component of the NAD field at the Earth's surface is less than −15 μT (Constable 2007). The *stars* indicate the core sites

55 m, the relative paleointensity becomes comparable to the maximum value observed pre-reversal.

Age model

To construct an age model for the interval spanning the upper Olduvai polarity reversal (1.8–1.7 Ma) in Core M, we correlated our $\delta^{18}O$ profile generated from tests of *G. inflata* in the core (Fig. 8g) with the LR04 global $\delta^{18}O$ stack (Lisiecki and Raymo 2005; Fig. 8k) and determined the peaks of marine isotope stages (MIS) 64 and 63.

The peak observed at 75.00 m (1.61 ‰) is comparable to the peak of MIS 64 because it is just below the Olduvai termination, and the broad low at approximately 58.00–52.00 m (0.78–0.83 ‰) can be correlated to MIS 63 because it is just above the Olduvai termination. The maximum amplitude of $\delta^{18}O$ in the core is 0.83 ‰. This is quite comparable to the amplitude of 0.71 ‰ observed in the LR04 stack between MIS 64 and MIS 63. We therefore assigned the peak at 75.00 m and the broad low at approximately 58.00–52.00 m to MIS 64 and to MIS 63, respectively.

It is, however, difficult to determine a particular position within the broad trough at approximately 58.00–52.00 m for correlation with the lowest point of MIS 63. To determine which position is suitable as the MIS 63 trough, we used Core I, which was drilled in the same study area (Fig. 1c). From Core I, the Olduvai termination boundary, magnetic susceptibility, and relative paleointensity were obtained, and a significant trough comparable to MIS 63 was recognized in the $\delta^{18}O$ curve (Kusu et al. 2014). The relative paleointensity and $\delta^{18}O$ curve for Core I were determined using the same methods as in this study for Core M. Figure 8 shows a comparison between the two cores and includes the horizons of MISs 64 and 63. The vertical axes are expressed as core depths. Because a coarse-grained tuff bed, a pumice-rich lapilli tuff bed, and a fine-grained tuff bed (A, B, C, respectively, in Fig. 8) are commonly recognized in both cores, these three tephra beds were used

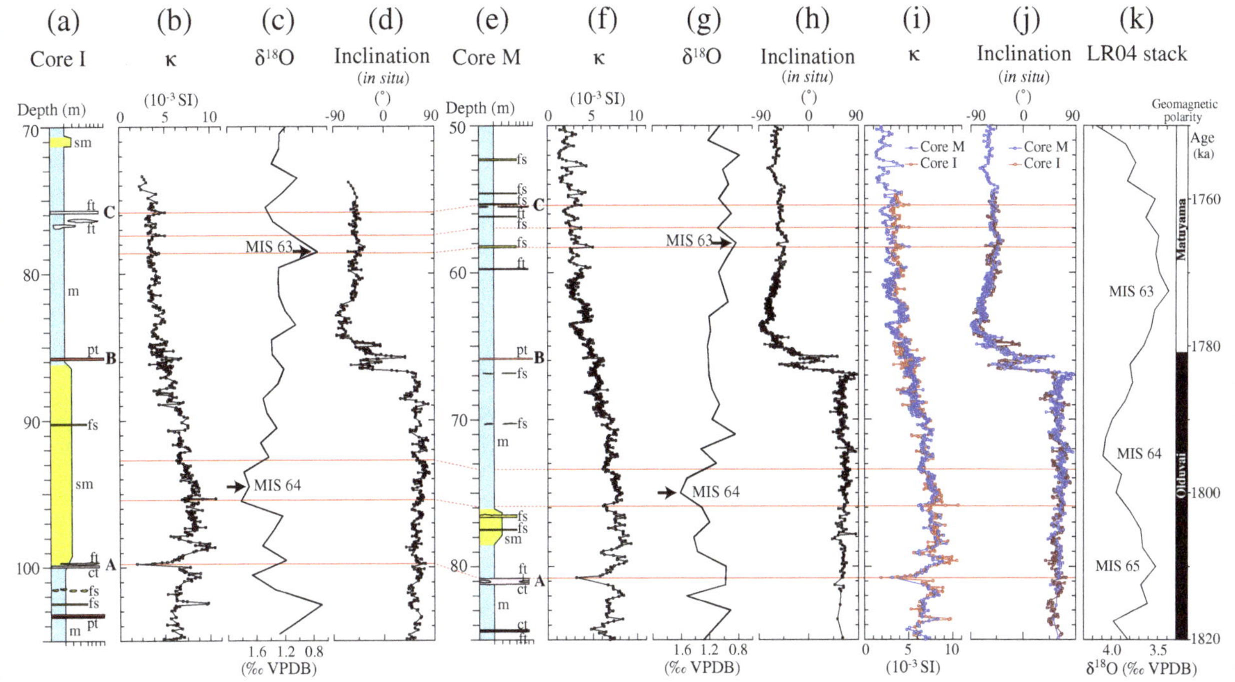

Fig. 8 Comparison between Cores I and M in the vicinity of the upper Olduvai polarity reversal. Lithofacies, volume magnetic susceptibility (κ), and oxygen isotope fluctuations based on measurements in *Globorotalia inflata* ($\delta^{18}O$) and *in situ* paleomagnetic inclinations are shown in **a**–**d** for Core I and **e**–**h** for Core M, respectively. **i** Correlation of the peaks of the magnetic susceptibility between the two cores, using the depths from Core M. **j** Correlation of *in situ* paleomagnetic inclinations between the two cores, using the same depth as **i**. **k** The LR04 benthic $\delta^{18}O$ stack (Lisiecki and Raymo 2005). Abbreviations shown on the lithology columns are the following: *m* mudstone (mud content ≥70 %), *sm* sandy mudstone (70 % > mud content ≥ 45 %), *fs* fine-grained sandstone, *ft* fine tuff, *ct* coarse tuff, *pt* pumice-rich lapilli tuff. For the *red lines*, see the text

to correlate the two cores. In the interval between the coarse-grained tuff bed (A) and the fine-grained tuff bed (C), the fluctuating patterns in the two cores' magnetic susceptibility records coincide well (Fig. 8i), as do the records of *in situ* paleomagnetic inclinations (Fig. 8j). We therefore used the magnetic susceptibility curves to more closely correlate the two cores. The red lines in Fig. 8 indicate the correlation between the two cores using representative tie horizons in the vicinities of MIS 64 and MIS 63.

From this correlation, we determined that the most likely position corresponding to the trough of MIS 63 in Core M is at a depth of 58.00 m. We assigned MIS 64 (1795.0 ka) to the peak at 75.00 m and MIS 63 (1772.5 ka) to the trough at 58.00 m, which serves as an age model for Core M. Based on the age model, the average sedimentation rate between MISs 64 and 63 is 73.9 cm/kyr.

Discussion

Our results from Cores M and I show that the fluctuating patterns in the records of magnetic susceptibility (Fig. 8i), *in situ* paleomagnetic inclinations (Fig. 8j), and relative paleointensities (Fig. 7c) coincide well with each other, and the ChRM inclinations of Core M are in close agreement with the value expected for a geocentric axial dipole (Fig. 6c). This supports the reliability of our results for the upper Olduvai polarity transition.

The VGP latitudes and relative paleointensities are plotted against age in Fig. 7b, c, respectively. If the polarity transition interval is defined as the interval when the VGPs cross the equator and are more than 45° apart from the geographical poles, the transitional interval for Core M lies between depths of 66.99 and 63.60 m. This interval is dated at between 1784.4 and 1779.9 ka, and the duration of the transition is thus estimated to be 4.5 kyr. The relative paleointensity evidently decreased rapidly approximately 1 kyr before the beginning of the polarity transition and gradually recovered to its original value after almost 12 kyr.

The VGP cluster over the southern South Atlantic has a similar location to VGP clusters seen in lava-derived records of polarity transitions spanning the last 10 Myr (Hoffman 1991, 1992). Previous studies suggest that the VGP cluster areas during polarity transitions are associated with the vertical flux patches in the present non-axial dipole (NAD) field, although the VGP records from different polarity transitions merge (e.g., Channell et al. 2003, 2004; Hoffman et al. 2008). In Fig. 7a, the blue-shaded areas indicate the downward flux patches of the

NAD, averaged for the last 400 years, where the vertical component is less than −15 μT (Constable 2007). This figure shows that our results of the VGPs from the upper Olduvai polarity transition match well with the NAD vertical flux patches.

The results for the upper Olduvai polarity transition obtained from this study were compared with records from ODP Site 983 (Mazaud and Channell 1999; Channell et al. 2002). At Site 983 (60.40° N, 23.64° W), detailed records of the upper Olduvai polarity transition were obtained by u-channel measurements at 1 cm intervals. Figure 7 compares the VGP path, VGP latitude, and relative paleointensity results of the two studies. The VGP position data and age model from Site 983 are based on Channell et al. (2002), and the relative paleointensity data from this site is based on Mazaud and Channell (1999). Note that the age models used in the two studies are different. The VGP path from the present study does not exhibit the large loop seen in the Site 983 record (Fig. 7d). However, the VGP cluster areas in eastern North America, the South Atlantic, and the Middle East obtained from Site 983 are similar to the areas reported in this study, whereas the VGP paths between the clusters are different. The bold letters in Fig. 7 represent the temporal order of the VGP clusters at each study site. As shown in Fig. 7, the temporal orders of the VGP clusters are different for the two studies, but the positions of the clusters are similar. The two records indicate that the VGP cluster moved to a location close to the study site at the beginning of the transition, relocated to eastern North America, and then relocated to the Southern Hemisphere. This suggests that the VGPs at the beginning of the transition could have been affected by one of the vertical flux patches of the NAD field located near the study site, and that, just before the polarity reversal, the VGPs were strongly affected by the NAD field near eastern North America.

Based on the ten most detailed records of reversals from volcanic materials, Valet et al. (2012) proposed that a geomagnetic reversal can be characterized by three successive phases: a precursory event (phase 1), a 180° polarity switch (phase 2), and a rebound (phase 3). Furthermore, they proposed that the maximum durations of phases 1 and 3 are ~2.5 kyr and the maximum duration of phase 2 is approximately ~1.0 kyr. Our records for VGP cluster A could correspond to phase 1, the VGP path between clusters C and D could correspond to phase 2, and VGP cluster E could correspond to phase 3.

A previous study that analyzed variations in field intensity across the five polarity reversals over the past 2 Myr suggested that the intensity records were characterized by a rapid recovery following the transitions (Valet et al. 2005). However, the relative paleointensity variations during the upper Olduvai polarity transition reconstructed in this study (Fig. 7c) as well as from Site 983 (Fig. 7f) show a slow and gradual recovery after the reversal. Thus, the results from this study and Site 983 suggest that the paleointensity variations associated with geomagnetic reversals could differ between each reversal.

Clement (2004) estimated that the average duration of a polarity transition is approximately 7 kyr, based on data from 30 selected sedimentary records spanning the four most recent polarity reversals (the Matuyama-Brunhes, the upper and the lower Jaramillo, and the upper Olduvai), and that this duration varies with site latitude: shorter durations were observed at low-latitude sites, and longer durations were observed at mid- to high-latitude sites. Our estimated duration of 4.5 kyr for the polarity transition closely matches the durations observed at mid-latitude sites, which range from 3 to 10 kyr, by Clement (2004).

At Site 983, the upper Olduvai polarity transition was recorded between 248 and 247 mcd (Mazaud and Channell 1999). The 248 mcd was dated at 1778 ka using the age model constructed by Channell et al. (2002) based on the astrochronology of Shackleton et al. (1990). For Core M, the upper Olduvai polarity transition was determined to be between depths of 66.99 and 63.60 m. The depth of 66.99 m is dated at 1773.4 ka using an age model also based on the astrochronology of Shackleton et al. (1990). This estimated age from Core M is close to but slightly younger (ca. 4 kyr) than the estimated age from Site 983. The reason for this discrepancy could be partly due to a lock-in depth inherent in the Site 983 sediments. However, we acknowledge that our age model, which consists of only two tie points, has to be improved to allow more precise comparison with other records in terms of reversal timing.

Conclusions

We present a nearly continuous paleomagnetic record of the upper Olduvai polarity reversal from a sediment core (Core M) drilled in southern Yokohama City, on the Pacific side of central Japan. The results of our study are summarized as follows.

(1) In Core M, the reversal boundary was observed between 65.69 and 65.67 m depth (dated to 1782.7 ka), the polarity transition was observed between 66.99 and 63.60 m depth (dated to between 1784.4 and 1779.9 ka), and the duration of the polarity transition is estimated to be 4.5 kyr, based on an age model derived from $\delta^{18}O$ measurements in the planktonic foraminifer *G. inflata* from the same core.

(2) In the polarity transition period, VGP clusters were observed in five areas: eastern Asia near Japan, the Middle East, eastern North America (the North

Atlantic), off southern Australasia, and the southern South Atlantic off South Africa. The primary locations of the observed VGP clusters are closely related to the vertical flux patches of the present NAD field.

(3) The relative paleointensity decreased rapidly approximately 1 kyr before the beginning of the polarity reversal and recovered to its original value after almost 12 kyr. Between depths of 67.03 and 63.46 m, where a large movement in the VGPs was observed, the relative paleointensity dropped to approximately 12 % of the maximum value observed pre-reversal.

Abbreviations

AFD: Alternating field demagnetization; AMS: Anisotropy of magnetic susceptibility; ARM: Anhysteretic remanent magnetization; ChRM: Characteristic remanent magnetization; MAD: Maximum angular dispersion; MD: Multidomain; MIS: Marine isotope stage; NAD: Non-axial dipole; NRM: Natural remanent magnetization; PSD: Pseudo-single domain; SD: Single domain; THD: Thermal demagnetization; VGP: Virtual geomagnetic pole; κ: Magnetic susceptibility; κ_{ARM}: Anhysteretic remanent magnetization susceptibility

Acknowledgements

We would like to thank Y. Yamamoto who supported the rock-magnetic measurements at the Center for Advanced Marine Core Research (CMCR), Kochi University under the cooperative research program (grants 15A046, 15B041, 14A017, and 14A018). We also thank T. Gokan (Moriya City Office) and Y. Haneda (Ibaraki University) for the cooperation on magnetic measurements. We would like to thank K. Kamoshida (Nitori Co., Ltd.) for the isotopic measurements. We would like to thank M. Arima, M. Ishikawa, R. Wani, and S. Kawagata (Yokohama National University) for the discussions in our seminars. We also would like to thank M. Utsunomiya (Geological Survey of Japan, AIST) for the helpful discussions. We would like to thank two reviewers and the editor of *Progress in Earth and Planetary Science* for the constructive comments that greatly improved the manuscript.

Funding

This study was supported by the joint research project of the Faculty of Environment and Information Sciences, Yokohama National University in fiscal 2010, by a grant from the Fujiwara Natural History Foundation in 2015 and in part by Grants-in-Aid for Scientific Research (A: no. 16204041 and B: no. 20403015) from the Japan Society for the Promotion of Science (JSPS).

Authors' contributions

MO proposed the paleomagnetic study of polarity transition. CK and MO wrote the manuscript and coordinated the structure of the paper. CK collected the paleomagnetic and rock-magnetic samples, carried out the experiments and measurements, and analyzed the data. AN carried out the geological surveys. RM conducted the drill core research project and contributed to the construction of the manuscript. HW contributed to the isotope analyses. All the authors read and approved the final manuscript.

Competing interests

The authors declare that they have no competing interests.

Author details

[1]Graduate School of Environment and Information Sciences, Yokohama National University, 79-7 Tokiwadai, Hodogaya-ku, Yokohama 240-8501, Japan. [2]College of Science, Ibaraki University, 2-1-1 Bunkyo, Mito, Ibaraki 310-8512, Japan. [3]Hiratsuka City Museum, 12-41 Sengencho, Hiratsuka, Kanagawa 254-0041, Japan. [4]Faculty of Environment and Information Sciences, Yokohama National University, 79-7 Tokiwadai, Hodogaya-ku, Yokohama 240-8501, Japan. [5]Faculty of Science, Shizuoka University, 836 Ohya, Suruga-ku, Shizuoka 422-8529, Japan.

References

Channell JET, Guyodo Y (2004) The Matuyama Chronozone at ODP Site 982 (Rockall Bank): evidence for decimeter-scale magnetization lock-in depths. In: Channell JET, Kent DV, Lowrie W, Meert JG (eds) Timescales of the paleomagnetic field, vol 145, Geophys Monogr Ser. AGU, Washington DC, pp 205–219. doi:10.1029/145GM15

Channell JET, Mazaud A, Sullivan P, Turner S, Raymo ME (2002) Geomagnetic excursions and paleointensities in the Matuyama Chron at Ocean Drilling Program Sites 983 and 984 (Iceland Basin). J Geophys Res 107(B6). doi:10.1029/2001JB000491

Channell JET, Labs J, Raymo ME (2003) The Réunion Subchronozone at ODP Site 981 (Feni Drift, North Atlantic). Earth Planet Sci Lett 215(1):1–12. doi:10.1016/S0012-821X(03)00435-7

Channell JET, Curtis JH, Flower BP (2004) The Matuyama-Brunhes boundary interval (500-900 ka) in North Atlantic drift sediments. Geophys J Int 158(2): 489–505. doi:10.1111/j.1365-246X.2004.02329.x

Channell JET, Xuan C, Hodell DA (2009) Stacking paleointensity and oxygen isotope data for the last 1.5 Myr (PISO-1500). Earth Planet Sci Lett 283(1):14–23. doi:10.1016/j.epsl.2009.03.012

Channell JET, Mazaud A, Sullivan P, Turner S, Raymo ME (2013) Declination and inclination of ODP Site., pp 162–983. doi:10.1594/PANGAEA.808978

Clement BM (1991) Geographical distribution of transitional VGPs: evidence for non-zonal equatorial symmetry during the Matuyama-Brunhes geomagnetic reversal. Earth Planet Sci Lett 104(1):48–58. doi:10.1016/0012-821X(91)90236-B

Clement BM (2004) Dependence of the duration of geomagnetic polarity reversals on site latitude. Nature 428(6983):637–640. doi:10.1038/nature02459

Constable C (2007) Non-dipole field. In: Gubbins D, Herrero-Bervera E (eds) Encyclopedia of geomagnetism and paleomagnetism. Springer, Netherlands, pp 701–704

Day R, Fuller M, Schmidt VA (1977) Hysteresis properties of titanomagnetites: grain size and composition dependence. Phys Earth Planet Int 13(4):260–267. doi:10.1016/0031-9201(77)90108-X

deMenocal PB, Ruddiman WF, Kent DV (1990) Depth of the post-depositional remanence acquisition in deep-sea sediments: a case study of the Brunches-Matuyama reversal and oxygen isotopic stage 19.1. Earth Planet Sci Lett 99:1–13. doi:10.1016/0012-821X(90)90066-7

Dunlop DJ (2002) Theory and application of the Day plot (Mrs/Ms versus Hcr/Hc), 1. Theoretical curves and tests using titanomagnetite data. J. Geophys Res 107(B3). doi:10.1029/2001JB000486

Eto T (1986) Stratigraphy of the Hayama Group in the Miura Peninsula, Japan. Sci Rep Yokohama National Univ (Sec II) 33:67–106

Hoffman KA (1991) Long-lived transitional states of the geomagnetic field and the two dynamo families. Nature 354(6351):273–277

Hoffman KA (1992) Dipolar reversal states of the geomagnetic field and core-mantle dynamics. Nature 359(6398):789–794

Hoffman KA, Singer BS, Camps P, Hansen LN, Johnson KA, Clipperton S, Carvallo C (2008) Stability of mantle control over dynamo flux since the mid-Cenozoic. Phys Earth Planet Int 169(1):20–27. doi:10.1016/j.pepi.2008.07.012

Hyodo M (1984) Possibility of reconstruction of the past geomagnetic field from homogeneous sediments. J Geomag Geoelectr 36:45–62. doi:10.5636/jgg.36.45

Irving E, Major A (1964) Post-depositional detrital remanent magnetization in a synthetic sediment. Sedimentology 3(2):135–143. doi:10.1111/j.1365-3091.1964.tb00638.x

Kent DV (1973) Post-depositional remanent magnetisation in deep-sea sediment. Nature 246(5427):32–34

King JW, Banerjee SK, Marvin J (1983) A new rock-magnetic approach to selecting sediments for geomagnetic paleointensity studies: application to paleointensity for the last 4000 years. J Geophys Res 88:5911–5921

Kirschvink JL (1980) The least-squares line and plane and the analysis of palaeomagnetic data. Geophys J Int 62:699–718. doi:10.1111/j.1365-246X.1980.tb02601.x

Kitazaki T, Majima R (2003) A slope to outer-shelf cold-seep assemblage in the Plio-Pleistocene Kazusa Group, Pacific side of central Japan. Paleontol Res 7:279–296

Kusu C, Nozaki A, Okada M, Wada H, Majima R (2014) Lithology and upper boundary of the Olduvai Subchronozone in a core recovered from the middle Kazusa Group (Lower Pleistocene) on the Miura Peninsula, Pacific side of central Japan. J Geol Soc Jpn 120:53–70. doi:10.5575/geosoc.2014.0002

Laj C, Mazaud A, Weeks R, Fuller M, Herrero-Bervera E (1991) Geomagnetic reversal paths. Nature 351:477. doi:10.1038/351447a0

Lisiecki LE, Raymo ME (2005) A Pliocene-Pleistocene stack of 57 globally distributed benthic $\delta^{18}O$ records. Paleoceanography 20:PA1003. doi:10.1029/2004PA001071

Mazaud A, Channell JET (1999) The top Olduvai polarity transition at ODP Site 983 (Iceland Basin). Earth Planet Sci Lett 166:1–13. doi:10.1029/2001JB000491

Mazaud A, Channell JET, Xuan C, Stoner JS (2009) Upper and lower Jaramillo polarity transitions recorded in IODP Expedition 303 North Atlantic sediments: implications for transitional field geometry. Phys Earth Planet Int 172(3):131–140. doi:10.1016/j.pepi.2008.08.012

Merrill RT, McFadden PL (1999) Geomagnetic polarity transitions. Rev Geophys 37(2):201–226

Mitsunashi T, Suda Y (1980) Geological map of Japan 1: 200 000, Otaki. Geol Surv Japan, Tsukuba

Mitsunashi T, Ono K, Suda Y (1980) Geological map of Japan 1:200 000, Yokosuka. Geol Surv Japan, Tsukuba

Nozaki A, Majima R, Kameo K, Sakai S, Kouda A, Kawagata S, Wada H, Kitazato H (2014) Geology and age model of the Lower Pleistocene Nojima, Ofuna, and Koshiba Formations of the middle Kazusa Group, a forearc basin-fill sequence on the Miura Peninsula, the Pacific side of central Japan. Island Arc 23:157–179. doi:10.1111/iar.12066

Oba T, Irino T, Yamamoto M, Murayama M, Takamura A, Aoki K (2006) Paleoceanographic change off central Japan since the last 144 000 years based on high-resolution oxygen and carbon isotope records. Glob Planet Change 53:5–20

Ohno M, Murakami F, Komatsu F, Guyodo Y, Acton G, Kanamatsu T, Evans HF, Nanayama F (2008) Paleomagnetic directions of the Gauss-Matuyama polarity transition recorded in drift sediments (IODP Site U1314) in the North Atlantic. Earth Planets Space 60:e13–e16

Okada M, Niitsuma N (1989) Detailed paleomagnetic records during the Brunhes-Matuyama geomagnetic reversal, and a direct determination of depth lag for magnetization in marine sediments. Phys Earth Planet Int 56:133–150

Passier HF, De Lange GJ, Dekkers MJ (2001) Magnetic properties and geochemistry of the active oxidation front and the youngest sapropel in the eastern Mediterranean Sea. Geophys J Int 145(3):604–614

Sakamoto T, Sakai A, Hata M, Unozawa A, Oka S, Hiroshima T, Komazawa M, Murata Y (1987) Geological Map of Japan 1: 200 000, Tokyo. Geol Surv Japan, Tsukuba

Seno T, Takano T (1989) Seismotectonics at the trench-trench-trench triple junction off central Honshu. Pure Appl Geophys 129:27–40

Shackleton NJ, Berger A, Peltier WR (1990) An alternative astronomical calibration of the lower Pleistocene timescale based on ODP Site 677. Trans R Soc Edinburgh Earth Sci 81:251–261

Suganuma Y, Yokoyama Y, Yamazaki T, Kawamura K, Horng CS, Matsuzaki H (2010) ^{10}Be evidence for delayed acquisition of remanent magnetization in marine sediments: implication for a new age for the Matuyama-Brunhes boundary. Earth Planet Sci Lett 296(3):443–450. doi:10.1016/j.epsl.2010.05.031

Suganuma Y, Okuno J, Heslop D, Roberts AP, Yamazaki T, Yokoyama Y (2011) Post-depositional remanent magnetization lock-in for marine sediments deduced from ^{10}Be and paleomagnetic records through the Matuyama-Brunhes boundary. Earth Planet Sci Lett 311(1):39–52. doi:10.1016/j.epsl.2011.08.038

Takahashi N, Mitsuoka T, Katoh A, Yokoyama K (2005) Correlation of the key tephra bed 'Kd38' occurring near the boundary between the Tertiary and Quaternary in the southern part of the Kanto Province, central Japan: correlation in the Kazusa Group and the Chikura Group, Boso Peninsula. J Geol Soc Jpn 111:371–388

Tate Y, Majima R (1998) A chemosynthetic fossil community related to cold seeps in the outer shelf environment: a case study in the Lower Pleistocene Koshiba Formation, Kazusa Group, central Japan. J Geol Soc Jpn 104:24–41

Tauxe L (1993) Sedimentary records of relative paleointensity of the geomagnetic field: theory and practice. Rev Geophys 31:319–354

Tauxe L, Yamazaki T (2007) Paleointensities. In: Schubert G (ed) Treatise on geophysics, vol 5, 1st edn. Elsevier, Amsterdam, pp 509–563

Tauxe L, Herbert T, Shackleton NJ, Kok YS (1996) Astronomical calibration of the Matuyama–Brunhes boundary: consequences for magnetic remanence acquisition in marine carbonates and the Asian loess sequences. Earth Planet Sci Lett 140:133–146. doi:10.1016/0012-821X(96)00030-1

Tauxe L, Steindor JL, Harris A (2006) Depositional remanent magnetization: toward an improved theoretical and experimental foundation. Earth Planet Sci Lett 244:515–529. doi:10.1016/j.epsl.2006.02.003

Tric E, Laj C, Jéhanno C, Valet JP, Kissel C, Mazaud A, Iaccarino S (1991) High-resolution record of the Upper Olduvai transition from Po Valley (Italy) sediments: support for dipolar transition geometry? Phys Earth Planet Int 65(3):319–336

Unozawa A, Oka S, Sakamoto T, Komazawa M (1983) Geological map of Japan 1: 200 000, Chiba. Geol Surv Japan, Tsukuba

Utsunomiya M, Majima R (2012) Paleobathymetries of the Plio-Pleistocene Urago and Nojima Formations, Kazusa Group, Miura Peninsula, central Japan: revision on the basis of molluscan fossils from new localities. Fossils (Palaeontological Soci Jpn) 91:5–14

Valet JP, Meynadier L, Guyodo Y (2005) Geomagnetic dipole strength and reversal rate over the past two million years. Nature 435(7043):802–805

Valet JP, Fournier A, Courtillot V, Herrero-Bervera E (2012) Dynamical similarity of geomagnetic field reversals. Nature 490(11491):89–93. doi:10.1038/nature11491

Verosub KL (1977) Depositional and postdepositional processes in the magnetization of sediments. Rev Geophys 15(2):129–143

Wada H, Fujii N, Niitsuma N (1984) Analytical method of stable isotope for ultra-small amounts of carbon dioxide with MAT250 mass-spectrometer. Geosci Repts Shizuoka Univ 10:103–112

Yamazaki T, Oda H (2005) A geomagnetic paleointensity stack between 0.8 and 3.0 Ma from equatorial Pacific sediment cores. Geochem Geophys Geosys 6(11):Q11H20. doi:10.1029/2005GC001001

Zijderveld JDA (1967) A. C. demagnetization of rocks: analysis of results. In: Collinson DW, Creer KM, Runcorn SK (eds) Methods in palaeomagnetism. Elsevier, Amsterdam, pp 254–286

11

Expected geoneutrino signal at JUNO

Virginia Strati[1,2*], Marica Baldoncini[1,3], Ivan Callegari[2], Fabio Mantovani[1,3], William F McDonough[4], Barbara Ricci[1,3] and Gerti Xhixha[2]

Abstract

Constraints on the Earth's composition and on its radiogenic energy budget come from the detection of geoneutrinos. The Kamioka Liquid scintillator Antineutrino Detector (KamLAND) and Borexino experiments recently reported the geoneutrino flux, which reflects the amount and distribution of U and Th inside the Earth. The Jiangmen Underground Neutrino Observatory (JUNO) neutrino experiment, designed as a 20 kton liquid scintillator detector, will be built in an underground laboratory in South China about 53 km from the Yangjiang and Taishan nuclear power plants, each one having a planned thermal power of approximately 18 GW. Given the large detector mass and the intense reactor antineutrino flux, JUNO aims not only to collect high statistics antineutrino signals from reactors but also to address the challenge of discriminating the geoneutrino signal from the reactor background. The predicted geoneutrino signal at JUNO is $39.7^{+6.5}_{-5.2}$ terrestrial neutrino unit (TNU), based on the existing reference Earth model, with the dominant source of uncertainty coming from the modeling of the compositional variability in the local upper crust that surrounds (out to approximately 500 km) the detector. A special focus is dedicated to the $6° \times 4°$ local crust surrounding the detector which is estimated to contribute for the 44% of the signal. On the basis of a worldwide reference model for reactor antineutrinos, the ratio between reactor antineutrino and geoneutrino signals in the geoneutrino energy window is estimated to be 0.7 considering reactors operating in year 2013 and reaches a value of 8.9 by adding the contribution of the future nuclear power plants. In order to extract useful information about the mantle's composition, a refinement of the abundance and distribution of U and Th in the local crust is required, with particular attention to the geochemical characterization of the accessible upper crust where 47% of the expected geoneutrino signal originates and this region contributes the major source of uncertainty.

Keywords: Geoneutrino flux; JUNO experiment; Earth reference model; Earth composition; Heat-producing elements; Reactor antineutrinos

Background

The first experimental evidence of geoneutrinos, i.e., electron antineutrinos produced in beta decays along the ^{238}U and ^{232}Th decay chains, was claimed by the Kamioka Liquid scintillator Antineutrino Detector (KamLAND) Collaboration in 2005 (KamLAND Collaboration 2005), which ushered in a new method for exploring the Earth's interior and provided constraints on the planet's composition and specifically its radiogenic element budget (Fiorentini et al. 2007). The geoneutrino energy spectrum contains in it distinctive contributions from U and Th, each one resulting from different rates and shapes of their decays (see Figures three and five of Fiorentini et al. 2007) and from concentrations and spatial distributions of these elements inside the Earth.

Geoneutrinos are measured in liquid scintillation detectors via the inverse beta decay (IBD) reaction on free protons:

$$\bar{\nu}_e + p \rightarrow e^+ + n$$

whose energy threshold of 1.806 MeV means that only a small fraction of the antineutrinos produced from the U and Th decay chains are detectable. The IBD detection event in a liquid scintillator produces two flashes of light: the annihilation flash, from electron-positron interaction, followed by the deuterium formation flash, which is 2.2 MeV of light that follows some 200 μs later. The delayed coincidence of these two flashes of light provides the critical identification of the antineutrino

* Correspondence: strati@fe.infn.it
[1]Department of Physics and Earth Sciences, University of Ferrara, Via Saragat 1, 44121 Ferrara, Italy
[2]Legnaro National Laboratories, INFN, Viale dell'Università, 2, 35020 Legnaro, Italy
Full list of author information is available at the end of the article

interaction and eliminates most background events. The KamLAND and Borexino experiments recently reported 116^{+28}_{-27} geoneutrino events over 2,991 days (KamLAND Collaboration 2013) and 14.3 ± 4.4 geoneutrino events in 1,353 days (Borexino Collaboration 2013), respectively. Differences in the detection rates reflect the detector sizes, with the KamLAND detector being approximately 1 kton and the Borexino detector 0.3 kton.

The most significant source of background for geoneutrino measurements is due to reactor antineutrinos, i.e., electron antineutrinos emitted during the beta decays of fission products from ^{235}U, ^{238}U, ^{239}Pu, and ^{241}Pu burning. Approximately 30% of the reactor antineutrino events are recorded in the geoneutrino energy window extending from the IBD threshold up to the endpoint of the ^{214}Bi beta decay spectrum (3.272 MeV) (Fiorentini et al. 2010). The Terrestrial Neutrino Unit (TNU), which is the signal that corresponds to one IBD event per 10^{32} free protons per year at 100% efficiency, is used to compare the different integrated spectral components (i.e., antineutrinos from U, Th, and reactors) measured by the detectors or just beneath the Earth's surface.

In the past decade, reactor antineutrino experiments played a decisive role in unraveling the neutrino puzzle, which currently recognizes three flavor eigenstates (ν_e, ν_μ, and ν_τ), each of which mixes with three mass eigenstates (ν_1, ν_2, and ν_3) via three mixing angles (θ_{12}, θ_{13}, and θ_{23}). The quantities that govern the oscillation frequencies are two differences between squared masses (i.e., $\delta m^2 = m_2^2 - m_1^2 > 0$ and $\Delta m^2 = m_3^2 - (m_1^2 + m_2^2)/2$). Central to neutrino studies is understanding the neutrino mass hierarchy (i.e., $\Delta m^2 > 0$ or $\Delta m^2 < 0$) (Capozzi et al. 2014; Ge et al. 2013).

Massive (>10 kton) detectors such as the Jiangmen Underground Neutrino Observatory (JUNO) (Li 2014) and Reno-50 (Kim 2013) experiments are being constructed at medium baseline distances (a few tens of kilometers) away from bright reactor antineutrino fluxes in order to assess significant physics goals regarding the neutrino properties, in the first place, the mass hierarchy. These experiments intend also to obtain subpercent precision measurements of neutrino oscillation parameters and along the way make observations of events of astrophysical and terrestrial origin.

JUNO is located (N 22.12°, E 112.52°) in Kaiping, Jiangmen, Guangdong Province (South China), about 53 km away from the Yangjiang and Taishan nuclear power plants, which are presently under construction. The combined thermal power of these two units is planned to be on the order of 36 GW (Li and Zhou 2014) (Figure 1). The JUNO experiment is designed as a liquid scintillator detector of 20 kton mass that will be built in a laboratory some 700 m underground (approximately 2,000 m water equivalent). This amount of overburden will attenuate the cosmic muon flux, which contributes to the overall detector background signal, but this overburden is significantly less than that at the KamLAND and Borexino experiments. The detector energy response and the spatial distribution of the reactor cores are the most critical features affecting the experimental sensitivity (Li et al. 2013) required to achieve the intended physics goals.

The goal of this present study is to predict the geoneutrino signal at JUNO on the basis of an existing reference Earth model (Huang et al. 2013), together with an estimate of the expected reactor antineutrino signal. Since a significant contribution to the expected geoneutrino signal comes from U and Th in the continental crust surrounding the site, we follow past approaches to study the local contribution (Coltorti et al. 2011; Fiorentini et al. 2012; Huang et al. 2013; Huang et al. 2014), with a particular interest in focusing on the closest 6° × 4° grid surrounding the detector. We define this latter region as the local crust (LOC) (Figure 1).

Methods

The geoneutrino signal expected at JUNO is calculated adopting the same methodology and the same inputs of the reference Earth model developed by Huang et al. (2013). It provides a description of the abundances and distribution of the heat-producing elements (HPEs; i.e., U, Th, and K) in the Earth's crust, along with their uncertainties. According to this model, the silicate portion of the Earth is composed of five dominant reservoirs: the depleted mantle (DM), the enriched mantle (EM), the lithospheric mantle (LM), the continental crust (CC), and the oceanic crust (OC). The continental crust is dominantly composed of the lower crust (LC), middle crust (MC), and upper crust (UC), and it is overlain by shallow layers of sediments (Sed) which also cover the OC.

The surface geoneutrino flux is calculated by dividing the Earth's surface in 1° × 1° tiles that are projected vertically into discrete volume cells, and each cell is assigned with physical and chemical states. Just for the sake of computing flux, the 1° × 1° tiles are further subdivided into many subcells with the same properties of the parent tile. The number of subcells is progressively bigger approaching the detector location with the aim of not introducing any bias due to discretization.

The total crustal thickness of each cell and its associated uncertainty correspond, respectively, to the mean and the half range of three crustal models obtained from different approaches: the global crustal model based on reflection and refraction data 'CRUST 2.0' (Bassin et al. 2000; Laske et al. 2001), the global shear velocity model of the crust and upper mantle 'CUB 2.0' (Shapiro and Ritzwoller 2002), and the high-resolution map of Moho (crust-mantle boundary) depth based on the gravity field data 'GEMMA' (Reguzzoni and Tselfes 2009; Reguzzoni

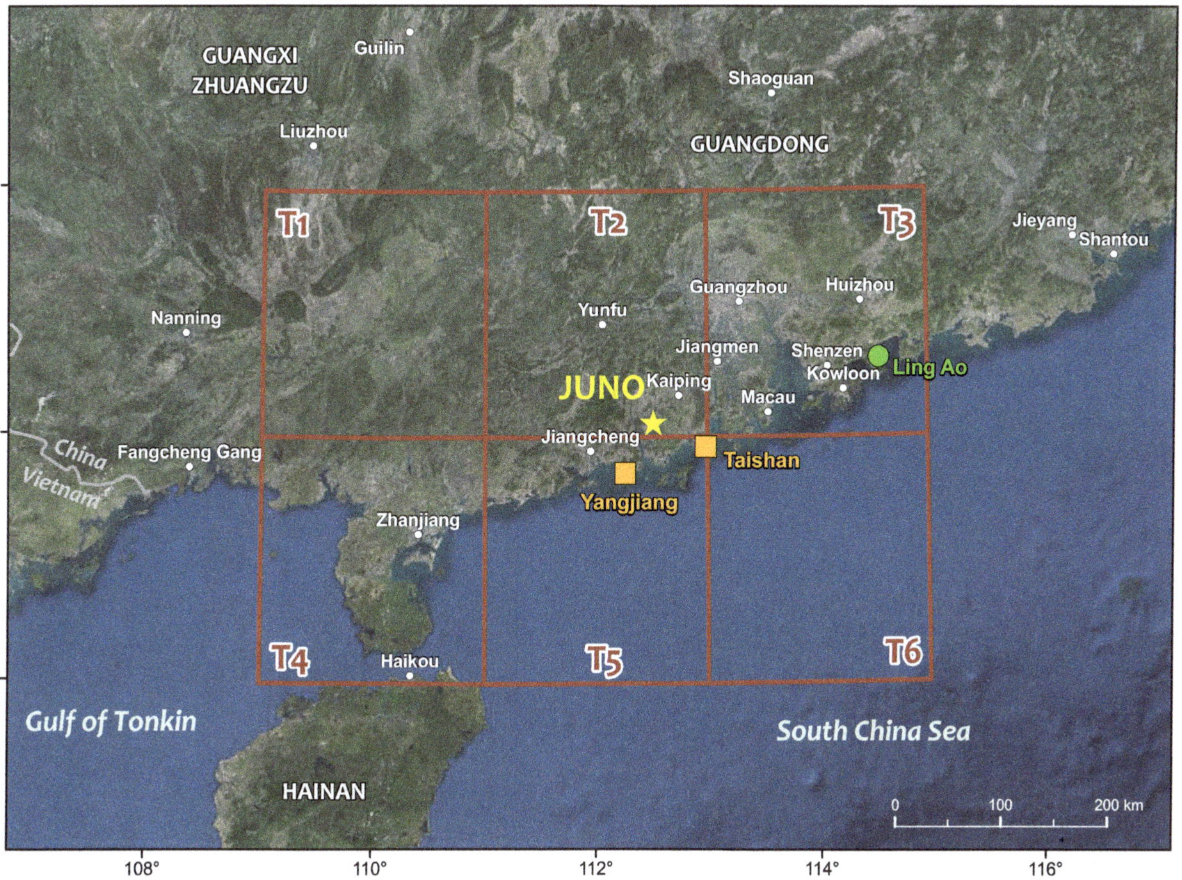

Figure 1 Map of LOC surrounding JUNO. JUNO (yellow star) located in Kaiping, Jiangmen, Guangdong Province (South China) and the planned (orange square) and operational (green circle) nuclear power plants. The six 2° × 2° tiles (dark red lines) define the LOC.

and Sampietro 2015). The reference model incorporates the relative proportional thickness of the crustal layers along with density and elastic properties (compressional and shear wave velocity) reported in CRUST 2.0. The same information is adopted for the Sed layer using the global sediment map of Laske and Masters (1997). In Figure 2, the thicknesses of the continental crust layers in the 24 cells constituting the LOC for JUNO are reported. Their total crustal thickness ranges between 26.3 and 32.3 km with an uncertainty for each cell of approximately 7%.

The HPE abundances in the Sed, OC, and UC layers are assumed to be relatively homogenous and correspond to the values reported in Table three of Huang et al. (2013). The ratio between the felsic and mafic components in the deep CC (MC and LC) is inferred from seismic velocity data, and these data are in turn used to estimate the U and Th content of each cell of the reference crustal model. Focusing on the LOC, the central values of U abundance in MC and LC vary in the range 0.8 to 1.2 μg/g and 0.3 to 0.1 μg/g, respectively. The Th/U ratio in the deep CC of the LOC is typically approximately 5 as compared to a bulk silicate Earth ratio of 3.9 or a bulk CC ratio just greater than 4.0; the higher Th/U ratio in the deep CC is likely due to the greater upward mobility of U during dehydration reactions that accompany granulite facies metamorphism of the deep CC.

In the reference model of Huang et al. (2013), the LM corresponds to the portion of the Earth between the Moho discontinuity and an assumed standard depth of 175 km beneath the surface. The thickness of this unit in the LOC ranges between 143 and 149 km, and its composition is modeled from the database reported in McDonough (1990) and the update in Huang et al. (2013). In our calculation, we adopt for the LM the U and Th abundances of $0.03^{+0.05}_{-0.02}$ and $0.15^{+0.28}_{-0.10}$ μg/g, respectively (Huang et al. 2013).

The sublithospheric mantle extends down from the base of the lithosphere to the core-mantle boundary and is divided in two spherically symmetric domains, the DM and the EM, whose density profiles are derived from the Preliminary Reference Earth Model (PREM) (Dziewonski and Anderson 1981). Adopting a mass ratio $M_{DM}/M_{EM} = 4.56$ (Huang et al. 2013), we calculate the

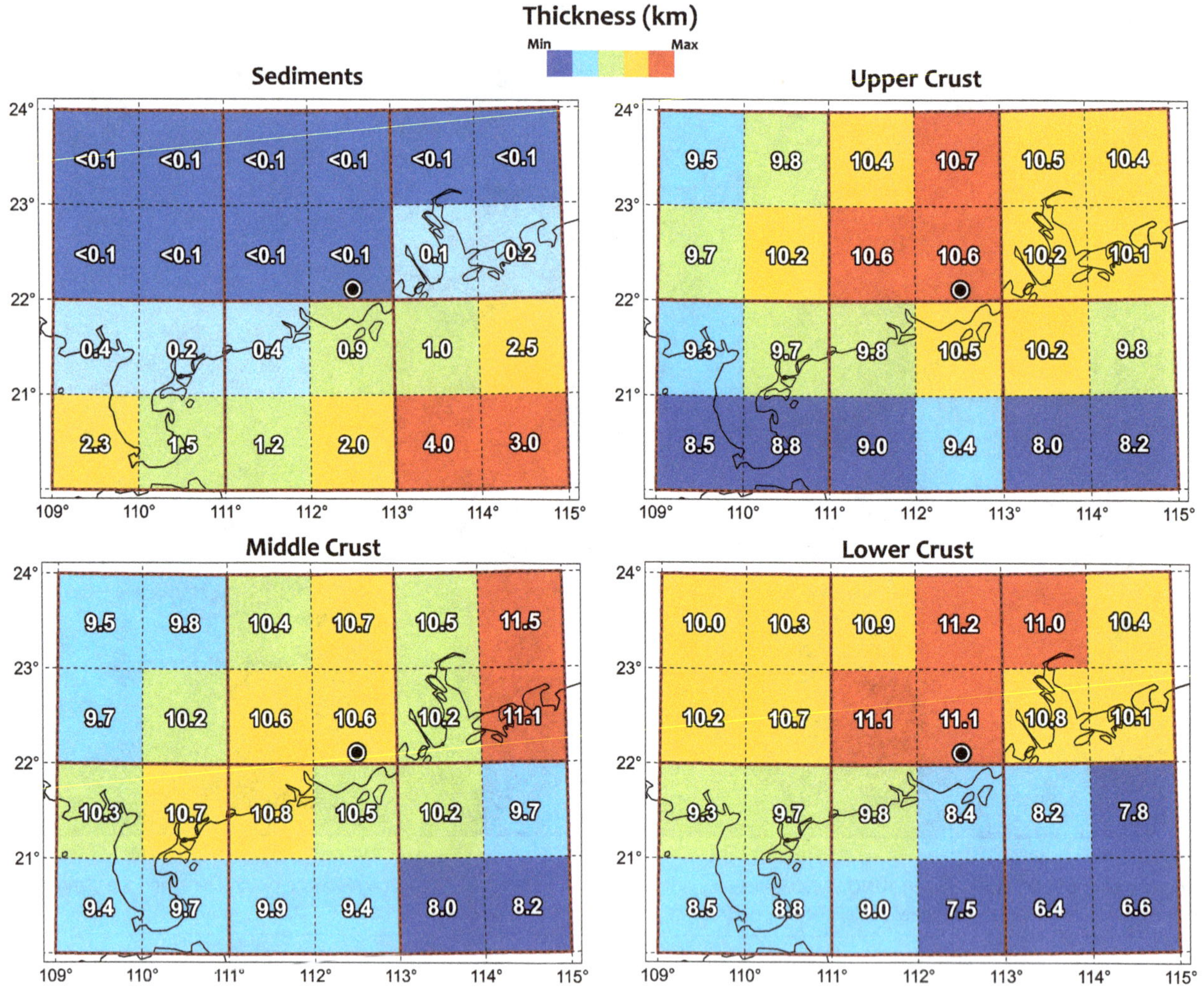

Figure 2 Thicknesses of the four crustal layers in the LOC. The thicknesses in kilometers of the Sed, UC, MC, and LC layers are reported for each of the 24 cells constituting the LOC surrounding JUNO (black circle), with color coding to illustrate gradients in thickness.

masses of these two reservoirs $M_{DM} = 3.207 \times 10^{24}$ kg and $M_{EM} = 0.704 \times 10^{24}$ kg. In a survey of the midocean ridge basalts (MORB), Arevalo and McDonough (2010) reported the lognormal-based average abundances of uranium ($U_{MORB} = 80$ ng/g) and thorium ($Th_{MORB} = 220$ ng/g) and, from this, calculated the $U_{DM} = 8$ ng/g and $Th_{DM} = 22$ ng/g based on a simple melting model. Based on these assumptions, the U_{EM} can be calculated:

$$U_{EM} = \frac{m_{BSE} - m_C}{M_{EM}} - U_{DM}\frac{M_{DM}}{M_{EM}},$$

where $m_{BSE} = 8.1 \times 10^{16}$ kg is the U mass in the bulk silicate earth (BSE) (McDonough and Sun 1995) and $m_C = 3.1 \times 10^{16}$ kg is the total U mass in the crust (Huang et al. 2013). The mantle geoneutrino signals reported in Table 1 are calculated with $U_{DM} = 8$ ng/g and $U_{EM} = 34$ ng/g together with $(Th/U)_{DM} = 2.8$ and $(Th/U)_{EM} = 4.8$.

Table 1 Geoneutrino signals from U and Th expected in JUNO

	S (U)	*S* (Th)	*S* (U + Th)
Sed CC	$0.48^{+0.06}_{-0.06}$	$0.16^{+0.02}_{-0.02}$	$0.64^{+0.06}_{-0.06}$
UC	$14.6^{+3.5}_{-3.4}$	$3.9^{+0.5}_{-0.5}$	$18.5^{+3.6}_{-3.4}$
MC	$4.7^{+3.0}_{-1.8}$	$1.7^{+1.6}_{-0.8}$	$6.8^{+3.6}_{-2.3}$
LC	$0.9^{+0.7}_{-0.4}$	$0.4^{+0.7}_{-0.2}$	$1.5^{+1.0}_{-0.6}$
Sed OC	$0.08^{+0.02}_{-0.02}$	$0.03^{+0.01}_{-0.01}$	$0.11^{+0.02}_{-0.02}$
OC	$0.05^{+0.02}_{-0.02}$	$0.01^{+0.01}_{-0.01}$	$0.06^{+0.02}_{-0.02}$
Bulk crust	$21.3^{+4.8}_{-4.2}$	$6.6^{+1.9}_{-1.2}$	$28.2^{+5.2}_{-4.5}$
CLM	$1.3^{+2.4}_{-0.9}$	$0.4^{+1.0}_{-0.3}$	$2.1^{+2.9}_{-1.3}$
Total lithosphere	$23.2^{+5.9}_{-4.8}$	$7.3^{+2.4}_{-1.5}$	$30.9^{+6.5}_{-5.2}$
DM	4.2	0.8	5.0
EM	2.9	0.9	3.8
Grand total	$30.3^{+5.9}_{-4.8}$	$9.0^{+2.4}_{-1.5}$	$39.7^{+6.5}_{-5.2}$

The inputs for the calculations are taken from Huang et al. (2013), and the signals from the different reservoirs indicated in the first column are in TNU.

Results and discussion

In Table 1, we summarize the contributions to the expected geoneutrino signal at JUNO produced by U and Th in each of the reservoirs identified in the model. The central value and the asymmetric uncertainties are, respectively, the median and 1σ errors of a positively skewed distribution obtained from Monte Carlo simulation. This approach was developed for the first time by Huang et al. (2013) in order to combine the Gaussian probability density function of geophysical and (some) geochemical inputs, together with the lognormal distributions of U and Th abundances observed in the felsic and mafic rocks of MC and LC.

The total geoneutrino signal at JUNO is $G = 39.7^{+6.5}_{-5.2}$ TNU where the 1σ error only recognizes the uncertainties of the inputs of the lithosphere, which are mainly due to the uncertainties in the composition of the rocks and subsequently to the geophysical inputs. The predicted mantle contribution at JUNO is assumed to be $S_M \approx 9$ TNU (Huang et al. 2013). The expected geoneutrino signal from the mantle is essentially model dependent, and it is estimated according to a mass balance argument. Uncertainty in the assumed mantle model is much less than the predicted for the lithosphere (e.g., $\delta G \approx \pm 6$ TNU). An extensive discussion of different mantle structures is described in Šrámek et al. (2013), which considers a range of geoneutrino signals for different mantle models.

Thus, a future refinement of the abundances and distribution of HPEs in the UC surrounding the JUNO detector is strongly recommended, as this region provides approximately 47% of the total geoneutrino signal (G) and is a significant contributor to the total uncertainty.

Plotting the cumulative geoneutrino signal as a function of the distance from JUNO for the different Earth reservoirs (Figure 3), we observe that half of the total signal comes from U and Th in the regional crust that lies within 550 km of the detector. Since the modeling of the geoneutrino flux is based on 1° × 1° cells, we study the signal produced in LOC subdivided in six 2° × 2° tiles (Figure 1).

The geoneutrino signals from U and Th in the lithosphere of each tile are reported in Table 2 with their uncertainties. The main contribution (27% of G) comes from tile T2 in which the JUNO experiment is located (Figure 1). The thick UC in this tile, which is covered by a very shallow layer of Sed (Figure 2), is predicted to give a signal of $7.6^{+1.5}_{-1.4}$ TNU. Therefore, a refined study of the U and Th content of the UC in tile T2 is a high-value target for improving the accuracy and precision of the predicted geoneutrino signal at JUNO. Evaluating the antineutrino signal requires knowledge of several ingredients necessary for modeling the three antineutrino life stages: production, propagation to the detector site, and detection in liquid scintillation detectors via the IBD reaction. The propagation and detection processes are independent from the source of the particles, and we modeled these two stages using the oscillation parameters from Ge et al. (2013) and the IBD cross section from Strumia and Vissani (2003). The spectral parameters for U and Th geoneutrinos are from Fiorentini et al. (2007), and the modulation of these fluxes are based on Huang et al. (2013). Reactor antineutrino production is calculated adopting the data from a worldwide reference model from Baldoncini et al. (2015). Reported in Figure 4 are the energy distributions of geoneutrinos and reactor antineutrino signals in two different scenarios: in the full energy region, $R_{OFF} = 95.3^{+2.6}_{-2.4}$ TNU is obtained with data from the worldwide commercial reactors operating in 2013 and $R_{ON} = 1,566^{+111}_{-100}$ TNU, including the Yangjiang (17.4 GW) and Taishan (18.4 GW) nuclear power plants operating at an 80% annual average load factor (Baldoncini et al. 2015). In the geoneutrino energy window (i.e., 1.806 to 3.272 MeV), the reactor signals are $S_{OFF} = 26.0^{+2.2}_{-2.3}$ and $S_{ON} = 354^{+45}_{-41}$ TNU (Table 3).

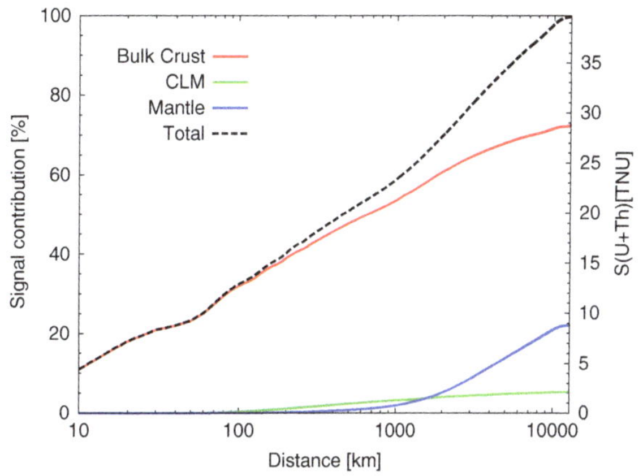

Figure 3 Geoneutrino signal contribution. The cumulative geoneutrino signal and the percentage contributions of the bulk crust, continental lithospheric mantle (CLM), and mantle are represented as functions of the distance from JUNO.

Table 2 Geoneutrino signals from six tiles of the LOC

Tile	S (U)	S (Th)	S (U + Th)	Percentage
T1	$0.4^{+0.1}_{-0.1}$	$0.1^{+0.1}_{-0.1}$	$0.5^{+0.1}_{-0.1}$	3.0
T2	$8.1^{+1.9}_{-1.7}$	$2.6^{+0.8}_{-0.5}$	$10.8^{+2.1}_{-1.8}$	62.1
T3	$1.1^{+0.3}_{-0.2}$	$0.4^{+0.2}_{-0.1}$	$1.5^{+0.3}_{-0.3}$	8.6
T4	$0.3^{+0.1}_{-0.1}$	$0.1^{+0.1}_{-0.1}$	$0.4^{+0.1}_{-0.1}$	2.2
T5	$2.5^{+0.5}_{-0.5}$	$0.7^{+0.2}_{-0.1}$	$3.2^{+0.6}_{-0.5}$	18.2
T6	$0.8^{+0.2}_{-0.2}$	$0.2^{+0.1}_{-0.1}$	$1.0^{+0.2}_{-0.2}$	5.9

The expected geoneutrino signals from U and Th contained in the lithosphere (CC+CLM) of the six tiles reported in Figure 1 are expressed in TNU. In the last column contributions in percentage are reported.

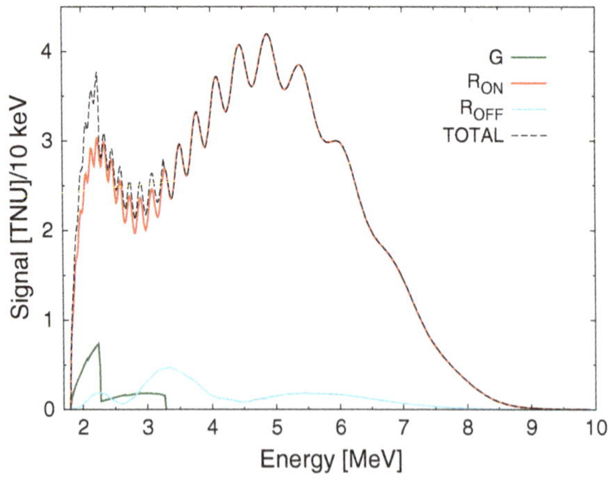

Figure 4 Antineutrino energy spectra expected at JUNO. Geoneutrino energy spectrum (green) is reported together with the energy reactor antineutrino spectra computed considering the commercial reactors operating all over the world in 2013 (cyan) and adding the contribution of the Yangjiang and Taishan nuclear power plants (red). The reactor antineutrino spectra are computed assuming normal hierarchy and neutrino oscillation. The total spectrum (black dashed lines) is obtained assuming the R_{ON} scenario.

Assuming a scenario whereby JUNO's signal does not have a background signal from Yangjiang and Taishan nuclear power plants, the ratio of $S_{OFF}/G = 0.7$, which compares to a value of 0.6 for the Borexino detector (Baldoncini et al. 2015). Considering only the statistical uncertainties, in the R_{OFF} scenario, JUNO is an excellent experiment for geoneutrino measurements reaching a 10% accuracy on the geoneutrino signal in approximately 105 days (assuming a $C_{17}H_{28}$ liquid scintillator composition, a 100% detection efficiency, and reactor antineutrinos as the sole source of background), given 576 geoneutrino events per year for a target mass of 14.5×10^{32} free protons. This optimistic expectation does not take into account the uncertainties of S_{OFF} and the background due to production of cosmic muon spallation, accidental coincidences, and radioactive contaminants in the detector.

Table 3 Geoneutrino and reactor antineutrinos signals at JUNO

	S (TNU)
Local contribution	$17.4^{+3.3}_{-2.8}$
Far-field crust	$13.4^{+3.3}_{-2.4}$
Mantle	8.8
Grand total of geoneutrinos	$39.7^{+6.5}_{-5.2}$
Reactors OFF	$26.0^{+2.2}_{-2.3}$
Reactors ON	354^{+45}_{-41}

The total geoneutrino signal (*G*) is the sum of the contributions from the local lithosphere (S_{LOC}), from the rest of the lithosphere (i.e., far-field crust, S_{FFC}), and from the mantle (S_M). The reactor antineutrino signal in the geoneutrino window is calculated from the data for commercial reactors operating all over the world in 2013 (S_{OFF}) and adding the contribution of the Yangjiang (17.4 GW) and Taishan (18.4 GW) nuclear power plants (S_{ON}) (Baldoncini et al. 2015). All the signals are expressed in TNU.

Conclusions

Designed as a 20 kton liquid scintillator detector, the JUNO experiment will collect high statistics for antineutrino signals from reactors and from the Earth. In this study, we focused on predicting the geoneutrino signal using the Earth reference model of Huang et al. (2013). The contribution originating from naturally occurring U and Th in the 6° × 4° LOC surrounding the JUNO detector (Figure 1) was determined. The main results of this study are summarized as follows:

- The thicknesses of the Sed, UC, MC, and LC layers of the 24 1° × 1° cells of the LOC are reported (Figure 2). The Moho depth of the continental LOC ranges between 26.3 and 32.3 km, and the uncertainty for each 1° × 1° cell is of the order of 7%.
- The total and local geoneutrino signals at JUNO are $G = 39.7^{+6.5}_{-5.2}$ and $S_{LOC} = 17.4^{+3.3}_{-2.8}$ TNU, respectively. The asymmetric 1σ errors are obtained from the Monte Carlo simulations and account only for uncertainties from the lithosphere. The major source of uncertainty comes from predicting the abundances and distribution of U and Th in local crustal rocks.
- High-resolution seismic data acquired in the LOC can improve the present geophysical model of the crust and CLM, of which the latter is assumed to have a homogenous depth of 175 ± 75 km. The CLM composition is derived from data for U and Th abundances inferred from the peridotite xenoliths, and its geoneutrino signal is of $2.1^{+2.9}_{-1.3}$ TNU.
- The HPEs in the regional crust extending out to 550 km from the detector produce half of the total expected geoneutrino signal (Figure 3). The U and Th in the 2° × 2° tile that hosts JUNO produces $10.8^{+2.1}_{-1.8}$ TNU corresponding to 27% of *G*. Since this region is characterized by a thick UC, which gives $7.6^{+1.5}_{-1.4}$ TNU, a refined geophysical and geochemical model of the UC of this tile is highly desired.
- The reactor signal in the geoneutrino window assuming two scenarios is $S_{OFF} = 26.0^{+2.2}_{-2.3}$ TNU with the 2013 reactor operational data only and $S_{ON} = 355^{+44}_{-41}$ TNU when the contributions of the Yangjiang and Taishan nuclear power plants are added. There is a potential to achieve up to 10% accuracy on geoneutrinos after 105 days of data accumulation, under conditions of the Yangjiang and Taishan nuclear power plants being off.

The JUNO experiment has the potential to reach a milestone in geoneutrino science, although some technical challenges must be addressed to minimize background (e.g., production of cosmic muon spallation, accidental coincidences, and radioactive contaminants in the detector). Assuming $S_{OFF}/G = 0.7$, JUNO can collect hundreds of low-background geoneutrino events in less than a year under optimal conditions. A future refinement of the U and Th distribution and abundance in the LOC is strongly recommended. Such data will lead to insights on the radiogenic heat production in the Earth, the composition of the mantle, and constraints on the chondritic building blocks that made the planet.

Abbreviations
BSE: bulk silicate Earth; CC: continental crust; CLM: continental lithospheric mantle; DM: depleted mantle; EM: enriched mantle; HPEs: heat-producing elements; IBD: inverse beta decay; JUNO: Jiangmen Underground Neutrino Observatory; LC: lower crust; LM: lithospheric mantle; MC: middle crust; OC: oceanic crust; Sed: sediments; TNU: terrestrial neutrino unit; UC: upper crust.

Competing interests
The authors declare that they have no competing interests.

Authors' contributions
VS and FM proposed and conceived the study and produced the geoneutrino data with WFM. MB and BR produced the reactor antineutrino data. IC and GX participated in the data analysis and interpretation of the results. VS took the lead in designing and composing the manuscript, and all the authors contributed to it. All authors read and approved the final manuscript.

Authors' information
The authors have a geological and physics background. VS and MB are graduate students attending PhD course in Physics at Physics and Earth Sciences Department of the University of Ferrara. FM and BR are researchers of Physics and Earth Sciences Department of the University of Ferrara and INFN. IC and GX are researchers at Legnaro National Laboratories. WFM is a professor of the Department of Geology at the University of Maryland.

Acknowledgements
We are grateful to R. L. Rudnick and Y. Huang for their fruitful discussions on crustal modeling of geoneutrino fluxes. We appreciate the observations on the geoneutrino signal predictions from S. Dye, G. Fiorentini, L. Ludhova, and H. Watanabe. We thank J. Mandula for the valuable help in compiling the nuclear reactor database. We wish to thank two anonymous reviewers for their detailed and thoughtful reviews. This work was partially supported by the Istituto Nazionale di Fisica Nucleare (INFN) through the ITALRAD Project, by the University of Ferrara through the research initiative 'Fondo di Ateneo per la Ricerca scientifica FAR 2014' and partially by the U.S. National Science Foundation Grants EAR 1067983/1068097.

Author details
[1]Department of Physics and Earth Sciences, University of Ferrara, Via Saragat 1, 44121 Ferrara, Italy. [2]Legnaro National Laboratories, INFN, Viale dell'Università, 2, 35020 Legnaro, Italy. [3]Ferrara Section, INFN, Via Saragat 1, 44121 Ferrara, Italy. [4]Department of Geology, University of Maryland, 237 Regents Drive, College Park, MD 20742, USA.

References

Arevalo R, McDonough WF (2010) Chemical variations and regional diversity observed in MORB. Chem Geol 271(1–2):70–85. doi:10.1016/j.chemgeo.2009.12.013

Baldoncini M, Callegari I, Fiorentini G, Mantovani F, Ricci B, Strati V, Xhixha G (2015) Reference worldwide model for antineutrinos from reactors. Physical Review D 91(6):065002

Bassin C, Laske G, Masters TG (2000) The current limits of resolution for surface wave tomography in North America. EOS Trans AGU 81:F897

Borexino Collaboration (2013) Measurement of geo-neutrinos from 1353 days of Borexino. Phys Lett B 722(4–5):295–300. doi:10.1016/j.physletb.2013.04.030

Capozzi F, Lisi E, Marrone A (2014) Neutrino mass hierarchy and electron neutrino oscillation parameters with one hundred thousand reactor events. Phys Rev D 89(1):013001

Coltorti M, Boraso R, Mantovani F, Morsilli M, Fiorentini G, Riva A, Rusciadelli G, Tassinari R, Tomei C, Di Carlo G, Chubakov V (2011) U and Th content in the Central Apennines continental crust: a contribution to the determination of the geo-neutrinos flux at LNGS. Geochim Cosmochim Acta 75(9):2271–2294. doi:10.1016/j.gca.2011.01.024

Dziewonski AM, Anderson DL (1981) Preliminary reference Earth model. Phys Earth Planet In 25:297–356. doi:10.1016/0031-9201(81)90046-7

Fiorentini G, Fogli G, Lisi E, Mantovani F, Rotunno A (2012) Mantle geoneutrinos in KamLAND and Borexino. Physical Review D 86 (3). doi:10.1103/PhysRevD.86.033004

Fiorentini G, Ianni A, Korga G, Lissia M, Mantovani F, Miramonti L, Oberauer L, Obolensky M, Smirnov O, Suvorov Y (2010) Nuclear physics for geo-neutrino studies. Physical Review C 81(3). doi:10.1103/PhysRevC.81.034602

Fiorentini G, Lissia M, Mantovani F (2007) Geo-neutrinos and Earth's interior. Phys Rep 453(5–6):117–172. doi:10.1016/j.physrep.2007.09.001

Ge S-F, Hagiwara K, Okamura N, Takaesu Y (2013) Determination of mass hierarchy with medium baseline reactor neutrino experiments. J High Energ Phys 2013(5):1–23. doi:10.1007/JHEP05(2013)131

Huang Y, Chubakov V, Mantovani F, Rudnick RL, McDonough WF (2013) A reference Earth model for the heat-producing elements and associated geoneutrino flux. Geochem Geophys Geosyst 14(6):2023–2029. doi:10.1002/ggge.20129

Huang Y, Strati V, Mantovani F, Shirey SB, McDonough WF (2014) Regional study of the Archean to Proterozoic crust at the Sudbury Neutrino Observatory (SNO+), Ontario: predicting the geoneutrino flux. Geochem Geophys Geosyst 15(10):3925–3944. doi:10.1002/2014gc005397

KamLAND Collaboration (2005) Measurement of neutrino oscillation with KamLAND: evidence of spectral distortion. Phys Rev Lett 94(8):081801

KamLAND Collaboration (2013) Reactor on-off antineutrino measurement with KamLAND. Phys Rev D 88(3):033001

Kim S-B (2013) Proposal for RENO-50. In: International workshop on RENO-50, Seoul National University, Korea, 13–14 June 2013

Laske G, Masters TG (1997) A global digital map of sediment thickness. Eos Trans AGU 78:F483

Laske G, Masters TG, Reif C (2001) CRUST 2.0: a new global crustal model at 2 × 2 degrees. http://igppweb.ucsd.edu/~gabi/crust2.html

Li Y-F, Cao J, Wang Y, Zhan L (2013) Unambiguous determination of the neutrino mass hierarchy using reactor neutrinos. Phys Rev D 88(1):013008

Li Y-F, Zhou Y-L (2014) Shifts of neutrino oscillation parameters in reactor antineutrino experiments with non-standard interactions. Nucl Phys B 888(0):137–153. doi:10.1016/j.nuclphysb.2014.09.013

Li YF (2014) Overview of the Jiangmen Underground Neutrino Observatory (JUNO). arXiv:1402.6143 [physics.ins-det] http://arxiv.org/pdf/1402.6143

McDonough WF (1990) Constraints on the composition of the continental lithospheric mantle. Earth Planet Sci Lett 101(1):1–18. doi:10.1016/0012-821x(90)90119-i

McDonough WF, Sun S-S (1995) The composition of the Earth. Chem Geol 120:223–253. doi:10.1016/0009-2541(94)00140-4

Reguzzoni M, Sampietro D (2015) GEMMA: an earth crustal model based on GOCE satellite data. Int J Appl Earth Observation Geoinform 35(Part A (0)):31–43. doi:10.1016/j.jag.2014.04.002

Reguzzoni M, Tselfes N (2009) Optimal multi-step collocation: application to the space-wise approach for GOCE data analysis. J Geodes 83(1):13–29. doi:10.1007/s00190-008-0225-x

Shapiro NM, Ritzwoller MH (2002) Monte-Carlo inversion for a global shear-velocity model of the crust and upper mantle. Geophys J Int 151:88–105. doi:10.1046/j.1365-246X.2002.01742.x

Šrámek O, McDonough WF, Kite ES, Lekić V, Dye ST, Zhong S (2013) Geophysical and geochemical constraints on geoneutrino fluxes from Earth's mantle. Earth Planet Sci Lett 361:356-366. doi:10.1016/j.epsl.2012.11.001

Strumia A, Vissani F (2003) Precise quasielastic neutrino/nucleon cross-section. Phys Lett B 564(1–2):42–54. doi:10.1016/s0370-2693(03)00616-6

12

Surface deformation and source modeling of Ayaz-Akhtarma mud volcano, Azerbaijan, as detected by ALOS/ALOS-2 InSAR

Kento Iio[1,3*] and Masato Furuya[2]

Abstract

Azerbaijan, located on the western edge of the Caspian Sea in Central Asia, has one of the highest populations of mud volcanoes in the world. We used satellite-based synthetic aperture radar (SAR) images derived from two L-band SAR satellites, ALOS/PALSAR along an ascending track from 2006 to 2011, and its successor ALOS-2/PALSAR-2 along both ascending and descending tracks from 2014 to 2017. First, we applied interferometric SAR (InSAR) technique to detect surface displacements at the Ayaz-Akhtarma mud volcano in Azerbaijan. The 35 derived interferograms indicate that the deformation of the mud volcano is largely characterized by horizontal displacement. Besides the InSAR technique, we also used multiple-aperture interferometry (MAI) to derive the surface displacements parallel to the satellite flight direction to complement the InSAR data. Using the InSAR and MAI data, we obtained 3D displacements, which indicate that the horizontal displacement is dominant relative to subsidence and possible uplift. To explain the displacements, we performed source modeling, based on the assumption of elastic dislocation theory in a half space. The derived model consists of a convex surface on which normal-fault-type slips are semi-radially distributed, causing the significant horizontal displacements with minor subsidence. The convex source surface suggests that a steady overpressure system would be maintained by constantly intruding mud and gas.

Keywords: Mud volcano, Surface deformation, Geodesy, Interferometric synthetic aperture radar, Multiple-aperture interferometry, Advanced Land Observing Satellite, Elastic dislocation model

Introduction

Mud volcanism is analogous to magmatic volcanism; however, the materials extruded to the surface are mud, gases (mostly methane), and saline water originating from deeper sediments. While mud volcanoes include a variety of surface features generated from the extruded materials, their morphology is basically a cone-shaped topographic high, though some can be relatively flat and can even include depressions or calderas. The size of mud volcanoes is generally smaller than that of magmatic volcanoes, but varies over a wide range, from tens of centimeters to several hundred meters in height and tens of kilometers in diameter (Kopf 2002; Dimitrov, 2002; Mazzini and Etiope 2017). Mud volcanoes are distributed both onshore and offshore, but are usually found in active tectonic settings, such as fold-and-thrust belts, accretionary complexes, and convergent plate margins (e.g., Milkov 2000; Kopf 2002; Dimitrov, 2002; Mazzini 2009; Bonini 2012; Mazzini and Etiope 2017), where we may expect compressive stress regimes and higher sedimentation rates. Also, the presence of mud volcanoes is often associated with petroleum systems (Dimitrov, 2002; Kopf, 2002; Mazzini and Etiope 2017). Multiple mechanisms are necessary to account for the formation of mud volcanoes. The first is the bulk density contrast between lighter clays and denser overburden in sedimentary layers, which will lead to the formation of mud diapirs. However, the buoyancy of mud diapirs is not strong enough alone for the formation of mud volcanoes, and additional overpressure is regarded as another requirement in their formation. The overpressure is

* Correspondence: knt-iio25@frontier.hokudai.ac.jp
[1]Department of Natural History Sciences, Graduate School of Science, Hokkaido University, N10W8, Kita-ku, Sapporo 060-0810, Japan
[3]Geospatial Information Authority of Japan, Kitasato 1, Tsukuba 305-0811, Japan
Full list of author information is available at the end of the article

generated not only from the compaction of initially retained water by the overburden layer but also from the biogenic gas produced at depth from organic matter (Dimitrov, 2002). Compressive tectonic stress also contributes to higher pore-fluid pressures. The higher content of water and gas will significantly reduce the bulk density, shear modulus, and viscosity, allowing the layer to flow. However, the actual subsurface geometry and locations of such fluid-rich muddy masses remain uncertain.

Using satellite-based synthetic aperture radar (SAR), we can image ground surfaces with a resolution on the order of ten meters or less, regardless of weather and sunlight. Taking the difference of the phase values of SAR images at different times, the interferometric SAR (InSAR) technique allows us to map surface displacements with unprecedented spatial resolution, with an accuracy of a few centimeters, and has been used to study earthquake faults and volcanic magma sources (e.g., Massonnet and Feigl, 1998; Bürgmann et al., 2000; Hanssen, 2001; Simons and Rosen, 2015). While there have been numerous applications of InSAR to surface deformation mapping at magmatic volcanoes, there are relatively few studies on its application to mud volcanoes, with the exception of the Lusi mud volcano eruption in Indonesia (e.g., Fukushima et al., 2009; Aoki and Sidiq, 2014).

Azerbaijan is located near the eastern edge of the Greater Caucasus and the western edge of the Caspian Sea and is one of the countries with the highest national populations of mud volcanoes (Fig. 1). Mud volcanoes in Azerbaijan are mostly located along the anticlinal structures found throughout the country (e.g., Bonini, 2012). While there have been a couple of pioneering studies that applied InSAR technique to detect the displacements due to mud volcanoes in Azerbaijan in the early 2000s (Hommels et al., 2003; Mellors et al., 2005), Antonielli et al. (2014) was the first to unequivocally reveal the surface displacements at four mud volcanoes in Azerbaijan with the uses of C-band Envisat/ASAR images gathered from 2003 to 2005. In particular, the study of the Ayaz-Akhtarma mud volcano (Fig. 2) revealed significant radar line-of-sight (LOS) changes, which were negative and positive in the eastern half and western half of the site, respectively. As the

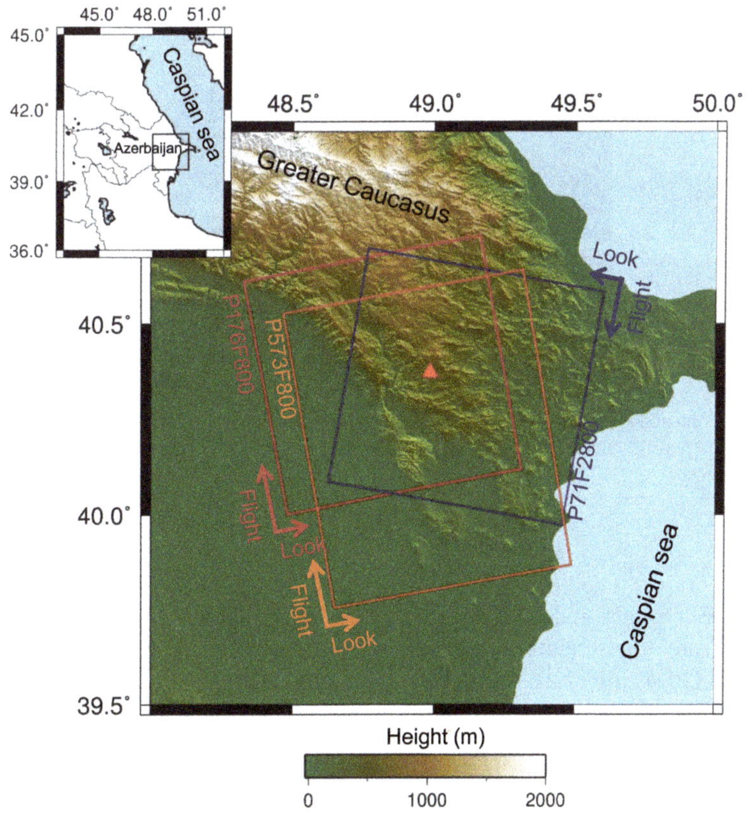

Fig. 1 Location of Ayaz-Akhtarma mud volcano, indicated with red triangle. Orange, red, and blue rectangles represent the imaging areas covered by ALOS/PALSAR ascending, ALOS-2/PALSAR-2 ascending, and ALOS-2/PALSAR-2 descending tracks, respectively. Upper-left panel indicates the location of Azerbaijan and Caspian Sea

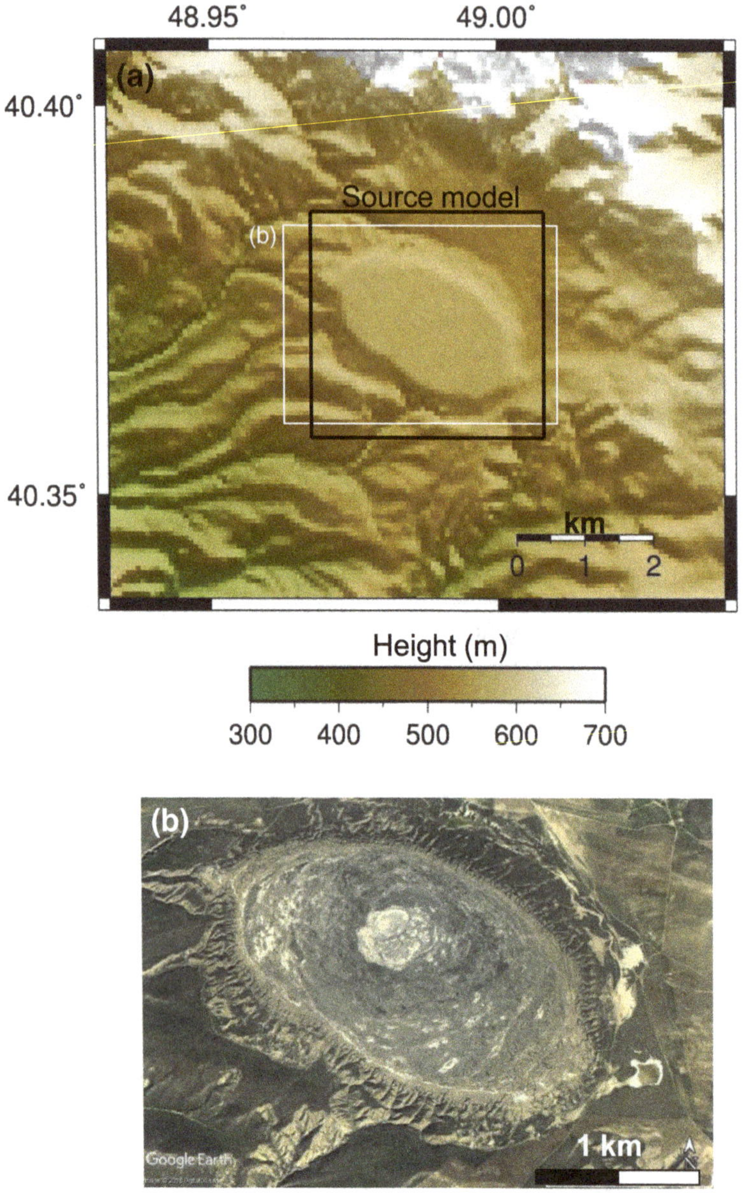

Fig. 2 **a** Elevation map of Ayaz-Akhtarma mud volcano and its surrounding area based on the ALOS World 3D-30 m (AW3D30) DEM. Plan view of the source model in Figs. 12 and 13 is also indicated with black line. **b** Google Earth image of Ayaz-Akhtarma mud volcano, acquired on 24 March 2004

negative and positive LOS changes indicate displacement toward and away from the satellite sensor, respectively, Antonielli et al. (2014) interpreted that these changes correspond to uplift and subsidence, respectively. However, the Envisat/ASAR images used in Antonielli et al. (2014) were acquired only from the descending track, with a fixed incidence angle. Because the radar LOS change derived by InSAR is a projection of the three-dimensional (3D) surface displacements onto the radar LOS direction, the observed LOS changes do not tell us the actual 3D displacement field, and the LOS is most sensitive to the vertical component of displacement because of its smaller incidence angle of ~ 20–40 degrees. The interpretations of the derived LOS changes by Antonielli et al. (2014) thus remain inconclusive.

Here, we also apply InSAR technique to detect the surface displacements at the Ayaz-Akhtarma mud volcano, but use L-band (wavelength 23.6 cm) images from Advanced Land Observing Satellite (ALOS)/

Phased Array type L-band Synthetic Aperture Radar (PALSAR) and its successor, ALOS-2/PALSAR-2, collected from 2006 to 2010 and 2014 to 2017, respectively. The L-band SAR is known to be more advantageous than shorter wavelength microwave bands in terms of interferometric coherence (Rosen et al., 1996), and hence allows improved unwrapping of the InSAR phase data. Moreover, we process the SAR images acquired from both ascending and descending tracks, which will help to judge if vertical displacements dominate over horizontal displacements. We also apply a multiple-aperture interferometry (MAI) technique to derive the surface displacements that are parallel to the satellite flight direction in order to complement the InSAR data (Bechor and Zebker, 2006), since InSAR phase is insensitive to these along-track (near north-south) displacements. We can thus infer the full 3D displacements at the Ayaz-Akhtarma mud volcano. Based on these displacements' data and analytical solutions of elastic dislocation theory in an elastic half-space, we derive a source model that consists of a convex fault surface with normal faulting and strike-slip, which we will use to investigate the on-going processes and their relation to the stress regime.

Methods/Experimental

Figure 1 shows the areas covered by ALOS/PALSAR for its ascending path and by ALOS-2/PALSAR-2 for both its ascending and descending paths; note that the look direction from the ascending tracks is opposite to that used by Antonielli et al. (2014). Details on the ALOS/ALOS-2 data used are listed in Tables 1, 2, 3, and 4. SAR data was processed using GAMMA software (Wegmüller and Werner, 1997). To acquire interferograms with high coherence, we formed interferometric pairs, making each temporal separation as shorter as possible. We analyzed these data such that the slave data become the next master data, and so on. We used the ALOS World 3D-30 m (AW3D30) digital elevation model (DEM) and the precision orbit data to eliminate topographic and orbital fringes, respectively. Since the interferograms still included long-wavelength phase trends, we removed these by fitting low-order polynomials (in view of the derived spatial scale of ground deformation, we feel that this removal does not affect the detection of the deformation signals). For the step of phase unwrapping, we used the minimum cost flow algorithm (Costantini, 1998).

The basic algorithm of MAI is described in Barbot et al. (2008). In the MAI method, we split the total aperture time into forward-looking and backward-look ing times and from these generate two single-look complex images. We then create forward and backward interferograms whose LOS directions are different. Taking the difference between the forward and backward interferograms, MAI provides us with the displacements projected along the satellite flight direction. Although the measurement precision is lower than that of InSAR, with a precision on the order of about 10 cm (Bechor and Zebker 2006; Barbot et al. 2008; Jung et al. 2009), the MAI data complements that obtained by InSAR and allows us to derive 3D displacements through the combination of both techniques. In the MAI processing, the multi-look size was set 2 and 4 in range and azimuth directions, respectively, and we also applied Goldstein-Werner's adaptive spectral filter with the exponent of 0.7 to smooth the signals (Goldstein and Werner 1998).

Table 1 ALOS/PALSAR datasets for interferograms in Fig. 3. All data sets are acquired from the ascending track (mostly from the south to the north) with an incidence angle 38.7° at the image center and a heading angle of − 10.1° measured clockwise from the north. The path and frame of all pairs are 573 and 800, respectively. Bperp means a perpendicular baseline

Pair no.	Date (yyyy/mm/dd)	Bperp (m)	Temporal baseline (days)
P1	2006/12/28–2007/02/12	1439	46
P2	2007/02/12–2007/06/30	614	138
P3	2007/06/30–2007/08/15	291	46
P4	2007/08/15–2007/09/30	271	46
P5	2007/09/30–2007/12/31	346	92
P6	2007/12/31–2008/04/01	1287	92
P7	2008/04/01–2008/05/17	199	46
P8	2008/05/17–2008/07/02	− 3292	46
P9	2008/07/02–2008/08/17	− 2829	46
P10	2008/08/17–2008/10/02	1015	46
P11	2008/10/02–2008/11/17	427	46
P12	2008/11/17–2009/01/02	181	46
P13	2009/01/02–2009/02/17	569	46
P14	2009/02/17–2009/08/20	754	184
P15	2009/08/20–2009/10/05	501	46
P16	2009/10/05–2010/01/05	488	92
P17	2010/01/05–2010/02/20	693	46
P18	2010/02/20–2010/04/07	152	46
P19	2010/04/07–2010/05/23	73	46
P20	2010/05/23–2010/07/08	52	46
P21	2010/07/08–2010/08/23	378	46
P22	2010/08/23–2011/01/08	880	138
P23	2011/01/08–2011/02/23	696	46

Results

Observation results

In Figs. 3 and 4, we show the observed ALOS and ALOS-2 InSAR data, both derived from the ascending path; details are shown in Table 1 for P1–P23 and

Table 2 ALOS-2/PALSAR-2 data sets for interferograms in Fig. 4. The A and D stand for ascending and descending (mostly from the north to the south) orbits. The path and frame of the ascending and descending datasets are 176–800 and 71–2800, respectively. The incidence angle of the ascending and descending datasets are 31.4° and 36.3°, respectively. Heading angles of ascending and descending datasets are – 10.8 and – 169.8°, respectively

Pair no.	Orbit	Date (yyyy/mm/dd)	Bperp (m)	Temporal baseline (days)
P24	A	2014/09/17–2014/11/26	– 16	70
P25	A	2014/11/26–2015/02/04	98	70
P26	A	2015/02/04–2015/04/29	49	84
P27	A	2015/04/29–2015/07/08	90	70
P28	A	2015/07/08–2015/09/16	– 125	70
P29	A	2015/09/16–2016/06/08	31	266
P30	A	2016/06/08–2016/09/14	– 88	98
P31	A	2016/09/14–2016/11/23	43	70
P32	A	2016/11/23–2017/07/05	– 104	224
P33	D	2016/03/23–2016/06/01	– 360	70
P34	D	2016/06/01–2017/03/22	490	294
P35	D	2017/03/22–2017/05/31	– 439	70

Table 2 for P24–P32. Although they show diverse patterns of LOS changes, which we will discuss later in this paper, all of the results indicate that the western and eastern sectors across a N-S trending boundary near the center of the mud volcano show the signals in opposite sign. Figure 3 indicates that the surface approached the satellite in the west sector and moved away from the satellite in the east sector, respectively. The 23 interferograms in Fig. 3 are stacked to represent cumulative LOS changes in Fig. 5a. The cumulative LOS displacements of the western and the eastern sectors are around – 2.5 m and + 3.5 m, respectively, during the period of December 28, 2006, to February 23, 2011. In Fig. 4, for the ALOS-2 data, we can see essentially the same

Table 3 ALOS/PALSAR datasets for MAI images in Fig. 6. Details of the orbit, incidence, and heading angle are the same as in Table 1

Pair no.	Date (yyyy/mm/dd)	Bperp (m)	Temporal baseline (days)
P36	2006/12/28–2007/06/30	2054	184
P37	2007/06/30–2007/12/31	908	184
P38	2007/12/31–2008/07/02	– 1813	184
P39	2008/07/02–2009/02/17	– 646	230
P40	2009/02/17–2009/08/20	754	184
P41	2009/08/20–2010/01/05	990	138
P42	2010/01/05–2010/08/23	1347	230
P43	2010/08/23–2011/02/23	1575	184

Table 4 ALOS-2/PALSAR-2 data sets for MAI images in Fig. 7. Details of the orbit, incidence, and heading angle are the same as in Table 2

Pair no.	Orbit	Date (yyyy/mm/dd)	Bperp (m)	Temporal baseline (days)
P44	A	2014/09/17–2015/02/04	83	140
P45	A	2015/02/04–2015/09/16	– 85	224
P46	A	2015/09/16–2016/06/08	31	266
P47	A	2016/06/08–2016/11/23	– 46	168
P48	A	2016/11/23–2017/07/05	– 104	224
P49	D	2016/03/23–2017/03/22	130	364

pattern of LOS changes as in the ALOS InSAR data. The cumulative interferogram for the data in Fig. 4 is shown in Fig. 5b, with minimum and maximum values of – 1.5 m and + 1.5 m for the western and eastern sectors, respectively, during the period of September 17, 2014, to July 5, 2017. Although there are only three interferometric pairs, the observed ALOS-2 InSAR data along the descending track (P33–P35 in Table 2) are shown in Fig. 4. Descending interferograms indicate that the spatial pattern of LOS changes is consistent with the previous study (Antonielli et al. 2014), which, as for the ascending track, show eastern and western sectors separated at the center of the mud volcano by a near N-S boundary (Figs. 3 and 4). The cumulative LOS changes based on the three interferograms are shown in Fig. 5c, up to + 25 cm and – 50 cm for the western and eastern sectors, respectively, during the period of March 23, 2016, to May 31, 2017.

The LOS changes in ascending and descending orbits indicate that the areas near and far from the satellite show negative and positive changes, respectively. This, in turn, suggests that the nearside areas are approaching the satellite and the farside areas are moving away from the satellite. Hence, we can conclude that the LOS changes are mostly dominated by horizontal, rather than vertical, displacements. Because InSAR LOS change is most sensitive to the vertical component, we may observe similar spatial patterns in the LOS changes with the same signs, regardless of ascending and descending orbits, when the surface displacements are dominated by subsidence or uplift signals over flat areas (e.g., Aoki and Sidiq 2014). Our observations, however, clearly demonstrate that the observed LOS changes are not entirely due to the vertical displacements and that it is critically important to view one area from various directions.

In Fig. 6, we show the LOS change time series for ALOS (Fig. 6a) and ALOS-2 (Fig. 6b) data along the ascending orbit, which were derived by averaging the LOS change data in the western and eastern areas marked in Fig. 6c. These results indicate that the LOS changes are largely linear with some temporal fluctuations. The

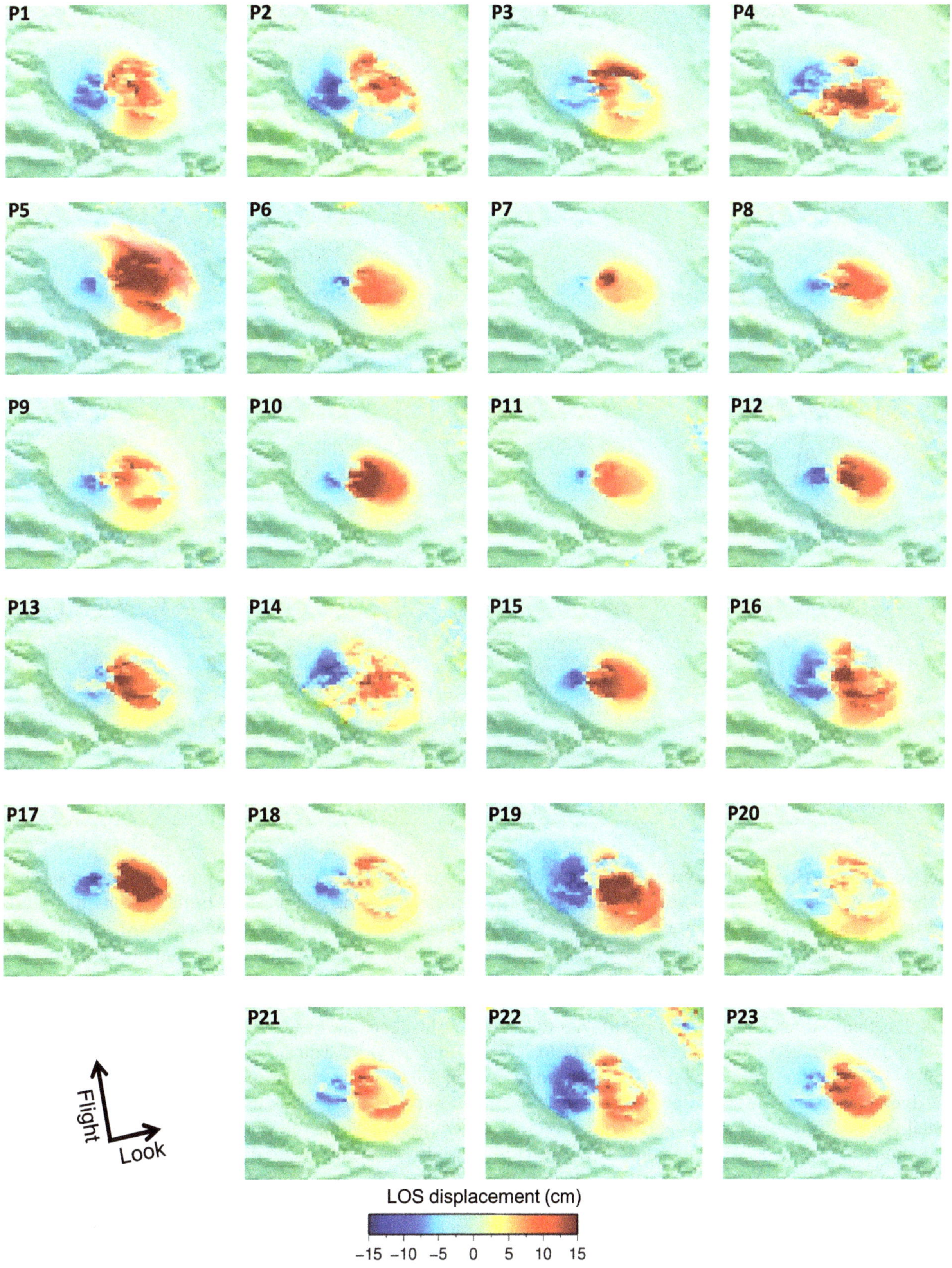

Fig. 3 Observed unwrapped InSAR images derived from ALOS/PALSAR ascending track data (path 573, frame 800). Positive and negative signals indicate the LOS changes away from and toward the satellite, respectively. Details of each data set are described in Table 1. Two arrows at the lower left indicate the satellite flight direction and beam radiation direction

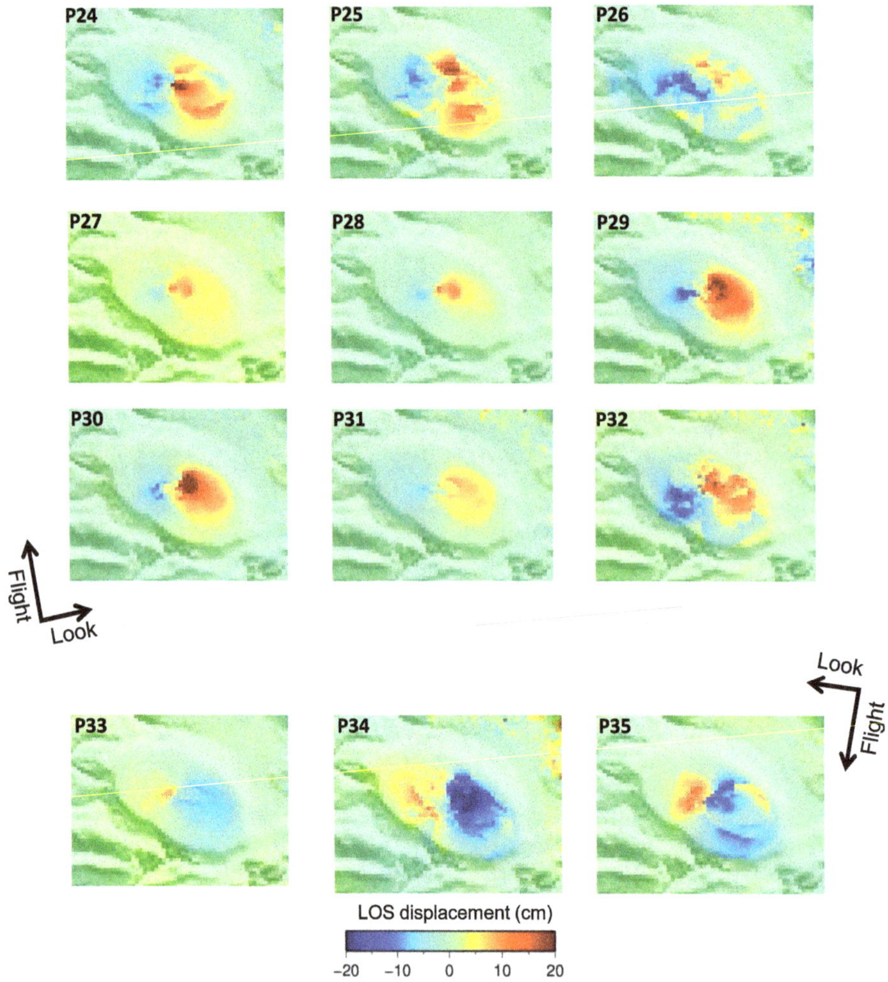

Fig. 4 Observed unwrapped InSAR images derived from ALOS-2/PALSAR-2 ascending track data (path 176, frame 800, P24–32) and descending track data (path 71, frame 2800, P33–35). Details of each data set are described in Table 2

inferred LOS velocities of the western and eastern parts of the ALOS data are − 31 cm/year and + 54 cm/year, and those of the ALOS-2 data are − 19 cm/year and + 35 cm/year, respectively. The larger velocities of the ALOS data compared to that of the ALOS-2 data are presumably due to the differences in the unit vectors of LOS directions, which are $(e_e, e_n, e_z) = (0.6157, 0.1085, -0.7804)$ for ascending ALOS and $(e_e, e_n, e_z) = (0.5120, 0.0974, -0.8535)$ for ascending ALOS-2. This indicates that the LOS vector of ALOS is more sensitive to horizontal displacement than that of the ALOS-2 data.

As noticed from the e_n component of the LOS vectors above, InSAR LOS changes are most insensitive to the north-south displacement because of the satellites near-polar orbit. To reveal the north-south displacement and thus to elucidate the 3D displacements, we employ MAI approach. The observed ALOS MAI data along the ascending path (Table 3) are shown in Fig. 7; the P38 data is much noisier due to lower coherence. The positive and negative signals indicate the displacements projected along and opposite to the satellite flight direction, respectively. The observed MAI displacements show that the southern part of the mud volcano is represented by significant large negative signals, indicating that the southern portion moved opposite to the satellite flight direction. In Fig. 8, we present the observed MAI using ALOS-2 data along the ascending track (P44–P48 in Table 4). As shown in the ALOS MAI data, we can observe a similar pattern of displacements in the southern sector. Moreover, we can identify the displacements along the satellite flight direction, which were not clearly detected by the ALOS data in the northern sector. In Fig. 9a, we show the cumulative along track displacements of the eight ALOS MAI data, which reached up to − 7 m in the southern sector during the same period as the ALOS InSAR data. The stacked image of the five ALOS-2 MAI data is shown in Fig. 9b and indicates that the displacements of the northern and southern sectors

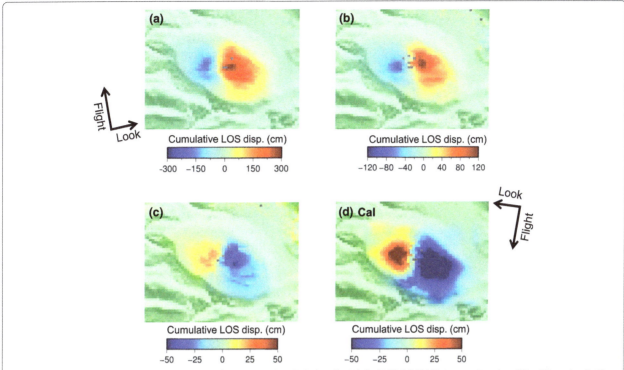

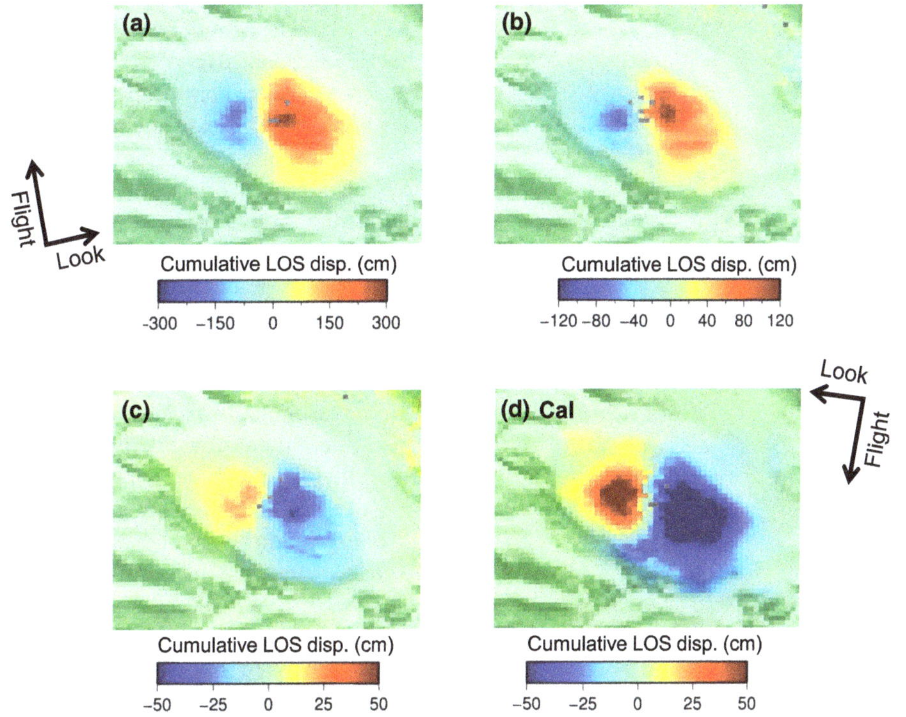

Fig. 5 Cumulative observed LOS changes derived from **a** ALOS/PALSAR data (P1-23), **b** ALOS-2/PALSAR-2 ascending data (P24–32), and **c** ALOS-2/PALSAR-2 descending data (P33–35). **d** Calculated ALOS-2/PALSAR-2 descending data, based on the slip distribution model in Fig. 12

reached up to + 1.5 m and − 4.0 m along the flight direction, respectively. The observed MAI data along the descending path (P49 in Table 4) is also shown at the bottom of Fig. 8; we show this pair because it covers a longer temporal period with shorter perpendicular baseline. Although there is only one descending MAI dataset, *which* is still a bit noisy, the overall spatial pattern is consistent with that detected in the ascending MAI data, namely, large positive and small negative signals, which are separated around the middle of the mud volcano, with a near east-west boundary.

We estimate the 3D surface displacements, using InSAR and MAI measurements acquired from both ascending and descending paths. Because the observation period is different for the ascending and descending datasets, and the surface displacements appear to have been constantly taking place in view of the present and previous observation results (Antonielli et al. 2014), we converted the cumulative LOS and MAI changes (Figs. 5b, c, 9b, and 8 (P49)) into average velocities for simplicity. The calculation method to map the 3D displacements from the observation data is basically the same as those used in several previous studies (e.g., Wright et al. 2004; Jung et al. 2011). The inferred 3D displacements are shown in Fig. 10, which consists of (a) north-south, (b) east-west, and (c) up-down displacements; we should note that each color scale in Fig. 10 is different. They indicate very large north-south horizontal displacements and smaller vertical displacements. We should note that the largest north-south displacements could only be inferred by using the MAI technique. In Fig. 10a, the broad area of the southern sector shows southward motion at a rate of ~ 1 m/year. In Fig. 10b, the east-west component indicates extension by up to 80 cm/year in both easterly and westerly directions in the eastern and western sectors, respectively. Considering all the components, the results indicate that this mud volcano has been extending horizontally from the center, with southward displacements being most significant, and with some minor subsidence and possible uplift in places. The data scatter outside the mud volcano suggests the errors by as much as 10 cm/year.

Source modeling

To account for and interpret the observed cumulative InSAR and MAI data, we develop a fault source model, based on the dislocation theory in an elastic half-space. Although the analytical solutions by Okada (1985) have been widely used to explain surface displacements, they are derived on the assumption of rectangular dislocation elements and thus will generate mechanically incompatible gaps or overlaps in the case of a non-planar fault plane. Whereas the analytical solutions for rectangular dislocation elements are useful, those due to triangular dislocation elements are more versatile and allow us to represent more complex fault geometries (Maerten et al.

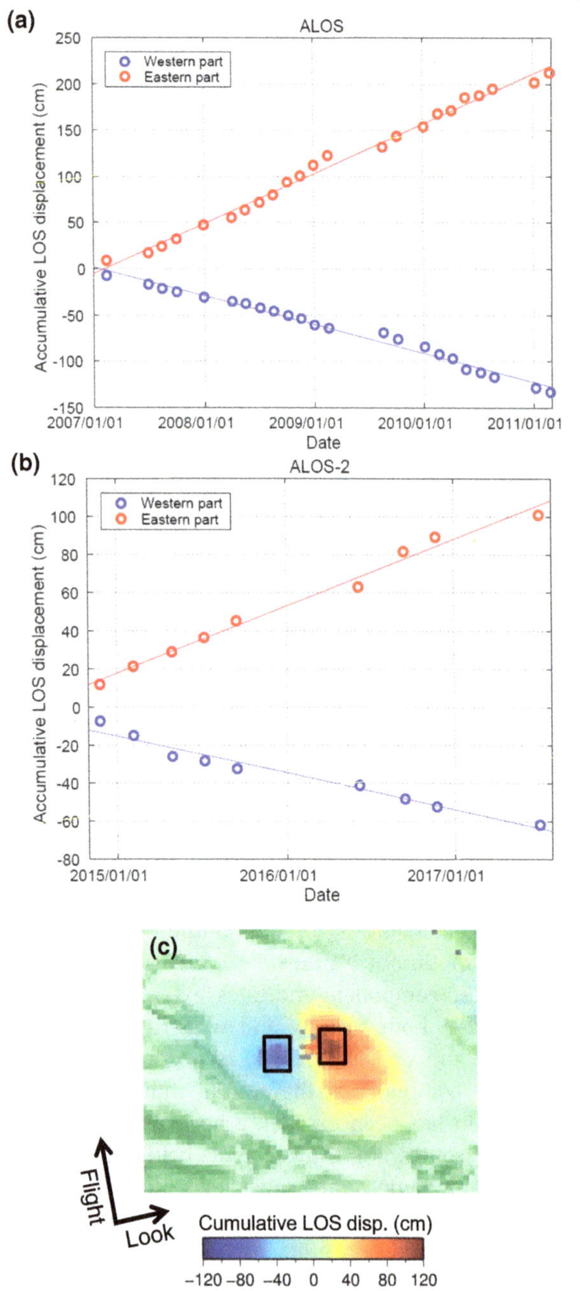

Fig. 6 History of the LOS changes derived from **a** ALOS and **b** ALOS-2 InSAR data along the ascending path. Red and blue circles indicate the cumulated value of LOS changes, derived by averaging the LOS changes within the black boxes of the western and eastern sections in **c**. Red and blue lines demonstrate the linear approximations of the cumulated LOS change described with circles

2005; Furuya and Yasuda 2011). We construct triangular meshes for the non-planar planes using Gmsh software (Geuzaine and Remacle 2009). The size of each side of the triangular mesh is approximately 150 m. To calculate Green's function of surface displacements due to triangular dislocation elements, we use the MATLAB code produced by Meade (2007), assuming a Poisson ratio of 0.3.

In order to derive physically plausible distributions, we use a non-negative least squares method that constrains the slip directions (e.g., Simons et al. 2002), whereas we allow distinct strike-slip directions, depending on the dipping direction of each segment as noted below. We also apply a constraint on the smoothness of the slip distributions, using an umbrella operator (Maerten et al. 2005; Furuya and Yasuda 2011). We use the cumulative displacements acquired from the ascending orbit in the inversion (Fig. 11a–d). We do not invert for the descending orbit data because of the lack of cumulative displacement data for this orbit compared to the ascending orbit. Instead, based on the derived source model, we can compute descending and 3D data, with which we will compare the observed ones so that we can check the consistency of source modeling.

The location and geometry of the source fault are the most important factors in reducing the misfit residuals and were derived by trial-and-error (e.g., Furuya and Yasuda 2011; Abe et al. 2013; Himematsu and Furuya 2015). To reproduce the large horizontal and minor vertical displacements, we set a convex source surface to be dipping most significantly toward the south with minor dip toward east, west, and north; the shallowest point of the convex surface is located beneath the center of the mud volcano. Thereby, we allow for radially distributed normal fault slip on the convex surface. To express oblique normal faulting, we imposed right lateral and left lateral strike-slip on the west-dipping and east-dipping segment, respectively; no such constraint was imposed on the north-south trending normal fault slip for simplicity. To generate the triangular meshes with Gmsh, we initially assigned the 3D coordinates at the 8 control points along the border of the surface. The plane and 3D views are shown in Figs. 2a, 12, and 13, respectively; it should be noted that the scales for the vertical and horizontal axes are different, and the slope of the source plane is exaggerated in Figs. 12 and 13. Although the source surface is convex, the height differences are at most 50 m over the entire surface extending ~3 km. Nonetheless, we consider that the curvature is important because the observed large southward horizontal displacements and minor subsidence are most significantly explained by the large southward normal fault slip.

In Figs. 12 and 13, we show the optimum fault slip amplitude distribution and slip vectors on each segment, assuming that the slip occurs at a constant rate; Figs. 12 and 13 are derived from the ascending path data from ALOS and ALOS-2, respectively. The shallowest point of the convex surface is located at a depth of 25 m. Normal dip slip is dominant on the southern part of the

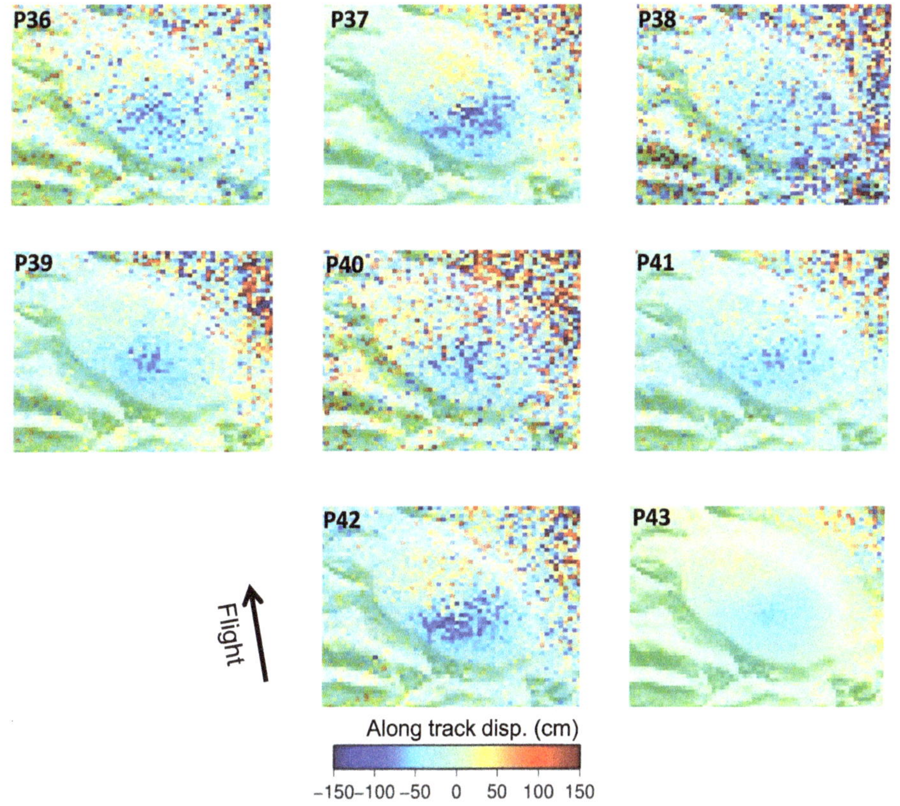

Fig. 7 Observed MAI data derived from ALOS/PALSAR ascending track data (path 573, frame 800). Positive and negative signals indicate the horizontal displacement projected in the direction of the satellite flight and the opposite direction, respectively. Details of each data set are described in Table 3

source plane and has maximum slip amplitudes around the depth of 35 m of up to 8 m (ALOS, Fig. 12c) and 4 m (ALOS-2, Fig. 13c). Moreover, we need to include "strike-slip" components on the western and eastern slope on the surface to represent the oblique normal dip slip. When we set the shallowest point further deeper by ~ 30 m, the estimated normal dip-slip amplitude was larger by ~ 1 m, whereas the misfit residuals became more significant.

Based on the estimated slip distributions, we compute the surface deformation projected onto the LOS and along the satellite track direction (Fig. 11e–h) and the misfit residuals (Fig. 11i–l). Although some residuals remain, the calculated surface displacements can largely reproduce the cumulative observations. Moreover, we show the computed 3D displacements (Fig. 10d–f) and descending InSAR data (Fig. 5d), both of which are based on the derived source model. Although we notice some differences particularly in the EW components (Fig. 10e), the calculated data turn out to largely reproduce the observed ones, which demonstrate the consistency of the source model. Although we tried setting a Mogi-type point source and a Yang-type spheroid source, we could not explain the observed data with physically plausible volume changes. We also tested tensile opening and closure, but could not consistently explain the estimated 3D and descending data. We consider that our source model is probably the simplest in terms of its overall geometry and slip distribution within a framework of elastic dislocation theory, whereas we do not preclude more sophisticated models. We interpret the inferred source model in the following section and discuss the implications for mud volcanism.

Discussion

Why are the LOS changes from ALOS1/2 data larger than those from Envisat data?

Comparing our observed interferograms with the results produced by Antonielli et al. (2014), we observe that the LOS changes in our study are much larger than those found in the previous study. The acquired cumulative LOS displacements of ALOS and ALOS-2 along the ascending track are up to 300 cm and 120 cm for about 4 and 2 years, respectively, and that of ALOS-2 along the descending track is up to 50 cm for about a year, whereas the cumulative LOS displacement calculated by

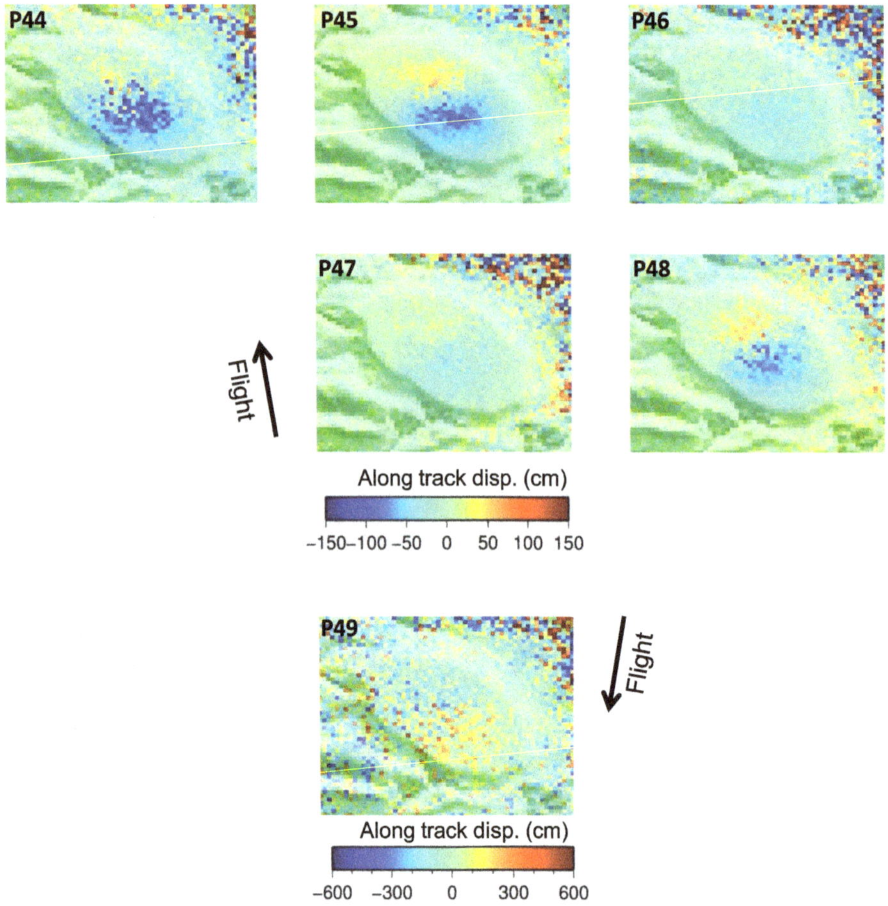

Fig. 8 Observed MAI data derived from ALOS-2/PALSAR-2 ascending track data (path 176, frame 800, P44–48) and descending track data (path 71, frame 2800, P49). Details of each data set are described in Table 4

Antonielli et al. (2014) is up to 20 cm for about 2 years. This notable difference may be caused by two possible factors. One possibility is an increase in the activity of the mud volcano, and the other is a difference in the local incidence angle between the data sets. The incidence angles at the image center of the ascending ALOS, ascending ALOS-2, and descending ALOS-2 are 38.7°, 31.4°, and 36.3°, respectively, whereas the angle of the Envisat is 23°, while the local heading angle does not differ significantly between the data sets. With the heading angle of 10° counter-clockwise at mid-latitude, the unitary LOS vectors for Envisat/ASAR are (e_e, e_n, e_z) = (0.416, 0.073, − 0.906). Compared to the LOS vectors for ALOS1/2 noted earlier, Envisat/ASAR is apparently more sensitive to vertical and less sensitive to horizontal displacements. As the inferred 3D displacements show,

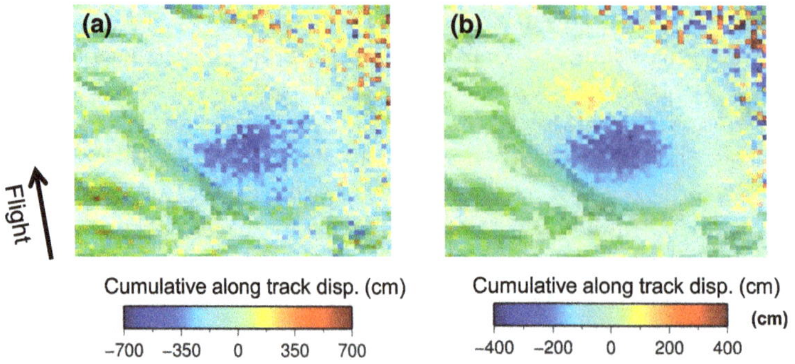

Fig. 9 Cumulative observed MAI data derived from **a** ALOS/PALSAR data (P36–43), and **b** ALOS-2/PALSAR-2 ascending data (P44–48)

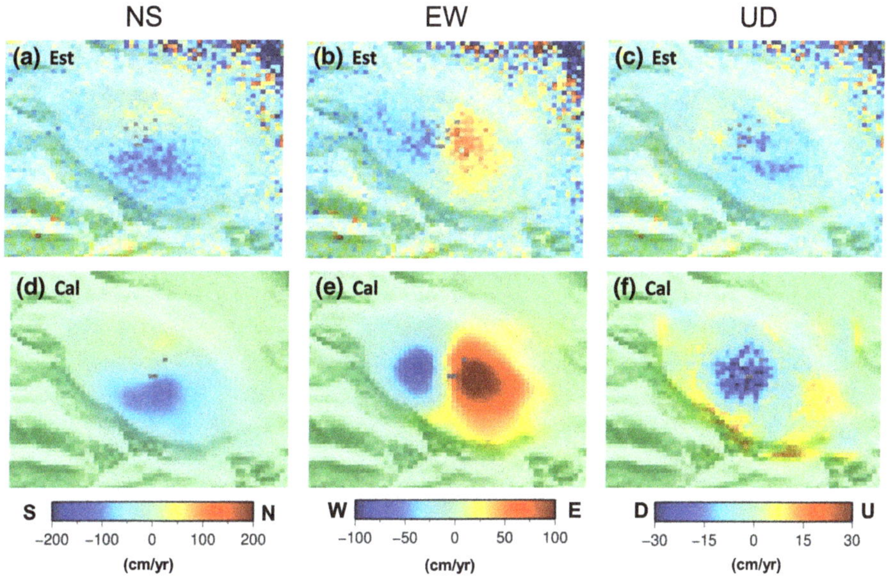

Fig. 10 3D displacements derived from both InSAR and MAI data, assuming constant velocity. **a** North-south displacement. **b** East-west displacement. **c** Up-down displacement. Also shown in **d**, **e**, and **f** are the predicted 3D displacements from the slip distribution model in Fig. 12. Positive signals indicate northward, eastward, and uplift movement, for **a**, **d**; **b**, **e**; and **c**, **f** respectively. Note the difference in the color bar scales

the largest displacements are due to north-south displacement, and the vertical displacement was the least significant, which is a rare occurrence to our knowledge. Using the derived 3D displacements in Fig. 10, we computed the LOS changes for the Envisat beam geometry. Indeed, the LOS changes for the Envisat beam turned out to be smaller than those due to the ALOS beam. However, the differences were not as striking as noted above. Thus, we cannot reject the first possibility.

Interpretations of the source model and the mud volcanism at Ayaz-Akhtarma

Although we could explain the cumulative LOS changes and MAI data with the use of the dislocation theory in elastic half-space, the inferred depth of the fault source is much shallower than those for earthquakes and magmatic eruptions, and the magnitudes of the estimated slip are so large that they may cast doubt on the validity of the use of elastic theory. We recognize that the elastic theory would be invalid in a strict sense for modeling of the observed large displacements, but we consider that the elastic dislocation theory is a helpful tool that can tell us the approximate location and geometry of the mechanical source. Moreover, because we inverted for the cumulative displacements instead of the short-term displacements, the inferred slip amplitude should be regarded as cumulative amplitude as well and may not be unreasonably large. Indeed, in view of each interferogram in Figs. 3 and 4, we can observe more complicated and non-smooth signals on the surface, which are rather localized and not clearly shown in the observed cumulative LOS changes; such localized signals in each interferogram cannot be reproduced in the calculated cumulative LOS (Fig. 5) and MAI (Fig. 9) data, either. We might be able to regard those small-scale signals in each interferogram as a surface expression of local elastic failure. We consider that the cumulative signals simply obscure those small-scale signals, which appear at each interferogram but do not persist over time.

The source model consists of a convex surface on which normal dip slips are semi-radially distributed from the top but most significantly on the south-dipping portion. Under such slip systems, it is reasonable to observe subsidence signals around the center of the mud volcano (Fig. 10c, f). Although the source model does not explicitly include any overpressure forcing elements, the convex geometry could be maintained by constant injection of mud and gas onto the slip surface. We consider that there would exist feeder channels or pipes from deeper mud diapirs (mud chambers) connected to the convex surface, which, however, do not exert sufficient stress to cause surface displacements. Meanwhile, the origin of the significant normal fault slip might seem to be puzzling, considering that the study area is under a compressive stress regime in terms of the global stress field. However, we should recall that, like many mud volcanoes in Azerbaijan, Ayaz-Akhtarma mud volcano is located along anticline axes formed by the regional compressive stress field. We interpret that the mud and gas have reached the anticlinal trap, where intruded

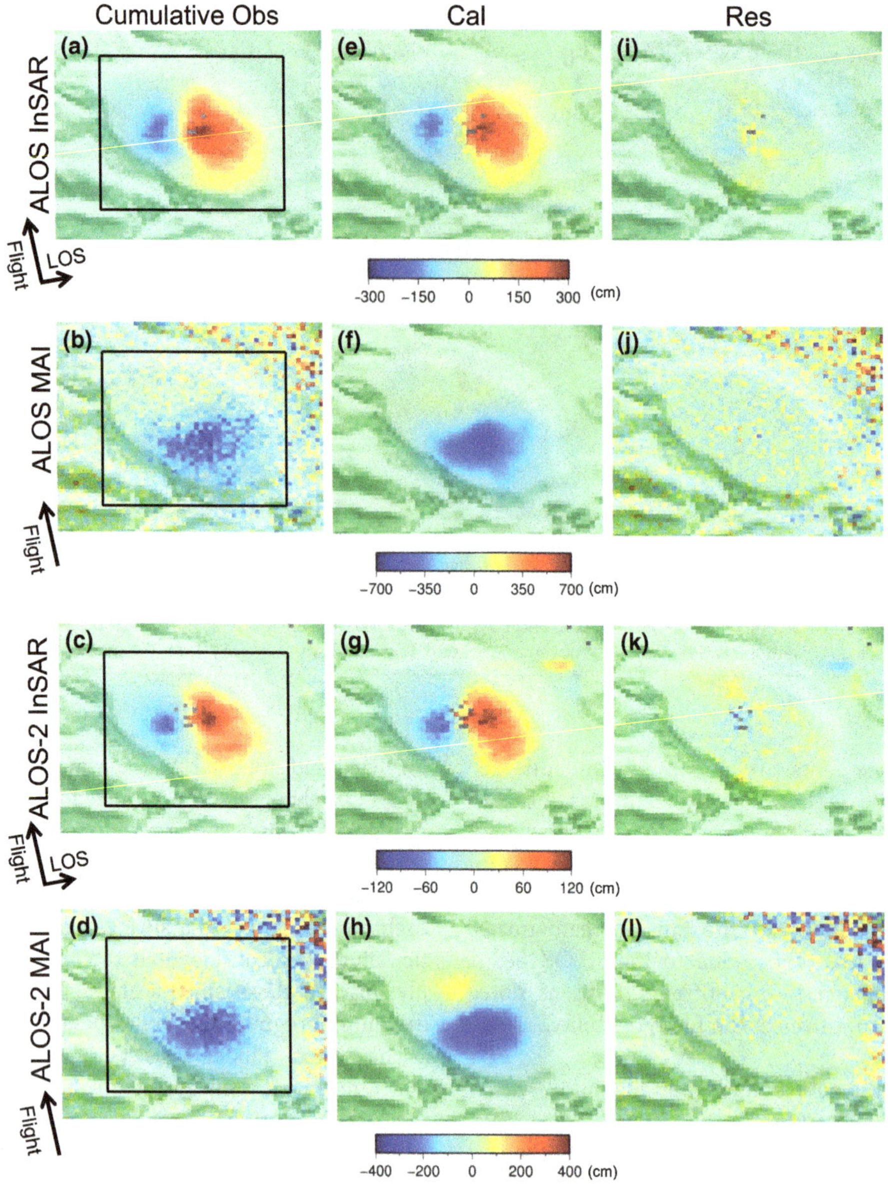

Fig. 11 Comparisons of cumulative observed data, calculated data, and the misfit residual. The left column (**a**–**d**) is the observed cumulative InSAR and MAI data, the middle one (**e–h**) is the calculated InSAR and MAI data derived from the source model in Figs. 12 and 13, and the right column (**i**–**l**) is the misfit between the observation and the calculation. The plan view of the source model in Figs. 11 and 12 is indicated with a black line

materials at very shallow depths are generating localized extensional stress and subsequent normal faulting.

Antonielli et al. (2014) performed a field survey and identified a 600 m-long fault or fracture, which we could also identify from a Google Earth image (Fig. 14b). Moreover, using Google Earth and USGS Landsat satellite images, we observed that the surface of the mud volcano is far from smooth, and such faults or fractures can be observed at many places on the surface (Fig. 14c). Our source model indicates that the largest normal fault slip amplitude lies roughly in a north-south direction. The dominant reported by Antonielli et al. (2014) are mechanically consistent with our source model. Our source model could be used as a guide for further field survey at unexplored areas.

Despite the large slip inferred from the source modeling, there is no evidence for major eruptive episodes during the analyzed period. If there were significant eruptions, we could not generate unwrapped interferograms because surface failures and eruption deposits

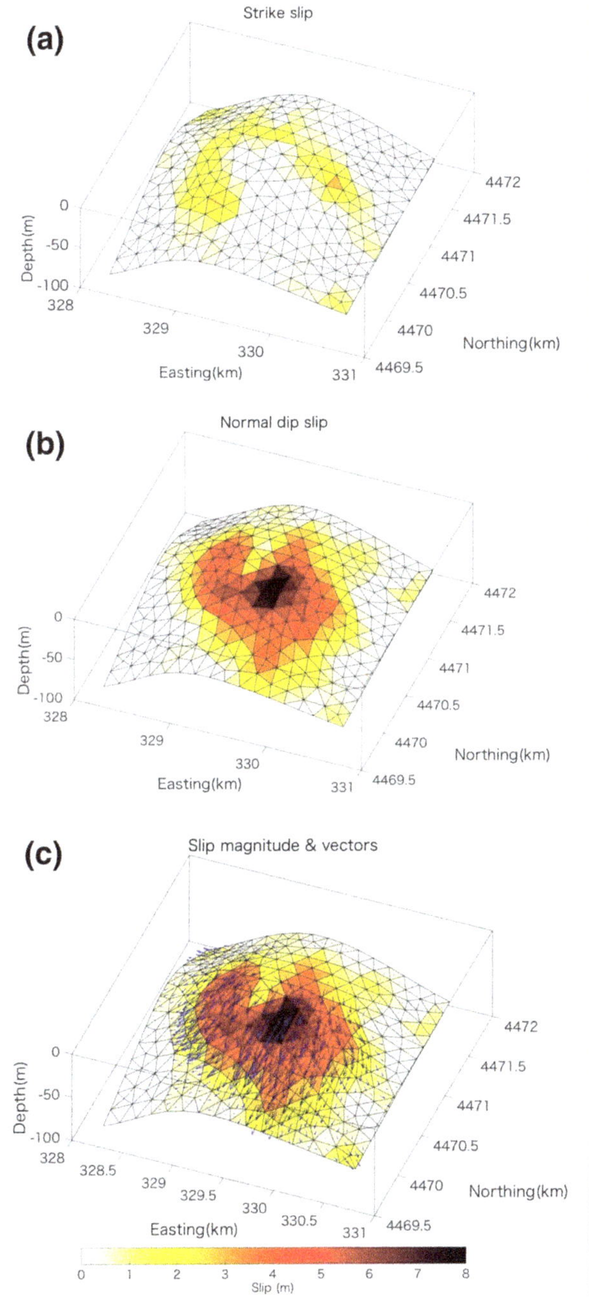

Fig. 12 Slip distributions of a source model inferred from both InSAR and MAI data derived by ALOS/PALSAR. **a** Strike-slip component. **b** Normal slip component. **c** Slip amplitude and vectors. Note that the scales for the vertical axes and horizontal axes are different

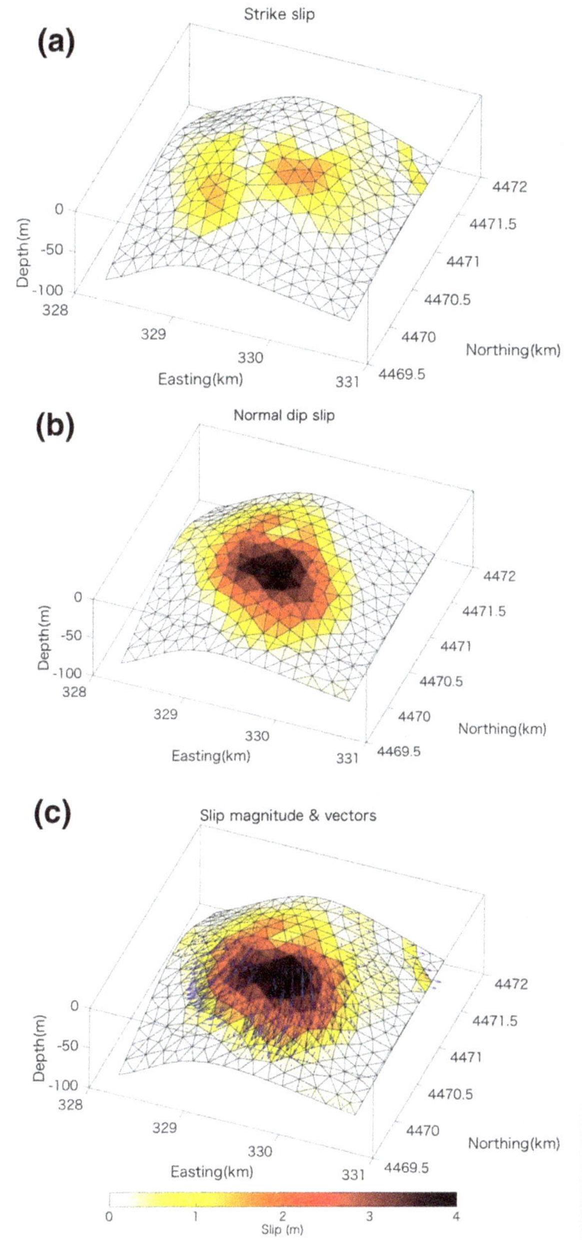

Fig. 13 Slip distributions of a source model inferred from both InSAR and MAI data derived by ALOS-2/PALSAR-2. **a** Strike-slip component. **b** Normal slip component. **c** Slip amplitude and vectors. Note that the scales for vertical axes and horizontal axes are different

from major mud eruptions would have caused the loss of interferometric coherence. However, small-scale eruptions and/or seepages of mud or gas seem to have been continuous (Dupuis et al., 2016). Those rather minor mud volcano activities presumably did not affect the present InSAR observations with a recurrence interval of ~50 days, whereas the large amplitude localized signals in Figs. 3 and 4 were probably due to minor eruptive episodes.

It is well known that the eruption of mud volcanoes is often triggered by earthquakes (e.g., Kopf 2002; Mellors et al. 2007; Manga et al. 2009 Bonini et al. 2016). Manga et al. (2009) proposed a relationship between an earthquake's magnitude and its hypocentral distance, which

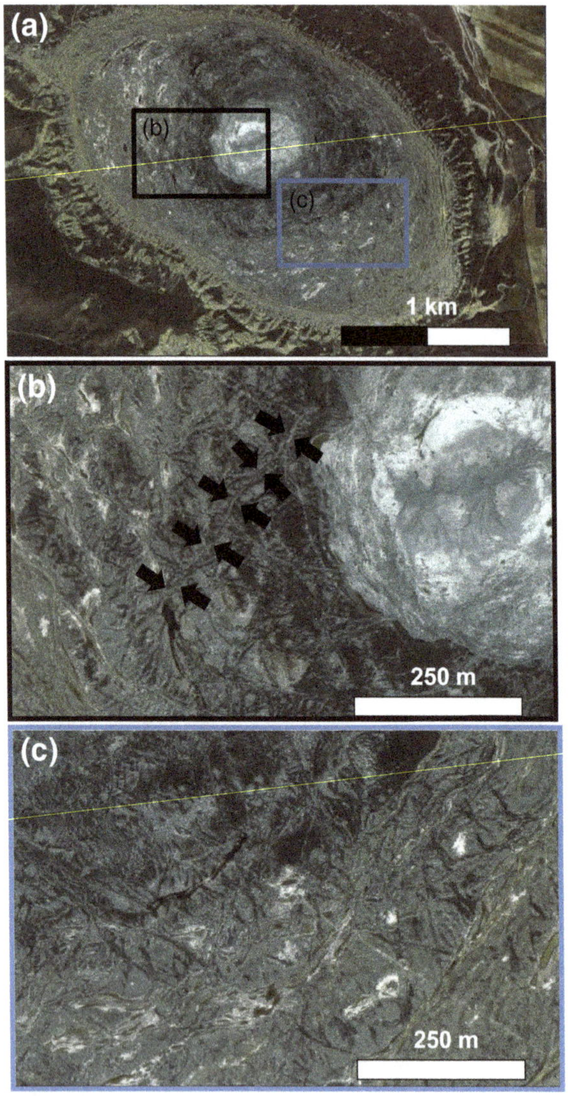

Fig. 14 a USGS Landsat satellite image of Ayaz-Akhtarma mud volcano available from the basemap gallery, https://go.usa.gov/xnhQV (accessed on Mar 13, 2018). **b** Expanded image of the black box in **a**. The lineament sandwiched by black arrows indicates the fault trace identified by Antonielli et al. (2014). **c** Expanded image of the blue box in **a**. Similar lines are observed in several locations

predicts that a mud volcano eruption can be triggered by an earthquake with M5 or above occurring within a distance of 20 km. Unfortunately, we do not have any local observation data to correlate seismic events with the triggered eruption(s), but even if they did occur, any eruptions must have been minor episodes, as speculated above.

Conclusions

Using ALOS/PALSAR and ALOS-2/PALSAR-2 InSAR and MAI data, we detected the detailed surface displacements associated with the activity of Ayaz-Akhtarma mud volcano. The MAI technique in particular turned out to be helpful in acquiring a more complete image of the surface deformation. By combining InSAR and MAI data, we were able to demonstrate that the actual displacements are dominated by horizontal, rather than vertical, components and that the north-south displacements are the largest. Our source model consists of semi-radially distributed normal dip-slip on a convex surface, on which mud and gas are continuously intruding.

Abbreviations

ALOS: Advanced Land Observing Satellite; InSAR: Interferometric synthetic aperture radar; LOS: Line of sight; MAI: Multiple-aperture interferometry; PALSAR: Phased Array type L-band Synthetic Aperture Radar; SAR: Synthetic aperture radar

Acknowledgements

All PALSAR1/2 data in this study are provided from PIXEL (PALSAR Interferometry Consortium to Study our Evolving Land Surface) under a cooperative research contract with the Earthquake Research Institute, The University of Tokyo. The ownership of ALOS/PALSAR and ALOS-2/PALSAR-2 data belong to JAXA/MITI and JAXA, respectively. We also thank Dr. Miho Asada for encouraging us to submit the manuscript. Comments from two anonymous reviewers were helpful to improve the original manuscript.

Funding

This work was supported, in part, by the Specific Research Project (B) 2015-B-02, "Studying surface changes using new generation synthetic aperture radar", Earthquake Research Institute, The University of Tokyo, and, in part, by the Ministry of Education, Culture, Sports, Science and Technology (MEXT) of Japan, under its Earthquake and Volcano Hazards Observation and Research Program.

Authors' contributions

KI processed and analyzed the ALOS-1/2 data. Both KI and MF have managed the research and drafted this manuscript. Both authors read and approved the final manuscript.

Competing interests

The authors declare that they have no competing interests.

Author details

[1]Department of Natural History Sciences, Graduate School of Science, Hokkaido University, N10W8, Kita-ku, Sapporo 060-0810, Japan. [2]Department of Earth and Planetary Sciences, Faculty of Science, Hokkaido University, N10W8, Kita-ku, Sapporo 060-0810, Japan. [3]Geospatial Information Authority of Japan, Kitasato 1, Tsukuba 305-0811, Japan.

References

Abe T, Furuya M, Takada Y (2013) Nonplanar fault source modeling of the 2008 Iwate-Miyagi inland earthquake (Mw 6.9) in Northeast Japan. Bull Seismo Soc Am 103:507–518

Antonielli B, Monserrat O, Bonini M, Righini G, Sani F, Luzi G, Feyzullayev AA, Allyev CS (2014) Pre-eruptive ground deformation of Azerbaijan mud volcanoes detected through satellite radar interferometry (DInSAR). Tectonophysics 637:163–177. https://doi.org/10.1016/j.tecto.2014.10.005

Aoki Y, Sidiq TP (2014) Ground deformation associated with the eruption of Lumpur Sidoarjo mud volcano, east Jana, Indonesia. J Volcanol Geotherm Res 278-279:96–102. https://doi.org/10.1016/j.jvolgeores.2014.04.012

Barbot S, Hamiel Y, Fialko Y (2008) Space geodetic investigation of the coseismic and postseismic deformation due to the 2003 M(w)7.2 Altai earthquake: implications for the local lithospheric rheology. J Geophys Res 113:B03403. https://doi.org/10.1029/2007JB005063

Bechor NBD, Zebker HA (2006) Measuring two-dimensional movements using a single InSAR pair. Geophys Res Lett 33:L16311. https://doi.org/10.1029/2006GL026883

Bonini M (2012) Mud volcanoes: indicators of stress orientation and tectonic controls. Earth Sci Rev 115:121–152. https://doi.org/10.1016/j.earscirev.2012.09.002

Bonini M, Rudolph ML, Manga M (2016) Long- and short-term triggering and modulation of mud volcano eruptions by earthquakes. Tectonophysics 672-673:190–211. https://doi.org/10.1016/j.tecto.2016.01.037

Bürgmann R, Rosen PA, Fielding EJ (2000) Synthetic aperture radar interferometry to measure Earth's surface topography and its deformation. Annu Rev Earth Planet Sci 28:169–209

Costantini M (1998) A novel phase unwrapping method based on network programming. IEEE Trans Geosci Remote Sens 36(3):813–821

Dimitrov LI (2002) Mud volcanoes—the most important pathway for degassing deeply buried sediments. Earth Sci Rev 59:49–76. https://doi.org/10.1016/S0012-8252(02)00069-7

Dupuis M, Odonne F, Imbert P, Abbasov O, Figarov T, Dofal A, Vendeville B (2016) The Ayaz-Akhtarma mud volcano: an actively growing mud pie in the foothills of the Greater Caucasus, Azerbaijan. Paper presented at the 13th International Conference on Gas in Marine Sediments, Tromsø, pp 19–22 Norway September 2016

Fukushima Y, Mori J, Hashimoto M, Kano Y (2009) Subsidence associated with the LUSI mud eruption, East Java, investigated by SAR interferometry. Mar Pet Geol 26:1740–1750. https://doi.org/10.1016/j.marpetgeo.2009.02.001

Furuya M, Yasuda T (2011) The 2008 Yutian normal faulting earthquake (Mw 7.1), NW Tibet: non-planar fault modeling and implications for the Karakax fault. Tectonophysics 511:125–133. https://doi.org/10.1016/j.tecto.2011.09.003

Geuzaine C, Remacle JF (2009) Gmsh: a 3-D finite element mesh generator with built-in pre- and post-processing facilities. Int J Numer Methods Eng 79:1309–1331. https://doi.org/10.1002/nme.2579

Goldstein RM, Werner CL (1998) Radar interferogram filtering for geophysical application. Geophys Res Lett 25:4035–4038. https://doi.org/10.1029/1998GL900033

Hanssen R (2001) Radar interferometry: data interpretation and error analysis Kluwer academic publishers, p 328 Netherlands pp

Himematsu Y, Furuya M (2015) Aseismic strike-slip associated with the 2007 dike intrusion episode in Tanzania. Tectonophysics 656:52–60. https://doi.org/10.1016/j.tecto.2015.06.005

Hommels A, Scholte KH, Munoz-Sabater J, Hanssen RF, Van der Meer FD, Kroonenberg SB, Aliyeva E, Huseynov D, Guliev I (2003) Preliminary ASTER and InSAR imagery combination for mud volcano dynamics, Azerbaijan. International Geoscience and Remote Sensing Symposium 3:1573–1575. https://doi.org/10.1109/IGARSS.2003.1294179

Jung HS, Lu Z, Won JS, Poland MP, Miklius A (2011) Mapping three-dimensional surface deformation by combining multiple-aperture interferometry and conventional interferometry: application to the June 2007 eruption of Kilauea volcano, Hawaii. IEEE Geosci Remote Sens Lett 8(1):34–38

Jung HS, Won JS, Kim SW (2009) An improvement of the performance of Multiple-Aperture SAR Interferometry (MAI). IEEE Trans Geosci Remote Sens 47(8):2859–2869

Kopf AJ (2002) Significance of mud volcanism. Rev Geophys 40:1–52. https://doi.org/10.1029/2000RG000093

Maerten F, Resor P, Pollard D, Maerten L (2005) Inverting for slip on three-dimensional fault surfaces using angular dislocations. Bull Seismol Soc Am 95(5):1654–1665. https://doi.org/10.1785/0120030181

Manga M, Brumm M, Rudolph ML (2009) Earthquake triggering of mud volcanoes. Mar Pet Geol 26:1785–1798. https://doi.org/10.1016/j.marpetgeo.2009.01.019

Massonnet D, Feigl KL (1998) Radar interferometry and its application to changes in the Earth's surface. Rev Geophys 36:441–500. https://doi.org/10.1029/97RG03139

Mazzini A (2009) Mud volcanism: processes and implications. Mar Pet Geol 26:1677–1688

Mazzini A, Etiope G (2017) Mud volcanism: an updated review. Earth Sci Rev 168:81–112. https://doi.org/10.1016/j.earscirev.2017.03.001

Meade BJ (2007) Algorithms for the calculation of exact displacements, strain, and stresses for triangular dislocation elements in a uniform elastic half space. Comput Geosci 33:1064–1075. https://doi.org/10.1016/j.cageo.2006.12.003

Mellors R, Kilb D, Aliyev A, Gasanov A, Yetirmishli G (2007) Correlations between earthquakes and large mud volcano eruptions. J Geophys Res 112:B04304. https://doi.org/10.1029/2006JB004489

Mellors RJ, Bunyapanasarn T, Panahi B (2005) InSAR analysis of the Absheron peninsula and nearby areas, Azerbaijan. Geodynamics and Seismicity NATO Science Series 51:201–209

Milkov AV (2000) Worldwide distribution of submarine mud volcanoes and associated gas hydrates. Mar Geol 167:29–42. https://doi.org/10.1016/S0025-3227(00)00022-0

Okada Y (1985) Surface deformation due to shear and tensile faults in a half-space. Bull Seismo Soc Am 75:1135–1154

Rosen PA, Hensley S, Zebker HA, Webb FH, Fielding EJ (1996) Surface deformation and coherence measurements of Kilauea volcano, Hawaii, from SIR-C radar interferometry. J Geophys Res 101(E10):23109–23125. https://doi.org/10.1029/96JE01459

Simons M, Fialko Y, Rivera L (2002) Coseismic deformation from the 1999 M_w 7.1 Hector Mine, California, earthquake as inferred from InSAR and GPS observations. Bull Seism Soc Am 92:1390–1402

Simons M, Rosen PA (2015) Interferometric synthetic aperture radar geodesy, in Schubert G (editor-in-chief): treatise in geophysics, vol 3, 2nd edn. Elsevier, Oxford, pp 339–385

Wegmüller U, Werner CL (1997) Gamma SAR processor and interferometry software. Proc. of the 3rd ERS symposium, European Space Agency Special Publication, ESA SP-414:1686–1692

Wright TJ, Parsons BE, Lu Z (2004) Toward mapping surface deformation in three dimensions using InSAR. Gephys Res Lett 31:L01607. https://doi.org/10.1029/2003GL018827

13

A numerical shallow-water model for gravity currents for a wide range of density differences

Hiroyuki A. Shimizu*, Takehiro Koyaguchi and Yujiro J. Suzuki

Abstract

Gravity currents with various contrasting densities play a role in mass transport in a number of geophysical situations. The ratio of the density of the current, ρ_c, to the density of the ambient fluid, ρ_a, can vary between 10^0 and 10^3. In this paper, we present a numerical method of simulating gravity currents for a wide range of ρ_c/ρ_a using a shallow-water model. In the model, the effects of varying ρ_c/ρ_a are taken into account via the front condition (i.e., factors describing the balance between the driving pressure and the ambient resistance pressure at the flow front). Previously, two types of numerical models have been proposed to solve the front condition. These are referred to here as the Boundary Condition (BC) model and the Artificial Bed (AB) model. The front condition is calculated as a boundary condition at each time step in the BC model, whereas it is calculated by setting a thin artificial bed ahead of the front in the AB model. We assessed the BC and AB models by comparing their numerical results with the analytical results for a simple case of homogeneous currents. The results from the BC model agree well with the analytical results when $\rho_c/\rho_a \lesssim 10^2$, but the model tends to overestimate the speed of the front position when $\rho_c/\rho_a \gtrsim 10^2$. In contrast, the AB model generates good approximations of the analytical results for $\rho_c/\rho_a \gtrsim 10^2$, given a sufficiently small artificial bed thickness, but fails to reproduce the analytical results when $\rho_c/\rho_a \lesssim 10^2$. Therefore, we propose a numerical method in which the BC model is used for currents with $\rho_c/\rho_a \lesssim 10^2$ and the AB model is used for currents with $\rho_c/\rho_a \gtrsim 10^2$.

Keywords: Gravity currents, Numerical model, Shallow-water model, Front condition

Introduction

Gravity currents are flows driven by density differences between the current and the ambient fluid. In geophysical settings, there are many types of high-Reynolds-number (typically $\gtrsim 10^3$) gravity currents that show a wide range of density ratios (ρ_c/ρ_a, where ρ_c and ρ_a are the densities of the current and ambient fluid, respectively), such as debris flows ($\rho_c/\rho_a \sim 10^3$; e.g., Iverson 1997), turbidity currents ($\rho_c/\rho_a \sim 10^0$; e.g., Meiburg and Kneller 2010), and pyroclastic density currents ($\rho_c/\rho_a = 10^0$–10^1 in the overlying parts and $\rho_c/\rho_a = 10^2$–10^3 in the underlying parts; e.g., Branney and Kokelaar 2002; Breard et al. 2016; Nield and Woods 2004). For the two extreme cases of $\rho_c/\rho_a \sim 10^0$ and 10^3, the fluid dynamical features of gravity currents (e.g., the shape of the interface and the propagation of the flow front) have been studied in detail using experimental investigations (e.g., Marino et al. 2005; Martin and Moyce 1952; Dressler 1954; Rottman and Simpson 1983), numerical investigations (e.g., Cantero et al. 2007; Ooi et al. 2009), and theoretical modeling (e.g., Benjamin 1968; Hogg and Pritchard 2004; Huppert and Simpson 1980; Stoker 1992; Ungarish and Zemach 2005). For intermediate density ratios ($10^0 < \rho_c/\rho_a < 10^3$), there have been some previous studies (e.g., Birman et al. 2005; Bonometti et al. 2011; Gröbelbauer et al. 1993; Hallworth and Huppert 1998; Härtel et al. 2000; Ungarish 2007), but the dynamics of gravity currents within this density range is less well understood than that of the extreme cases.

The purpose of this study is to develop a numerical model of gravity currents for a wide range of ρ_c/ρ_a based on a shallow-water model. The shallow-water model is an efficient mathematical model that captures the essential features of the vertically averaged motion of gravity currents with free surfaces (see

*Correspondence: s-hiro@eri.u-tokyo.ac.jp
Earthquake Research Institute, The University of Tokyo, 1-1-1 Yayoi, 113-0032 Bunkyo-ku, Tokyo, Japan

Ungarish 2009 for an extensive review). For simple initial and boundary conditions, analytical solutions of the shallow-water model for propagating gravity currents are available for a wide range of ρ_c/ρ_a (Ungarish 2007), and these analytical solutions have been verified by experimental measurements and direct numerical simulations using the Navier–Stokes equation (Bonometti and Balachandar 2010; Ungarish 2007). However, geophysical conditions of interest generally have rather complex initial and boundary conditions, so such analytical solutions are not always available. A numerical model that is applicable for complex initial and boundary conditions is highly desirable for simulations of gravity currents for a wide range of ρ_c/ρ_a.

This study is particularly concerned with a numerical treatment of the flow front of gravity currents. In the following sections, we formulate the mathematical problem and show that the numerical treatment at the flow front is key to correctly solving the dynamics of gravity currents for a wide range of ρ_c/ρ_a within the framework of the shallow-water model. We also assess previous numerical methods that have been used to calculate the behavior of the flow front by comparing numerical and analytical results, and we propose a numerical method to simulate the dynamics of gravity currents for a wide range of ρ_c/ρ_a under various geophysical conditions. Finally, as a geophysical application of our results, we develop a numerical model of a pyroclastic density current with strong density stratification.

Methods

Formulation

We consider a planar, inviscid, incompressible, immiscible gravity current of density ρ_c in a deep ambient fluid of density ρ_a, as shown in Fig. 1. The current propagates along a smooth horizontal bottom in the positive x^* direction in time t^*, and gravitational acceleration g acts in the negative z^* direction, where asterisks denote dimensional variables. The propagating current is initially stationary in a reservoir of length x_0 and height h_0, and propagation occurs after a dam at $x^* = x_0$ is rapidly removed at $t^* = 0$. The boundary at $x^* = 0$ is a rigid wall. The flow front at $x^* = x_N^*(t^*)$ is affected by the resistance of the ambient fluid, where N denotes the front. This problem is referred to as the "dam-break problem" (e.g., Ungarish 2009), and is a simple geophysical scenario.

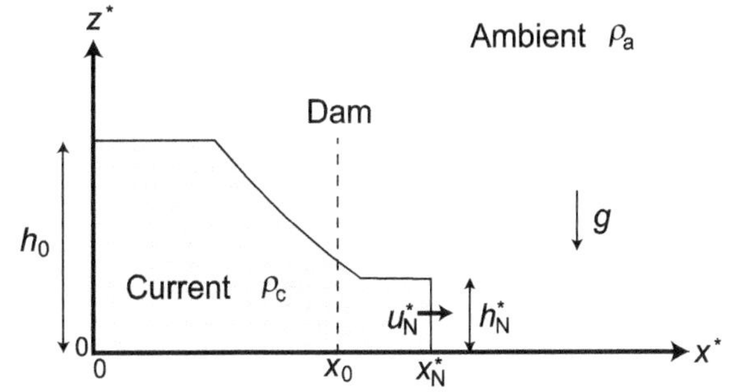

Fig. 1 Schematic of the gravity current released from a dam in a deep ambient fluid

We assume that the current is shallow, with $h_0/x_0 \ll 1$, and is in hydrostatic equilibrium in the vertical direction (i.e., the shallow-water approximation). In the shallow-water approximation, we can obtain the vertically averaged conservation equations of mass and momentum for the flow interior $x^* < x_N^*$ (e.g., Ungarish 2007) as follows:

$$\frac{\partial h^*}{\partial t^*} + \frac{\partial}{\partial x^*}(u^* h^*) = 0, \tag{1}$$

$$\frac{\partial}{\partial t^*}(u^* h^*) + \frac{\partial}{\partial x^*}\left(u^{*2} h^* + \frac{1}{2}\frac{\rho_c - \rho_a}{\rho_c} g h^{*2}\right) = 0, \tag{2}$$

where $h(x,t)$ is the local height and $u(x,t)$ is the local horizontal velocity.

At the flow front $x^* = x_N^*(t^*)$, the kinematic condition ($dx_N^*/dt^* = u_N^*$) and the mass and momentum equations should be taken into account. In addition, to describe realistic gravity current dynamics, we must consider a quasi-steady balance between the buoyancy pressure driving the current front ($\sim (\rho_c - \rho_a) g h_N^*$) and the resistance pressure caused by the acceleration of the ambient fluid around the front ($\sim \rho_a u_N^{*2}$). This condition is known as the front condition, and can be written as follows (e.g., Ungarish 2007):

$$u_N^* = Fr\sqrt{\frac{\rho_c - \rho_a}{\rho_a} g h_N^*} \quad \text{at} \quad x^* = x_N^*(t^*), \tag{3}$$

where Fr, which is an imposed frontal Froude number, is assumed to be a constant of order 10^0 (e.g., $\sqrt{2}$; Benjamin 1968).

Here, using x_0 as the length scale and h_0 as the height scale, we rewrite all dimensional variables to dimensionless variables as follows:

$$x = x^*/x_0, \quad h = h^*/h_0, \quad u = u^*/U, \quad t = t^*/T, \tag{4}$$

with

$$U = \sqrt{\frac{\rho_c - \rho_a}{\rho_c} g h_0}, \quad T = x_0/U. \tag{5}$$

Applying this scaling to Eqs. (1)–(3), we obtain

$$\frac{\partial}{\partial t}\boldsymbol{q} + \frac{\partial}{\partial x}\boldsymbol{f} = \boldsymbol{0} \tag{6}$$

$$u_N = Fr\sqrt{\rho_c/\rho_a}\sqrt{h_N} \quad \text{at} \quad x = x_N(t) \tag{7}$$

with

$$\boldsymbol{q} = \begin{pmatrix} h \\ uh \end{pmatrix}; \quad \boldsymbol{f} = \begin{pmatrix} uh \\ u^2 h + \frac{1}{2}h^2 \end{pmatrix}. \tag{8}$$

Note that the density ratio ρ_c/ρ_a is included only in the front condition (7). Hence, to capture the effects of ρ_c/ρ_a, it is important to calculate the front condition correctly (Ungarish 2007).

The behavior of the analytical solutions for the above equations depends on ρ_c/ρ_a (Fig. 2; Ungarish 2007). The analytical solutions of the dam-break problem consist of an initial "slumping" stage and a subsequent "self-similar" stage (Fig. 2a; e.g., Hogg 2006). During the slumping stage, the front moves with a constant speed and height. During this stage, an initial backward-propagating rarefaction wave arises from the rapidly removed dam, and then a wave arises from the reflection of this rarefaction wave at the back wall $x = 0$ at $t = 1$. The slumping stage continues until the front is caught by this reflection wave. After the slumping stage, the solution is asymptotic to a self-similar solution as time tends to infinity (i.e., the self-similar stage). During this stage, the velocity and height of the front decrease with time. The dependence of the solution on ρ_c/ρ_a is clearly observed in the behavior of the flow front. When $\rho_c/\rho_a \sim 10^0$, the front height h_N is on the order of 10^{-1} during the slumping stage and in the early self-similar stage (Fig. 2a). On the other hand, when $\rho_c/\rho_a \sim 10^3$, h_N is much smaller than 10^{-1}, even from the beginning, the front velocity u_N is substantially greater than u_N for $\rho_c/\rho_a \sim 10^0$ (Fig. 2b). These differences can be interpreted as follows: the momentum lost due to the resistance of the ambient fluid at the front becomes less significant with respect to the momentum of the current as ρ_c/ρ_a increases. We aim to numerically reproduce these features of the analytical solution below.

Numerical methods

In this study, we developed a numerical method for modeling gravity currents for a wide range of ρ_c/ρ_a by discretizing the dimensionless mass and momentum conservation equations (Eqs. (6) and (8)). As these equations are nonlinear and hyperbolic, shocks may develop in the currents. Consequently, we used a finite volume method with shock-capturing capability (e.g., LeVeque 2002; Toro 2001). The finite volume method updates a piecewise constant function $\boldsymbol{Q}_i^n$ that approximates the average value of the solution $\boldsymbol{q}$ in each grid cell i at time step n, using the expression

$$\boldsymbol{Q}_i^{n+1} = \boldsymbol{Q}_i^n - \frac{\Delta t}{\Delta x}(\boldsymbol{F}_{i+1/2} - \boldsymbol{F}_{i-1/2}), \tag{9}$$

where Δx is the constant cell length and Δt is the time interval. $\boldsymbol{F}_{i+1/2}$, which is the intercell flux between cells i and $i+1$, is obtained by using an exact Riemann solver or an approximate Riemann solver, such as the Roe scheme (e.g., LeVeque 2002; Toro 2001). The time interval Δt is

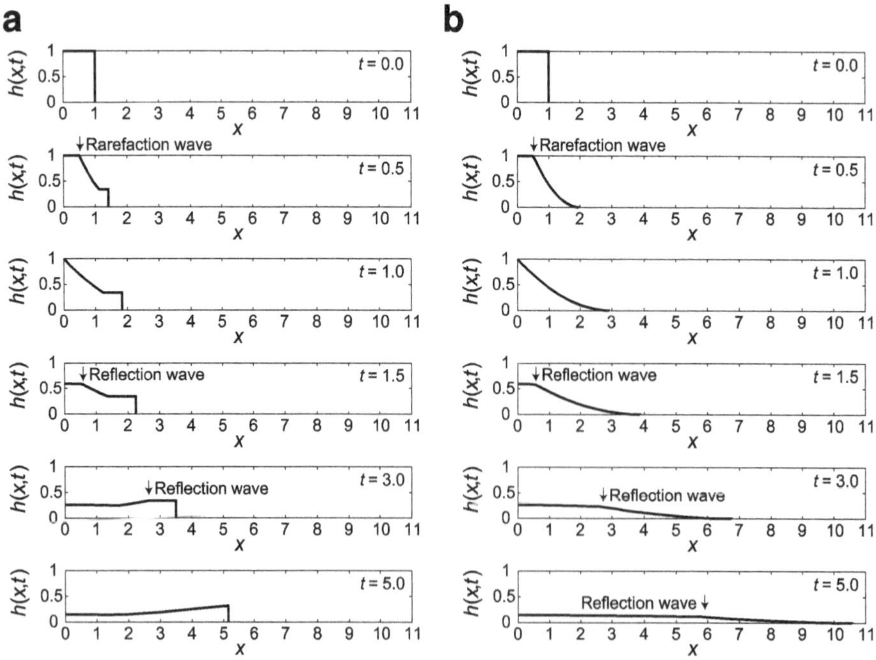

Fig. 2 Analytical solutions of $h(x, t)$ for the dam-break problem. Here, $Fr = \sqrt{2}$ (Benjamin 1968) is used. **a** $\rho_c/\rho_a = 1.01$. **b** $\rho_c/\rho_a = 1000$. In (**a**), the currents at $t = 0.5$, 1.0, 1.5, and 3.0 are in the slumping stage, and the current at $t = 5.0$ is in the self-similar stage. The initial backward-propagating rarefaction wave arising from the rapidly removed dam travels toward the back wall $x = 0$ (see the profile at $t = 0.5$), reaching the wall at $t = 1.0$. Then, a wave arises from the reflection of the rarefaction wave and travels toward the front (see the profiles at $t = 1.5$ and 3.0). After the front is caught by this reflection wave, the current is in the self-similar stage (see the profile at $t = 5.0$). In (**b**), all the currents are in the slumping stage. In this case (i.e., $\rho_c/\rho_a = 1000$), the slumping stage continues until $t \sim 226$ (see Hogg 2006 for details)

limited by the Courant–Friedrichs–Lewy condition (e.g., LeVeque 2002; Toro 2001).

As mentioned above, if we are to capture the effects of ρ_c/ρ_a, it is important to calculate the front condition (7) correctly. Previously, two types of numerical models have been proposed to calculate the front condition. In one, the front condition is calculated as a boundary condition at each time step (e.g., Ungarish 2009). We refer to this model as the Boundary Condition (BC) model (Fig. 3a). In the other, the front condition is calculated by setting a thin artificial bed ahead of the front (e.g., Toro 2001). We refer to this as the Artificial Bed (AB) model (Fig. 3b). In the AB model, the resistance of the ambient fluid at the flow front is modeled by the reaction of the force pushing the artificial bed at the flow front. These models will be described below.

Boundary Condition (BC) model

In the BC model, three quantities at the flow front (x_N, h_N, and u_N) are calculated as boundary conditions of the current from the three equations (mass and momentum conservation equations and front condition) at each time step. In the present numerical method, because we apply a fixed spatial coordinate with constant Δx, the front position $x = x_N(t)$ generally does not coincide with the margins of the grid cells. We therefore define the cell that includes the front as the front cell ($i = FC(t)$, where $FC(t)$ is an integer), and the width of the region that the current occupies in the front cell as $\Delta x_{FC}(t)$ ($0 \leq \Delta x_{FC}(t) < \Delta x$; see Fig. 4). Using $FC(t)$ and $\Delta x_{FC}(t)$, we can write the front position as

$$x_N(t) = (FC(t) - 1)\Delta x + \Delta x_{FC}(t). \tag{10}$$

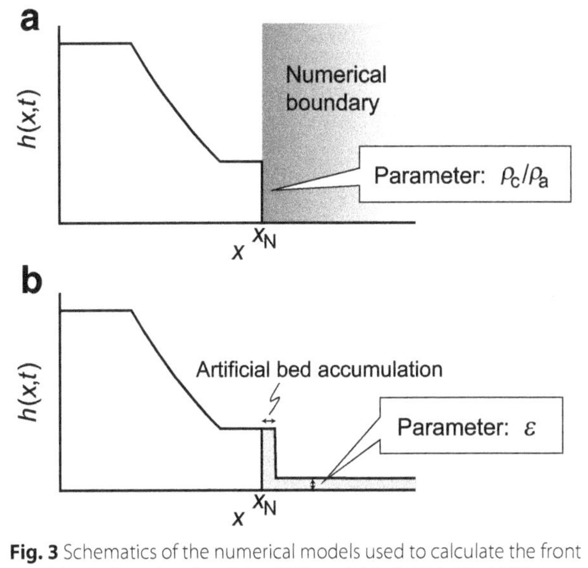

Fig. 3 Schematics of the numerical models used to calculate the front condition. **a** Boundary Condition (BC) model. **b** Artificial Bed (AB) model

The values of h_N and u_N are approximated by the values of h and u at the front cell (i.e., h_{FC} and u_{FC}).

When the kinematic condition ($dx_N/dt = u_N$) is taken into account, the discretized equations for mass and momentum conservation at the flow front are given by

$$\Delta x_{FC}^{n+1} h_{FC}^{n+1} = \Delta x_{FC}^{n} h_{FC}^{n} + \Delta t f_1 \tag{11}$$

and

$$\Delta x_{FC}^{n+1} (uh)_{FC}^{n+1} = \Delta x_{FC}^{n} (uh)_{FC}^{n} + \Delta t \left(f_2 - \frac{1}{2} \left(h_{FC}^{n+1} \right)^2 \right), \tag{12}$$

respectively, where $(f_1, f_2)^T$ represents the intercell flux $\boldsymbol{F}_{FC-1/2}$. From the front condition (i.e., Eq. (7)), we obtain

$$\frac{(uh)_{FC}^{n+1}}{h_{FC}^{n+1}} = Fr\sqrt{\rho_c/\rho_a}\sqrt{h_{FC}^{n+1}}. \tag{13}$$

Solving these three equations analytically (e.g., using Ferrari's method for the solution of the quartic equation) or numerically (e.g., using the Newton–Raphson iteration method), we obtain h_{FC}^{n+1}, u_{FC}^{n+1}, and Δx_{FC}^{n+1}, and hence, h_N, u_N, and x_N at each time step.

Artificial Bed (AB) model

In the AB model, the conservation equations (Eqs. (6) and (8)) are numerically solved using a shock-capturing method for not only the interior, but also the outside of the current by a priori setting a thin artificial bed ahead of the front. Through this numerical procedure, the flow front is generated as the flow following a shock formed ahead of the front without any additional calculation (see Fig. 3b). In this model, the thickness of the artificial bed (ε in Fig. 3b) is the parameter that controls the front condition (i.e., the values of h_N and u_N for different values of ρ_c/ρ_a; see section 10.8 in Toro 2001).

Here, we analytically determined the relationship between ε and ρ_c/ρ_a, as well as that between u_N and ε, on the basis of the analytical solution for the slumping stage of the dam-break problem (e.g., LeVeque 2002; Toro 2001; Ungarish 2009). The initial conditions are $h = 1$ and $u = 0$ in the domain $0 \leq x \leq 1$, and $h = \varepsilon$ and $u = 0$ in the domain $x > 1$, at $t = 0$. Let us consider the time evolution of the current before the rarefaction wave reaches the back wall $x = 0$ (i.e., $0 < t \leq 1$).

For hyperbolic equations such as those used in the present system (i.e., Eqs. (6) and (8)), the relationships between the variables (i.e., h and u) on the characteristics $c_\pm = u \pm \sqrt{h}$ are represented as follows:

$$\Gamma_\pm = u \pm 2\sqrt{h} = \text{const} \qquad \text{on} \quad \frac{dx}{dt} = c_\pm, \tag{14}$$

where $\Gamma_\pm$ are the "Riemann Invariants". Considering that c_+ characteristics from the domain with one initial con-

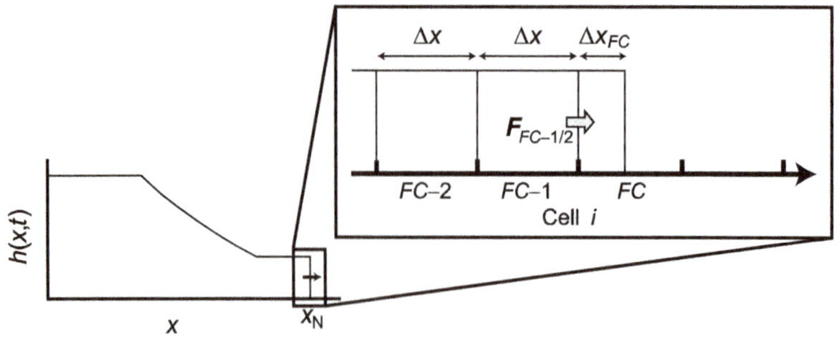

Fig. 4 Schematic of the computational domain of the BC model

dition ($h = 1, u = 0$) enter the front domain ($h = h_N$, $u = u_N$), we can obtain

$$u_N = 2\left(1 - \sqrt{h_N}\right) \tag{15}$$

from Eq. (14). The equation provides the relationship between $h = h_N$ and $u = u_N$ inside the current.

On the other hand, when an artificial bed with $h = \varepsilon$ and $u = 0$ is set, a shock wave traveling with speed S occurs ahead of the front. Across this shock wave, the Rankine–Hugoniot condition,

$$\boldsymbol{f}(\boldsymbol{q}_R) - \boldsymbol{f}(\boldsymbol{q}_L) = S(\boldsymbol{q}_R - \boldsymbol{q}_L) \tag{16}$$

should hold. Here, the subscript R denotes the state on the right side of the shock and L denotes the state on the left side. From Eq. (16), we obtain the state of the front domain behind the shock (i.e., the relationship between $h = h_N$ and $u = u_N$) as

$$u_N = (h_N - \varepsilon)\sqrt{\frac{1}{2}\left(\frac{h_N + \varepsilon}{h_N \varepsilon}\right)}, \tag{17}$$

and the shock speed as

$$S = \sqrt{\frac{1}{2}\frac{(h_N + \varepsilon)h_N}{\varepsilon}}. \tag{18}$$

Eliminating h_N from Eqs. (15), (17), and (18), we obtain u_N and S as a function of ε (Fig. 5a). Using the front condition (7) as well as these equations, we also obtain the relationship between the artificial bed thickness ε and the density ratio ρ_c/ρ_a (Fig. 5b) as

$$\left(1 - \frac{2}{Fr\sqrt{\rho_c/\rho_a} + 2}\right)\frac{4\sqrt{2\varepsilon}}{Fr\sqrt{\rho_c/\rho_a} + 2} = \left\{\left(\frac{2}{Fr\sqrt{\rho_c/\rho_a} + 2}\right)^2 - \varepsilon\right\}\sqrt{\left(\frac{2}{Fr\sqrt{\rho_c/\rho_a} + 2}\right)^2 + \varepsilon}. \tag{19}$$

Note that because we use Eq. (15) here, these relationships (Fig. 5) are in the slumping stage.

In Fig. 5a, S is larger than the front velocity, u_N, because of the accumulation of the artificial bed at the flow front (see Fig. 3b). This deviation of S from u_N is substantial for $\varepsilon \gtrsim 10^{-3}$. This implies that the position of the shock does not always approximate the flow front. If we are to extract the correct position of the flow front, we must calculate an advection equation for a passive tracer concentration, ϕ ($\phi = 1$ for $0 \le x \le 1$, and $\phi = 0$ for $x > 1$, at $t = 0$):

$$\frac{\partial \phi}{\partial t} + u\frac{\partial \phi}{\partial x} = 0 \tag{20}$$

after solving the equations of fluid motion (see section 13.12 in LeVeque 2002 for details).

Results and discussion

In this section, we compare the numerical results obtained from the BC and AB models with the analytical results, and assess the applicability of these models. Subsequently, as a geophysical application of our results, we develop a numerical model of pyroclastic density currents.

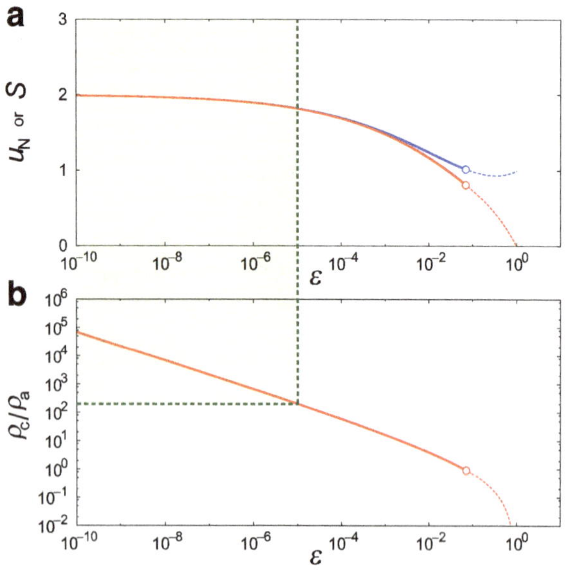

Fig. 5 Analytical solutions for the AB model during the slumping stage. **a** Front velocity u_N (*red curve*) and shock speed S (*blue curve*), as functions of ε. **b** Relationship between ε and ρ_c/ρ_a, in which $Fr = \sqrt{2}$ (Benjamin 1968) is used. *Dashed curves* represent the solutions for $\rho_c/\rho_a < 1$

Comparison of analytical and numerical results

Figure 6 shows the numerical results from the BC model along with the analytical results for the cases of $\rho_c/\rho_a = 1.01$ (a) and 1000 (b). The numerical results for $\rho_c/\rho_a = 1.01$ agree well with the analytical results from the early slumping stage to the late self-similar stage. The numerical results for $\rho_c/\rho_a = 1000$ also appear to agree with the analytical results, but the speed of the front position, $\dot{x}_N$, shows a numerical oscillation that is not observed in the analytical result (Fig. 7a). In particular, in the initial stage ($t \lesssim 0.0002$ in Fig. 7a), $\dot{x}_N$ tends to be overestimated. These oscillation and overestimation are caused by the assumption that the values of h_{FC} and u_{FC} are uniform across the width of the front cell Δx_{FC} in the present numerical method at first-order accuracy. For a large ρ_c/ρ_a, because h_N has a small value, the value of Δx_{FC} tends to be overestimated when a constant h_{FC} is assumed (Fig. 7b). We suggest, therefore, that the BC model is favorable for simulating gravity currents with relatively low ρ_c/ρ_a.

Figure 8 shows the numerical results from the AB model along with the analytical results. In these calculations, the values of ε for given values of ρ_c/ρ_a are set based on the relationship of Eq. (19) (see Fig. 5b). In Fig. 8b, the numerical results for $\rho_c/\rho_a = 1000$ ($\varepsilon = 4.58 \times 10^{-7}$) agree well with the analytical results. The numerical oscillations observed in the BC model do not occur with the AB model (Fig. 7a). In Fig. 8a, on the other hand, the numerical results for $\rho_c/\rho_a = 1.01$ ($\varepsilon = 6.58 \times 10^{-2}$) agree well with the analytical results only during the slumping stage ($t \lesssim 4.5$), but deviate from the analytical results during the self-similar stage ($t \gtrsim 4.5$). This agreement during the slumping stage and deviation during the self-similar stage occurs because ε is set using the analytical relationship (Eq. (19)) for the slumping stage of the dam-break problem. During the slumping stage, h_N and u_N are constant so that ε based on Eq. (19) provides the correct front condition. During the self-similar stage, on the other hand, the driving pressure, and hence h_N and u_N, decrease with time; therefore, the assumed value of ε is no longer consistent with the front condition Eq. (7).

The good agreement in the results of the AB model for $\rho_c/\rho_a = 1000$ reflects the fact that the dynamics of the gravity current becomes insensitive to the front condition for large values of ρ_c/ρ_a. In Fig. 5b, ε approaches 0 as ρ_c/ρ_a increases. In the limit as $\rho_c/\rho_a \to \infty$ and $\varepsilon \to 0$, u_N asymptotically approaches its maximum value, 2, and h_N asymptotically approaches 0. For sufficiently small ε, the solution converges to that in the limit as $u_N \to 2$ and $h_N \to 0$, and it becomes insensitive to the value of ε (see Fig. 5a). Indeed, as shown in Fig. 8b, we can confirm that the result of the AB model with a very small ε ($\varepsilon = 1.0 \times 10^{-10}$) is indistinguishable from that for $\rho_c/\rho_a = 1000$ ($\varepsilon = 4.58 \times 10^{-7}$). According to Fig. 5, the results of the AB model for the dam-break problem are insensitive to ε when $\varepsilon \lesssim 10^{-5}$, which corresponds to $\rho_c/\rho_a \gtrsim 10^2$ (Fig. 5b). Consequently, we suggest that the AB model is favorable for simulating gravity currents with high ρ_c/ρ_a for which the dynamics of the current is insensitive to the assumed value of ε.

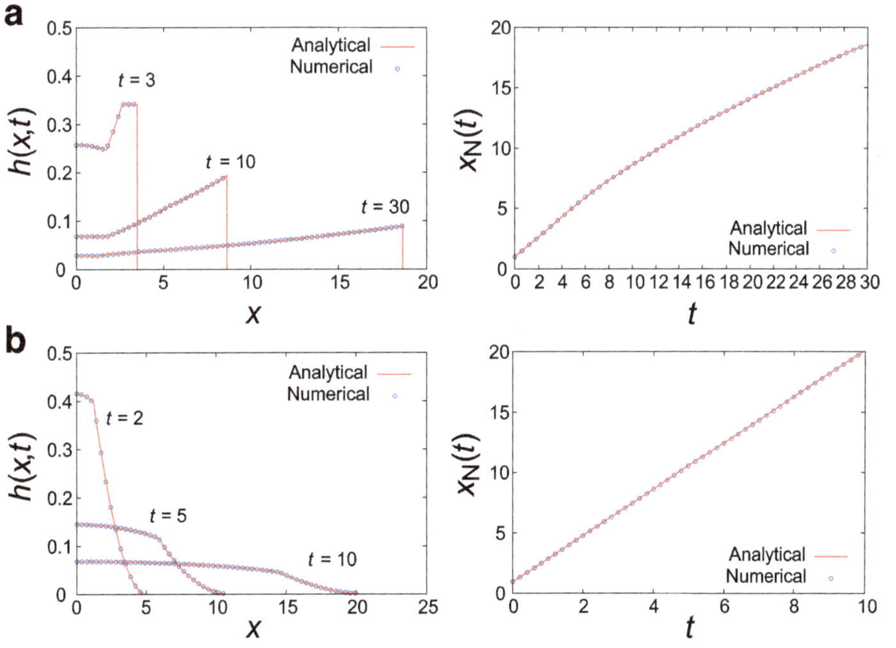

Fig. 6 Analytical results (*red curves*) and numerical results from the BC model (*blue symbols*) for $h(x, t)$ and $x_N(t)$. **a** $\rho_c/\rho_a = 1.01$. **b** $\rho_c/\rho_a = 1000$

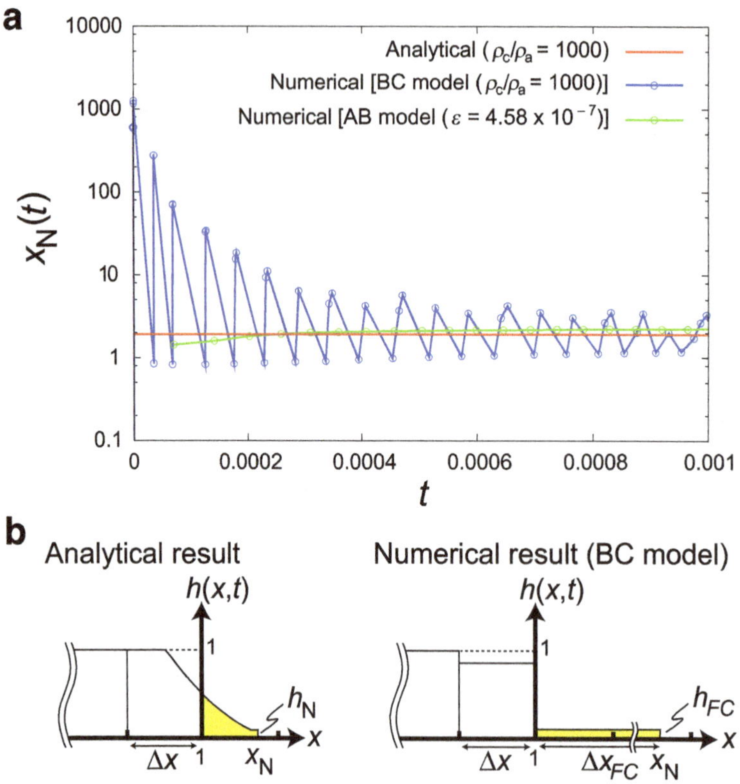

Fig. 7 Speed of the front position, $\dot{x}_N$, in the early time steps. **a** Comparisons between the analytical result with $\rho_c/\rho_a = 1000$ (*red line*) and numerical results of the BC model with $\rho_c/\rho_a = 1000$ (*blue symbols*) and of the AB model with $\varepsilon = 4.58 \times 10^{-7}$ (*green symbols*). In the numerical calculations, $\Delta x = 1.0 \times 10^{-4}$. **b** Illustrations of the overestimation of $\dot{x}_N$ by the BC model in $t \lesssim 0.0002$ in (**a**)

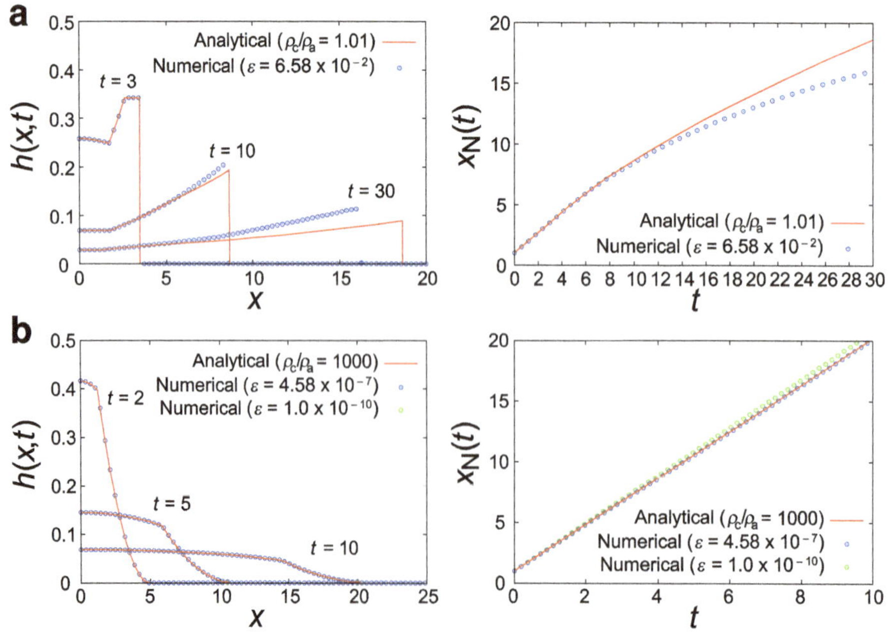

Fig. 8 Analytical results (*red curves*) and numerical results of the AB model (*symbols*) for $h(x, t)$ and $x_N(t)$. **a** $\rho_c/\rho_a = 1.01$. **b** $\rho_c/\rho_a = 1000$. In the numerical results, *blue symbols* represent the numerical results given ε based on the analytical solution during the slumping stage (**a** $\varepsilon = 6.58 \times 10^{-2}$; **b** $\varepsilon = 4.58 \times 10^{-7}$). *Green symbols* represent the numerical results given $\varepsilon = 1.0 \times 10^{-10}$

Applicability of the BC and AB models

Our results indicate that the BC and AB models each have their own advantages and disadvantages. The results obtained from the BC model agree well with the analytical results when $\rho_c/\rho_a \lesssim 10^2$ (Fig. 6a), whereas they show a numerical oscillation at the flow front and tend to overestimate the front speed when $\rho_c/\rho_a \gtrsim 10^2$ (Fig. 7). No such numerical oscillation nor overestimation is observed in the results from the AB model. For currents with $\rho_c/\rho_a \gtrsim 10^2$, the AB model provides good approximations of the analytical results, given a sufficiently small ε (Figs. 5 and 8b). For currents with $\rho_c/\rho_a \lesssim 10^2$, however, the AB model may fail to reproduce the analytical results for currents where the height and speed of the front change with time (Fig. 8a). Accordingly, we propose that the BC model should be used for currents with $\rho_c/\rho_a \lesssim 10^2$ and the AB model is applicable only to currents with $\rho_c/\rho_a \gtrsim 10^2$.

Because of its simple coding and numerical stability, the AB model with an arbitrarily small ε is commonly used for simulations of gravity currents in many geophysical situations (e.g., Denlinger and Iverson 2004; Doyle et al. 2007, 2008, 2011; Larrieu et al. 2006). This model would be applicable in simulating gravity currents with high values of ρ_c/ρ_a, such as debris flows (e.g., Denlinger and Iverson 2004). However, our results suggest that it may provide inaccurate results for gravity currents with $\rho_c/\rho_a \lesssim 10^2$, such as turbidity currents and dilute pyroclastic density currents. Numerical results for $\rho_c/\rho_a = 10$ show that the problem arises mainly from the behavior of the flow front (Fig. 9). Generally, a gravity current with a relatively low value of ρ_c/ρ_a is characterized by the formation of a large front height, which is caused by the resistance of the ambient fluid. This large front height is successfully reproduced by the BC model (Fig. 9a), while the AB model fails to capture it. The results from the AB model with $\varepsilon = 10^{-10}$ (Fig. 9b) show that the resistance at the front is too small to develop a large front height; consequently, the flow speed is substantially overestimated.

Geophysical application to pyroclastic density currents

Pyroclastic density currents (PDCs) are characterized by strong density stratification due to particle settling (e.g., Branney and Kokelaar 2002), whereby a dilute gravity current (particle suspension flow) with $\rho_c/\rho_a = 10^0$–10^1 overrides the dense basal gravity current (fluidized granular flow) with $\rho_c/\rho_a = 10^2$–10^3. The dynamics of PDCs is complex because the dilute and dense currents are influenced by a number of physical processes such as particle settling (e.g., Bonnecaze et al. 1993), entrainment of ambient air (e.g., Johnson and Hogg 2013; Sher and Woods 2015), and basal resistance (e.g., Roche et al. 2008). In addition to the effects of these processes, our results suggest that the application of the correct numerical model to the flow front is important if we are to understand the dynamics and sedimentation of PDCs. The BC model should be applied to the overlying dilute current, while the AB model is applicable to the underlying dense current. Here, we discuss how the resistance at the flow front of the dilute part influences the dynamics of PDCs as a whole.

Before discussing the complex dynamics of PDCs with strong density stratification, we briefly assess the effects of some important physical processes on the dilute and dense currents. Figure 10 shows the results of simulations using the BC model for a dilute gravity current generated by an instantaneous release (i.e., the dam-break problem) of an initially homogeneous particle suspension with $\rho_c/\rho_a = 8.495$. This model accounts for the effects of particle settling and entrainment of the ambient fluid following Bonnecaze et al. (1993) and Johnson and Hogg (2013), respectively. As with a homogeneous gravity current with a low density ratio (Fig. 6a), a thick front develops in the dilute gravity current of the particle suspension, suggesting that the resistance of the ambient fluid at the front plays a significant role in the dynamics of the dilute part of PDCs regardless of the presence or absence of the effects of particle settling or entrainment.

Figure 11 compares our numerical results obtained using the AB model with experimental results for a

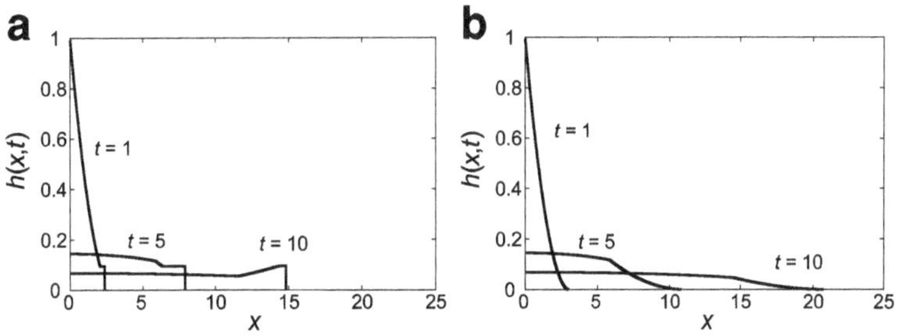

Fig. 9 Numerical results of $h(x, t)$ for $\rho_c/\rho_a = 10$. **a** BC model. **b** AB model with $\varepsilon = 1.0 \times 10^{-10}$. In (**b**), $\rho_c/\rho_a = 10$ is given when the basic equations (Eqs. (1) and (2)) are non-dimensionalized using Eqs. (4) and (5)

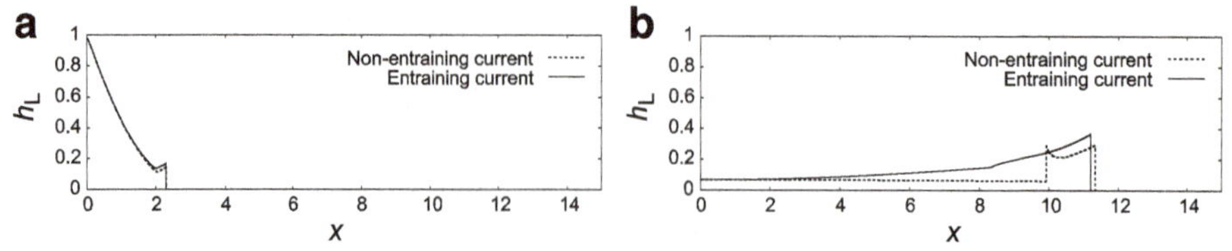

Fig. 10 Numerical results of a one-layer dilute PDC model for the dam-break problem. Heights h_L at **a** $t = 1$ and **b** $t = 10$ are shown for cases of entraining (*solid curves*) and non-entraining (*dotted curves*) currents. The initial density ratio (ρ_L^*/ρ_a) is set to 8.495, and the BC model is applied. The variables are non-dimensionalized using the initial values (e.g., Eqs. (4) and (5)). Given parameters: aspect ratio (initial height/initial length) = 1; initial particle concentration = 0.005; density ratio (particles/air) = 1500; density ratio (volcanic gas/air) = 1; and non-dimensional particle settling speed = 0.05

granular flow with $\rho_c/\rho_a = 10^2$–10^3 generated by an instantaneous release of an initially fluidized bed (Roche et al. 2008). Because the initial acceleration regime of the experimental setup (i.e., just after the release; (i) in Fig. 11b) is beyond the applicable range of the shallow-water model, we focus on the subsequent regimes here. The numerical results obtained using the AB model show the formation of a wedge-shaped flow front (Fig. 11a), which is in agreement with the experimental results of Roche et al. (2008). The results also indicate that the dynamics of the dense fluidized granular flow can be quantitatively simulated by the AB model when a suitable rheological model is applied for the basal resistance (Fig. 11b). The shear drag model explains the behavior during the constant velocity regime ((ii) in Fig. 11b; see Roche et al. 2008), whereas the Coulomb friction model better reproduces the features of the stopping regime ((iii) in Fig. 11b; see Roche et al. 2008).

The interplay between the dilute and dense parts may be crudely simulated by a two-layer model comprising a dilute layer and a dense layer (Doyle et al. 2008, 2011). The previous two-layer models for PDCs apply the AB model with a small ε to both layers. Here, we follow Doyle et al. (2011) for the basic formulation of the two-layer system, but apply the BC model to the dilute layer and the AB model to the dense layer. We also consider the effects of basal shear drag in the dense layer and entrainment of ambient air into the dilute layer (see Appendix for details). Figure 12 shows a representative result of our two-layer model for a PDC generated by the instantaneous release of an initially homogeneous dilute particle suspension with $\rho_c/\rho_a = 8.495$. When the BC model is applied, a thick front head develops in the dilute gravity current because of the resistance of the ambient fluid at the front (see Fig. 10). On the other hand, a dense gravity current with $\rho_c/\rho_a = 600.6$ is generated by particle settling from the overlying dilute gravity current. Because the rate at which particles are supplied to the dense layer is controlled by the conditions of the overlying dilute layer (e.g., thickness and particle concentration), the evolution and dynamics of the dense layer are critically dependent on those of the dilute layer. This suggests that the behavior of the flow

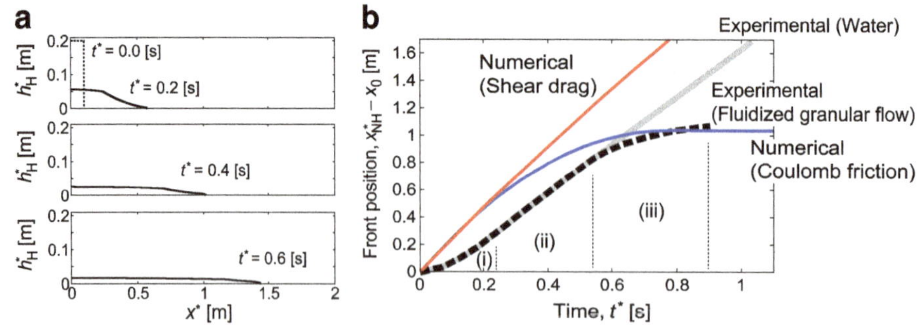

Fig. 11 Numerical results of a one-layer dense PDC model for the dam-break problem. The density ratio (ρ_H/ρ_a) is set to 600.6, and the AB model with $\varepsilon = 10^{-10}$ is applied. Given parameters: initial height $h_0 = 0.2$ m, initial length $x_0 = 0.1$ m, density of particles = 1500 kg/m^3; density of volcanic gas = 1 kg/m^3; and density of air = 1 kg/m^3. **a** Heights h_H^* of the dense PDC with shear drag at $t^* = 0.0$, 0.2, 0.4, and 0.6 s. **b** Front positions of the dense PDC with shear drag (*red solid curve*) and Coulomb friction (*blue solid curve*). In (**b**), the numerical results are compared with the experimental results of an initially fluidized granular flow (*black dashed curve*) and water (*gray solid curve*) (Roche et al. 2008). The experimental result of the initially fluidized granular flow has three distinct regimes: (i) initial acceleration, (ii) constant velocity, and (iii) stopping (see Roche et al. 2008 for details). Note that during (i) and (ii), the initially fluidized granular flow behaves as water. The numerical result with shear drag reproduces the slope (i.e., the constant velocity) of (ii). The numerical result with Coulomb friction reproduces the features of (iii)

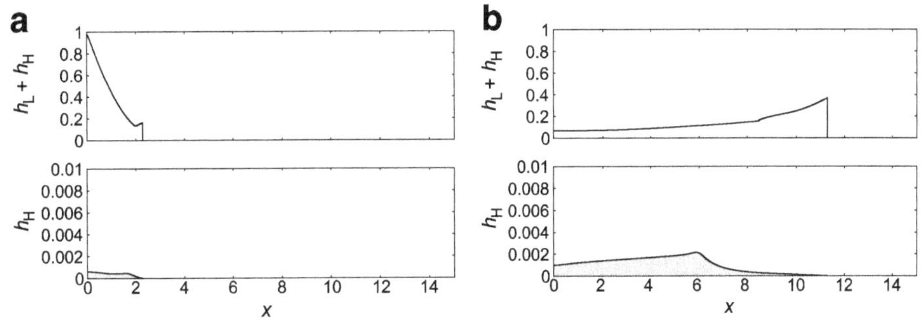

Fig. 12 Numerical results of a two-layer PDC model for the dam-break problem. The dense current height (h_H) and dilute current height ($h_L + h_H$) at **a** $t = 1$ and **b** $t = 10$ are shown. A PDC is generated by the instantaneous release of an initially homogeneous dilute layer. The overlying dilute layer (initially $\rho_L^*/\rho_a = 8.495$) is calculated using the BC model, and the underlying dense layer ($\rho_H/\rho_a = 600.6$) is calculated using the AB model with $\varepsilon = 10^{-10}$. Shear drag is applied for the basal resistance of the dense layer. The variables are non-dimensionalized using the initial values of the dilute layer (e.g., Eqs. (4) and (5)). Given parameters: aspect ratio (initial height/initial length) = 1; initial particle concentration of the dilute layer = 0.005; density ratio (particles/air) = 1500; density ratio (volcanic gas/air) = 1; and the non-dimensional particle settling speed = 0.05

front of the dilute layer controls not only the dynamics of the dilute layer but also the dynamics of the stratified PDC as a whole.

The results in Fig. 12 are preliminary, and a comprehensive understanding of the dynamics of PDCs should consider many other effects, such as the expansion of entrained air due to heating by pyroclasts, density stratification inside the overlying dilute layer, diffusion of the pore pressure and entrainment of air in the underlying dense layer, and the transport of particles from the underlying dense layer to the overlying dilute layer (e.g., Andrews 2014; Breard and Lube 2017; Bursik and Woods 1996; Dufek and Bergantz 2007; Esposti Ongaro et al. 2016; Ishimine 2005; Roche et al. 2008; Wilson and Walker 1982). Nevertheless, preliminary results (not shown here) have already indicated the diversity of the interplay between the dilute and dense layers, which depends on the initial particle concentration and grain size. The interaction also influences the sedimentation process from the PDCs (Fujii and Nakada 1999). A systematic parametric study of the two-layer PDC model using the BC model is in progress, with the aim of accounting for the diversity of PDC deposits.

Conclusion

A numerical shallow-water model of simulating gravity currents for a wide range of ρ_c/ρ_a has been proposed. In the model, the effects of varying ρ_c/ρ_a are taken into account via the front condition. We have assessed two types of numerical models for the front condition (the Boundary Condition (BC) model and the Artificial Bed (AB) model) by comparing their numerical results with the analytical results. The results from the BC model agree well with the analytical results when $\rho_c/\rho_a \lesssim 10^2$. In contrast, the AB model generates good approximations of the analytical results for $\rho_c/\rho_a \gtrsim 10^2$. On the basis of these results, we have developed a two-layer model of pyroclastic density currents (PDCs), in which the BC model is used for the overlying dilute part ($\rho_c/\rho_a = 10^0$–10^1) and the AB model is used for the underlying dense part ($\rho_c/\rho_a = 10^2$–10^3). This two-layer model successfully simulates some essential features of PDCs with strong density stratification.

Appendix: two-layer model of pyroclastic density currents

The dynamics of PDCs in which a strong density stratification develops is described here by a two-layer shallow-water model with overlying dilute and underlying dense layers (Fig. 13). The dilute layer is modeled as a dilute particle suspension, which is a continuum where monodisperse solid particles are suspended in an incompressible gas phase (volcanic gas and entrained air). Its vertically averaged mass and momentum conservation equations are respectively

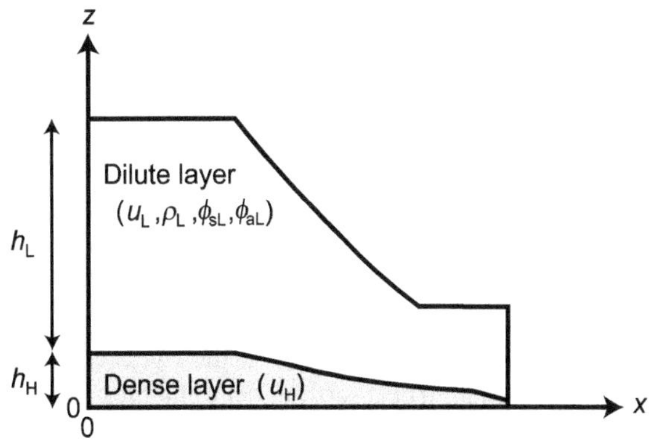

Fig. 13 Two-layer PDC model schematic. In the overlying dilute layer, the thickness h_L, velocity u_L, bulk density ρ_L, particle volume concentration ϕ_{sL}, and air volume concentration ϕ_{aL} evolve both temporally and spatially. In the underlying dense layer, the thickness h_H and velocity u_H evolve both temporally and spatially

$$\frac{\partial}{\partial t^*}\left(\rho_L^* h_L^*\right) + \frac{\partial}{\partial x^*}\left(\rho_L^* u_L^* h_L^*\right) = \rho_a E|u_L^*| - \rho_H \frac{\phi_{sL}}{\phi_{sH}} W_s, \quad (21)$$

$$\frac{\partial}{\partial t^*}\left(\rho_L^* u_L^* h_L^*\right) + \frac{\partial}{\partial x^*}\left(\rho_L^* u_L^{*2} h_L^* + \frac{\rho_L^* - \rho_a}{2} g h_L^{*2}\right) = -\left(\rho_L^* - \rho_a\right) g h_L^* \frac{\partial h_H^*}{\partial x^*} - \rho_H \frac{\phi_{sL}}{\phi_{sH}} u_L^* W_s - \tau_m^*. \quad (22)$$

The dense layer, modeled as a fluidized granular flow consisting of solid particles and volcanic gas, has the following vertically averaged mass and momentum conservation equations:

$$\frac{\partial h_H^*}{\partial t^*} + \frac{\partial}{\partial x^*}\left(u_H^* h_H^*\right) = \frac{\phi_{sL}}{\phi_{sH}} W_s, \quad (23)$$

$$\frac{\partial}{\partial t^*}\left(u_H^* h_H^*\right) + \frac{\partial}{\partial x^*}\left(u_H^{*2} h_H + \frac{1}{2}\frac{\rho_H - \rho_a}{\rho_H} g h_H^{*2}\right) = -\frac{\tau_b^*}{\rho_H} + \frac{\tau_m^*}{\rho_H} + \frac{\phi_{sL}}{\phi_{sH}} u_L^* W_s - \frac{h_H^*}{\rho_H}\frac{\partial}{\partial x^*}\left(\left(\rho_L^* - \rho_a\right) g h_L^*\right). \quad (24)$$

Here, $\phi(x^*, t^*)$ is volumetric concentration, and the subscripts L, H, s, g, and a denote the dilute (i.e., low-particle concentration) and dense (i.e., high-particle concentration) layers, solid particles, volcanic gas, and air, respectively. W_s is the settling velocity of the particles from the base of the dilute layer, E is the entrainment coefficient, τ_m^* is the interfacial shear drag, and τ_b^* is basal resistance of the dense layer.

The bulk density of the dilute layer is denoted by $\rho_L^* = \rho_s\phi_{sL} + \rho_a\phi_{aL} + \rho_g(1 - \phi_{sL} - \phi_{aL})$, in which the volume concentrations of the solid particles, ϕ_{sL}, and of the entrained air, ϕ_{aL}, evolve both temporally and spatially on the basis of their mass conservation equations:

$$\frac{\partial}{\partial t^*}\left(\phi_{sL} h_L^*\right) + \frac{\partial}{\partial x^*}\left(\phi_{sL} u_L^* h_L^*\right) = -\phi_{sL} W_s, \quad (25)$$

$$\frac{\partial}{\partial t^*}\left(\phi_{aL} h_L^*\right) + \frac{\partial}{\partial x^*}\left(\phi_{aL} u_L^* h_L^*\right) = E|u_L^*|. \quad (26)$$

In the dilute layer, it is assumed that turbulent mixing is sufficiently intense to maintain vertically uniform volumetric concentrations (e.g., Bonnecaze et al. 1993; Bursik and Woods 1996; Johnson and Hogg 2013). The dense layer is assumed to have a constant bulk density $\rho_H = \rho_s\phi_{sH} + \rho_g(1 - \phi_{sH})$, where the particle volumetric concentration ϕ_{sH} is set to 0.4 (Breard et al. 2016).

Interactions between the two layers are treated in the source terms of Eqs. (21)–(26) (i.e., the right-hand sides of the equations). Particle settling from the dilute layer to the dense layer is taken into account in the second source terms in Eqs. (21) and (22), the source terms in Eqs. (23) and (25), and the third source term in Eq. (24). The acceleration of the dilute layer over the basal contact is taken into account in the first source term in Eq. (22). The pressure gradient on the dense layer exerted by variations in the height of the dilute layer is taken into account in the fourth source term in Eq. (24).

The entrainment of ambient air into the dilute layer is taken into account in the first source term in Eq. (21) and the source term in Eq. (26). Thermal expansion of the entrained air is neglected here for the sake of ease. Air entrainment is also assumed to occur on the upper surface of the dilute layer (e.g., Bursik and Woods 1996; Johnson and Hogg 2013), although a different process for entrainment was recently proposed by Sher and Woods (2015). We adopted the entrainment coefficient proposed by Johnson and Hogg (2013): i.e., $E = 0.075/(1+27Ri)$, where $Ri \equiv \left(\rho_L^* - \rho_a\right) g h_L^* / \left(\rho_L^* u_L^{*2}\right)$.

The interfacial drag τ_m^* and the basal resistance of the dense layer τ_b^* are treated in the source terms in Eqs. (22) and (24). The interfacial shear drag τ_m^* is given by (Doyle et al. 2008, 2011):

$$\tau_m^* = \rho_L^* C_d \left(u_L^* - u_H^*\right) |u_L^* - u_H^*|, \quad (27)$$

where the drag coefficient C_d is set to 0.001 (Hogg and Pritchard 2004). The basal resistance τ_b^* is modeled such that shear drag is adopted during the constant velocity regime ((ii) of Fig. 11b) and Coulomb friction is adopted during the stopping regime ((iii) of Fig. 11b):

$$\tau_b^* = \begin{cases} \rho_H C_d u_H^* |u_H^*| & \text{(Shear drag)}, \\ \tan\delta\ (\rho_H - \rho_a) g h_H^* u_H^* / |u_H^*| & \text{(Coulomb friction)}, \end{cases} \quad (28)$$

(Figure 11b; see Roche et al. 2008 for details), where the dynamic basal friction angle δ is set to 20° (Doyle et al. 2008).

The front condition of the dilute layer is given by (Ungarish 2007):

$$u_{NL}^* = Fr\sqrt{\frac{\rho_{NL}^* - \rho_a}{\rho_a} g h_{NL}^*} \quad \text{at} \quad x^* = x_{NL}^*, \quad (29)$$

which is numerically treated by the BC model. The front condition of the dense layer, given by Eq. (3), is numerically treated by the AB model with $\varepsilon = 10^{-10}$.

In calculating the two-layer PDC model (Fig. 12), the conservation Eqs. (21)–(26) were numerically solved. In calculating the one-layer dilute PDC model (Fig. 10), we solved conservation Eqs. (21), (22), (25), and (26), where $h_H^* \approx 0$ and $u_H^* \approx 0$. In calculating the one-layer dense PDC model (Fig. 11), we solved conservation Eqs. (23) and (24). For numerical simulations, we used a fractional-step method to solve the conservation equations with source terms (e.g., LeVeque 2002), and the HLL approximate Riemann solver to calculate the intercell flux of the equations (e.g., Toro 2009).

Abbreviations

AB model: Artificial Bed model; BC model: Boundary Condition model; PDC: Pyroclastic density current

Acknowledgements
We thank Editor Colin J. N. Wilson, Reviewer Eric C. P. Breard, and an anonymous reviewer for their thoughtful comments.

Funding
Not applicable.

Authors' contributions
HAS carried out this study under the supervision of TK and YJS and prepared the first draft of the manuscript. All authors contributed to writing the manuscript and approved the final version.

Authors' information
HAS is a Ph.D. candidate supervised by TK and YJS. TK is a Professor at the Earthquake Research Institute, the University of Tokyo. YJS is an Assistant Professor at the Earthquake Research Institute, the University of Tokyo.

Competing interests
The authors declare that they have no competing interests.

References
Andrews BJ (2014) Dispersal and air entrainment in unconfined dilute pyroclastic density currents. Bull Volcanol 76:1–14. doi:10.1007/s00445-014-0852-4

Benjamin TB (1968) Gravity currents and related phenomena. J Fluid Mech 31:209–48. doi:10.1017/S0022112068000133

Birman V, Martin J, Meiburg E (2005) The non-boussinesq lock-exchange problem. part 2. High-resolution simulations. J Fluid Mech 537:125–44. doi:10.1017/S0022112005005033

Bonnecaze RT, Huppert HE, Lister JR (1993) Particle-driven gravity currents. J Fluid Mech 250:339–69. doi:10.1017/S002211209300148X

Bonometti T, Balachandar S (2010) Slumping of non-boussinesq density currents of various initial fractional depths: a comparison between direct numerical simulations and a recent shallow-water model. Comput Fluids 39:729–34

Bonometti T, Ungarish M, Balachandar S (2011) A numerical investigation of high-reynolds-number constant-volume non-boussinesq density currents in deep ambient. J Fluid Mech 673:574–602. doi:10.1017/S0022112010006506

Branney MJ, Kokelaar BP (2002) Pyroclastic density currents and the sedimentation of ignimbrites. Geol Soc, London

Breard ECP, Lube G (2017) Inside pyroclastic density currents—uncovering the enigmatic flow structure and transport behaviour in large-scale experiments. Earth Planet Sci Lett 458:22–36. doi:10.1016/j.epsl.2016.10.016

Breard ECP, Lube G, Jones JR, Dufek J, Cronin SJ, Valentine GA, Moebis A (2016) Coupling of turbulent and non-turbulent flow regines within pyroclastic density currents. Nat Geosci 9:767–71. doi:10.1038/NGEO2794

Bursik MI, Woods AW (1996) The dynamics and thermodynamics of large ash flows. Bull Volcanol 58:175–193. doi:10.1007/s004450050134

Cantero MI, Lee J, Balachandar S, Garcia MH (2007) On the front velocity of gravity currents. J Fluid Mech 586:1–39. doi:10.1017/S0022112007005769

Denlinger RP, Iverson RM (2004) Granular avalanches across irregular three-dimensional terrain: 1. theory and computation. J Geophys Res 109:F01014. doi:10.1029/2003JF000085

Doyle EE, Hogg AJ, Mader HM, Sparks RSJ (2008) Modeling dense pyroclastic basal flows from collapsing columns. Geophys Res Lett 35:L04305. doi:10.1029/2007GL032585

Doyle EE, Hogg AJ, Mader HM (2011) A two-layer approach to modelling the transformation of dilute pyroclastic currents into dense pyroclastic flows. Proc R Soc A 467:1348–71. doi:10.1098/rspa.2010.0402

Doyle EE, Huppert HE, Lube G, Mader HM, Sparks RSJ (2007) Static and flowing regions in granular collapses down channels: insights from a sedimenting shallow-water model. Phys Fluids 106601:19. doi:10.1063/1.2773738

Dressler RF (1954) Comparison of theories and experiments for the hydraulic dam-break wave. Int Assoc Sci Hydrol 3:319–28

Dufek J, Bergantz G (2007) Suspended load and bed-load transport of particle-laden gravity currents: the role of particle-bed interaction. Theor Comput Fluid Dyn 21:119–45. doi:10.1007/s00162-007-0041-6

Esposti Ongaro T, Orsucci S, Cornolti F (2016) A fast, calibrated model for pyroclastic density currents kinematics and hazard. J Volcanol Geotherm Res 327:257–72. doi:10.1016/j.jvolgeores.2016.08.002

Fujii T, Nakada S (1999) The 15 September 1991 pyroclastic flows at Unzen Volcano (Japan): a flow model for associated ash-cloud surges. J Volcanol Geotherm Res 89:159–72. doi:10.1016/S0377-0273(98)00130-9

Gröbelbauer H, Fanneløp T, Britter R (1993) The propagation of intrusion fronts of high density ratios. J Fluid Mech 250:669–87. doi:10.1017/S0022112093001612

Hallworth MA, Huppert HE (1998) Abrupt transitions in high-concentration, particle-driven gravity currents. Phys Fluids 10:1083. doi:10.1063/1.869633

Härtel C, Meiburg E, Necker F (2000) Analysis and direct numerical simulation of the flow at a gravity-current head. part 1. flow topology and front speed for slip and no-slip boundaries. J Fluid Mech 418:189–212. doi:10.1017/S0022112000001221

Hogg AJ (2006) Lock-release gravity currents and dam-break flows. J Fluid Mech 569:61–87. doi:10.1017/S0022112006002588

Hogg AJ, Pritchard D (2004) The effects of hydraulic resistance on dam-break and other shallow inertial flows. J Fluid Mech 501:179–212. doi:10.1017/S0022112003007468

Huppert HE, Simpson JE (1980) The slumping of gravity currents. J Fluid Mech 99:785–99. doi:10.1017/S0022112080000894

Ishimine Y (2005) Numerical study of pyroclastic surges. J Volcanol Geotherm Res 139:33–57. doi:10.1016/j.jvolgeores.2004.06.017

Iverson RM (1997) The physics of debris flows. Rev Geophys 35:245–96. doi:10.1029/97RG00426

Johnson CG, Hogg AJ (2013) Entraining gravity currents. J Fluid Mech 731:477–508. doi:10.1017/jfm.2013.329

Larrieu E, Staron L, Hinch E (2006) Raining into shallow-water as a description of the collapse of a column of grains. J Fluid Mech 554:259–70. doi:10.1017/S0022112005007974

LeVeque RJ (2002) Finite volume methods for hyperbolic problems. Cambridge University Press, Cambridge

Marino B, Thomas L, Linden P (2005) The front condition for gravity currents. J Fluid Mech 536:49–78. doi:10.1017/S0022112005004933

Martin JC, Moyce WJ (1952) An experimental study of the collapse of liquid columns on a rigid horizontal plane. Phil Trans R Soc Lond A 244:312–24. doi:10.1098/rsta.1952.0006

Meiburg E, Kneller B (2010) Turbidity currents and their deposits. Annu Rev Fluid Mech 42:135–56. doi:10.1146/annurev-fluid-121108-145618

Nield SE, Woods AW (2004) Effects of flow density on the dynamics of dilute pyroclastic density currents. J Volcanol Geotherm Res 132:269–81. doi:10.1016/S0377-0273(03)00314-7

Ooi SK, Constantinescu G, Weber L (2009) Numerical simulations of lock-exchange compositional gravity current. J Fluid Mech 635:361–88. doi:10.1017/S0022112009007599

Roche O, Montserrat S, Niño Y, Tamburrino A (2008) Experimental observations of water-like behavior of initially fluidized, dam break granular flows and their relevance for the propagation of ash-rich pyroclastic flows. J Geophys Res 113:B12203. doi:10.1029/2008JB005664

Rottman JW, Simpson JE (1983) Gravity currents produced by instantaneous releases of a heavy fluid in a rectangular channel. J Fluid Mech 135:95–110. doi:10.1017/S0022112083002979

Sher D, Woods AW (2015) Gravity currents: Entrainment, stratification and self-similarity. J Fluid Mech 784:130–62. doi:10.1017/jfm.2015.576

Stoker JJ (1992) Water waves: the mathematical theory with applications. Wiley, New York

Toro EF (2001) Shock-capturing methods for free-surface shallow flows. Wiley, Chichester

Toro, EF (2009) Riemann solvers and numerical methods for fluid dynamics: a practical introduction. Springer, Berlin

Ungarish M, Zemach T (2005) On the slumping of high reynolds number gravity currents in two-dimensional and axisymmetric configurations. European J Mech B/Fluids 24:71–90. doi:10.1016/j.euromechflu.2004.05.006

Ungarish M (2007) A shallow-water model for high-reynolds-number gravity currents for a wide range of density differences and fractional depths. J Fluid Mech 579:373–82. doi:10.1017/S0022112007005484

Ungarish M (2009) An introduction to gravity currents and intrusions. CRC Press, Boca Raton

Wilson CJN, Walker GPL (1982) Ignimbrite depositional facies: the anatomy of a pyroclastic flow. J Geol Soc 139:581–92. doi:10.1144/gsjgs.139.5.0581

Complex inner core boundary from frequency characteristics of the reflection coefficients of PKiKP waves observed by Hi-net

Satoru Tanaka[1*] and Hrvoje Tkalčić[2]

Abstract

Frequency-dependent reflection coefficients of P waves at the inner core boundary (ICB) are estimated from the spectral ratios of PKiKP and PcP waves observed by the high-sensitivity seismograph network (Hi-net) in Japan. The corresponding PKiKP reflection locations at the ICB are distributed beneath the western Pacific. At frequencies where noise levels are sufficiently low, spectra of reflection coefficients show four distinct sets of characteristics: a flat spectrum, a spectrum with a significant spectral hole at approximately 1 or 3 Hz, a spectrum with a strong peak at approximately 2 or 3 Hz, and a spectrum containing both a sharp peak and a significant hole. The variety in observed spectra suggests complex lateral variations in ICB properties. To explain the measured differences in frequency characteristics of ICB reflection coefficients, we conduct 2D finite difference simulations of seismic wavefields near the ICB. The models tested in our simulations include a liquid layer and a solid layer above the ICB, as well as sinusoidal and spike-shaped ICB topography with varying heights and scale lengths. We find that the existence of a layer above the ICB can be excluded as a possible explanation for the observed spectra. Furthermore, we find that an ICB topographic model with wavelengths and heights of several kilometers is too extreme to explain our measurements. However, restricting the ICB topography to wavelengths and heights of 1.0–1.5 km can explain the observed frequency-related phenomena. The existence of laterally varying topography may be a sign of lateral variations in inner core solidification.

Keywords: Inner core boundary; Topography; PKiKP; Finite difference modeling

Background

The inner core boundary (ICB) is one of the vital regions for understanding the Earth's core dynamics (Loper and Roberts 1981; Loper 1983; Bergman and Fearn 1994; Shimizu et al. 2005; Deguen et al. 2007; Sumita and Bergman 2009; Deguen 2012). Seismological studies of the ICB and its inferred characteristics, such as the density jump between the inner and the outer cores, the shear-wave velocity at the top of the inner core, and scattering of seismic energy from small-scale topography at the inner core surface, are important in elucidating the growth mechanism of the inner core and the source of the geodynamo (Souriau 2007; Sumita and Bergman 2009; Deuss 2014; Tkalčić 2015).

The hypothesis of a hemispherical structure in the upper inner core (Tanaka and Hamaguchi 1997) has been widely accepted because it is supported by seismic observations of body waves and free oscillations (Creager 1999; Deuss et al. 2010). Hemispherical dichotomy is recognized as a global phenomenon near the ICB (Niu and Wen 2001; Waszek et al. 2011) and possibly near the center of the inner core (Lythgoe et al. 2014). To explain the hemispherical structure of the inner core, two models have been proposed: a large-scale asymmetric flow in the outer core (Sumita and Olson 1999; Aubert et al. 2008; Gubbins et al. 2011) and translational convection in the inner core (Alboussiere et al. 2010; Monnereau et al. 2010). The nature of these models is that they allow diametrically opposite scenarios of freezing and melting, i.e., one of the two hemispheres solidifies faster than the other. This has raised further questions about whether the surface of the inner core in

* Correspondence: stan@jamstec.go.jp
[1]Department of Deep Earth Structure and Dynamics Research, Japan Agency for Marine-Earth Science and Technology, Yokosuka 237-0061, Japan
Full list of author information is available at the end of the article

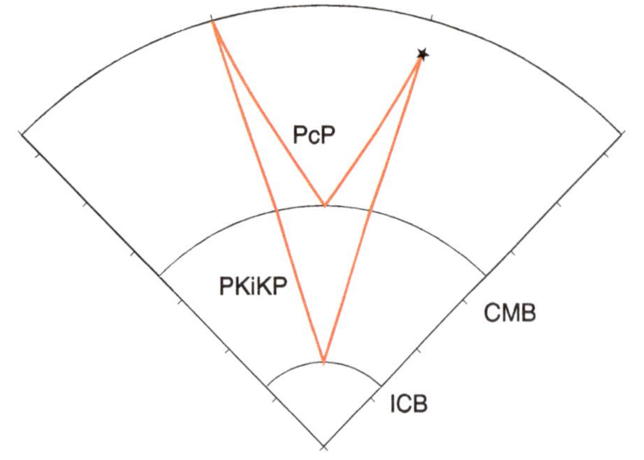

Fig. 1 Ray paths of PcP and PKiKP seismic phases. A cross section of Earth with ray paths of PcP and PKiKP seismic phases originating at the source (*star*) and recorded by a station at the Earth's surface

the eastern hemisphere is melting or freezing. Although translational convection models predict that the eastern hemisphere is melting, outer-core large asymmetric flow models disagree about which hemisphere is melting, due to the problem setting, e.g., heat flux at the core–mantle boundary (CMB) and outer core dynamics.

Measurements of PKiKP/PcP amplitude ratios have been used to infer the density jump at the ICB, as well as to determine the shear velocity at the top of the inner core. A sketch of their ray paths is shown in Fig. 1. Studies in the twentieth century usually analyzed PKiKP recorded on short-period seismographs with a predominant frequency of 1 Hz (Engdahl et al. 1970; Buchbinder 1972; Engdahl et al. 1974; Souriau and Souriau 1989; Shearer and Masters 1990). These pioneering studies may have been hampered by the small number of observations and the large scatter in PKiKP/PcP amplitude ratios due to the nature of noisy amplitude measurements (Tkalčić et al. 2009). Another factor influencing the amplitude ratios, and possibly the anti-correlation between the two phases, is the occurrence of near-receiver crustal and mantle heterogeneity (Tkalčić et al. 2010).

The new era of modern instruments, dense networks, and improved global coverage enables observation of a large number of PKiKP and PcP phases on the same seismogram. For example, Koper et al. (2003) analyzed a significant number of PKiKP phases at shorter distances recorded by small aperture arrays of the International Monitoring System. This study was followed by a large number of new studies (Cao and Romanowicz 2004; Koper et al. 2004; Koper and Pyle 2004; Poupinet and Kennett 2004; Koper and Dombrovskaya 2005; Krasnoshchekov et al. 2005; Leyton et al. 2005; Kawakatsu 2006; Leyton and Koper 2007a, b; Peng et al. 2008).

More specifically, Koper and Pyle (2004) measured PKiKP/PcP amplitude ratios from seismograms filtered between 1 and 3 Hz. Their analysis did not reveal any differences between the eastern and western hemispheres of the inner core. Using phases from a Mariana event observed in Japan, whose raypaths sampled the ICB beneath the western Pacific, Kawakatsu (2006) found little scattering energy in the PKiKP coda. However, Leyton and Koper (2007b) analyzed the coda of PKiKP and suggested the existence of small-scale heterogeneities in the uppermost inner core. This result indicates that the strong scattering region is located beneath the Pacific Ocean and Asia, which covers parts of the eastern and western hemispheres. On the basis of scattering properties and Q structure, Cormier (2007) inferred textural differences between the eastern and western hemispheres near the surface of the inner core, including vertically oriented structures in the eastern hemisphere.

Interestingly, the existence of a high-frequency PKiKP phase (up to 5 Hz) with steep incidence angles at the ICB was first observed by Poupinet and Kennett (2004) using phases recorded by narrow-aperture arrays and temporal broadband networks on the Australian continent, whose reflection points were in the eastern hemisphere. Recently, the same class of PKiKP waves was observed in the eastern hemisphere on Chinese and Japanese short-period and broadband stations (J-array) (Tkalčić et al., 2009, 2010) and by the high-sensitivity seismograph network (Hi-net) in Japan (Dai et al. 2012, Jiang and Zhao 2012).

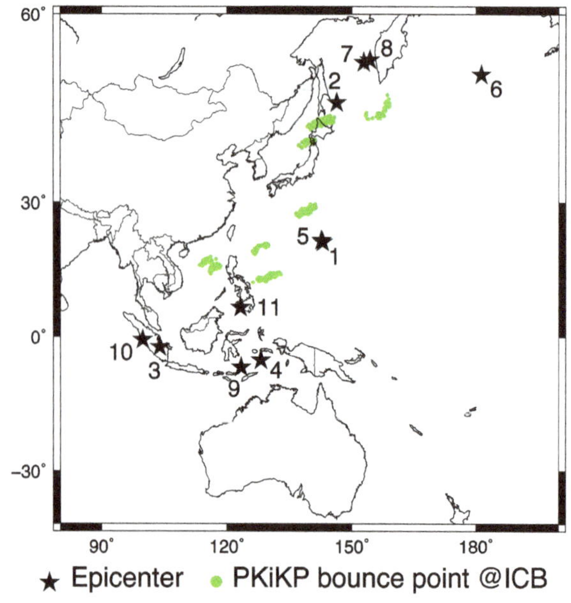

Fig. 2 Study area. Map view of epicenters (*stars*) and PKiKP bounce points at the ICB (*green circles*). Event numbers are taken from Table 1

Table 1 Event list

Event	Date	Time (UTC)	Lat. (°)	Lon. (°)	Depth (km)	m_b	Remarks	No. of used records
1	20010703	13:10:42.60	21.641	142.984	290.0	6.0	DC	73
2	20021117	04:53:48.46	47.946	146.419	470.2	5.8	Sm	58
3	20040725	14:35:19.06	−2.427	103.981	582.1	6.8	DC	83
4	20060127	16:58:53.67	−5.473	128.131	397.0	7.0	Sm	77
5	20070928	13:38:57.88	22.013	142.668	260.0	6.7	Sm	41
6	20080322	21:24:11.27	52.176	−178.716	132.0	5.8	Sm	65
7	20080705	02:12:04.48	53.882	152.886	632.8	6.8	DC	49
8	20081124	09:02:58.76	54.203	154.322	492.3	6.5	DC	83
9	20090828	01:51:20.40	−7.146	123.427	642.4	6.3	DC	31
10	20090930	10:16:09.25	−0.720	99.867	81.0	7.1	DC	33
11	20100723	23:15:10.19	6.780	123.260	640.0	7.4	Sm	53

Hypocenters are taken from the Earthquake data report provided by U.S.G.S.
DC double couple, *Sm* smoothed radiation pattern

To contribute to a better understanding of inner core dynamics and to constrain ICB structure in the eastern hemisphere, here, we collect an extensive dataset of PKiKP waves recorded by Hi-net in Japan. Our aim is to shift focus from analyzing a single value of the PKiKP/PcP amplitude ratio to evaluating its broad frequency characteristics, which is philosophically similar to how Cummins and Johnson (1988) evaluated pre-critical PKiKP waveforms and spectra by using a hybrid full wave-reflectivity algorithm. A dense configuration of borehole seismograms with high SNR observations of PKiKP waves over an unusually broad range of frequencies facilitates this new approach to estimate ICB properties. Thus, we examine data in the frequency domain and investigate possible broader implications for Earth's core dynamics.

Methods

Hi-net comprises approximately 700 short-period seismographs placed at the bottoms of individual boreholes (Okada et al. 2004). However, even for large-volume datasets such as Hi-net, good records of PKiKP and PcP are not frequently observed. We have detected PcP and PKiKP waveforms from 11 earthquakes with body wave magnitude ≥5.8 and focal depths of ≥80 km around Japan before the 2011 Tohoku earthquake (Fig. 2, Table 1). This covers epicentral distances from 15° to 50°. The reflection points of PKiKP at the ICB are distributed beneath the western Pacific, which is part of the eastern hemisphere (Fig. 2).

After visual examination of waveforms, amplitude spectra of PKIKP phases, and corresponding waveforms and spectra of pre-arrival noise, we applied a zero-phase Butterworth band-pass filter with corner frequencies of 2 and 5 Hz to 60 s record segments centered around theoretical PKiKP arrival times (Fig. 3a, b). This was done for all records except the Mariana event (event 1), which was examined by Kawakatsu (2006) and observed to have large PKiKP signals around 1 Hz. Note that we use band-pass filtering only for initial identification of PKiKP.

After applying the band-pass filter described above and retrieving record segments of ±10 s length around PKiKP arrivals, seismograms were divided into sub-groups comprising 70–230 stations with traces sorted by increasing epicentral distance. We then selected subsets of coherent waveforms by using the cross-correlation matrix method (Tkalčić et al. 2011), which retained 46–220 records per event. To find coherent PKiKP arrivals in each sub-group, we empirically determined the minimum percentage (τ) of all waveform pairs that should cross-correlate in such a way that the average cross-correlation coefficient equals or exceeds a threshold β. The algorithm calculates the cross-correlation coefficients for each pair of waveforms and counts the total percentage of pairs with cross-correlation coefficient exceeding β. For example, for $\beta > 0.4$, $\tau > 10$ %, and 130 total waveforms, we found that 77 waveforms satisfied these criteria and were consequently selected as "mutually coherent" (Fig. 3c). Incoherent waveforms were not used in further analyses. Subsequently, we visually checked the waveforms that satisfied the above criteria to find possible PKiKP signals with high signal-to-noise ratios, which resulted in a station list of "good" sites for PKiKP observations.

In the next step, we prepared a spectrogram of 90 s length from the unfiltered seismograms, with a sampling interval of 1 s and a lapse time of 50 s from the theoretical arrival times of PcP or PKiKP (Fig. 4). Each Fourier spectrum is calculated for a Welch tapered 10 s window (Press et al. 1988), then smoothed using a three-point moving average. Using the station list obtained with the

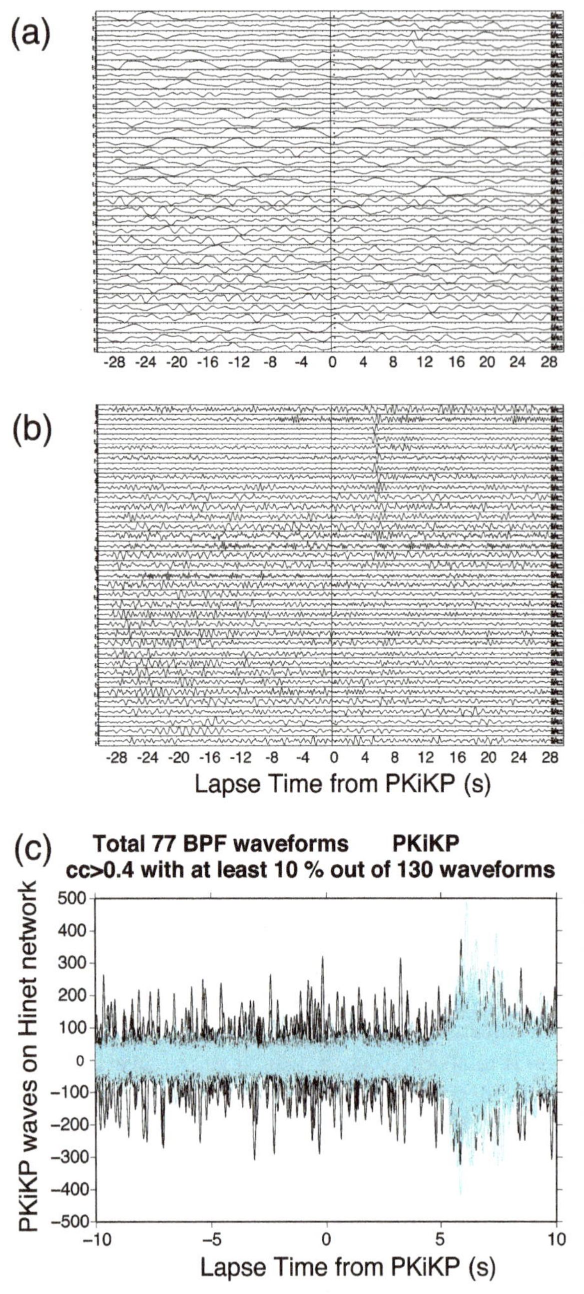

Fig. 3 Selection of coherent waveforms. **a** Raw seismograms of event 3 recorded by Hi-net stations whose names start with letter "I." *Horizontal axis* is the lapse time from the theoretical arrival times of PKiKP. **b** Seismograms from **a** are band-pass filtered with cutoff frequencies of 2 and 5 Hz. **c** The selected set of coherent waveforms after employing the cross-correlation matrix method (Tkalčić et al., 2011) is shown by *cyan lines*

cross-correlation matrix method, we used both the seismograms and spectrograms of PKiKP and PcP to identify acceptable data for calculating spectral ratios of PKiKP/PcP (Fig. 5). For each event, between 33 and 83 station records were retained (Table 1). Finally, the mean PKiKP/PcP spectral ratio was computed at each frequency that exceeded the mean of the noise/PcP ratio (Fig. 5). The spectra of PcP and PKiKP contained several peaks and holes whose central frequencies were not uniform for all events. For example, for event 3, the spectral ratio of PKiKP/PcP exceeded the mean noise/PcP ratio in the range 0.8 to 3.0 Hz (Fig. 5a). In the case of other events, however, valid frequency ranges were somewhat different, e.g., 2–3 Hz for event 6 (Fig. 5b) and 1–4 Hz for event 11 (Fig. 5c). Despite the above restrictions, the relatively high-frequency content of the PcP and PKiKP waves means that these high-quality simultaneous observations over a broad frequency interval are unprecedented.

To correct the amplitude ratios, we determined focal mechanisms using the program of Kikuchi and Kanamori (2003), rather than cataloged Global CMT solutions, USGS Moment tensor, and double-couple solutions. The broadband displacements of P and SH waves in the frequency range 0.002–1 Hz were used for the inversion. Furthermore, we compared the short-period (SP) P wave amplitudes (1–5 Hz) with the radiation pattern predicted from each focal mechanism solution. Although the observed SP amplitudes showed a high degree of scatter, we found that some events had clear energy even near the nodal plane and null axis, suggesting a smoothed radiation pattern. This may be due to scattering near the source region, as discussed in previous studies that determined magnitudes from SP data (e.g., (Schweitzer and Kværna 1999; Takemura et al. 2015). These results are summarized in the Remarks column of Table 1. The smoothed radiation pattern for the short period is a likely explanation for the observation of PcP and PKiKP in the events from Kuril (event 2) and Mindanao (event 11), in which the take-off azimuths and angles of PcP and PKiKP were located near the nodal plane.

Attenuation factors for the mantle and core at arbitrary frequencies were normalized to a reference frequency of 1 Hz. We then obtained the corrected relationship between spectral and theoretical amplitude ratios from the following relationship:

$$\left(\frac{A_{\mathrm{PKiKP}}}{A_{\mathrm{PcP}}}\right)_{\mathrm{corrected}} = \left(\frac{A_{\mathrm{PKiKP}}}{A_{\mathrm{PcP}}}\right)_{\mathrm{observed}} \frac{F_{\mathrm{PcP}}}{F_{\mathrm{PKiKP}}} \frac{\exp(-\pi f_0 \Delta t^*)}{\exp(-\pi f \Delta t^*)} = \frac{T_{\mathrm{PK}} R_{\mathrm{KK}} T_{\mathrm{KP}}}{R_{\mathrm{PP}}} \frac{G_{\mathrm{PKiKP}}}{G_{\mathrm{PcP}}} \frac{\exp(-\pi f_0 t^*{}_{\mathrm{PKiKP}})}{\exp(-\pi f_0 t^*{}_{\mathrm{PcP}})}, \quad (1)$$

where the subscripts PKiKP and PcP denote the corresponding seismic phases. On the left side of the equation, A is the spectral amplitude at an arbitrary frequency, F is the focal mechanism radiation pattern, t^* is the anelastic parameter (Lay and Wallace 1995), and Δt^* is the differential anelastic parameter between PKiKP

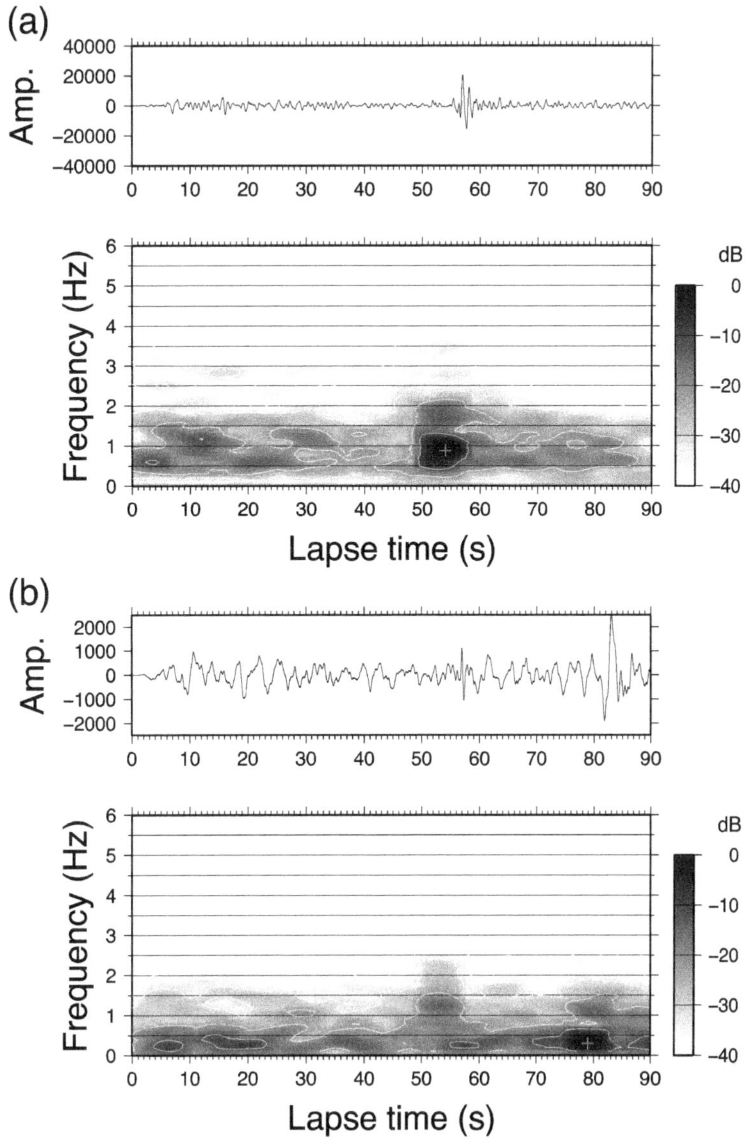

Fig. 4 Spectrograms including PcP and PKiKP. Raw seismograms and spectrograms of event 3 for **a** PcP and **b** PKiKP recorded at station IKTH. Spectrogram scaling is in decibels (dB). The predicted arrival times of PcP and PKiKP are located at a lapse time of 50 s

and PcP. On the right side, G represents the geometrical spreading factor; f_0 (=1 Hz) and f are reference and observed frequencies, respectively; T_{PK} and T_{KP} are the transmission coefficients at the CMB; and R_{PP} and R_{KK} denote reflection coefficients at the CMB and ICB, respectively. The reflection coefficient R_{KK} at the ICB is estimated from the corrected spectral ratios by considering the terms of the reflection and transmission coefficients at the CMB, the geometrical spreading factor, and anelastic parameters, which are calculated using ak135 (Kennett et al. 1995):

$$(R_{\mathrm{KK}})_{\mathrm{estimated}} = \left[\left(\frac{A_{\mathrm{PKiKP}}}{A_{\mathrm{PcP}}}\right)_{\mathrm{corrected}}\right] \Big/ \left[\frac{T_{\mathrm{PK}}T_{\mathrm{KP}}}{R_{\mathrm{PP}}}\frac{G_{\mathrm{PKiKP}}}{G_{\mathrm{PcP}}}\frac{\exp(-\pi f_0 t^*{}_{\mathrm{PKiKP}})}{\exp(-\pi f_0 t^*{}_{\mathrm{PcP}})}\right]. \tag{2}$$

We converted the spectral ratios to reflection coefficients as a function of frequency. To the first order, geometric spreading is constant with respect to frequency under a ray-theoretical assumption. The reflection and transmission coefficients were then calculated for planar boundaries. The correction of the attenuation factor using t^* gives a smooth exponential variation with frequency.

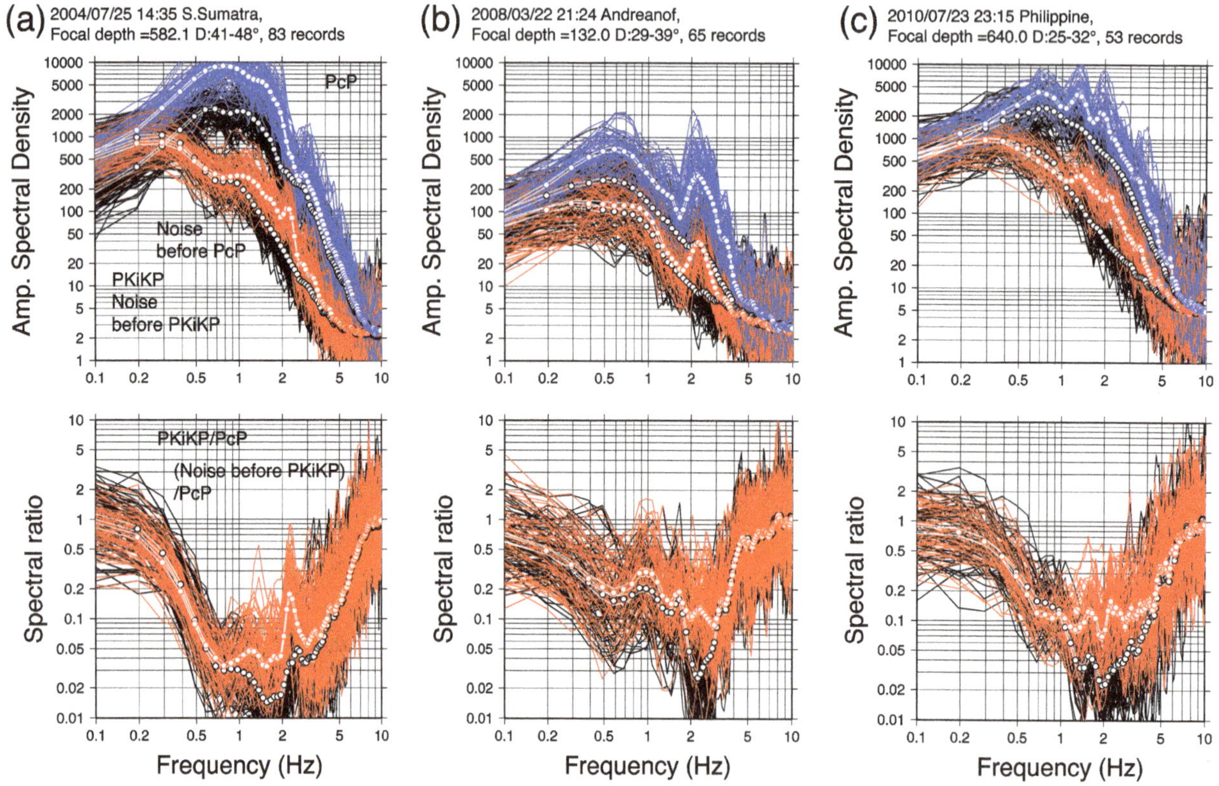

Fig. 5 Spectra of PcP and PKiKP and spectral ratios of PKiKP/PcP. Individual PKiKP, PcP, noise spectra, and PKiKP/PcP spectral ratios for **a** event 3, **b** event 6, and **c** event 11 (see Fig. 2 for the event location). (*Top row*) Spectra of PcP (*blue lines*) and PKiKP (*red lines*), noise in the 20 s window prior to PcP (*gray lines*), and noise in the 20 s window prior to PKiKP (*black lines*). *Open circles* are average spectral amplitudes for PcP, PKiKP, noise prior to PcP, and noise prior to PKiKP at each frequency. (*Bottom row*) Spectral ratios of PKiKP/PcP (*red lines*) and noise before PKiKP/PcP (*black lines*). *Open circles* are average values of spectral ratios at each frequency

Thus, these corrections will not result in any spectral holes or peaks.

To reduce the unwanted effect of the CMB on the PcP spectra, the spectral ratios of PcP/P were examined. Similar to the above procedure, we estimated the P wave reflection coefficients at the CMB as a function of frequency. The obtained reflection coefficients did not always coincide with theoretical values due to the uncertainty in focal mechanisms, large differences in P and PcP take-off angles, and other unknown causes. Thus, we corrected only the fluctuations in the reflection coefficients around the average values in the frequency range 1–3 Hz. The reflection coefficients at the ICB are multiplied by the fluctuations in reflection coefficients at the CMB, which can result in either amplifying an apparently small PKiKP/PcP due to a large peak in the PcP spectrum or reducing a large PKiKP/PcP due to a small peak in the PcP.

Results

A summary of frequency-dependent ICB P wave reflection coefficients, as derived from Eqs. (1) and (2), is shown in Fig. 6. Panels are sorted by increasing incidence angle at the ICB; we refer to each panel hereafter as result *n*, where *n* stands exclusively for panel number. Figure 7 summarizes frequency-dependent P wave reflection coefficients at the CMB derived from PcP/P spectral ratios. In Fig. 8, these are used to correct the reflection coefficients at the ICB (Fig. 6). These corrected values are used in finite difference modeling and subsequent interpretation.

The frequency characteristics of the ICB reflection coefficients are quite complex, even in a narrow effective signal band (Fig. 8). Roughly speaking, peaks in reflection coefficients appear around 2 Hz (results 8, 9, and 10) and 3 Hz (result 3), and holes are observed around 1 Hz (results 2, 3, and 4) and 3 Hz (result 10). Although the discrimination is still qualitative, we recognize four general categories of frequency-dependent characteristics: (i) a flat variation, where fluctuations in the relative strengths of peaks and holes are between half and double those of the theoretical reflection coefficients (results 1, 5, 6, 7, and 11); (ii) a distinct single hole in each reflection coefficient spectrum (results 2 and 4); (iii) a strong single peak in each spectrum (results 8 and 9);

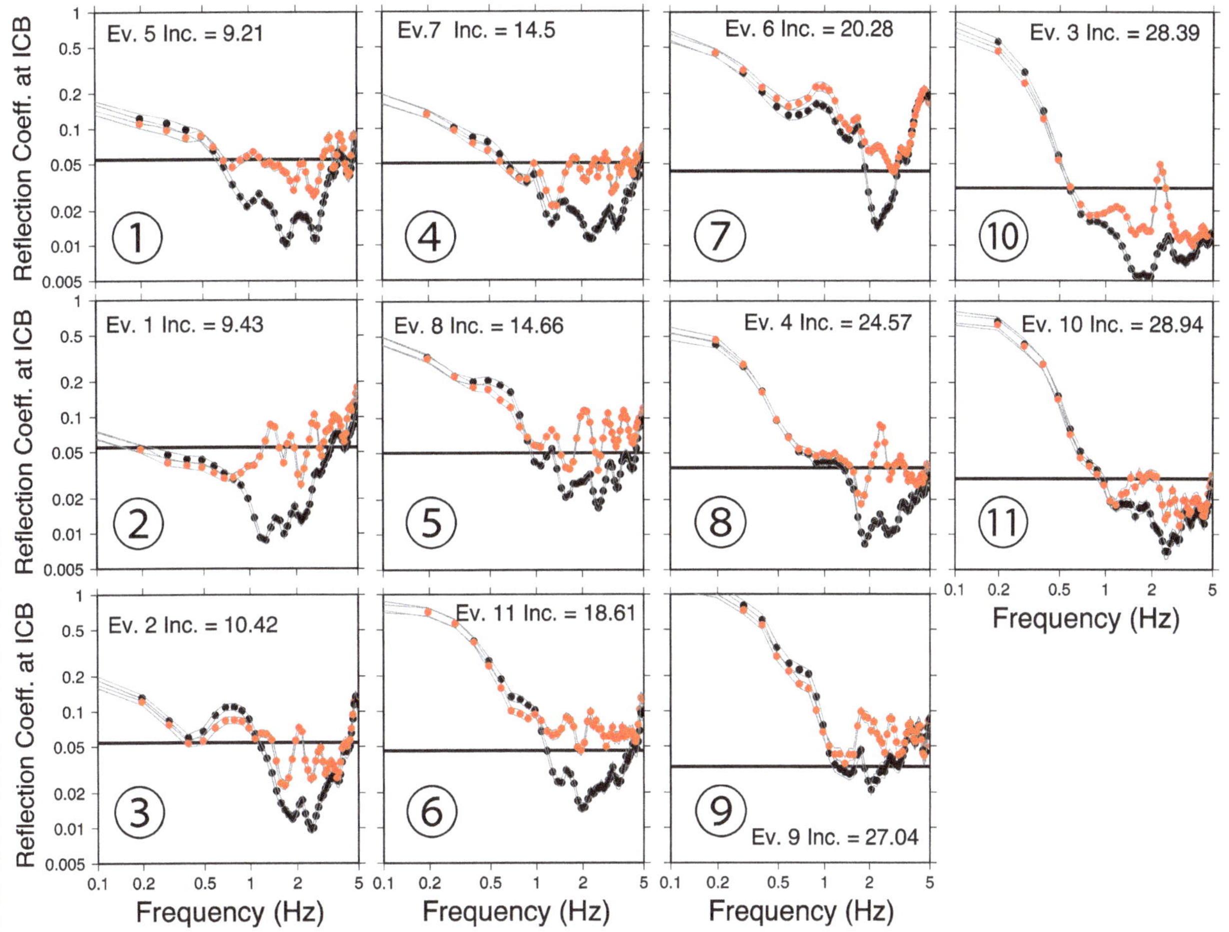

Fig. 6 P wave reflection coefficients at the ICB. P wave reflection coefficients at the ICB as a function of frequency, estimated from PKiKP/PcP spectral ratios (*red circles*) and noise/PcP (*black circles*). *Thin lines* show the upper and lower bounds of standard errors. *Thick lines* are theoretical values of reflection coefficients calculated using ak135 (Kennett et al., 1995). *Encircled numbers* are "result numbers," arranged by ascending incidence angle

and (iv) a strong peak and hole in the same spectrum (results 3 and 10).

The geographical distribution of these categories is plotted in Fig. 9. The diameters of the 1 and 2 Hz Fresnel zones for PKiKP are approximately 80 and 60 km at the ICB, respectively. This is equivalent to the area of PKiKP reflection points at the ICB that is covered by a single event and all stations that detect PKiKP. This result suggests that our averaging within each region number is meaningful, but it is not appropriate in discussing lateral variations in individual measurements within each region. Although the sampling areas are sparse, we note a tendency of frequency peaks in reflection coefficient spectra to be most observable at low latitudes (results 8, 9, and 10). Frequency holes show no such latitudinal trend (results 2, 3, 4, and 10).

Discussion

Effects of the CMB

Regarding the effects of the CMB on PKiKP spectra during transmission through the CMB, we address this issue in the context of the results of previous studies. Using amplitude of precursors to PKIKP, Dai et al. (2012) and Yao and Wen (2014) showed that several regions exhibit weak scattering in the lowermost mantle beneath the southwestern Pacific. PKiKP phases from events that occurred in the Banda Sea (events 4 and 9), Sumatra (events 3 and 10), and the Philippines (event 11) enter a "normal" CMB. According to Hedlin and Shearer (2000), there is a relatively weak scattering region in the lowermost mantle beneath the Philippine Sea, which corresponds to the CMB entry points of PKiKP for events 1 and 5 and the CMB exit points for events 1, 3, 4, 5, 9,

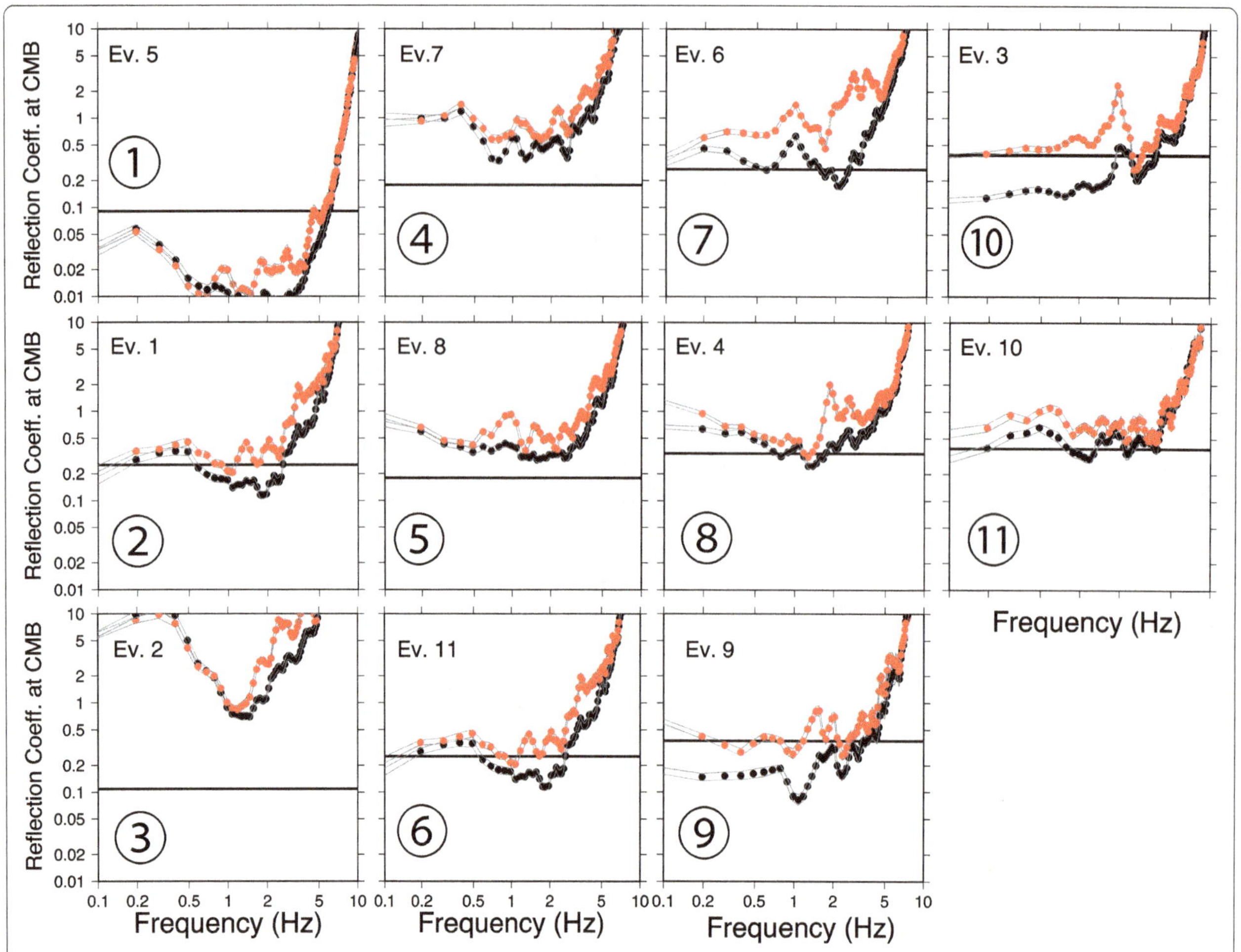

Fig. 7 P wave reflection coefficients at the CMB. P wave reflection coefficients at the CMB as a function of frequency, estimated from PcP/P spectral ratios (*red circles*) and noise/P (*black circles*). *Thin lines* show the upper and lower bounds of standard errors. *Thick lines* are theoretical values of the reflection coefficient calculated using ak135 (Kennett et al., 1995). *Encircled numbers* have the same meaning as in Fig. 6

10, and 11. Thus, we have reason to believe that the CMB effects on the estimated reflection coefficients for results 1, 2, 6, 8, 9, 10, and 11 will be negligible. However, a strong scattering area near the CMB exists beneath north Japan and the northwestern Pacific, which includes the PKiKP CMB entry and exit points for events 2, 6, 7, and 8. Thus, we cannot rule out the possibility that the frequency characteristics of calculated reflection coefficients for results 3, 4, 5, and 7 are CMB effects, e.g., the high-frequency components of PKiKP may be lost by scattering at the CMB.

Numerical simulations

Problem setup

To explain our observations, we use the 2D finite difference program *e3d* (Larsen and Shultz 1995; Rodgers et al. 2006) to simulate wave propagation by solving the full wave equation on a staggered grid. The solutions are fourth order accurate in space and second order accurate in time. Figure 10a shows the configuration of the simulation, for which the grid spacing is 70 m. P- and S-wave velocities and densities above and below the ICB are taken from ak135 (Kennett et al. 1995). We place a sequence of point sources with 1 km spacing on a straight line 100 km long and generate a plane wave. The incidence angle θ is a control parameter. As the input, the representative P wave waveform is taken from the Mariana event (event 1). The calculation is valid for frequencies up to 5 Hz. We examine the observed spectral ratios between incident and reflected waves at three points (triangles in Fig. 10a). To verify the configuration and boundary conditions, and to ensure that there are no unwanted numerical effects from the edges of the box, we conducted a test run of a simple, flat ICB, with elastic parameters on both sides of the ICB taken from the ak135 model (Fig. 10b). The spectral ratios between upgoing and downgoing waves for incidence angles of 10°, 20°, and 30° correspond to epicentral distances of approximately 16°, 32°, and 48°, respectively. Although there are small frequency-dependent fluctuations (within

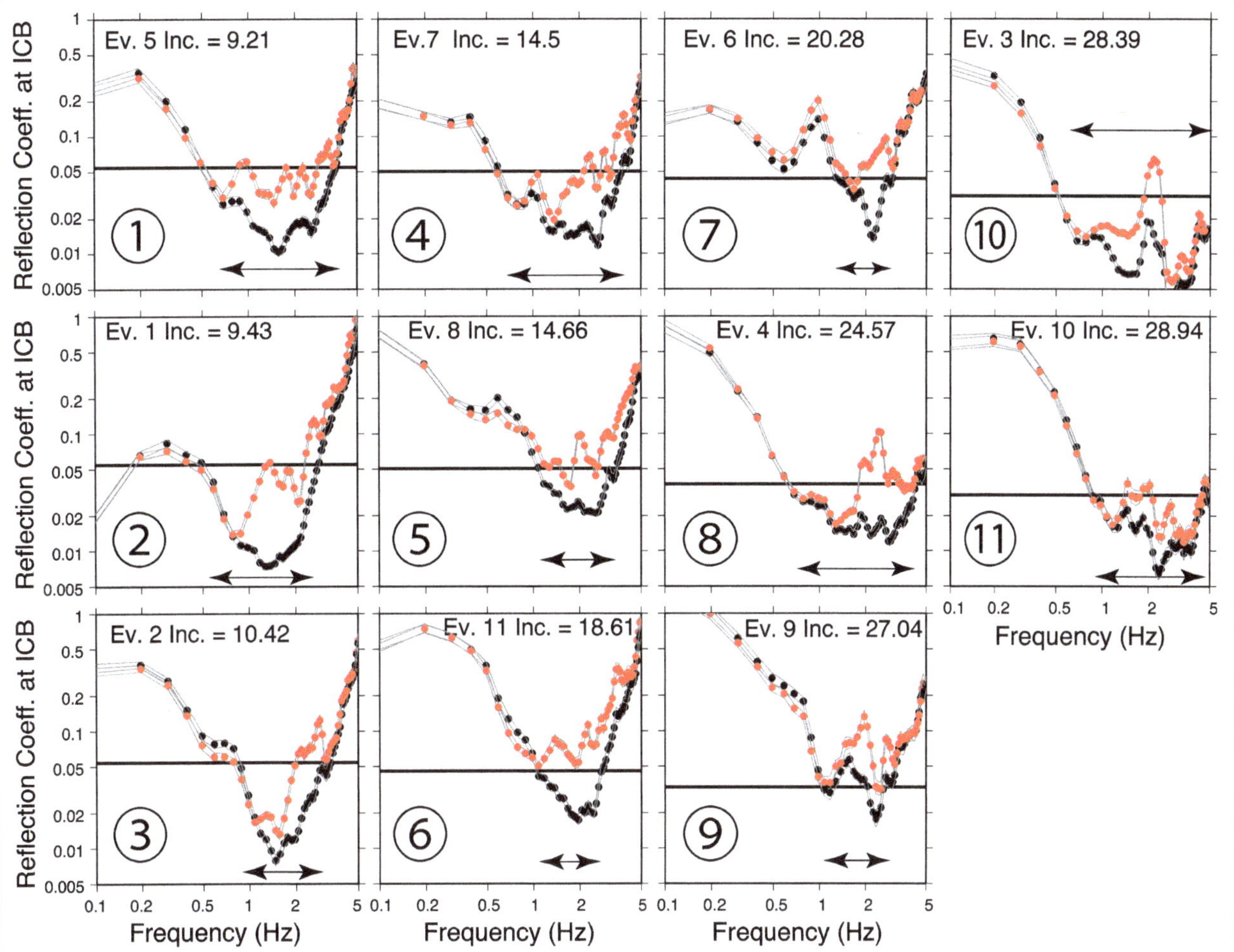

Fig. 8 Corrected P wave reflection coefficients at the ICB. P wave reflection coefficients at the ICB as a function of frequency, estimated from PKiKP/PcP spectral ratios (*red circles*) and noise/PcP (*black circles*). Ratios are corrected for fluctuations estimated from the P wave reflection coefficients at the CMB, obtained from the PcP/P ratios in Fig. 7. *Thin lines* are upper and lower bounds by standard errors. *Thick lines* are theoretical values of the reflection coefficient calculated using ak135 (Kennett et al., 1995). *Horizontal arrows* indicate the effective frequency ranges. *Encircled numbers* are explained in Fig. 6

a factor of 2, at most), their average values coincide well with the theoretical reflection coefficients at the ICB.

Simulations for a thin layer above the ICB

As discussed in the "Background" section, there are multiple conceptual models that can explain freezing and melting of the quasi-eastern hemisphere. To investigate the possibility of a thin layer above the ICB, we run a series of simulations with layer thicknesses of 0.1, 0.25, 0.5, 0.75, 1.0, 3.0, and 5.0 km. For a solid layer above the ICB with finite shear-wave velocity and slight variations in compressional wave velocity and density (Vp = 10.6–10.9 km/s, Vs = 1–3 km/s, $\rho = 12.4–12.5$ g/cm^3), all the resultant reflected wave amplitudes are small. However, a light or heavy liquid layer (Vp = 9.6–10.3 km/s, $\rho = 12.5$ g/cm^3 for heavy liquid, $\rho = 11.5$ g/cm^3 for light liquid) results in large amplitudes. In particular, we find that the amplitudes of seismograms filtered with central frequencies of 1 and 2 Hz become large when we insert a heavy liquid layer above the ICB with Vp = 9.6 km/s, $\rho = 12.5$ g/cm^3, and a thicknesses of 0.75 or 1 km (Fig. 11a, b). This finding is qualitatively consistent with the results of Krasnoshchekov et al. (2005). However, we also find that spectral ratios continuously increase as a function of frequency for any incidence angle (Fig. 11c), without distinct peaks or holes in the frequency range 0.5 to 3 Hz. Since this does not explain our observations, we reject the model of a thin layer above the ICB as a possible explanation for the observed frequency-domain characteristics of PKiKP.

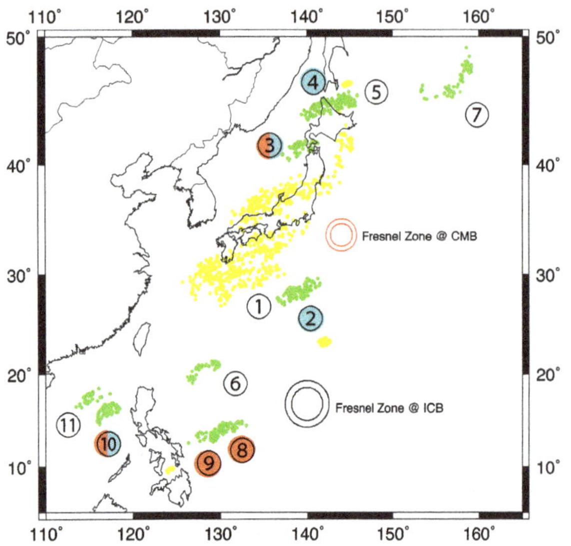

Fig. 9 Summary map of frequency characteristics of P wave reflection coefficients at the ICB. Distribution of the reflection points of PKiKP waves at ICB (*green dots*), and "result numbers" indicating their frequency characteristics by colors. *White*: almost flat spectrum; *blue*: spectrum with a spectral hole; *red*: spectrum with a spectral peak; *half red and blue*: both a hole and a peak. *Yellow dots* indicate the exit and entrance points of PKiKP at the CMB. *Double black and red circles* are Fresnel zones at the ICB and CMB, respectively. *Inner and outer circles* are Fresnel zone estimates for 2 and 1 Hz, respectively

Simulations for topography at the ICB

Earlier PKiKP and PcP amplitude ratio analyses made sporadic seismological observations of a lower density contrast at the ICB than predicted values for spherically symmetric Earth models, as low as 200–300 g/m^3(Koper and Pyle 2004; Tkalčić et al. 2009). In addition, Gubbins et al. (2008) inferred a low density contrast from geodynamical considerations. These results indirectly support the existence of a dense layer at the top of the inner core (F-layer) to explain the smaller density differences between the outer and inner cores. However, there is no proof from seismology that such a layer is a global feature. Masters and Gubbins (2003) noted that even the relatively large density jump inferred from free oscillation analyses is consistent with an F-layer having a strong density gradient: the free oscillation data would average over thick layers below and above the ICB, while the body wave would be sensitive to variations across the ICB. A lower density contrast across the ICB would permit larger topography (Buffett 1997). Of particular interest is the possibility of sharp edges between the solidification and melting areas. Recent waveform modeling suggests significant topography (Dai et al. 2012); however, determination of the amplitude is a more difficult problem than determination of the wavelength.

One possibility of most simplified geometry is that a sinusoidal topographic structure might develop at the largest scales (λ = 10–100 km) (Buffett 1997). At these length scales, there is an inverse relationship between relaxation time scale and wavelength (Turcotte and Schubert 2002). In addition, the time scale required for topography to relax varies inversely with density contrast and linearly with viscosity. As the viscosity of the outer core is effectively zero (de Wijs et al. 1998), the rate of relaxation is thus entirely controlled by deformation in the inner core.

On the other hand, spike-shaped topographic structures might develop as a result of dendritic growth, likely at smaller scales (λ = 10–several 100 m) (Bruce Buffett, pers. comm.). Such topography could be relaxed through melting and freezing (thermal relaxation). As the temperature gradients are steeper at short wavelengths, this can drive the heat flow needed to melt or freeze.

Given the above, we test two different classes of topography at the ICB. In the first scenario, we test a regular sinusoidal topography with wavelength λ and height H (Fig. 12a). The values of (λ, H) tested, in kilometers, are (0.1, 0.1), (0.2, 0.2), (0.5, 0.5), (1.0, 1.0), (1.5, 1.5), (2.5, 2.5), (2.5, 2.6), (2.5, 2.7), (5.0, 2.5), and (10.0, 5.0). Figure 13 shows the results of selected combinations of λ and H. The topographies with $\lambda = 0.1$ and 0.2 km and $H = 0.1$ and 0.2 km did not reproduce any spectral peaks and holes and yielded spectra similar to those obtained from a simple discontinuity model (Fig. 10b). However, for a slightly more prominent topography of $\lambda = 0.5$ km and $H = 0.5$ km, we find a clear peak in spectral ratios around 1.2 Hz (Fig. 13a). The ICB topography with $\lambda = 1.0$ and $H = 1.0$ km results in an increased number of spectral peaks (Fig. 13b). The topographic model with $\lambda = 1.5$ and $H = 1.5$ km produces distinct peaks around 2 Hz for incidence angles of 20° and 30° and remarkable holes around 1.2 Hz for incidence angles of 10° and 30° (Fig. 13c). The best results are obtained for the topography characterized by $\lambda = 1.5$ and $H = 1.5$ km. The simulated spectra contain similar characteristics to those observed in the reflection coefficient profiles of results 8, 9, and 10. However, when the longer λ and larger H are used in simulations (e.g., λ = 5 and $H = 2.5$ km; $\lambda = 10$ and $H = 5$ km), the reflection coefficients overall become smaller than those for the flat ICB. The spectra are devoid of peaks at higher frequencies (Fig. 13d, e).

Finally, we introduce "spiky" topography to address the possibility of dendritic growth of the inner core (Sumita and Bergman 2009), which is mathematically expressed in our simulations by a reversed cycloid with

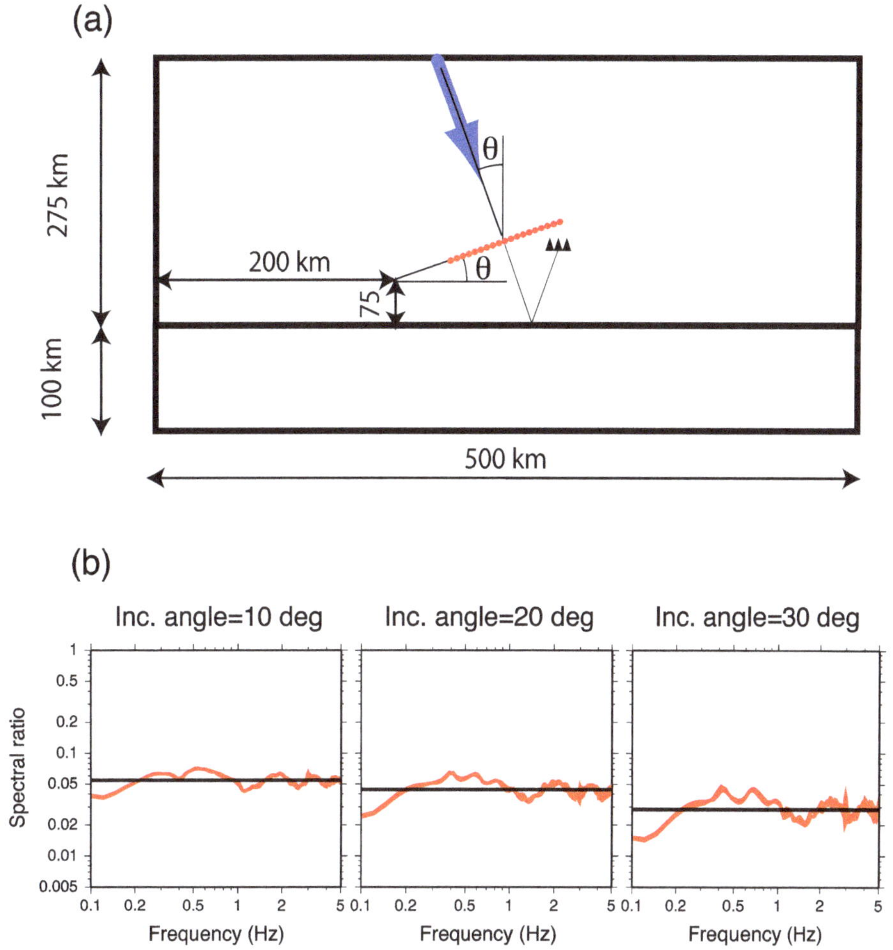

Fig. 10 Configuration of 2D finite difference simulations and a test run. **a** Configuration of 2D finite difference simulations using e3d (Larsen and Shultz 1995). *Small red circles* represent multiple sources used to simulate a plane wave. *Triangles* mark the locations of virtual receivers used to record the reflected waves. Incidence angles θ are a control parameter. **b** Spectral ratios between upward and downward waves using a flat discontinuity in simulations for incidence angles of 10°, 20°, and 30°. *Thick lines* are theoretical values of the reflection coefficient calculated using ak135 (Kennett et al. 1995)

upward sharp tips (Fig. 12b). We tested several (λ, H) combinations for this conceptual model: (0.1, 0.1), (0.2, 0.2), (0.5, 0.2), (0.5, 0.1), (0.5, 0.2), (1.0, 0.2), (1.0, 0.5), (1.0, 1.0), (1.0, 2.0), (1.2, 1.7), (1.5, 1.5), (1.5, 2.0), (2.0, 1.0), (2.0, 1.5), (2.0, 2.0), and (5.0, 2.0). The overall characteristics of the spectral ratios for spiky topography, with $\lambda \le 2$ km and $H \le 2$ km (Fig. 14a–d), can be summarized as a gradual decrease in spectral ratio as frequency increases from 0.5 to 5 Hz. This topographic model can explain several peaks or holes for incidence angles of $\theta = 10°$ and 20° and many peaks at frequencies around 2, 3, and 4 Hz for $\theta = 30°$. The topography with $\lambda = 5$ km and $H = 2$ km results in relatively flat and small spectral ratios in the frequency range 0.7–5 Hz. Such a model can explain many holes at frequencies larger than 2.5 Hz for $\theta = 10°$–20° and peaks for $\theta = 30°$. However, these results cannot adequately explain our observations because the overall reflection coefficients are too small.

The resultant characteristics are slightly different from those produced by the sinusoidal topographies, even though their structural dimensions are the same. The frequencies of the distinct spectral peaks decrease with increasing wavelength and height (Fig. 14b–d), whereas no distinct peaks are observed for the topography with $\lambda = H = 0.5$ km. The spiky topography with $\lambda = H =$ 1.0 km results in a distinct peak around 1.7 Hz for $\theta =$ 10°–20° and a large hole around 1.5 Hz for $\theta = 30°$ (Fig. 14b). The topography with $\lambda = 1.5$ and $H = 1.5$ km yields a distinct single peak at roughly $f = 1.2$ Hz. There are several spectral holes for $\theta = 10°$–20° and peaks for $\theta = 30°$ at $f > 2$ Hz (Fig. 14c). Of all the test cases with spiky topography, the case with $\lambda = H = 1.0$ km most closely matches the observed spectral ratios.

In summary, our observations and numerical simulations suggest that the inner core boundary is a sharp boundary without transitional layers. The hypothesis of

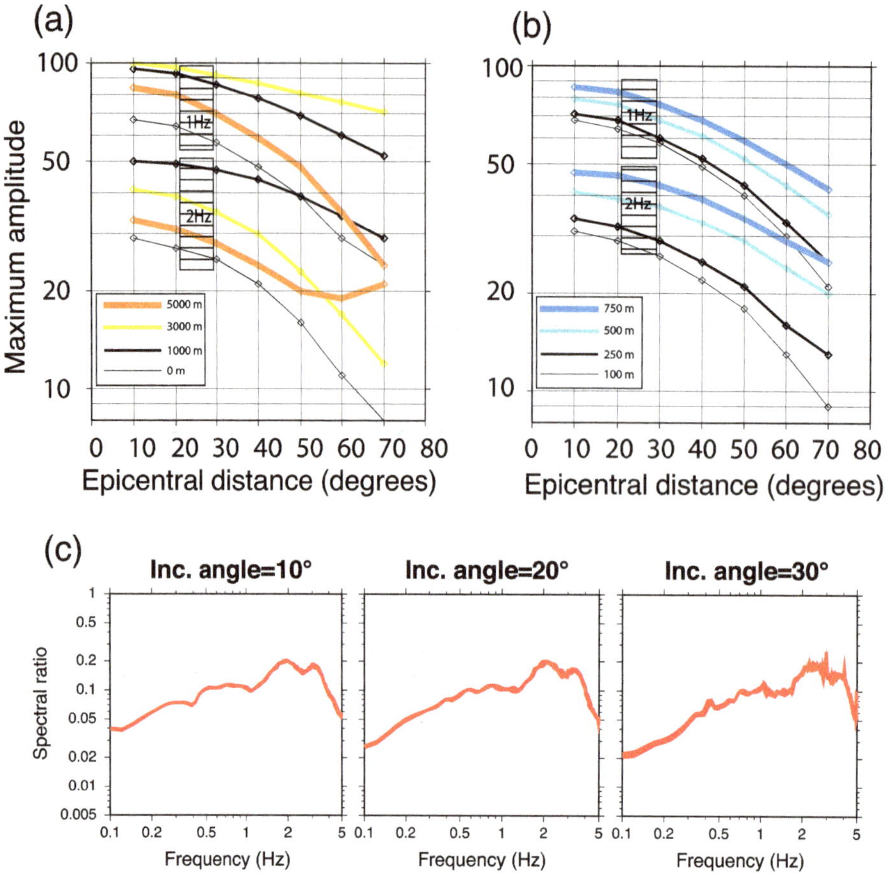

Fig. 11 Results of numerical simulations for a liquid thin layer above the ICB. Maximum amplitudes of the reflected waves measured on seismograms filtered around 1 and 2 Hz as a function of increasing angular distance at the ICB. Synthetic seismograms are calculated by including a heavy liquid layer (Vp = 9.6 km/s, $\rho = 12.5$ g/cm^3) overlying the ICB with varying thicknesses: **a** 0, 1000, 3000, and 5000 m and **b** 100, 250, 500, and 750 m. **c** Spectral ratios for the 1000 m-thick heavy liquid layer for incidence angles of 10°, 20°, and 30°

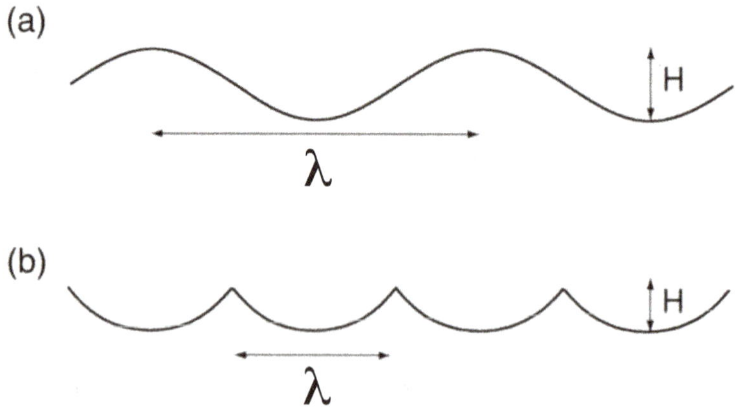

Fig. 12 Diagram of ICB topography. Diagram of ICB topography used in numerical simulations with wavelength and height defined as λ and H, respectively. **a** Sinusoidal topography; **b** "spiky" topography. The meaning of each topographic model is explained in the text

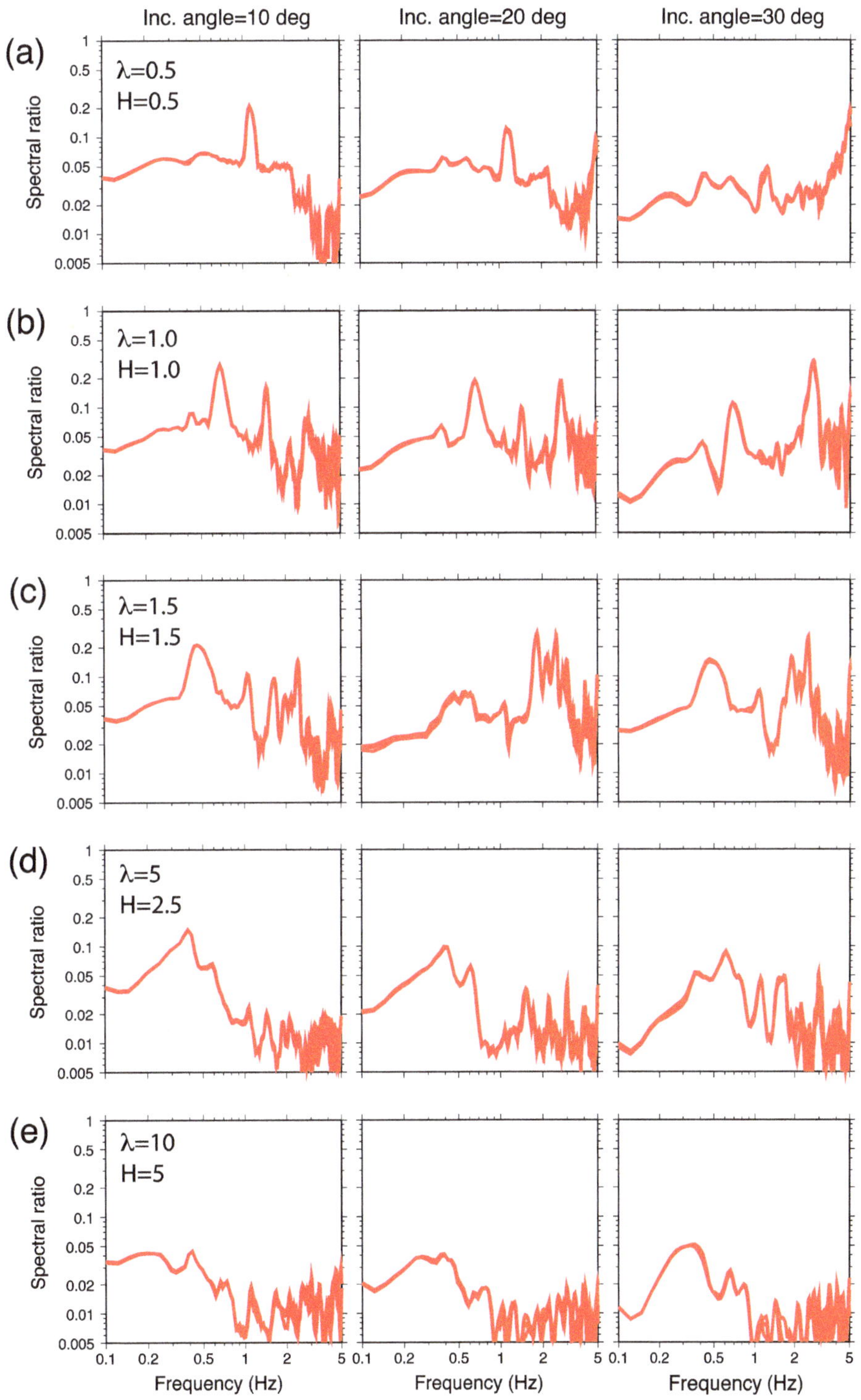

Fig. 13 Results for sinusoidal topographies. Representative results, characterized by height *H* and length *λ*, for incidence angle 10 degrees (*left*), 20 degrees (*middle*) and 30 degrees (*right*). The values of *H* and lambda are as follows: **a**) $\lambda = 0.5$, $H = 0.5$, **b**) $\lambda = 1.0$, $H = 1.0$, **c**) $\lambda = 1.5$, $H = 1.5$, **d**) $\lambda = 5$, $H = 2.5$, **e**) $\lambda = 10$ $H = 5$

melt at the surface of the inner core in the quasi-eastern hemisphere is not supported by our simulations. The most likely scenario to explain some of the observed spectral characteristics is the existence of topography at the ICB; however, more than one class of topography must be invoked to explain all observations. We therefore conclude that the topography characteristics of the ICB vary laterally. These variations may result from

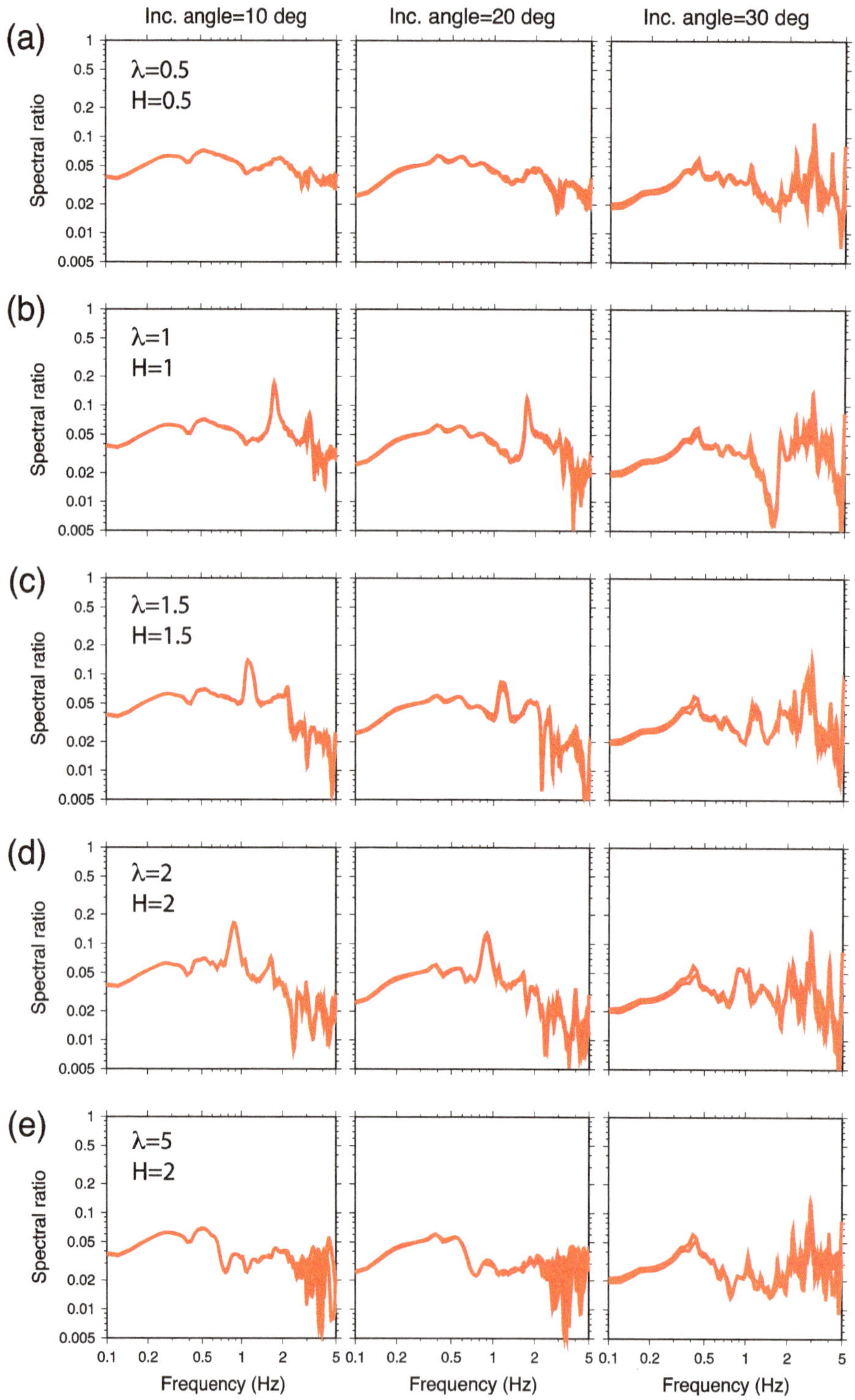

Fig. 14 Results for spiky topographies. Same as Fig. 13, but "spiky" topography was used in simulations. The values of H and λ are as follows: **a**) $\lambda = 0.5$, $H = 0.5$, **b**) $\lambda = 1$, H=1, **c**) $\lambda = 1.5$, $H = 1.5$, **d**) $\lambda = 2$, $H = 2$, **e**) $\lambda = 5$ $H = 2$

lateral variations of inner core solidification. If solidification is dynamically driven from top to bottom, its geographical pattern will be controlled by the pattern of outer core convection (Bergman et al. 2002; Aubert et al. 2008; Gubbins et al. 2011). If the solidification is instead driven from the bottom up, the pattern will be affected by variations in inner core convection (Deguen and Cardin 2011). Furthermore, small-scale variations in topographic characteristics suggest small-scale convection in a mushy zone at the ICB (Bergman and Fearn

1994; Deguen et al. 2007). While we cannot distinguish between these hypotheses in the present study, largely due to the fact that we sample only sparse and limited areas of the ICB, further observations of PKiKP and PcP will improve our understanding of large-scale ICB structure and dynamics.

Conclusions

Frequency characteristics of ICB reflection coefficients were investigated for the area around Japan using Hi-net vertical component seismograms. We found four patterns in the frequency-dependent behavior of reflection coefficients: (a) a nearly flat spectrum (little variation), (b) a significant hole at a frequency of approximately 1 or 3 Hz, (c) a peak at a frequency of approximately 2 or 3 Hz, and (d) the existence of a hole and a peak. The variety in observed spectra reflects the complex nature of the ICB. To interpret these observations, we conducted 2D finite difference simulations. Since we tested only limited cases with planar geometry, further simulations are required. Our modeling results suggest that holes and peaks in the spectra of reflection coefficients can be qualitatively explained by a sinusoidal or spike-like topography at the ICB, with wavelengths and heights ~1–1.5 km, whereas a liquid or solid layer overlying the ICB does not reproduce any of the observed spectral features.

Abbreviations

CMB: core–mantle boundary; ICB: inner core boundary.

Competing interests

The authors declare that they have no competing interests.

Authors' contributions

ST analyzed waveform data. HT performed finite difference simulations. The manuscript was written by ST and HT. Both authors read and approved the final manuscript.

Authors' information

ST is Deputy Director and Senior Scientist at the Department of Deep Earth Structure and Dynamics Research, Japan Agency for Marine-Earth Science and Technology.
HT is Associate Professor and Senior Fellow at Seismology and Mathematical Geophysics, Research School of Earth Sciences, The Australian National University.

Acknowledgements

The authors are grateful to the National Research Institute for Earth Science and Disaster Prevention (NIED), Japan, for providing high-quality seismograms recorded by the high-sensitivity seismograph network (Hi-net). The authors thank two anonymous reviewers who give valuable comments. S.T. was supported in part by MEXT KAKENHI grant number 15H05832. H.T. was supported by the Japan Society for the Promotion of Science and the Research School of Earth Science, The Australian National University during his stay in Japan. Most of the figures were drawn using GMT (Wessel and Smith 1998).

Author details

[1]Department of Deep Earth Structure and Dynamics Research, Japan Agency for Marine-Earth Science and Technology, Yokosuka 237-0061, Japan.
[2]Research School of Earth Sciences, The Australian National University, Canberra ACT 2601, Australia.

References

Alboussiere T, Deguen R, Melzani M (2010) Melting-induced stratification above the Earth's inner core due to convective translation. Nature 466:744–9
Aubert J, Amit H, Hulot G, Olson P (2008) Thermochemical flows couple the Earth's inner core growth to mantle heterogeneity. Nature 454:758–62
Bergman MI, Fearn DR (1994) Chimneys on the Earth's inner-outer core boundary. Geophys Res Lett 21:477–80
Bergman MI, Cole DM, Jones JR (2002) Preferred crystal orientations due to melt convection during directional solidification. J. Geophys. Res., 107: doi:10.1029/2001JB000601
Buchbinder GGR (1972) Travel times and velocities in the outer core from PmKP. Earth Planet Sci Lett 14:161–8
Buffett BA (1997) Geodynamic estimates of the viscosity of the Earth's inner core. Nature 388:571–3
Cao A, Romanowicz B (2004) Constraints on density and shear velocity contrast at the inner core boundary. Geophys J Int 157:1146–51
Cormier VF (2007) Texture of the uppermost inner core from forward- and back-scattered seismic waves. Earth Planet Sci Lett 258:442–53
Creager KC (1999) Large-scale variations in inner core anisotropy. J Geophys Res 104:23127–39
Cummins P, Johnson L (1988) Synthetic seismograms for an inner core transition of finite thickness. Geophys J 94:21–34
Dai Z, Wang W, Wen L (2012) Irregular topography at the Earth's inner core boundary. Proc Natl Acad Sci 109:7654–8
de Wijs GA, Kresse G, Vocadlo L, Dobson DP, Alfè D, Gillan M, Price GD (1998) The viscosity of liquid iron at the physical conditions of the Earth's core. Nature 392:805–7
Deguen R (2012) Structure and dynamics of Earth's inner core. Earth Planet Sci Lett 333–334:211–25
Deguen R, Cardin P (2011) Thermochemical convection in Earth's inner core. Geophys J Int 187:1101–18
Deguen R, Alboussiere T, Brito D (2007) On the existence and structure of a mush at the inner core boundary of the Earth. Phys Earth Planet Inter 164:36–49
Deuss A (2014) Heterogeneity and anisotropy of Earth's inner core. Ann Rev Earth Planet Sci 42:103–26
Deuss A, Irving JCE, Woodhouse JH (2010) Regional variation of inner core anisotropy from seismic normal mode observations. Science 328:1018–20
Engdahl ER, Flinn EA, Romney CF (1970) Seiemic waves reflected from the Earth's inner core. Nature 228:852–3
Engdahl ER, Flinn EA, Massé RP (1974) Differential PKiKP travel times and the radius of the inner core. Geophys J R Astron Soc 39:457–63
Gubbins D, Masters G, Nimmo F (2008) A thermochemical boundary layer at the base of Earth's outer core and independent estimate of core heat flux. Geophys J Int 174:1007–18
Gubbins D, Sreenivasan B, Mound J, Rost S (2011) Melting of the Earth's inner core. Nature 473:361–3
Hedlin MAH, Shearer PM (2000) An analysis of large-scale variations in small-scale mantle heterogeneity using global seismographic network recordings of precursors to PKP. J Geophys Res 105:13655–73
Jiang G, Zhao D (2012) Observation of high-frequency PKiKP in Japan: insight into fine structure of inner core boundary. J Asian Earth Sci 59:167–84
Kawakatsu H (2006) Sharp and seismically transparent inner core boundary region revealed by an entire network observation of near-vertical PKiKP. Earth Planets Space 58:855–63
Kennett BLN, Engdahl ER, Buland R (1995) Constraints on seismic velocities in the Earth from travel-times. Geophys J Int 122:108–24
Kikuchi M, Kanamori H (2003), Note on teleseismic body-wave inversion program, http://www.eri.u-tokyo.ac.jp/ETAL/KIKUCHI/
Koper KD, Dombrovskaya M (2005) Seismic properties of the inner core boundary from PKiKP/P amplitude ratios. Earth Planet Sci Lett 237:680–94
Koper KD, Pyle ML (2004) Observations of PKiKP/PcP amplitude ratios and implications for Earth structure at the boundaries of the liquid core. J. Geophys. Res., 109: doi:10.1029/2003JB002750
Koper KD, Pyle ML, Franks JM (2003) Constraints on aspherical core structure from PKiKP-PcP differential travel times. J Geophys Res 108:2168, doi:2110.1029/2002JB001995
Koper KD, Franks JM, Dombrovskaya M (2004) Evidence for small-scale heterogeneity in Earth's inner core from a global study of PKiKP coda waves. Earth Planet Sci Lett 228:227–41

Krasnoshchekov DN, Kaazik PB, Ovtchinnikov VM (2005) Seismological evidence for mosaic structure of the surface of the Earth's inner core. Nature 435: 483–7

Larsen SC, Shultz CA (1995), E3D:2D/3D Elastic finite-difference wave propagation code, Lawrence Livermore National Laboratory,Livermore, CA, USA. 1–18.

Lay T, Wallace T (1995) Modern global seismology. Academic, San Diego

Leyton F, Koper KD (2007a) Using PKiKP coda to determine inner core structure: 1. Synthesis of coda envelopes using single-scattering theories. J. Geophys. Res., 112: doi:10.1029/2006JB004369.

Leyton F, Koper KD (2007b) Using PKiKP coda to determine inner core structure: 2. Determination of Q_C. J. Geophys. Res., 112: doi:10.1029/2006JB004370.

Leyton F, Koper KD, Zhu L, Dombrovskaya M (2005) On the lack of seismic discontinuities within the inner core. Geophys J Int 162:779–86

Loper DE (1983) Structure of the inner core boundary. Geophys Astrophys Fluid Dyn 25:139–55

Loper DE, Roberts PH (1981) A study of conditions at the inner core boundary of the Earth. Phys Earth Planet Inter 24:302–7

Lythgoe KH, Deuss A, Rudge JF, Neufeld JA (2014) Earth's inner core: Innermost inner core or hemispherical variations? *Earth Planet.* Sci Lett 385:181–9

Masters G, Gubbins D (2003) On the resolution of density within the Earth. Phys Earth Planet Inter 140:159–67

Monnereau M, Calvet M, Margerin L, Souriau A (2010) Lopsided growth of Earth's inner core. Science 328:1014–7

Niu FL, Wen LX (2001) Hemispherical variations in seismic velocity at the top of the Earth's inner core. Nature 410:1081–4

Okada Y, Kasahara K, Hori S, Obara K, Sekiguchi S, Fujiwara H, Yamamoto A (2004) Recent progress of seismic observation networks in Japan –Hi-net, F-net, K-NET and KiK-net–. Earth Planets Space, 56: xv-xxviii

Peng ZG, Koper KD, Vidale JE, Leyton F, Shearer P (2008) Inner-core fine-scale structure from scattered waves recorded by LASA. J. Geophys. Res., 113: doi:10.1029/2007jb005412

Poupinet G, Kennett BLN (2004) On the observation of high frequency PKiKP and its coda in Australia. Phys Earth Planet Inter 146:497–511

Press WH, Teukolsky SA, Vetterling WT, Flannery BP (1988) Numerical recipes in C: the art of scientific computing. Cambridge University Press, Cambridge

Rodgers A, Tkalčić H, McAllen D (2006) Seismic ground motion and site response in Las Vegas Valley, Nevada from NTS explosions and earthquake data. Pure Appl Geophys 163:55–80

Schweitzer J, Kværna T (1999) Influence of source radiation patterns on globally observed short-period magnitude estimates (m_b). Bull Seism Soc Am 89: 342–7

Shearer P, Masters G (1990) The density and shear velocity contrast at the inner core boundary. Geophys J Int 102:491–8

Shimizu H, Poirier JP, Le Mouël JL (2005) On crystallization at the inner core boundary. Phys Earth Planet Inter 151:37–51

Souriau A (2007) Deep Earth structure—the Earth's cores. In: Romanowicz B, Dziewonski AM (eds) Treatise on geophysics, vol. 1, seismology and structure of the Earth. Elsevier, Amsterdam, pp 655–93

Souriau A, Souriau M (1989) Ellipticity and density at the inner core boundary from subcritical PKiKP and PcP data. Geophys J Int 98:39–54

Sumita I, Bergman M (2009) Inner-core dynamics. In: Olson P (ed) Treatise on geophysics, vol. 8, core dynamics. Elsevier, Amsterdam, pp 299–318

Sumita I, Olson P (1999) A laboratory model for convection in Earth's core driven by a thermally heterogeneous mantle. Science 286:1547–9

Takemura S, Furumura T, Maeda T (2015) Scattering of high-frequency seismic waves caused by irregular surface topography and small-scale velocity inhomogeneity. Geophys J Int 201:459–74

Tanaka S, Hamaguchi H (1997) Degree one heterogeneity and hemispherical variation of anisotropy in the inner core from PKP(BC)-PKP(DF) times. J Geophys Res 102:2925–38

Tkalčić H (2015) Complex inner core of the Earth: the last frontier of global seismology. Rev. Geophys., 53: doi:10.1002/2014RG000469

Tkalčić H, Kennett BLN, Cormier VF (2009) On the inner-outer core density contrast from PKiKP/PcP amplitude ratios and uncertainties caused by seismic noise. Geophys J Int 179:425–43

Tkalčić H, Cormier VF, Kennett BLN, He K (2010) Steep reflections from the earth's core reveal small-scale heterogeneity in the upper mantle. Phys Earth Planet Inter 178:80–91

Tkalčić H, Chen Y, Liu R, Huang Z, Sun L, Chan W (2011) Multistep modelling of teleseismic receiver functions combined with constraints from seismic tomography: crustal structure beneath southeast China. Geophys J Int 187:303–26

Turcotte D, Schubert G (2002) Geodynamics, 2nd edn. Cambridge University Press, Cambridge

Waszek L, Irving J, Deuss A (2011) Reconciling the hemispherical structure of Earth's inner core with its super-rotation. Nat Geosci 4:264–7

Wessel P, Smith WHF (1998) New improved version of generic mapping tools released. EOS Trans Am Geophys Un 79:579

Yao J, Wen L (2014) Seismic structure and ultra-low velocity zones at the base of the Earth's mantle beneath Southeast Asia. Phys Earth Planet Inter 233: 103–11

15

Deep magnetic field stretching in numerical dynamos

Diego Peña[1*], Hagay Amit[2] and Katia J. Pinheiro[1,2]

Abstract

The process of magnetic field stretching transfers kinetic energy to magnetic energy and thereby maintains dynamos against ohmic dissipation. Stretching at depth may play an important role in shaping the field morphology and in the dynamo action. Here, we analyze snapshots from self-consistent 3D numerical dynamos to unravel the nature of field-flow interactions that induces stretching secular variation of the radial magnetic field at mid-depth of the shell. We search for roots of intense flux patches identified at the outer boundary. The deep radial field structures exhibit a position shift with respect to the locations of the outer boundary patches, consistent with a mixed effect of tangent cylinder rim and plume-like dynamics. A global stretching/advection rms ratio is ~ 1.5–3 times larger than that of poloidal/toroidal flows. In addition, local stretching is often more effective than advection, in particular at regions of significant field-aligned flow. On average at roots of high-latitude flux patches, total stretching is 1.1 times larger than total advection despite the poloidal flow being only 0.37 of the toroidal flow. Radial stretching secular variation acts as an effective dynamo mechanism at regions where laterally varying radial flow shears toroidal field lines to generate a poloidal magnetic field. Stretching at depth exhibits similar parameter dependence as that of stretching at the outer boundary, with the strongest dependence being on the magnetic Prandtl number in both cases. Our results provide insights into the underlying deep dynamo mechanisms that sustain intense magnetic flux patches at the outer boundary.

Keywords: Magnetic field, Dynamo, Stretching, Flux patch, Secular variation, Core stratification

Introduction

The geomagnetic field is generated by rapidly rotating convective motions of an electrically conductive fluid in Earth's outer core. Temporal changes in the geomagnetic field termed secular variation (SV) may provide constraints on the fluid dynamics at the top of the core and possibly on the dynamo action. Indeed, geomagnetic field and SV models based on surface observations and satellite data (e.g. Jackson et al. 2000; Olsen and Mandea 2008) have been used to characterize Earth's core dynamics (e.g. Finlay and Jackson 2003), in particular the fluid flow at the top of the core (for a review, see Holme 2015), or as constraints on numerical dynamo simulations (e.g. Christensen et al. 1998, 2010; Aubert et al. 2013). In contrast, detailed morphological analyses of spatial patterns of the SV at depth have not been conducted.

*Correspondence: dpena@on.br
[1]Geophysics Department, Observatório Nacional, Rio de Janeiro 20921-400, Brazil
Full list of author information is available at the end of the article

According to the magnetic induction equation, the SV is comprised of magnetic advection, stretching, and diffusion. Magnetic field advection transfers magnetic energy from one degree to another, whereas magnetic field stretching transfers kinetic energy to magnetic energy and by that maintains dynamo action against ohmic dissipation (e.g. Moffatt 1978; Mininni 2011). Kageyama and Sato (1997a) found that in an α-dynamo mechanism, axial convective cylinders generate a poloidal field from a basic toroidal field and vice versa (for illustration, see, e.g., Figure 5 of Olson et al. 1999). In these dynamo models, azimuthal magnetic field lines at the equatorial plane are transformed into a poloidal field in the form of axial field lines along the column, while axial magnetic field lines along the column are transformed into a toroidal field at the equatorial plane (see also Aubert et al. 2008b).

Morphological criteria for characterizing the observed geomagnetic field on the core-mantle boundary (CMB) include concentrated flux patches (Christensen et al. 2010). High-latitude normal flux patches (i.e., where the local field has the same sign as the axial dipole field) contribute significantly to the dominant axial dipole, while intensification and expansion of reversed flux patches (i.e., where the sign of the local field is opposite to that of the axial dipole field) diminish it (Gubbins 1987; Olson and Amit 2006). In addition, these patches have a particular signature on the SV. For example, an advected patch leads to a bipolar SV pattern (Livermore et al. 2017), whereas a patch intensified by downwelling gives a same-sign SV structure (Amit 2014). Due to the importance of these patches in defining the field morphology, it is essential to explore their origin. Local analysis of field-flow interactions may provide a detailed interpretation of the SV in the vicinity of these robust field features.

In regions of high-latitude intense geomagnetic flux patches, stretching may play an important role. These robust non-axisymmetric features are typically observed near the edge of the inner core tangent cylinder (Jackson et al. 2000), possibly due to a flow barrier and surface convergence at these regions (Olson et al. 1999). In dipole-dominated numerical dynamo models, surface convergence is correlated with columnar cyclones (Kageyama and Sato 1997b; Olson et al. 2002; Amit et al. 2007), so the flow near these patches has a large field-aligned component and produces little magnetic advection (Finlay and Amit 2011). Magnetic field stretching may also be the underlying mechanism for regions of weak field intensity at Earth's surface. Strong deviations of the geomagnetic field from axial dipolarity appear in the form of reversed flux patches which may reflect expulsion of a toroidal field (Bloxham 1986) advected from depth to the CMB by fluid upwelling (e.g., Aubert et al. 2008b). The current weak surface field intensity in Brazil (Hartmann and Pacca 2009) is related to reversed flux patches on the CMB below the Atlantic (Aubert 2015; Tarduno et al. 2015), though this relation is not trivial (Terra-Nova et al. 2017).

The existence of stretching SV just below the CMB depends on whether a stably stratified layer is present (e.g., Whaler 1980). Stable stratification at the top of Earth's core might inhibit penetration of radial motion. Under such conditions, the tangential flow is purely toroidal. Some seismic studies (e.g., Helffrich and Kaneshima 2010) and mineral physics models (e.g., de Koker et al. 2012; Pozzo et al. 2012) suggest that the top of the core is indeed stably stratified (Gubbins and Davies 2013). In contrast, other studies claimed that the thermal conductivity of the core is as low as previously estimated (Konôpková et al. 2016; Ohta et al. 2016) and thus that the whole of the outer core convects. Even if the thermal conductivity is high, exsolution of mantle material (Badro et al. 2016; O'Rourke and Stevenson 2016) may destabilize the top of the core. Regional interpretations of the geomagnetic SV also suggest some local upwelling/downwelling (Olson and Aurnou 1999; Chulliat et al. 2010; Amit 2014). Quasi-geostrophic core flow models rely on surface poloidal flow to infer the flow at depth (Pais and Jault 2008; Gillet et al. 2009, 2011, 2015). Lesur et al. (2015) argued that the geomagnetic data could not be adequately explained by a purely toroidal flow, but inclusion of a weak poloidal flow is sufficient to explain the SV, suggesting that the upper part of the core is weakly stratified.

Peña et al. (2016) studied global and local stretching at the top of the shell of numerical dynamos in detail. They found that stretching has a significant influence on the SV despite the relatively weak poloidal flow. In addition, their analysis showed that local stretching is often more effective than advection, in particular at regions of significant field-aligned flow as well as in intensifying magnetic flux patches. Morphological resemblance between local stretching in the dynamo models of Peña et al. (2016) and local observed geomagnetic SV (Amit 2014) may indicate the presence of stretching at the top of the Earth's core.

Of course dynamo action might not necessarily occur in the entire outer core. The dynamo may be generated exclusively at depth due to stable stratification at the top of the core. For example, the very weak, large-scale, and axisymmetric magnetic field of Mercury (Anderson et al. 2011; Oliveira et al. 2015) can be explained by a skin effect across a stably stratified layer (Christensen 2006; Christensen and Wicht 2008; Wicht and Heyner 2014). As mentioned above, such a layer was also proposed for the Earth (Pozzo et al. 2012; Gubbins and Davies 2013). In this paper, we analyze output from the same set of numerical dynamos as Peña et al. (2016), but deep in the shell, to understand the field-flow interactions and the contribution of magnetic field stretching to the SV there. In particular, we search for the roots of the intense magnetic flux patches, identified at the outer boundary by Peña et al. (2016), to explore their kinematic origins. We focus on the contribution of stretching as well as other mechanisms to the SV of the radial component of the field for comparison with the radial field on the outer boundary, which is the component accessible from observations. The contributions of the different SV mechanisms at depth and their dependence on the dynamo control parameters are explored. The results for the SV of the radial magnetic field at depth are discussed in the context of inductive effects and the dynamo process.

As mentioned above, the mechanism of magnetic field generation by convection of an electrically conductive fluid in a rotating spherical shell has been extensively studied using numerical dynamos (e.g., Kageyama and Sato 1997a; Olson et al. 1999; Aubert et al. 2008b; Takahashi and Shimizu 2012). Here, we focus on analyzing exclusively the different terms of the radial component of the magnetic induction equation and only at a specific spherical surface situated at depth. For practical reasons, we sample the deep shell only at mid-depth. We compare our findings at depth with the corresponding results just below the outer boundary (Peña et al. 2016).

The paper is outlined as follows. In the "Theory" section, we describe the theoretical basis of our study, including the different terms in the radial induction equation at depth, the effect of the tangent cylinder at depth, and an assessment of the image distortion effects from the CMB to mid-depth. The methods are described in the "Methods/Experimental" section, including numerical dynamos, the criterion for identifying roots of outer boundary patches, and the statistical tools used to evaluate the results. In the "Results" section, we report the global and local stretching contributions as well as their dependence on the dynamo control parameters. We discuss our main findings in the "Discussion" section.

Methods/Experimental

Theory

Radial magnetic induction equation far from the boundaries

The magnetic induction equation in a 3D vectorial form is given by

$$\frac{\partial \vec{B}}{\partial t} = \nabla \times (\vec{u} \times \vec{B}) + \lambda \nabla^2 \vec{B} \tag{1}$$

where $\vec{B}$ is the magnetic field, $\vec{u}$ is the fluid velocity, and λ is the magnetic diffusivity. For comparison with information accessible from observations, we focus on the non-diffusive (or frozen-flux, denoted by superscript "ff") part of (1):

$$\frac{\partial \vec{B}^{ff}}{\partial t} = \nabla \times (\vec{u} \times \vec{B}) \tag{2}$$

Using a vector identity for the right-hand side of (2) and considering the non-divergence of the flow and the field, (2) can be rewritten as

$$\frac{\partial \vec{B}^{ff}}{\partial t} = (\vec{B} \cdot \nabla)\vec{u} - (\vec{u} \cdot \nabla)\vec{B} \tag{3}$$

We now consider the radial component of (3):

$$\hat{r} \cdot \frac{\partial \vec{B}^{ff}}{\partial t} = \hat{r} \cdot (\vec{B} \cdot \nabla)\vec{u} - \hat{r} \cdot (\vec{u} \cdot \nabla)\vec{B} \tag{4}$$

The first term of (4) is simply

$$\hat{r} \cdot \frac{\partial \vec{B}^{ff}}{\partial t} = \frac{\partial B_r{}^{ff}}{\partial t} \tag{5}$$

where the subscript r denotes the radial component. The second term of (4) is

$$\hat{r}\cdot(\vec{B}\cdot\nabla)\vec{u} = B_r\frac{\partial u_r}{\partial r} + \frac{B_\theta}{r}\frac{\partial u_r}{\partial \theta} + \frac{B_\varphi}{r\sin\theta}\frac{\partial u_r}{\partial \varphi} - \frac{B_\theta u_\theta}{r} - \frac{B_\varphi u_\varphi}{r} \tag{6}$$

where θ and φ are colatitude and longitude, respectively. Considering $\nabla \cdot \vec{u} = 0$

$$\frac{\partial u_r}{\partial r} = -\nabla_h \cdot \vec{u}_h - \frac{2}{r}u_r \tag{7}$$

and substituting (7) into (6) gives

$$\hat{r}\cdot(\vec{B}\cdot\nabla)\vec{u} = -B_r\nabla_h\cdot\vec{u}_h + \vec{B}_h\cdot\nabla_h u_r - 2\frac{B_r u_r}{r} - \frac{B_\theta u_\theta}{r} - \frac{B_\varphi u_\varphi}{r} \tag{8}$$

where the subscript h denotes the component tangential to a spherical surface of constant radial distance. The third term of (4) is

$$\begin{aligned}\hat{r}\cdot(\vec{u}\cdot\nabla)\vec{B} &= u_r\frac{\partial B_r}{\partial r} + \frac{u_\theta}{r}\frac{\partial B_r}{\partial \theta} + \frac{u_\varphi}{r\sin\theta}\frac{\partial B_r}{\partial \varphi} - \frac{B_\theta u_\theta}{r} - \frac{B_\varphi u_\varphi}{r} \\ &= \vec{u}_h \cdot \nabla_h B_r + u_r\frac{\partial B_r}{\partial r} - \frac{B_\theta u_\theta}{r} - \frac{B_\varphi u_\varphi}{r}\end{aligned} \tag{9}$$

Finally, by substituting (5), (8), and (9) into (4), the radial component of the magnetic induction equation at a radius r far from the boundaries, i.e., without assuming $u_r = 0$, can be written as

$$\frac{\partial B_r{}^{ff}}{\partial t} = -\vec{u}_h\cdot\nabla_h B_r - B_r\nabla_h\cdot\vec{u}_h - \frac{u_r}{r^2}\frac{\partial(r^2 B_r)}{\partial r} + \vec{B}_h\cdot\nabla_h u_r \tag{10}$$

The term on the left-hand side of (10) is the frozen-flux SV, i.e., the change in the radial field due to the flow. The first and third terms on the right-hand side of (10) represent the tangential and radial magnetic advection, respectively. The second and fourth terms on the right-hand side of (10) represent the tangential and radial magnetic field stretching, respectively.

We note that it is possible to rewrite the radial advection term using $\nabla\cdot\vec{B} = 0$ as $u_r\nabla_h\cdot\vec{B}_h$. However, in fluid dynamics, advection is affiliated with the flow interacting with a gradient of a tracer (in this case the magnetic field). In our radial advection term that is precisely the case, whereas expressing this term using a divergence is less intuitive.

Figure 1 illustrates the four SV mechanisms at depth. Tangential advection Ad_h occurs due to the displacement of the radial magnetic field on the spherical surface when a tangential flow has a component perpendicular to the B_r-isolines (Fig. 1a). This process produces bipolar features in the SV (Amit 2014; Livermore et al. 2017). Tangential stretching St_h occurs due to the interaction of B_r with downwelling or upwelling structures (Fig. 1b). Downwelling intensifies the radial magnetic field (Christensen et al. 1998), resulting in same-sign polarity features in the SV (Amit 2014; Peña et al. 2016). Similar to Ad_h, radial advection Ad_r occurs due to the radial displacement of B_r by the radial flow (Fig. 1c). Radial stretching St_r occurs due

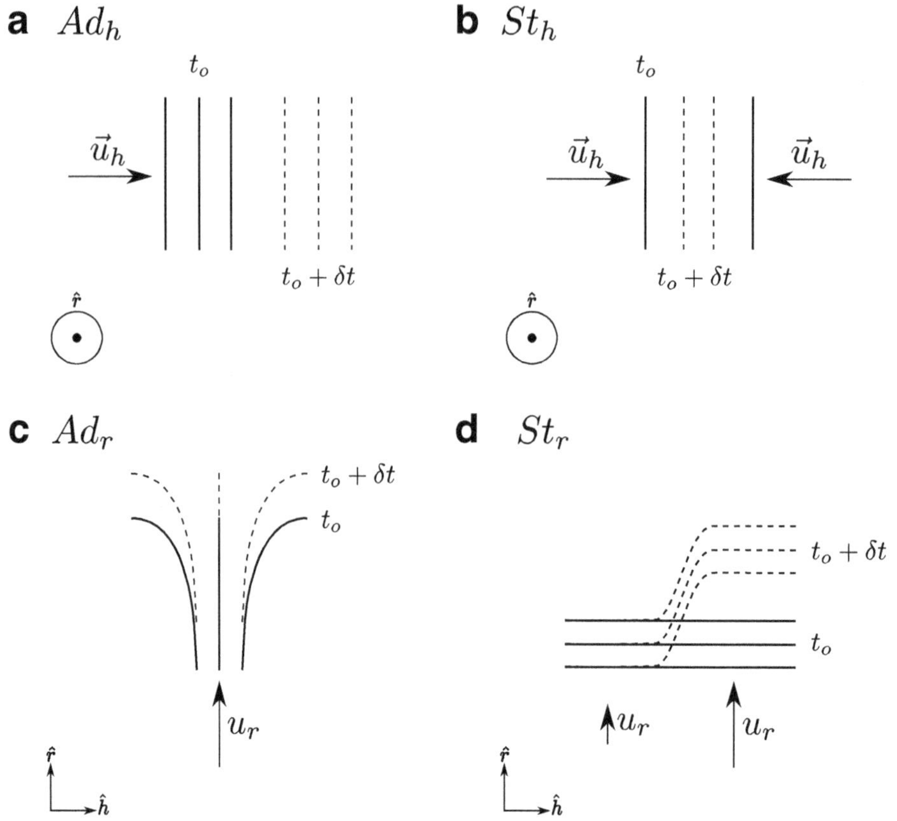

Fig. 1 Schematic illustrations of the radial frozen-flux SV mechanisms. **a** Tangential advection. **b** Tangential stretching. **c** Radial advection. **d** Radial stretching. The solid lines represent the initial configuration at time t_o, and the dashed lines represent an evolved configuration due to the fluid flow action after some time δt. The lines represent either radial field contours (**a** and **b**) or magnetic field lines (**c** and **d**). Spatial orientations are given at the bottom left of each subplot

to the interaction of tangential magnetic field lines with tangentially varying radial flow (Fig. 1d). In this dynamo mechanism, the laterally varying radial flow shears the toroidal field lines to generate a poloidal magnetic field (Olson et al. 1999).

At the CMB ($r = r_o$, where r_o is the outer core radius), the radial velocity vanishes ($u_r = 0$). Then, the radial advection and radial stretching terms (third and fourth terms on the right-hand side of (10), respectively) also vanish, and the radial component of the magnetic induction equation becomes (e.g. Bloxham and Jackson 1991; Holme 2015)

$$\frac{\partial B_r^{ff}}{\partial t} = -\vec{u}_h \cdot \nabla_h B_r - B_r \nabla_h \cdot \vec{u}_h \tag{11}$$

Tangent cylinder at depth

The presence of the inner core affects the geometry of the fluid flow in the outer core. The tangent cylinder—a hypothetical cylinder coaxial with Earth's rotation axis and tangential to the inner core at the equatorial plane (Fig. 2)—separates the outer core into two regions in which the fluid flow and the resulting magnetic field are expected to be quite different (e.g. Aurnou et al. 2003). In particular, the tangent cylinder rim acts as a flow barrier, resulting in surface convergence and intense magnetic flux patches in its intersection with the CMB (e.g. Olson et al. 1999).

Because the inner to outer radius ratio is $r_i/r_o = 0.35$ (Dziewonski and Anderson 1981), the intersection of the tangent cylinder with the CMB occurs at colatitude $\theta_o^c \sim 20.5°$ $\left(\sin\theta_o^c = 0.35/1,\text{ see Fig. 2}\right)$. At mid-depth ($r_{1/2}/r_o = 0.675$, where $r_{1/2}$ is the radius at mid-depth), the intersection occurs at colatitude $\theta_{1/2}^c \sim 31.2°$ $\left(\sin\theta_{1/2}^c = 0.35/0.675\right)$. The difference between these two angles implies that robust features on the CMB (e.g., high-latitude intense flux patches) may have their roots at mid-depth at lower latitudes if these features are related to the tangent cylinder rim effect.

Image distortion

When considering the same longitudinal and latitudinal extents for both the outer boundary and mid-depth patches, a root (at mid-depth) of an intense flux patch (at the CMB) would exhibit an apparent distortion due to

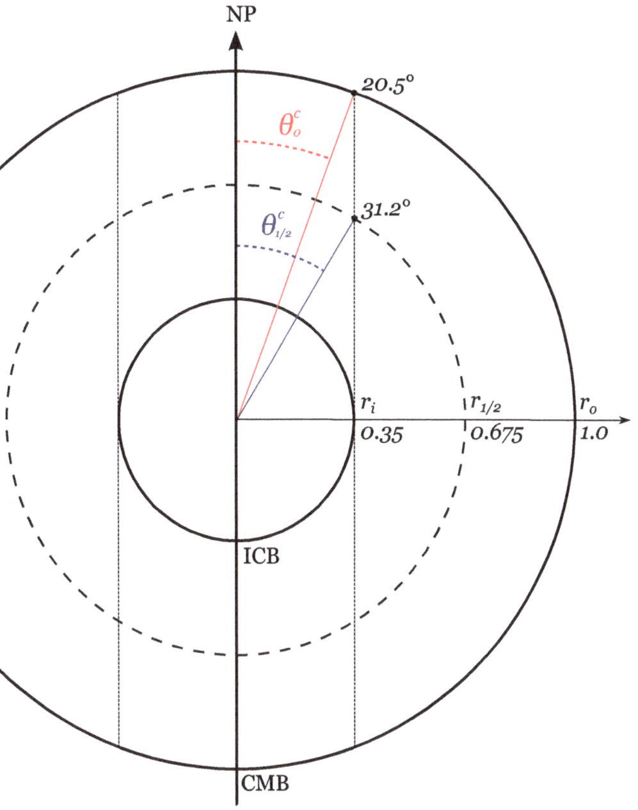

Fig. 2 Schematic illustration of the inner core tangent cylinder and its intersection with the CMB (red) and with the spherical surface at mid-shell (blue)

the difference between the corresponding CMB and mid-depth areas of equal angular extents. This difference is associated with the change in radial level from the CMB to mid-depth as well as a possible change in latitude. The root occupies a larger portion of the angular surface than a CMB structure of the same size. This radial distortion ratio is given by

$$\left(\frac{r_o}{r_{1/2}}\right)^2 = \left(\frac{1}{0.675}\right)^2 \tag{12}$$

Such a distortion, which we will refer to as plume-like, should conserve the spatial angle, i.e., $\delta\vartheta = \delta\varphi = 0$, where $\delta\vartheta$ and $\delta\varphi$ are the latitudinal and longitudinal shifts, respectively, between the root and the outer boundary patches. In contrast, if the roots reside at lower latitudes than the CMB structures, e.g., due to a tangent cylinder rim effect, the roots appear relatively smaller. The purely latitudinal distortion from one CMB patch centered at the intersection of the tangent cylinder and the CMB to another CMB patch centered at the intersection of the tangent cylinder and the mid-depth spherical surface is given by $\frac{\sin\theta_o^c}{\sin\theta_{1/2}^c} = 0.675$.

The ratio between the areas of a structure at the CMB S_o and its root at mid-depth $S_{1/2}$ with both structures centered at the tangent cylinder, which we will refer to as the tangent cylinder rim effect, is

$$\frac{S_o}{S_{1/2}} = \left(\frac{r_o}{r_{1/2}}\right)^2 \cdot \frac{\sin\theta_o^c}{\sin\theta_{1/2}^c} = \frac{1}{0.675} \tag{13}$$

For a general latitudinal distortion combined with the radial distortion, we calculated the correlation between a pair of CMB and mid-depth field structures for a range of $S_o/S_{1/2}$ (Fig. 3). In Fig. 3a, we show two examples of these distortions: a root with a larger area and a root with a smaller area representing a purely radial and a purely latitudinal distortion, respectively. Note that the latter is hypothetical since radial distortion is always present. In practice, for a latitudinal distortion that ranges from zero to a tangent cylinder rim effect (as we will later show is the case), the combined radial and latitudinal distortion is represented by the gap between the green and red vertical lines in Fig. 3b. According to these results, the correlation between a flux patch at the CMB and its root at mid-depth is high, in the range 0.77—0.93. This indicates that it is possible to identify a root despite the geometrical

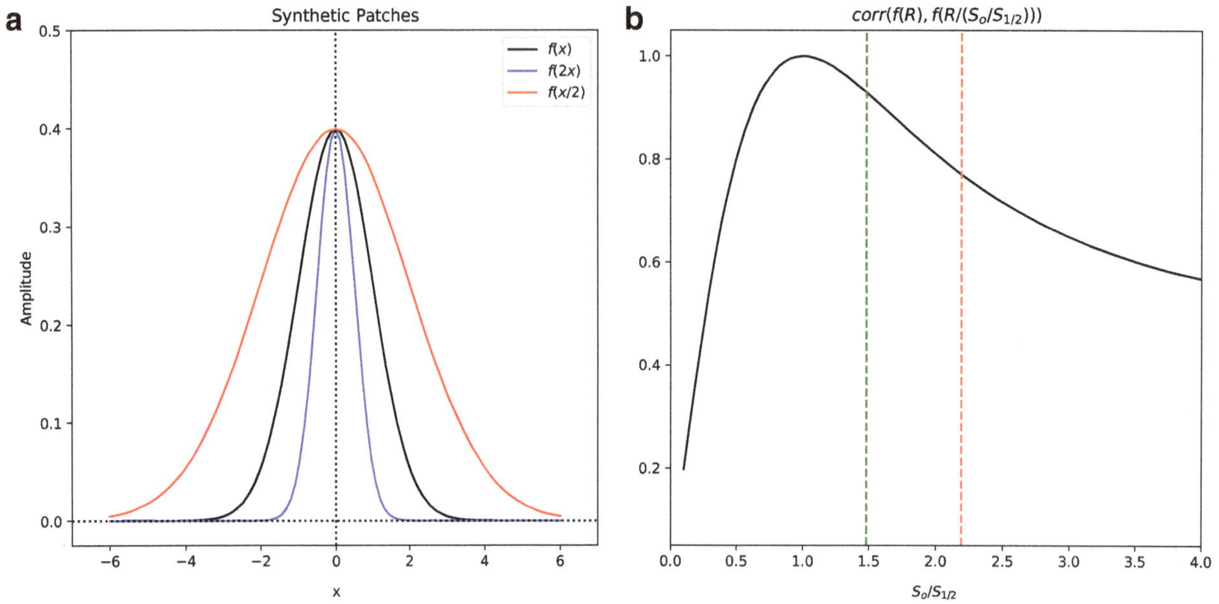

Fig. 3 Quantification of geometrical distortion. **a** Examples of a root with a larger area (red) and a smaller area (blue) compared to the CMB patch (black). All structures f are modeled by Gaussians with standard deviations representing the width. The horizontal axis represents either longitude or colatitude cross sections. **b** Correlation of a flux patch (R is the radial distance from the center of the patch) and its mid-depth root for a range of geometrical distortion ratios $S_o/S_{1/2}$. The red and green vertical dashed lines represent the radial distortion and the tangent cylinder rim effect, respectively

distortions, as long as the patch is not too distorted by the dynamics in the shell.

Numerical dynamos

We use the same dynamo models as in Peña et al. (2016). Here, we briefly recall the description of these models (for further details, see Peña et al. 2016). Numerical dynamos are solutions to the magnetohydrodynamics equations: Navier-Stokes, magnetic induction, conservation of energy, conservation of mass (continuity for an incompressible fluid), and no magnetic monopoles. In non-dimensional form, these equations can be written (e.g. Olson et al. 1999) as follows:

$$E\left(\frac{\partial \vec{u}}{\partial t} + \vec{u}\cdot\nabla\vec{u} - \nabla^2\vec{u}\right) + 2\hat{z}\times\vec{u} + \nabla P = Ra\frac{\vec{r}}{r_o}T + \frac{1}{Pm}\left(\nabla\times\vec{B}\right)\times\vec{B} \tag{14}$$

$$\frac{\partial \vec{B}}{\partial t} = \nabla\times(\vec{u}\times\vec{B}) + \frac{1}{Pm}\nabla^2\vec{B} \tag{15}$$

$$\frac{\partial T}{\partial t} + \vec{u}\cdot\nabla T = \frac{1}{Pr}\nabla^2 T + \epsilon \tag{16}$$

$$\nabla\cdot\vec{u} = 0 \tag{17}$$

$$\nabla\cdot\vec{B} = 0 \tag{18}$$

where $\vec{u}$ is the fluid velocity, $\vec{B}$ is the magnetic field, T is the temperature (or more generally co-density), t is the time, $\hat{z}$ is a unit vector in the direction of the rotation axis, P is the pressure, $\vec{r}$ is the position vector, and ϵ is a heat (or buoyancy) source or sink.

Four non-dimensional parameters in (14)–(16) control the dynamo action. The heat flux Rayleigh number (Olson and Christensen 2002) represents the strength of the buoyancy force driving the convection relative to retarding forces

Table 1 Dynamo models control parameters: Rayleigh Ra, Ekman E, and magnetic Prandtl Pm

Model	Ra	E	Pm	Rm	$\bar{\delta\tau}$
1	2×10^5	1×10^{-3}	5	137	14.75
2	2×10^5	1×10^{-3}	10	255	11.9
3	4×10^5	1×10^{-3}	5	219	33.07
4	5×10^5	3×10^{-4}	3	82	22.51
5	1×10^6	3×10^{-4}	3	125	14.60
6	3×10^6	3×10^{-4}	3	234	78.98
7	1×10^7	1×10^{-4}	1.3	126	9.42
8	1×10^7	1×10^{-4}	2	218	5.50
9	3×10^7	1×10^{-4}	2	446	16.66

For all models, we set the Prandtl number to $Pr = 1$. $\bar{\delta\tau}$ denotes the average time difference between successive snapshots in units of magnetic advection time

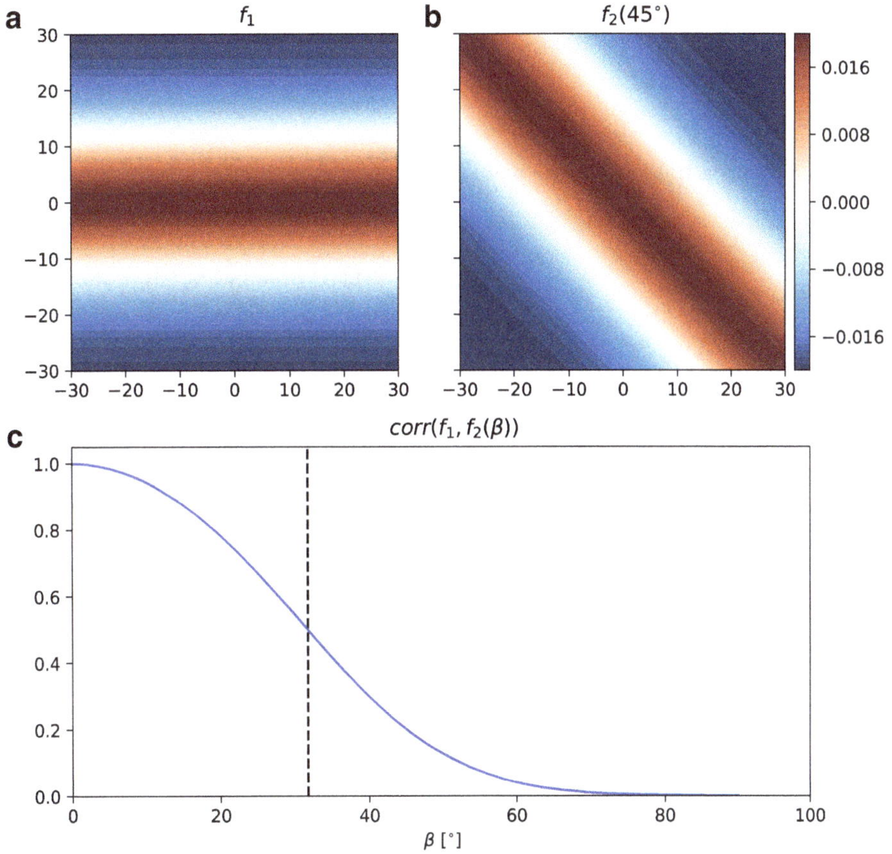

Fig. 4 Same polarity image distortion. **a** An example of a synthetic intense flux patch at the CMB. **b** Its root at mid-depth rotated by 45°. The patch is modeled by a Gaussian with standard deviation equals to 1. **c** Correlation between the intense flux patch in **a** and its root rotated by α. The black dashed line shows the maximum acceptable rotation $\beta \approx 31.7°$ for correlations higher than 0.5

$$Ra = \frac{\alpha g_o q_o D^4}{k\kappa\nu} \quad (19)$$

where α is the thermal expansivity, g_o is the gravitational acceleration on the outer boundary at radius r_o, q_o is the mean heat flux across the outer boundary, D is the shell thickness, k is the thermal conductivity, κ is the thermal diffusivity, and ν is the kinematic viscosity. The Ekman number represents the ratio of viscous to Coriolis forces

$$E = \frac{\nu}{\Omega D^2} \quad (20)$$

Table 2 Global statistics at mid-depth: dynamo models time average and standard deviation values

Model	Ad_r/Ad_h	St_r/St_h	St_h/Ad_h	St/Ad	$\mathcal{P}/\mathcal{T}$	$\|u_r\|/\|\vec{u}_h\|$
1	0.63 ± 0.05	1.33 ± 0.22	0.44 ± 0.05	0.78 ± 0.16	0.37 ± 0.04	0.69 ± 0.06
2	0.60 ± 0.10	1.26 ± 0.20	0.41 ± 0.07	0.66 ± 0.10	0.38 ± 0.04	0.71 ± 0.04
3	0.63 ± 0.08	1.29 ± 0.18	0.38 ± 0.06	0.65 ± 0.12	0.43 ± 0.03	0.76 ± 0.10
4	0.50 ± 0.07	1.76 ± 0.22	0.43 ± 0.08	0.90 ± 0.12	0.32 ± 0.04	0.60 ± 0.06
5	0.52 ± 0.07	1.45 ± 0.17	0.48 ± 0.07	0.85 ± 0.09	0.38 ± 0.03	0.65 ± 0.06
6	0.64 ± 0.07	1.25 ± 0.15	0.50 ± 0.05	0.81 ± 0.10	0.42 ± 0.05	0.75 ± 0.05
7	0.57 ± 0.08	1.27 ± 0.11	0.64 ± 0.11	1.00 ± 0.14	0.38 ± 0.03	0.71 ± 0.06
8	0.59 ± 0.05	1.29 ± 0.13	0.56 ± 0.04	0.88 ± 0.13	0.42 ± 0.05	0.74 ± 0.07
9	0.65 ± 0.04	1.14 ± 0.07	0.52 ± 0.05	0.75 ± 0.06	0.44 ± 0.04	0.79 ± 0.05

St_r, St_h, and St are the radial, tangential, and total stretching rms, respectively; Ad_r, Ad_h, and Ad are the radial, tangential, and total advection rms, respectively; $\mathcal{P}/\mathcal{T}$ is the poloidal/toroidal flow rms ratio, and $\|u_r\|/\|\vec{u}_h\|$ is the radial/tangential flow rms ratio where Ω is the rotation rate. The Prandtl number is the ratio of kinematic viscosity to thermal diffusivity

$$Pr = \frac{\nu}{\kappa} \tag{21}$$

and the magnetic Prandtl number is the ratio of kinematic viscosity to magnetic diffusivity

$$Pm = \frac{\nu}{\lambda} \tag{22}$$

The most important output parameter is the magnetic Reynolds number Rm, which represents the scaled ratio between magnetic advection and diffusion and is given by

$$Rm = \frac{UD}{\lambda} \tag{23}$$

where U is a typical velocity scale.

We used the numerical implementation MagIC (Wicht 2002). Due to computational limitations, dynamo simulations use control parameters very far from Earth-like conditions, therefore relating the results to the real core conditions is challenging. Our chosen control parameters (Table 1) are moderate compared to what modern computers are capable of. The reason is that smaller E

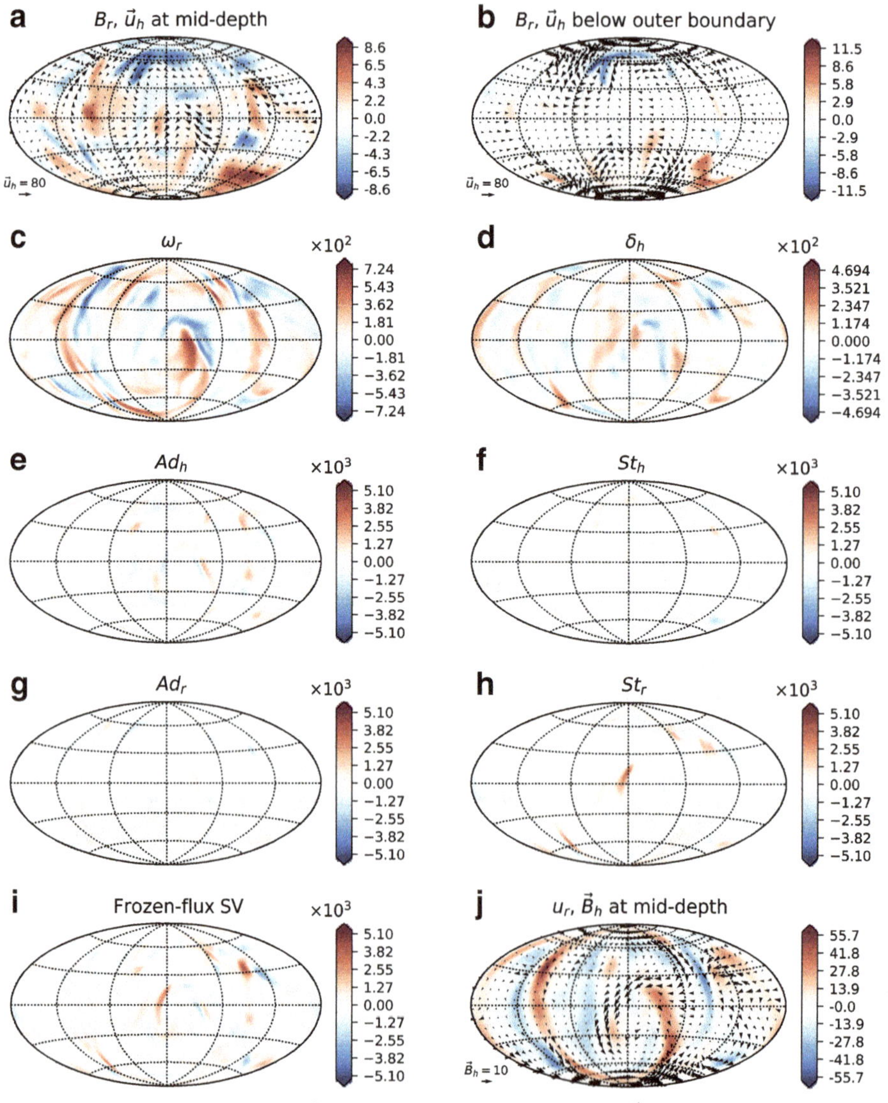

Fig. 5 A snapshot from dynamo model 1: radial magnetic field B_r (colors) and the tangential flow $\vec{u}_h$ (arrows) at **a** mid-depth and at **b** the top of the free stream just below the Ekman boundary layer. The velocity arrows in **a** and **b** have the same scale. The following quantities are all at mid-depth: **c** radial vorticity ω_r, **d** tangential divergence δ_h, **e** tangential advection SV, **f** tangential stretching SV, **g** radial advection SV, **h** radial stretching SV, and **i** total frozen-flux SV. Also in **j**, the radial flow u_r (colors) and the tangential magnetic field $\vec{B}_h$ (arrows) are shown. All variables are non-dimensional. The global statistics for this snapshot are as follows: $Ad_r/Ad_h = 0.60$; $St_r/St_h = 1.91$; $St_h/Ad_h = 0.51$; $St/Ad = 1.22$; $\mathcal{P}/\mathcal{T} = 0.39$; and $\|u_r\|/\|\vec{u}_h\| = 0.64$

values produce such small-scale structures that the local relations between the field and the flow would become difficult to interpret. We focus on dynamos in the non-reversing dipole-dominated regime (e.g. Christensen and Aubert 2006; Kutzner and Christensen 2002).

The shell geometry is identical to Earth's core, with an inner to outer boundary radius ratio of $r_i/r_o = 0.35$. The inner and outer boundaries of the shell are set to be insulating and rigid. To simulate generic thermo-chemical convection (e.g. Aubert et al. 2008a), on the inner core boundary, a fixed co-density is set; on the outer boundary, a fixed heat flux is prescribed; and the source/sink term in (15) is set to $\epsilon = 0$. The number of radial grid points N_r is chosen to accommodate at least five grid points across the Ekman boundary layer. In our models, N_r varies from 49 for the larger $E = 1 \times 10^{-3}$ cases to 61 for the smaller $E = 1 \times 10^{-4}$ cases. Horizontal resolution is also increased with a decreasing Ekman number, from a maximum degree and order $\ell_{\max} = 64$ for the $E = 1 \times 10^{-3}$ cases to $\ell_{\max} = 96$ for the $E = 1 \times 10^{-4}$ cases.

Criterion for root identification

In order to identify deep roots of intense flux patches on the outer boundary, we used the identified structures at the outer boundary which were studied by Peña et al. (2016). These outer boundary B_r structures were compared with B_r structures at mid-depth using an auto-correlation function for the same longitudinal and latitudinal extent. Our algorithm calculates the Pearson correlation (e.g. Press et al. 1992) between local B_r on the outer boundary and local B_r at depth

$$\frac{\sum_i \left(B^i_{r,o} - \overline{B^i_{r,o}}\right)\left(B^i_{r,1/2} - \overline{B^i_{r,1/2}}\right)}{\sqrt{\sum_i \left(B^i_{r,o} - \overline{B^i_{r,o}}\right)^2}\sqrt{\sum_i \left(B^i_{r,1/2} - \overline{B^i_{r,1/2}}\right)^2}} \tag{24}$$

In (24), i denotes a set of points (θ_i, φ_i) defined at both r_0 and $r_{1/2}$. We consider a displacement of the center of the latter of up to 15° in colatitude and longitude while conserving the longitudinal and latitudinal extents. The best correlation determines the location of the patch at depth. Structures with correlations higher than a critical threshold are defined as roots.

As shown in Fig. 3, the deterioration of the correlation due to geometrical distortion is expected to be minor. The opposite problem of biased high correlations might occur merely because same-polarity CMB and deep structures are considered. For example, deep structures rotated with respect to a CMB patch are nevertheless positively correlated. However, a large angle representing a false root would result in a reduced correlation coefficient between the two radial field structures (see Fig. 4a, b for an example with an angle of 45°). For this study, we accept roots with a correlation higher than an arbitrary threshold value of 0.5 (black dashed line in Fig. 4c) that corresponds to an angle of $\sim 31.7°$. With this selection process, we avoid falsely attributing roots to positive correlations that originate from merely same-sign structures.

Statistics

As in Peña et al. (2016), we calculated global and local rms ratios ($\|X\|/\|Y\|$) between pairs of quantities X and Y. The rms $\|X\|$ is obtained by integration of X over the spherical surface at mid-shell. In addition to the statistical measures calculated by Peña et al. (2016), we also quantified SV terms that appear at depth but vanish on approach to the boundary. These include the radial stretching to tangential stretching rms ratio St_r/St_h and radial advection to tangential advection rms ratio Ad_r/Ad_h. We also quantified the rms ratio of radial to tangential velocity $\|u_r\|/\|\vec{u}_h\|$.

We used the same local classification by polarity and by latitude as in Peña et al. (2016). Polarity is defined as normal and reversed with respect to the sign of the axial dipole. High latitudes are arbitrarily defined by patches that are centered at latitudes higher than 45°. Classified this way, four types of patches are possible: normal polarity at high latitudes (HN), normal polarity at low latitudes (LN), reversed polarity at high latitudes (HR), and reversed polarity at low latitudes (LR). Note that we classified the deep structures according to the latitudes of the correlated outer boundary structures even if the roots are found on the other side of latitude 45°.

To examine a possible tangent cylinder rim effect, we calculated the dislocation of pairs of outer boundary and mid-depth patches. We defined $\delta\varphi$ and $\delta\vartheta$ as the longitudinal and latitudinal dislocation of patches

Table 3 Identified roots: detection fraction in percentage of roots at mid-depth of the intense magnetic flux patches at the top of the shell reported by Peña et al. (2016) in each dynamo model

Model	Total	HN	LN	LR
1	82.4	77.5	100	80
2	65.3	70.8	40	57.9
3	60.6	60.7	–	60
4	44.4	55.2	0	0
5	71.4	80	0	0
6	62.5	64.5	66.7	50
7	85.2	95.2	66.7	33.3
8	38.7	37.9	–	50
9	20	15.4	25	40

The detected roots are classified on the basis of the latitudes of the magnetic flux patches on the outer boundary: high-latitude normal intense flux patches (HN), low-latitude normal intense flux patches (LN), and low-latitude reversed flux patches (LR)

from the outer boundary to mid-depth, respectively. Positive/negative $\delta\vartheta$ corresponds to lower/higher latitudes of mid-depth patches, respectively.

Finally, we examined the dependence of the statistical quantities on the non-dimensional control parameters of the dynamo models. Each quantity may be expressed as a generic power law (as in Peña et al. 2016):

$$f = C \cdot E^a \cdot Ra^b \cdot Pm^c \tag{25}$$

where f is the statistical quantity and C, a, b, and c are fitting coefficients. The relative misfit σ_r of the power law is given by

$$\sigma_r = \sqrt{\sum_{i=1}^{n} \left(f_i^{\text{dyn}} - f_i\right)^2 / \sum_{i=1}^{n} \left(f_i^{\text{dyn}}\right)^2} \tag{26}$$

where f^{dyn} is the statistical quantity obtained from the dynamo models and n is the number of dynamo

Table 4 Latitudinal shift: mid-depth root latitudinal shift $\delta\vartheta$ with respect to the coordinates of the correlated outer boundary patch

Model	PT	Ad_h	St_h	Ad_r 666666	St_r	All
	HN	3.37 ± 5.06	7.50 ± 0.00	5.62 ± 0.00	3.44 ± 6.71	3.58 ± 5.31
1	LN	−15.00 ± 0.00	–	–	–	−15.00 ± 0.00
	LR	−3.00 ± 6.67	5.62 ± 0.00	−1.88 ± 1.88	–	−1.64 ± 6.04
	HN	4.31 ± 4.32	−1.88 ± 0.00	5.62 ± 0.00	5.63 ± 3.75	4.25 ± 4.75
2	LN	−7.50 ± 1.88	–	–	–	−7.50 ± 1.88
	LR	−0.19 ± 7.48	−15.00 ± 0.00	–	–	−1.53 ± 8.30
	HN	2.46 ± 6.91	5.62 ± 0.00	–	–	2.65 ± 6.75
3	LN	–	–	–	–	–
	LR	3.75 ± 0.00	5.62 ± 7.50	–	–	5.00 ± 6.19
	HN	11.25 ± 5.81	–	10.00 ± 3.85	5.86 ± 4.44	8.32 ± 5.42
4	LN	–	–	–	–	–
	LR	–	–	–	–	–
	HN	8.04 ± 2.18	–	7.50 ± 1.33	10.21 ± 2.67	8.91 ± 2.58
5	LN	–	–	–	–	–
	LR	–	–	–	–	–
	HN	5.25 ± 5.82	11.25 ± 0.00	11.25 ± 0.00	−1.87 ± 8.53	4.78 ± 6.87
6	LN	13.13 ± 1.88	–	–	–	13.13 ± 1.88
	LR	−5.00 ± 5.79	–	–	–	−5.00 ± 5.79
	HN	3.75 ± 0.00	–	8.44 ± 0.94	5.62 ± 4.77	5.81 ± 4.51
7	LN	0.94 ± 0.94	–	–	–	0.94 ± 0.94
	LR	1.88 ± 0.00	–	–	–	1.88 ± 0.00
	HN	10.31 ± 7.08	–	0.00 ± 0.00	4.69 ± 4.81	6.31 ± 6.46
8	LN	–	–	–	–	–
	LR	−7.50 ± 0.00	–	–	–	−7.50 ± 0.00
	HN	−1.41 ± 5.83	–	–	–	−1.41 ± 5.83
9	LN	8.44 ± 2.82	–	–	–	8.44 ± 2.82
	LR	−1.88 ± 0.00	–	5.62 ± 0.00	–	1.87 ± 3.75
	HN	4.52 ± 5.99	5.62 ± 4.78	7.64 ± 3.41	5.66 ± 5.70	5.11 ± 5.80
$\bar{x}$	LN	1.67 ± 9.62	–	–	–	1.67 ± 9.62
	LR	−1.62 ± 6.82	0.47 ± 10.38	0.62 ± 3.85	–	−1.10 ± 7.25

Average and standard deviation values are given for each patch type (PT) and SV dominant mechanism. Positive values represent roots at lower latitudes, and negative values represent roots at higher latitudes. $\bar{x}$ denotes averages over all dynamo models, and "All" denotes averages over roots of all patch types models analyzed. Relative misfits larger than an arbitrary threshold value of 0.2 were considered inadequate, and in these cases, the fits were not interpreted.

The power law fits (25) obtained by the misfit minimization (26) were applied to time-average statistical quantities (Tables 2 and 5). The time dependence was expressed by the standard deviation σ. Note that σ was not used to obtain the fits.

Results

Global stretching

Several statistical quantities were analyzed at mid-depth globally, i.e., over the spherical surface, for the nine

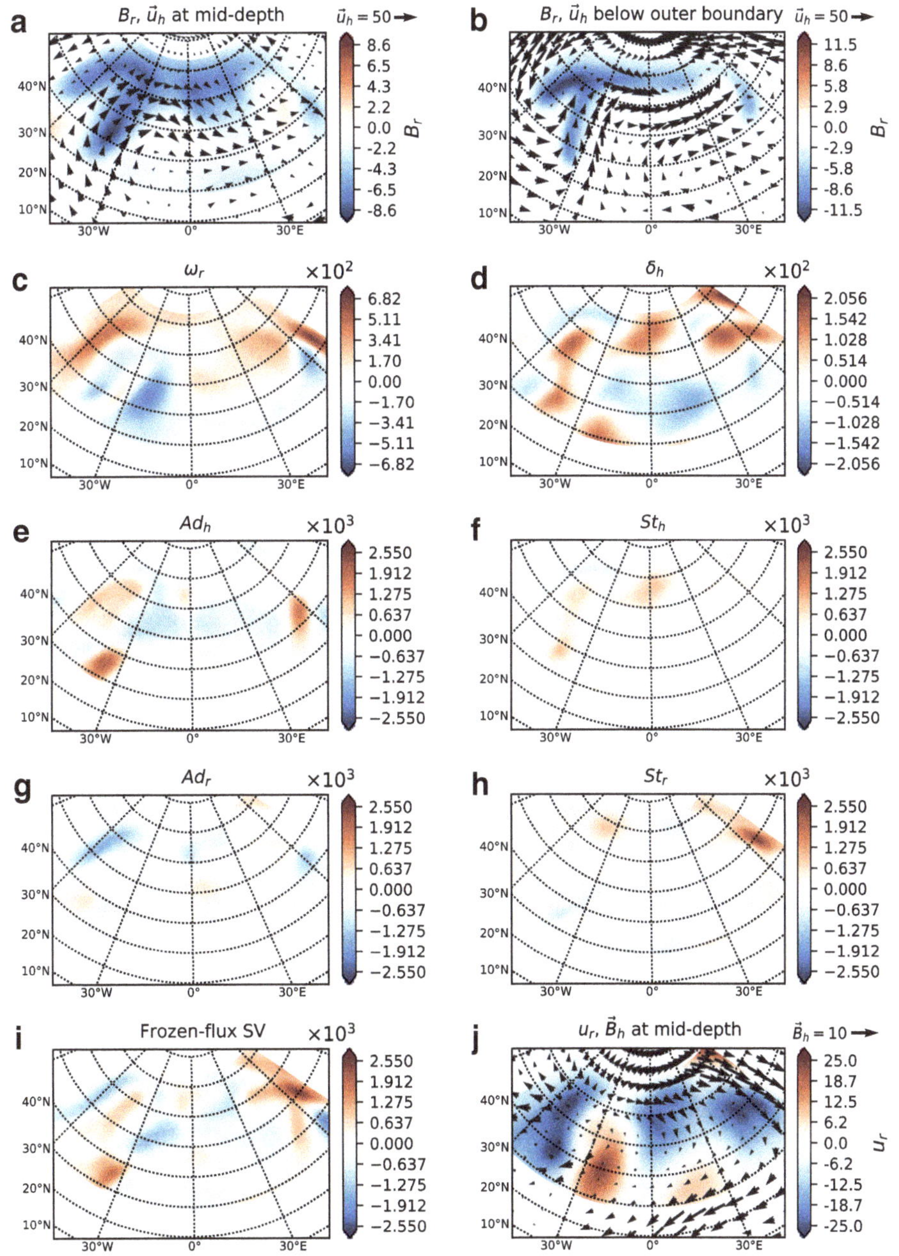

Fig. 6 Intense high-latitude normal polarity magnetic flux patch (HN) in dynamo model 1 (Fig. 5): radial magnetic field B_r (colors) and the tangential flow $\vec{u}_h$ (arrows) at **a** mid-depth and at **b** the top of the free stream just below the Ekman boundary layer. The velocity arrows in **a** and **b** have the same scale. The following quantities are all at mid-depth: **c** radial vorticity ω_r, **d** tangential divergence δ_h, **e** tangential advection SV, **f** tangential stretching SV, **g** radial advection SV, **h** radial stretching SV, and **i** total frozen-flux SV. Also in **j**, the radial flow u_r (colors) and the tangential magnetic field $\vec{B}_h$ (arrows) are shown. All variables are non-dimensional. The mid-depth patch in **a** is shifted with respect to the outer boundary patch in **b** by $\delta\vartheta = 1.88°$ and $\delta\varphi = -1.88°$ with an auto-correlation of 0.68. The local statistics for this patch are as follows: $Ad_r/Ad_h = 0.62$; $St_r/St_h = 1.69$; $St_h/Ad_h = 0.39$; $St/Ad = 0.78$; $\mathcal{P}/\mathcal{T} = 0.41$; and $\|u_r\|/\|\vec{u}_h\| = 0.62$

dynamo models described in Table 1. For each dynamo model, ten arbitrary snapshots well separated in time (see $\delta\bar{\tau}$ in Table 1) were analyzed.

We decomposed the tangential flow into toroidal and poloidal parts using $\vec{u}_{\text{tor}} = \nabla \times \mathcal{T}\hat{r}$ and $\vec{u}_{\text{pol}} = \nabla_h \mathcal{P}$, where $\mathcal{T}$ and $\mathcal{P}$ are the respective flow potentials. Note that in this decomposition, the radial poloidal flow is not considered. The ratio of rms flows $||\vec{u}_{\text{pol}}||/||\vec{u}_{\text{tor}}||$ is denoted by $\mathcal{P}/\mathcal{T}$.

The global statistics for each dynamo model are shown in Table 2. All cases exhibit a dominance of tangential advection Ad_h over radial advection Ad_r and a dominance of radial stretching St_r over tangential stretching St_h. In addition, all models exhibit a total stretching/advection ratio of $St/Ad \sim 1.5$–3 times larger than the poloidal/toroidal flow ratio $\mathcal{P}/\mathcal{T}$ (Table 2).

Figure 5a, b shows the radial field at mid-depth and at the outer boundary in a snapshot from dynamo model

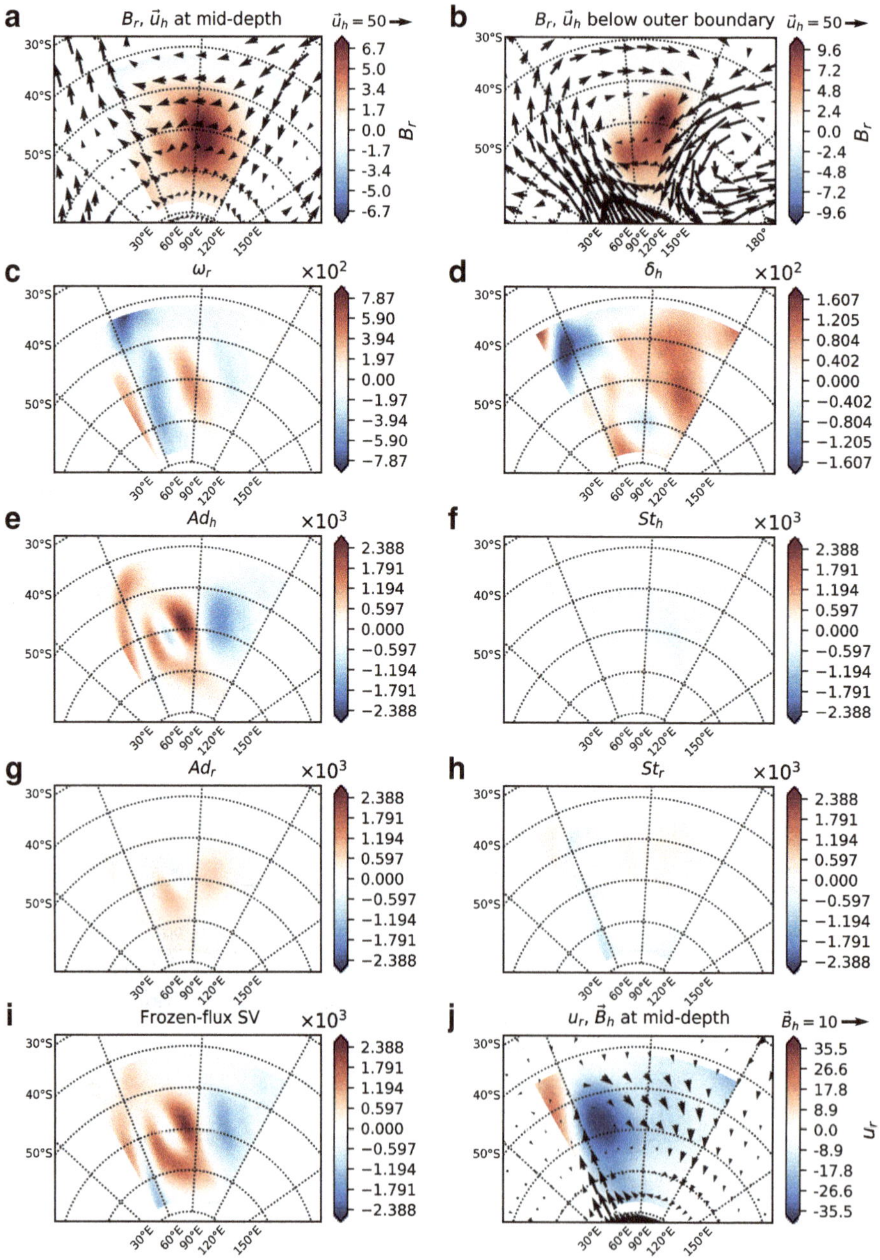

Fig. 7 As in Fig. 6 for another intense high-latitude normal polarity magnetic flux patch (HN). This patch presents a dominant Ad_h in dynamo model 3. The mid-depth patch in **a** is shifted with respect to the outer boundary patch in **b** by $\delta\vartheta = 1.88°$ and $\delta\varphi = -9.38°$ with an auto-correlation of 0.82. The local statistics for this patch are as follows: $Ad_r/Ad_h = 0.31$; $St_r/St_h = 1.36$; $St_h/Ad_h = 0.21$; $St/Ad = 0.29$; $\boldsymbol{\mathcal{P}/\mathcal{T}} = 0.25$; and $||u_r||/||\vec{u}_h|| = 0.50$

1. The field is much less concentrated with many more intense patches at mid-depth than at the outer boundary. Consequently, only a few intense patches previously detected at the outer boundary by Peña et al. (2016) could be identified at mid-depth. The radial vorticity $\omega_r = \hat{r} \cdot \nabla \times \vec{u}$ (Fig. 5c) and the tangential divergence $\delta_h = \nabla_h \cdot \vec{u}_h$ (Fig. 5d) are weakly correlated, in contrast to the helical flow correlation (Amit and Olson 2004) observed at the top of the shell of numerical dynamos (Olson et al. 2002; Amit et al. 2007; Peña et al. 2016). The toroidal flow dominates over the poloidal, and the tangential flow dominates over the radial. Nevertheless, due to particular field-flow interactions, in this snapshot, the total stretching is larger than the total advection by a factor of 1.22 (see Fig. 5e–h).

Kinematics of roots of intense magnetic flux patches

Using the criteria described in the "Criterion for root identification" section, we identified and analyzed 213 mid-depth roots of outer boundary patches. Table 3

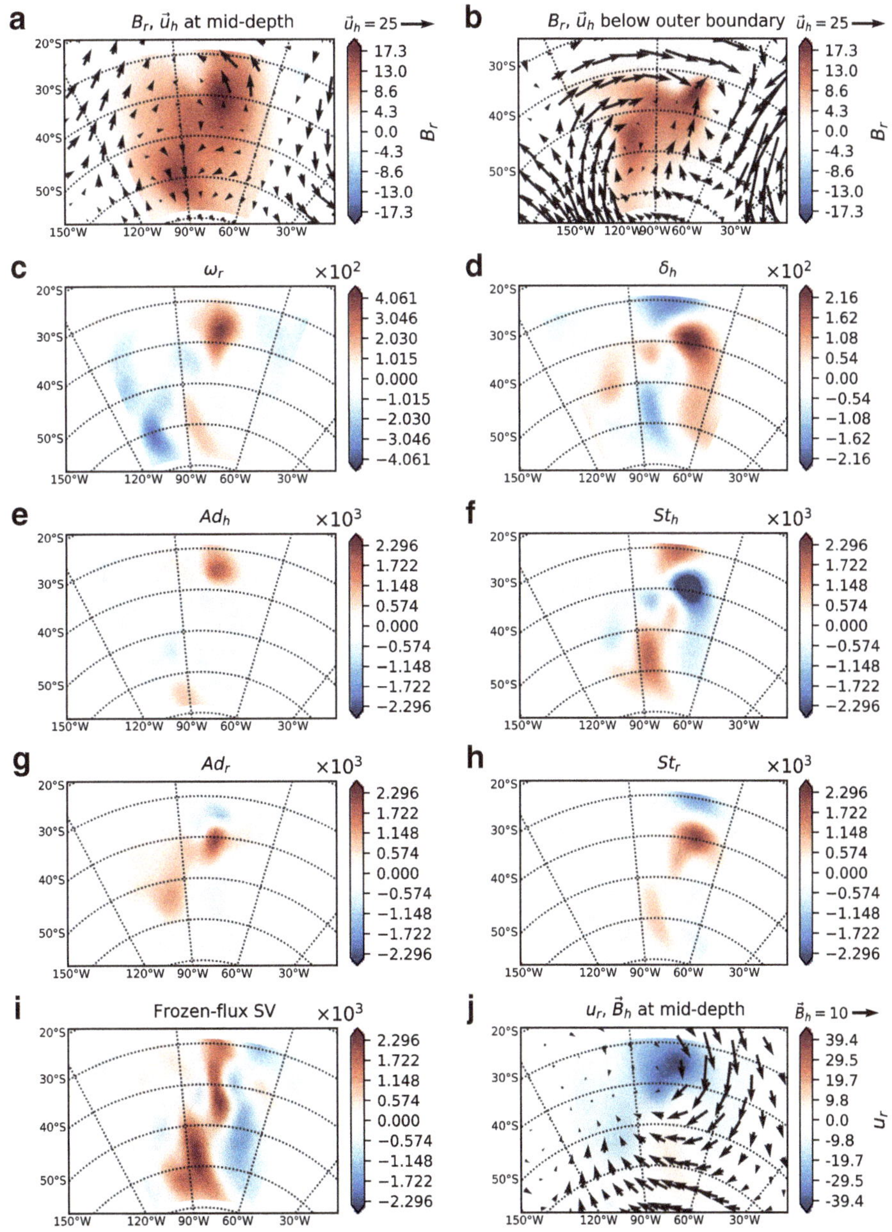

Fig. 8 As in Fig. 6 for another intense high-latitude normal polarity magnetic flux patch (HN). This patch presents a dominant St_h in dynamo model 2. The mid-depth patch in **a** is shifted with respect to the outer boundary patch in **b** by $\delta\vartheta = 5.62°$ and $\delta\varphi = 7.50°$ with an auto-correlation of 0.69. The local statistics for this patch are as follows: $Ad_r/Ad_h = 1.16$; $St_r/St_h = 0.65$; $St_h/Ad_h = 2.10$; $St/Ad = 1.47$; $\mathcal{P}/\mathcal{T} = 1.07$; and $\|u_r\|/\|\vec{u}_h\| = 1.62$

shows the proportion of outer boundary patches for which roots were detected for each dynamo model. Dynamo models with stronger magnetic advection effects exhibit more distortion from the outer boundary to depth. Indeed, dynamo model 9 has the lowest percentage of roots and the highest Rm while dynamo models 1 and 7 have the highest percentage of roots and a fairly low Rm. In addition, we separated the percentage of detected roots by patch type. We note that in the strong advection dynamo model 9, the percentage of detected HN roots is lower than the percentage of detected LN and LR roots, whereas in the weaker advection dynamo models 1 and 7, the percentage of HN roots is either comparable to or larger than that of LN and LR roots.

In general, the mid-depth roots appear with some shift in latitude and longitude with respect to the location of the outer boundary patches. Table 4 shows the average

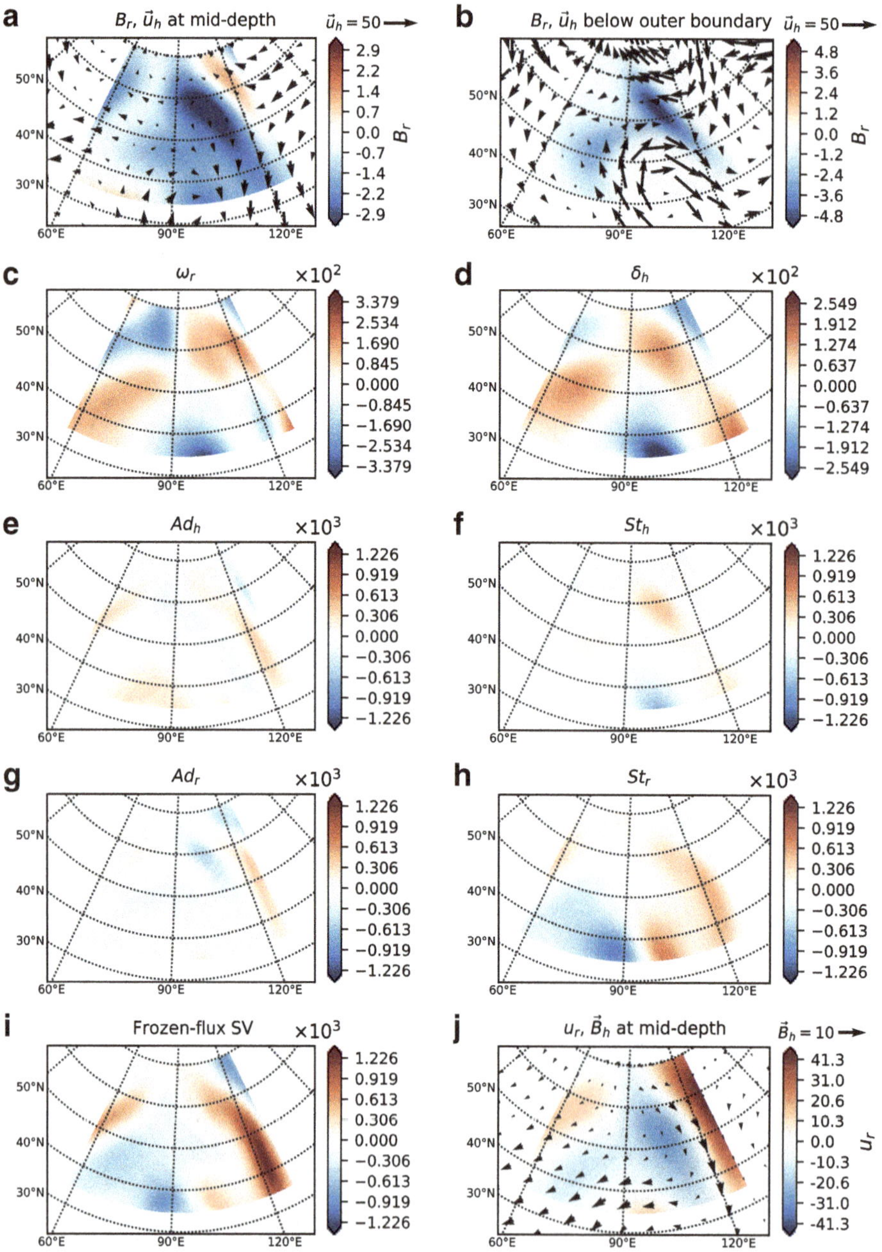

Fig. 9 As in Fig. 6 for another intense high-latitude normal polarity magnetic flux patch (HN). This patch presents a dominant St_r in dynamo model 1. The mid-depth patch in **a** is shifted with respect to the outer boundary patch in **b** by $\delta\vartheta = 5.62°$ and $\delta\varphi = -1.88°$ with an auto-correlation of 0.53. The local statistics for this patch are as follows: $Ad_r/Ad_h = 0.87$; $St_r/St_h = 2.42$; $St_h/Ad_h = 0.98$; $St/Ad = 1.74$; $\mathcal{P}/\mathcal{T} = 0.57$; and $\|u_r\|/\|\vec{u}_h\| = 1.00$

latitudinal shift $\delta\vartheta$ of the identified mid-depth roots. In some cases, the shift is rather close to the 10.7° value expected from the tangential cylinder rim effect, e.g., in dynamo models 4 and 8 with tangential advection dominance and in dynamo model 5 with radial stretching dominance. Overall, the positive $\delta\vartheta$ for HN (Table 4 and Figs. 6, 7, 8, 9, and 10) is consistent with some tangent cylinder rim effect at high latitudes. In contrast, the generally lower $\delta\vartheta$ for LN and LR is suggestive of more plume-like dynamics at low latitudes. Lower standard deviations were found for HN, supporting the robustness of lower latitude roots for these features. In contrast, the values for LN and LR are more dispersed.

In some cases, small $\delta\varphi$ values were found (Figs. 6, 9, and 10). This is consistent with both a tangent cylinder rim effect and a plume-like effect. However, the larger $\delta\varphi$ values in other cases (Figs. 7 and 8) are indicative of more complex kinematics.

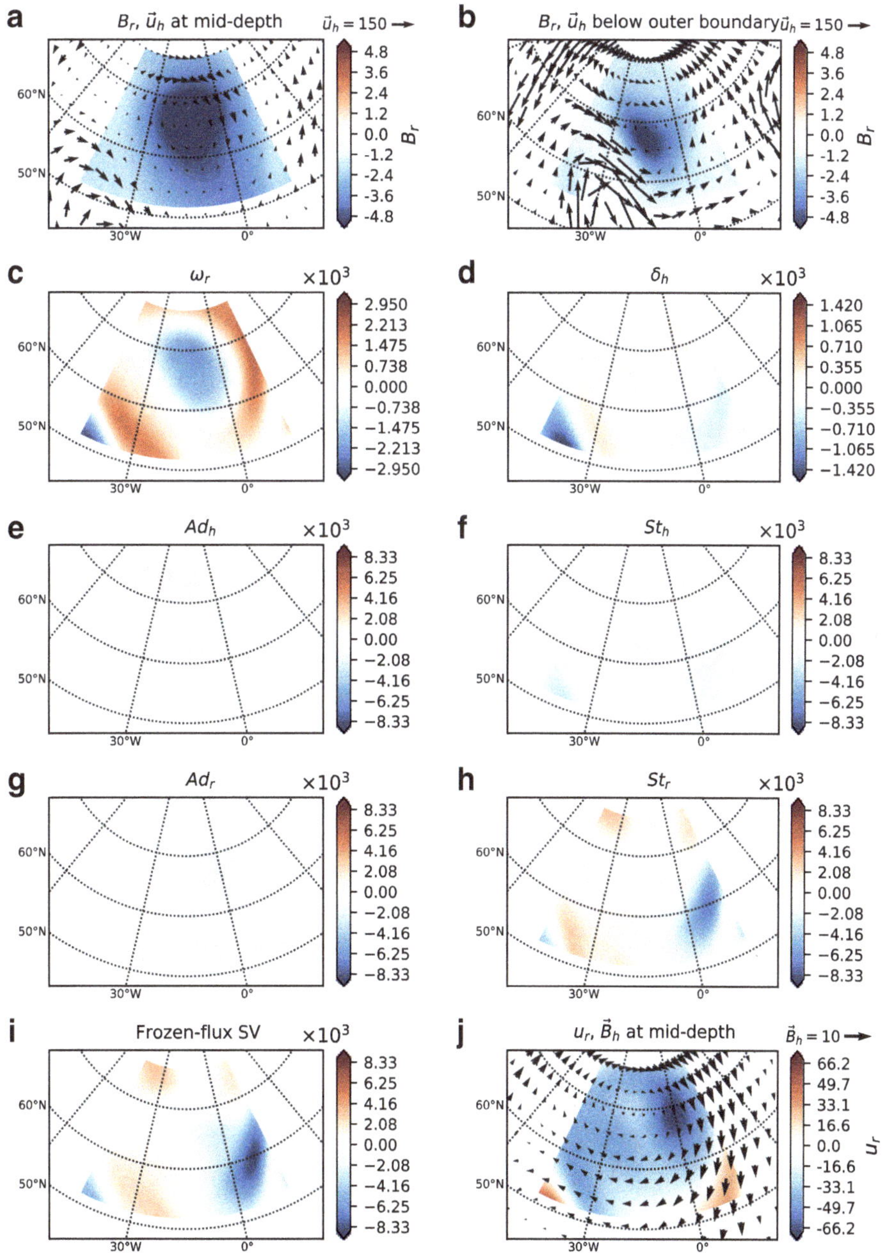

Fig. 10 As in Fig. 6 for another intense high-latitude normal polarity magnetic flux patch (HN). This patch presents a dominant St_r in dynamo model 7. The mid-depth patch in **a** is shifted with respect to the outer boundary patch in **b** by $\delta\vartheta = 5.62°$ and $\delta\varphi = 0°$ with an auto-correlation of 0.80. The local statistics for this patch are as follows: $Ad_r/Ad_h = 1.26$; $St_r/St_h = 2.77$; $St_h/Ad_h = 2.52$; $St/Ad = 4.86$; $\mathcal{P}/\mathcal{T} = 0.25$; and $\|u_r\|/\|\vec{u}_h\| = 0.71$

The local statistics for each dynamo model and root type are given in Table 5. The statistics for the roots are similar to the global statistics, with a dominance of Ad_h over Ad_r and of St_r over St_h. In the HN roots, the contribution of the total stretching to the frozen-flux SV is larger than the total advection contribution despite the low and moderate values of $\mathcal{P}/\mathcal{T}$ and $\|u_r\|/\|\vec{u}_h\|$, respectively. In contrast, the roots of LN and LR patches exhibit a larger advection contribution to the frozen-flux SV. Also, the poloidal part of the flow is stronger in LN and LR than in HN.

Figure 6 is an example of a very clear root at mid-depth from dynamo model 1. The outer boundary HN patch in Fig. 6b is highly auto-correlated (0.68) with the deep structure detected in Fig. 6a. The flow pattern also exhibits some similarities between mid-depth and the outer boundary. The tangential flow is dominantly toroidal, and the radial component of the flow is only 0.62 of the tangential component. Significant field-flow alignment seen in Fig. 6a produces a relatively weak Ad_h (Fig. 6e) in some parts of the area. However, Ad_h is

Table 5 Local statistics at mid-depth: dynamo models time average and standard deviation values for each patch type (PT)

Model	PT	Ad_r/Ad_h	St_r/St_h	St_h/Ad_h	St/Ad	$\mathcal{P}/\mathcal{T}$	$\|u_r\|/\|\vec{u}_h\|$
1	HN	0.69 ± 0.29	1.31 ± 0.50	0.65 ± 0.28	0.90 ± 0.46	0.36 ± 0.12	0.74 ± 0.24
	LN	0.46 ± 0.00	2.87 ± 0.00	0.20 ± 0.00	0.56 ± 0.00	0.27 ± 0.00	0.45 ± 0.00
	LR	0.77 ± 0.38	0.84 ± 0.21	0.63 ± 0.32	0.71 ± 0.28	0.41 ± 0.14	0.97 ± 0.21
2	HN	0.71 ± 0.29	1.49 ± 0.67	0.52 ± 0.37	0.81 ± 0.46	0.37 ± 0.16	0.82 ± 0.32
	LN	0.55 ± 0.11	0.77 ± 0.18	0.56 ± 0.02	0.67 ± 0.11	0.35 ± 0.05	0.98 ± 0.08
	LR	0.55 ± 0.28	1.15 ± 0.51	0.38 ± 0.15	0.55 ± 0.18	0.37 ± 0.15	0.90 ± 0.40
3	HN	0.61 ± 0.27	1.53 ± 0.80	0.42 ± 0.13	0.63 ± 0.26	0.32 ± 0.11	0.72 ± 0.18
	LN	–	–	–	–	–	–
	LR	2.23 ± 1.69	0.87 ± 0.44	0.99 ± 0.72	0.56 ± 0.25	1.26 ± 0.63	1.65 ± 1.29
4	HN	0.77 ± 0.33	1.53 ± 0.82	0.92 ± 0.38	1.35 ± 0.66	0.33 ± 0.08	0.80 ± 0.24
	LN	–	–	–	–	–	–
	LR	–	–	–	–	–	–
5	HN	0.68 ± 0.24	1.29 ± 0.45	1.02 ± 0.38	1.50 ± 0.91	0.40 ± 0.12	0.71 ± 0.20
	LN	–	–	–	–	–	–
	LR	–	–	–	–	–	–
6	HN	0.65 ± 0.30	1.44 ± 0.79	0.54 ± 0.25	0.81 ± 0.33	0.37 ± 0.17	0.95 ± 0.50
	LN	0.52 ± 0.18	1.10 ± 0.13	0.55 ± 0.09	0.54 ± 0.24	1.18 ± 0.62	0.62 ± 0.08
	LR	0.78 ± 0.14	0.91 ± 0.41	0.71 ± 0.15	1.03 ± 0.26	0.53 ± 0.14	1.21 ± 0.20
7	HN	0.78 ± 0.19	1.37 ± 0.47	1.39 ± 0.50	1.75 ± 1.03	0.40 ± 0.21	0.82 ± 0.28
	LN	0.48 ± 0.22	1.44 ± 0.70	0.53 ± 0.31	0.99 ± 0.48	0.24 ± 0.01	0.66 ± 0.19
	LR	0.56 ± 0.00	1.25 ± 0.00	0.38 ± 0.00	0.51 ± 0.00	0.33 ± 0.00	0.66 ± 0.00
8	HN	0.70 ± 0.20	1.33 ± 0.58	1.18 ± 0.49	1.68 ± 0.92	0.47 ± 0.17	0.85 ± 0.22
	LN	–	–	–	–	–	–
	LR	0.68 ± 0.00	1.42 ± 0.00	0.43 ± 0.00	0.80 ± 0.00	0.41 ± 0.00	0.59 ± 0.00
9	HN	0.64 ± 0.21	1.03 ± 0.47	0.64 ± 0.26	0.77 ± 0.27	0.35 ± 0.12	1.18 ± 0.79
	LN	0.74 ± 0.17	0.99 ± 0.39	0.71 ± 0.00	1.12 ± 0.20	0.52 ± 0.01	1.14 ± 0.04
	LR	0.87 ± 0.34	0.57 ± 0.15	0.80 ± 0.51	1.01 ± 0.64	0.45 ± 0.05	1.37 ± 0.16
$\bar{x}$	HN	0.70 ± 0.28	1.40 ± 0.65	0.77 ± 0.47	1.10 ± 0.75	0.37 ± 0.15	0.81 ± 0.33
	LN	0.56 ± 0.19	1.28 ± 0.73	0.54 ± 0.21	0.80 ± 0.36	0.54 ± 0.46	0.80 ± 0.26
	LR	0.83 ± 0.76	0.98 ± 0.43	0.58 ± 0.39	0.68 ± 0.33	0.50 ± 0.36	1.04 ± 0.57

Quantities are the same as in Table 2. $\bar{x}$ denotes averages over all dynamo models the dominant SV mechanism due to other highly advective regions where the flow is nearly perpendicular to the B_r-isolines. The contribution of Ad_r is low (see Fig. 6g). St_r is the second dominant mechanism (Fig. 6h), with a strong feature due to a strong meridional magnetic field B_θ sheared by the edge of an intense descending flow (Fig. 6j). Also, this St_r feature is located near an intense tangentially diverging structure (Fig. 6d), but because the local radial field is weak, no significant St_h contribution appears there (Fig. 6f). In this root, the total stretching SV amounts to 0.78 of the total advective SV.

The root shown in Fig. 7 is an example of a strong Ad_h contribution from dynamo model 3. This mid-depth patch is highly auto-correlated with the HN patch detected at the outer boundary (Fig. 7b), shifted by $\delta\vartheta = 1.88°$ (i.e. to lower latitudes) and $\delta\varphi = -9.38°$ (i.e. to the west). Westward flow roughly perpendicular to the B_r-isolines (Fig. 7a) produces a characteristic bipolar SV pattern centered at the center of the patch (Fig. 7e) with its axis parallel to the direction of the flow (Amit 2014; Peña et al. 2016; Livermore et al. 2017). The low value of St_h is due to the low correlation between δ_h (Fig. 7d) and the B_r feature (Fig. 7a). The intense u_r in Fig. 7j is not well correlated with the B_r feature in Fig. 7a, resulting in a weak Ad_r (Fig. 7g). In addition, the significant field-flow alignment between u_r and $\vec{B}_h$ (Fig. 7j) results in a weak St_r (Fig. 7h).

Figure 8 shows a mid-depth root with St_h dominance from dynamo model 2. The high auto-correlation (0.69) between the deep structure detected in Fig. 8a and the outer boundary HN patch in Fig. 8b is very clear. The root is shifted by $\delta\vartheta = 5.62°$ (i.e., to lower latitudes) and $\delta\varphi = 7.50°$ (i.e., to the east). The tangential flow has a large field-aligned component which produces a weak Ad_h (Fig. 8e). In contrast, the overlap of the tangential divergence/convergence structures (Fig. 8d) with the B_r root (Fig. 8a) produces the strong St_h contribution in Fig. 8f. For example, the northeast peak of B_r (Fig. 8a) is dispersed by tangential divergence (a positive δ_h in the same area in Fig. 8d) giving a strong negative St_h there (Fig. 8f). In addition, both Ad_r and St_r are locally strong due to the strong descending flow in the northern part (Fig. 8j). The Ad_r structure (Fig. 8g) appears near the peak of u_r where the radial field is advected from above, whereas the St_r bipolar structure (Fig. 8h) appears at the edges of the u_r structure where B_θ is sheared by a variable convective flow in the north-south direction (Fig. 8j). In this root, the poloidal flow is slightly larger than the toroidal and the radial flow is significantly larger than the tangential. Overall, in this patch, stretching mechanisms dominate over advection mechanisms in the frozen-flux SV by a significant factor of 1.47.

Mid-depth roots dominated by Ad_r are rare (Table 6). On average, Ad_r is significantly weaker than Ad_h (Table 5).

Table 6 Dominant SV mechanisms: portion of the dominant SV mechanisms (in percentage) in all the identified roots for each dynamo model

Model	Ad_h	St_h	Ad_r	St_r
1	71.4	7.1	7.1	14.3
2	80.9	4.3	10.6	4.3
3	80	0	20	0
4	31.3	18.8	6.3	43.8
5	35	20	0	45
6	76	4	8	12
7	17.4	8.7	0	73.9
8	41.7	8.3	0	50
9	87.5	12.5	0	0

Over all nine dynamo models for HN patches, the ratio Ad_r/Ad_h has a rather narrow range of 0.61–0.78.

Finally, the root shown in Fig. 9 presents a dominant St_r contribution from dynamo model 1. This root is again well auto-correlated with the HN patch detected at the outer boundary (Fig. 9b), shifted by $\delta\vartheta = 5.62°$ (again to lower latitudes) and $\delta\varphi = -1.88°$ (i.e., to the west). It is located in the middle of a flow saddle (Fig. 9c) that generates a relatively weak Ad_h (Fig. 9e). This flow is predominantly toroidal and with comparable radial and tangential components. The tangential divergence (Fig. 9d) overlaps with the intense B_r, resulting in some St_h contribution (Fig. 9f). However, the strong total stretching SV mostly originates from the dominant St_r contribution (Fig. 9h) due to the strong azimuthal field B_φ in the southern part (Fig. 9j). Note that in the eastern part, strong u_r and B_θ features yield a weak St_r due to the north-south orientation of the radial flow structure. Similar St_r dominance can also be found in lower E dynamo models. Figure 10 focuses on an HN patch from our lowest E and Pm dynamo model 7. Although the tangential field is in general nearly parallel to the u_r isolines (Ferraro's law; see Aubert 2005), small deviations from this field-flow alignment suffice to produce intense St_r contributions at the edges of a descending flow structure where the tangential field is strong.

In general, we found a predominance of Ad_h (more than $\sim$70% of the studied roots in five dynamos; see Table 6). The other four dynamos exhibit a larger number of roots with St_r dominance, but only in dynamo model 7 the number of roots with St_r dominance is significantly larger than the other mechanisms. Radial advection is the least dominant mechanism.

Parameter dependence

In order to examine quantitatively the dependence of the statistical measures on the non-dimensional control parameters, we used a generic power law (25) as in

Peña et al. (2016). Power law fits were applied for global and local measures. Significantly small powers were omitted. Also, we assigned the same value to powers with similar values. Finally, we approximated powers by discrete values. These steps were monitored using the relative misfit (26). The resulting power laws are shown in Table 7.

Figure 11 shows the parameter dependence of Ad_r/Ad_h and St_r/St_h. Globally, relative radial advection in the dynamo models increases with increasing convection but decreases when rotation increases (Fig. 11a). In the roots of high-latitude normal polarity patches (HN), we find an opposite behavior: relative radial advection decreases with increasing convection but increases when rotation increases (Fig. 11b). In both cases, global and local, the dependence is stronger on E. Global relative radial stretching increases with increasing rotation but decreases when convection and electrical conductivity increase. In this case, the dependence is strongest on E (Fig. 11c). Relative radial stretching on HN increases with increasing rotation but decreases when convection and electrical conductivity increase. This is in qualitative agreement with the global case. Relative radial stretching exhibits a strong dependence on Ra in HN roots (Fig. 11d).

We also compared the parameter dependence of St_h/Ad_h and St/Ad (Fig. 12). Globally, relative tangential stretching is mostly influenced by Pm, decreasing with increasing electrical conductivity (Fig. 12a). In HN roots, relative tangential stretching increases with increasing rotation but decreases when convection and electrical conductivity increase. The dependence is strongest on E (Fig. 12b). Both global and local total relative stretching also increase with increasing rotation and decrease when convection and electrical conductivity increase (Fig. 12c, d). However, the global St/Ad dependence is strongest on Pm (Fig. 12c) whereas in HN roots, St/Ad exhibits a strong dependence on E (Fig. 12d).

Finally, Fig. 13 shows the parameter dependence of $\mathcal{P}/\mathcal{T}$ and $\|u_r\|/\|\vec{u}_h\|$. Globally, relative poloidal flow decreases with increasing rotation but increases when convection increases. The dependence is strongest on E (Fig. 13a). In contrast, relative poloidal flow in HN roots is equally influenced by E and Pm, increasing when rotation and electrical conductivity increase (Fig. 13b). Similar to global relative poloidal flow, global relative radial flow exhibits a strong dependence on E, decreasing with increasing rotation and with decreasing convection (Fig. 13c). This similarity is expected since the radial flow is strictly poloidal. Relative radial flow in HN roots increases when convection and electrical conductivity increase, with a strongest dependence on Pm (Fig. 13d).

Discussion

The radial magnetic induction equation at the CMB in the frozen-flux limit (11) has been thoroughly explored in the context of core flow inversions from geomagnetic SV data (e.g. Bloxham and Jackson 1991; Holme 2015). According to this equation, tangential advection and tangential stretching mechanisms (Ad_h and St_h) induce the SV at

Table 7 Parameter dependence: power law fits and relative misfit σ_r

Quantity	Type	Mid-shell	σ_r	$\delta\sigma_r$	Outer boundary
Ad_r/Ad_h	Global	$0.738 \cdot (E^2 \cdot Ra)^{\frac{1}{8}}$	0.029	0.004	–
	HN	$0.580 \cdot (E^2 \cdot Ra)^{-\frac{1}{11}}$	0.042	0.000	–
St_r/St_h	Global	$2.081 \cdot (E^3 \cdot Ra^2 \cdot Pm)^{-\frac{1}{12}}$	0.054	0.007	–
	HN	$2.982 \cdot (E \cdot Ra^2 \cdot Pm)^{-\frac{1}{28}}$	0.080	0.000	–
St_h/Ad_h	Global	$0.628 \cdot Pm^{-\frac{1}{4}}$	0.071	0.004	$2.996 \cdot (E \cdot Ra \cdot Pm^3)^{-\frac{1}{6}}$
	HN	$0.765 \cdot (E^6 \cdot Ra^3 \cdot Pm^5)^{-\frac{1}{11}}$	0.140	0.005	$10.489 \cdot (E \cdot Ra \cdot Pm^2)^{-\frac{1}{3}}$
St/Ad	Global	$1.666 \cdot (E^4 \cdot Ra^3 \cdot Pm^6)^{-\frac{1}{24}}$	0.036	0.004	$2.996 \cdot (E \cdot Ra \cdot Pm^3)^{-\frac{1}{6}}$
	HN	$0.425 \cdot (E^9 \cdot Ra^4 \cdot Pm^3)^{-\frac{1}{13}}$	0.184	0.016	$10.489 \cdot (E \cdot Ra \cdot Pm^2)^{-\frac{1}{3}}$
$\mathcal{P}/\mathcal{T}$	Global	$0.302 \cdot (E^3 \cdot Ra^2)^{\frac{1}{16}}$	0.035	0.002	$0.319 \cdot Pm^{\frac{1}{6}}$
	HN	$0.105 \cdot (E^{-1} \cdot Pm)^{\frac{1}{8}}$	0.089	0.002	$0.261 \cdot Pm^{\frac{1}{6}}$
$\|u_r\|/\|\vec{u}_h\|$	Global	$0.556 \cdot (E^3 \cdot Ra^2)^{\frac{1}{16}}$	0.025	0.005	–
	HN	$0.117 \cdot (Ra^3 \cdot Pm^5)^{\frac{1}{20}}$	0.078	0.003	–

The fit deterioration is defined by $\delta\sigma_r = \sigma_r - \sigma_{ro}$ where σ_{ro} is the initial best fit relative misfit. Quantities are the same as in Table 2. Also given are the power laws for the outer boundary (Peña et al. 2016)

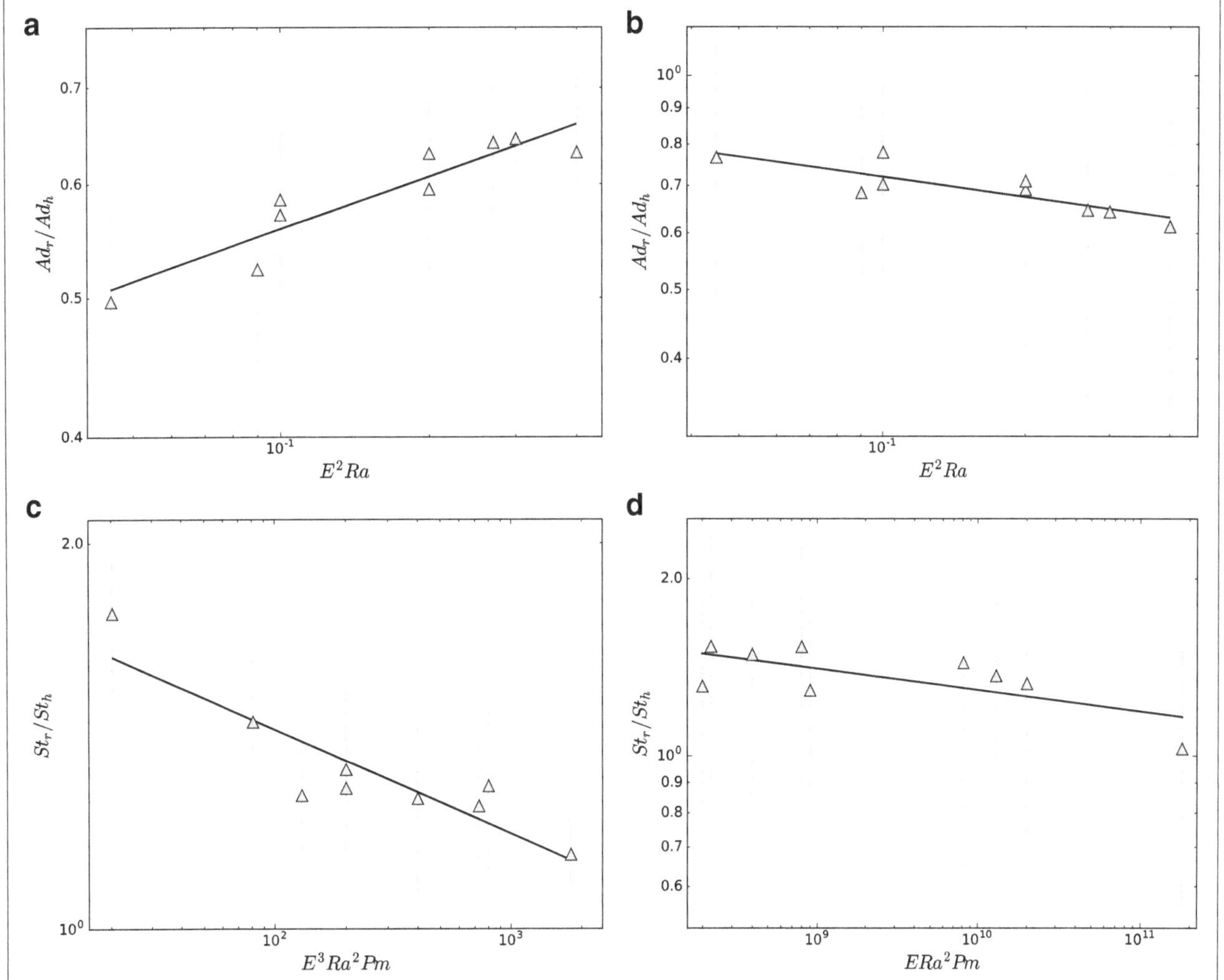

Fig. 11 Parameter dependence of Ad_r/Ad_h and St_r/St_h rms ratios. Each point represents **a**, **c** global mean values and **b**, **d** HN mean values of each dynamo simulation. Standard deviations are denoted by error bars which represent time dependence

the top of the core. The much-less explored radial induction equation away from the boundaries where $u_r \neq 0$ includes additional radial advection and radial stretching mechanisms (Ad_r and St_r; see Fig. 1) which lead to richer kinematic scenarios.

At mid-shell, global tangential advection dominates the SV in all dynamo models. Total stretching SV varies between two thirds of total advection to comparable, whereas the toroidal flow is 2–3 times larger than the tangential poloidal flow (Table 2). Similar results were reported by Peña et al. (2016) at the top of the shell. Thus, the contribution of stretching mechanisms to the SV at mid-depth is more significant than expected based on the relative strength of the poloidal flow at the top of the shell.

The radial magnetic field is concentrated in robust intense flux patches, as observed at the top of the shell of numerical dynamos as well as in the geomagnetic field models on the CMB (Christensen et al. 2010). In contrast, we found that at depth, the radial magnetic field is more distributed over the spherical surface (Fig. 5). This weaker flux concentration at mid-depth renders challenging the identification of deep roots of the intense flux patches at the top of the shell. Nevertheless, proper accounting for various image distortion effects led us to identify such roots.

Different portions of detected roots were found for the different dynamo models (see Table 3). The number of identified roots is roughly inversely proportional to the magnetic Reynolds number *Rm*. Stronger magnetic mixing effects at larger *Rm* correspond to more vigorous magnetic advection effects between the top and mid-shell, which distort the patches and reduce the possibility of identifying their roots. The identified mid-depth roots were classified by polarity and by latitude according to their correlated outer boundary structures. Roots of high-latitude intense normal flux patches (HN) were identified

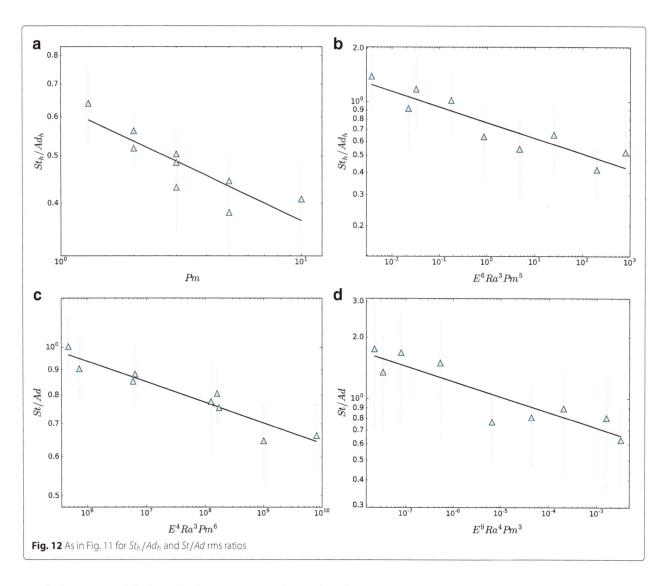

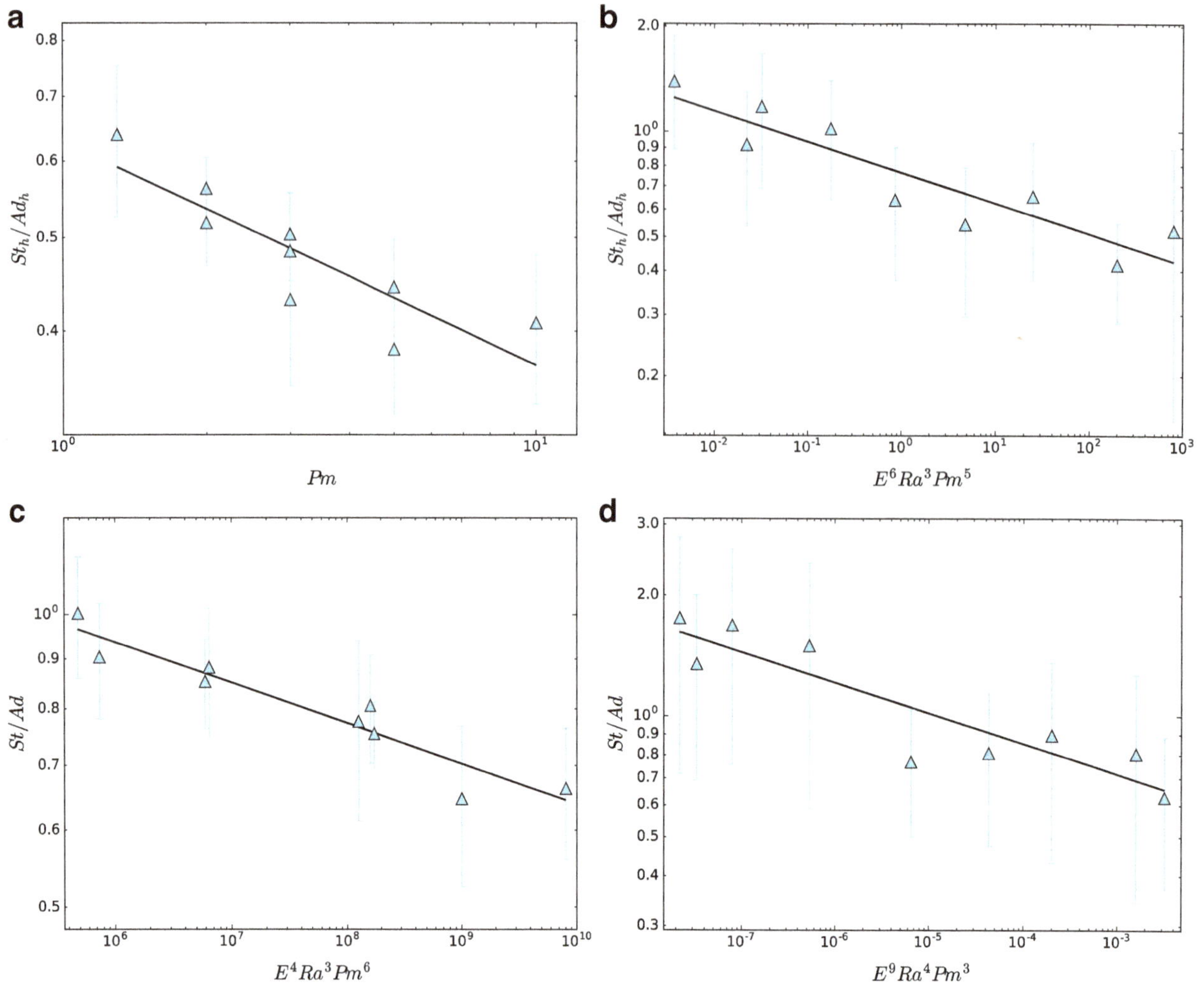

Fig. 12 As in Fig. 11 for St_h/Ad_h and St/Ad rms ratios

in all dynamo models, but the fraction of HN detected roots is also dependent on the strength of magnetic advection at each dynamo model. Stronger advection models exhibit a lower HN detection than that of the roots of low-latitude normal and reversed flux patches (LN and LR, respectively), whereas in weaker advection models, the HN detection is either comparable to or larger than that of LN and LR (Table 3).

The roots exhibit a position shift with respect to the location of the outer boundary patches. Generally, the latitudinal shift $\delta\vartheta > 0$, i.e., the roots are at lower latitudes than the outer boundary patches, in qualitative agreement with a tangent cylinder rim effect (Fig. 2). However, the mean latitudinal shift is lower than expected from a pure tangent cylinder rim effect ($\delta\vartheta < 10.7°$; see Table 4), indicating a mixed effect of a tangent cylinder rim and plume-like dynamics.

Relative stretching varies for different root types. In the HN roots, stretching dominates the SV in field-aligned flow regions where advection is not sufficiently effective, despite Ad_h being the single dominant SV mechanism (Table 6). On average in HN, total stretching SV exceeds total advection by a factor of 1.1, with the poloidal flow being only 0.37 of the toroidal flow (Table 5). These results are consistent with those found in the HN outer boundary patches reported by Peña et al. (2016). The contribution of radial stretching is significant for total stretching SV dominance at depth. In contrast, the structures at low latitudes (LN and LR) exhibit lower values of total relative stretching and higher values of relative poloidal flow.

At mid-shell, the ratio of global tangential stretching over tangential advection increases with decreasing electrical conductivity. In HN roots, relative tangential stretching increases with decreasing E, Ra, and Pm. The strongest dependence is on E, but Pm also exhibits a strong influence (Fig. 12a, b). The ratio of total stretching over total advection also increases with decreasing E, Ra, and Pm in both the global and HN roots

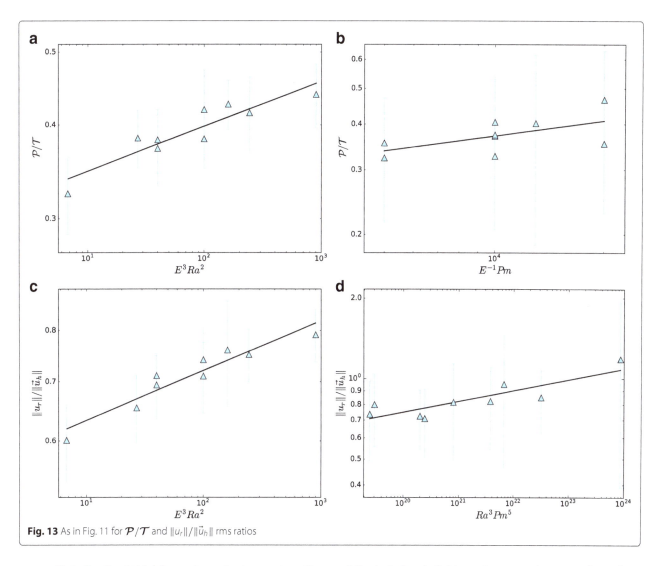

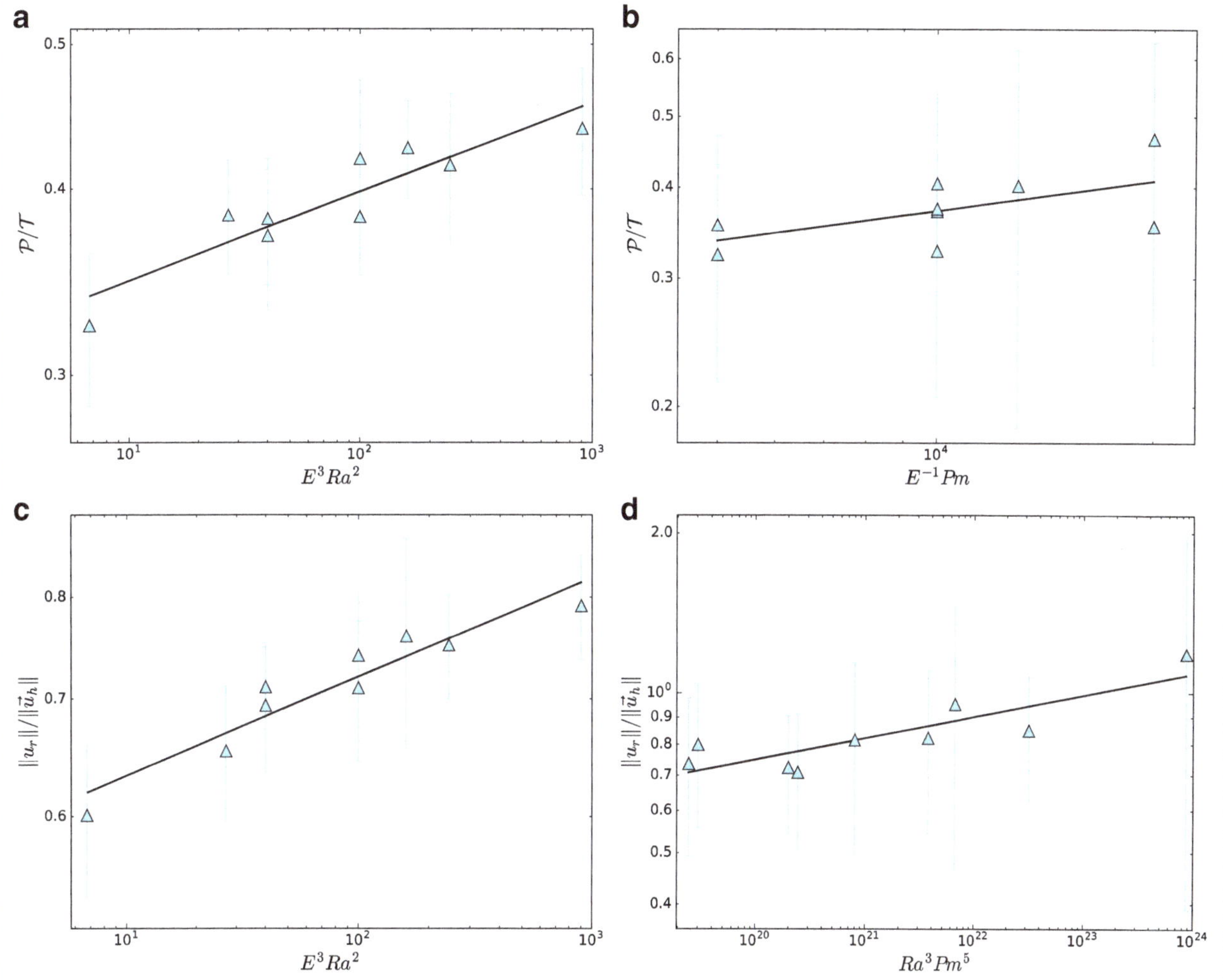

Fig. 13 As in Fig. 11 for $\mathcal{P}/\mathcal{T}$ and $\|u_r\|/\|\vec{u}_h\|$ rms ratios

cases. Globally, the St/Ad dependence is strongest on Pm while the dependence in the HN roots is strongest on E (Fig. 12c, d). These behaviors are in qualitative agreement with the decrease with E, Ra, and Pm found by Peña et al. (2016) for the ratio of stretching over advection on the outer boundary (see Table 7).

The relative global poloidal flow at mid-shell is strongly influenced by E, increasing with increasing E and Ra. In contrast, the relative poloidal flow in the HN roots increases with increasing E and decreasing Pm (Fig. 13a, b). These results contradict the behavior on the outer boundary, where both global and HN $\mathcal{P}/\mathcal{T}$ only depend on Pm, increasing with increasing Pm (Peña et al. 2016). The parameter dependence of the ratio of global radial flow over global tangential flow is practically identical to that of $\mathcal{P}/\mathcal{T}$. The relative radial flow at HN also has the strongest dependence on Pm, increasing with increasing Ra and Pm (Fig. 13c, d).

It is worth elaborating on the radial stretching SV mechanism because it contributes significantly at mid-shell while it is by definition absent at the outer boundary where the radial flow vanishes. Of the four radial field SV mechanisms at depth (Fig. 1), radial stretching is the only dynamo mechanism (Oslon et al. 1999, Aubert et al. 2008b). When the tangential magnetic field lines $\vec{B}_h$ are perpendicular to the u_r-isolines, strong St_r structures are produced (as in Fig. 9h, j). These strong St_r structures are characterized by an intense tangential field residing at the edges of radial plumes where the tangential gradient of u_r is large. As illustrated in Fig. 1d, this process transforms a (tangential) toroidal magnetic field into a (radial) poloidal one.

Takahashi and Shimizu (2012) performed a detailed local analysis of the field generation process in numerical dynamos. They found four typical stretching scenarios, two shallow and two deep. One deep stretching mechanism is associated with downwellings inside columnar vortices, which is related to the tangential stretching SV at high latitudes in our radial field analysis. The other deep stretching mechanism found by Takahashi and

Shimizu (2012) is associated with alternating flows in the direction perpendicular to columnar vortices, which is related to the radial stretching SV at low latitudes in our radial field analysis.

The significant contribution of radial stretching in the mid-shell roots indicates a possible underlying deep dynamo mechanism that sustains intense magnetic flux patches at the outer boundary. Because these high-latitude intense flux patches contribute significantly to the dominant geomagnetic axial dipole (Gubbins 1987; Olson and Amit 2006; Finlay 2008), radial stretching SV at depth may be considered as a mechanism to sustain the dipole.

Abbreviations

CMB: Core-mantle boundary; HN: Intense high-latitude normal polarity magnetic flux patch; HR: Intense high-latitude reversed polarity magnetic flux patch; LN: Intense low-latitude normal polarity magnetic flux patch; LR: Intense low-latitude reversed polarity magnetic flux patch; SV: Secular variation

Acknowledgements

We are grateful to two anonymous reviewers for their important comments. This work was partly carried out during several visits by D.P. to LPG Nantes.

Funding

This study was supported by the Centre National d'Etudes Spatiales (CNES) and the Observatório Nacional (ON/MCTI). D.P. was supported by a Ph.D. research grant from Coordenação de Aperfeiçoamento de Pessoal de Nível Superior (CAPES) and a grant by LPG-Nantes. K.P. was supported by la Région des Pays de la Loire and Coordenação de Aperfeiçoamento de Pessoal de Nível Superior (CAPES- Proc no BEX 2498/13-8) and Fundação de Amparo à Pesquisa do Estado do Rio de Janeiro (FAPERJ- E06/2015). This work acknowledges financial support from Région Pays de la Loire, project GeoPlaNet (convention no. 2016-10982).

Authors' contributions

HA ran the dynamo simulations. DP analyzed the models' output, produced the graphics, calculated the statistics, and wrote the paper. All authors read and improved the text and approved the final manuscript.

Competing interests

The authors declare that they have no competing interests.

Author details

[1]Geophysics Department, Observatório Nacional, Rio de Janeiro 20921-400, Brazil. [2]CNRS, Université de Nantes, Nantes Atlantiques Universités, UMR CNRS 6112, Laboratoire de Planétologie et de Géodynamique, 2 rue de la Houssinière, 44000 Nantes, France.

References

Amit H (2014) Can downwelling at the top of the Earth's core be detected in the geomagnetic secular variation? Phys Earth Planet Int 229:110–121

Amit H, Olson P (2004) Helical core flow from geomagnetic secular variation. Phys Earth Planet Int 147:1–25

Amit H, Olson P, Christensen UR (2007) Tests of core flow imaging methods with numerical dynamos. Geophys J Int 168:27–39

Anderson BJ, Johnson CL, Korth H, Purucker ME, Winslow RM, Slavin JA, Solomon SC, McNutt RL, Raines JM, Zurbuchen TH (2011) The global magnetic field of Mercury from MESSENGER orbital observations. Science 333(6051):1859–1862

Aubert J (2005) Steady zonal flows in spherical shell dynamos. J Fluid Mech 542:53–67

Aubert, J (2015) Geomagnetic forecasts driven by thermal wind dynamics in the Earth's core. Geophys J Int 203(3):1738–1751

Aubert J, Amit H, Hulot G, Olson P (2008a) Thermochemical flows couple the Earth's inner core growth to mantle heterogeneity. Nature 454(7205):758–761

Aubert J, Aurnou J, Wicht J (2008b) The magnetic structure of convection-driven numerical dynamos. Geophys J Int 172(3):945–956

Aubert J, Finlay CC, Fournier F (2013) Bottom up control of geomagnetic secular variation by the Earth's inner core. Nature 502:219–223

Aurnou J, Andreadis S, Zhu L, Olson P (2003) Experiments on convection in Earth's core tangent cylinder. Earth Planet Sci Lett 212(1): 119–134

Badro J, Siebert J, Nimmo F (2016) An early geodynamo driven by exsolution of mantle components from Earth's core. Nature 536:326–328

Bloxham J (1986) The expulsion of magnetic flux from the Earth's core. Geophys J R Astr Soc 87:669–678

Bloxham J, Jackson A (1991) Fluid flow near the surface of Earth's outer core. Rev Geophys 29(1):97–120

Christensen UR (2006) A deep dynamo generating Mercury's magnetic field. Nature 444:1056–1058

Christensen UR, Aubert J (2006) Scaling properties of convection-driven dynamos in rotating spherical shells and application to planetary magnetic fields. Geophys J Int 166:97–114

Christensen UR, Wicht J (2008) Models of magnetic field generation in partly stable planetary cores: Applications to Mercury and Saturn. Icarus 196:16–34

Christensen UR, Olson P, Glatzmaier G (1998) A dynamo model interpretation of geomagnetic field structures. Geophys Res Lett 25(10):1565–1568

Christensen UR, Aubert J, Hulot G (2010) Conditions for Earth-like geodynamo models. Earth Planet Sci Lett 296(3):487–496

Chulliat A, Hulot G, Newitt LR (2010) Magnetic flux expulsion from the core as a possible cause of the unusually large acceleration of the north magnetic pole during the 1990s. J Geophys Res 115(B7):1978–2012

Dziewonski AM, Anderson DL (1981) Preliminary reference Earth model. Phys Earth Planet Int 25(4):297–356

Finlay CC (2008) Historical variation of the geomagnetic axial dipole. Phys Earth Planet Int 170(1–2):1–14

Finlay CC, Amit H (2011) On flow magnitude and field flow alignment at Earth's core surface. Geophys J Int 186:175–192

Finlay CC, Jackson A (2003) Equatorially dominated magnetic field change at the surface of Earth's core. Science 300(5628):2084–2086

Gillet N, Pais MA, Jault D (2009) Ensemble inversion of time dependent core flow models. Geochem Geophys Geosyst 10(Q06004)

Gillet N, Schaeffer N, Jault D (2011) Rationale and geophysical evidence for quasi-geostrophic rapid dynamics within the Earth's outer core. Phys Earth Planet Int 187(3–4):380–390

Gillet N, Jault D, Finlay CC (2015) Planetary gyre, time-dependent eddies, torsional waves, and equatorial jets at the Earth's core surface. J Geophys Res Solid Earth 120(6):3991–4013

Gubbins D (1987) Mechanism for geomagnetic polarity reversals. Nature 326(6109):167–169

Gubbins D, Davies CJ (2013) The stratified layer at the core-mantle boundary caused by barodiffusion of oxygen, sulphur and silicon. Phys Earth Planet Int 215:21–28

Hartmann GA, Pacca IG (2009) Time evolution of the south atlantic magnetic anomaly. An Acad Bras Ciênc 81(2):243–255

Helffrich G, Kaneshima S (2010) Outer-core compositional stratification from observed core wave speed profiles. Nature 468:807–810

Holme R (2015) Large-scale flow in the core. In: Schubert G (ed). Treatise on Geophysics, 2nd edn. Elsevier, Oxford. pp 91–113. chap 4

Jackson A, Jonkers A, Walker M (2000) Four centuries of geomagnetic secular variation from historical records. Phil Trans R Soc Lond A 358:957–990

Kageyama A, Sato T (1997a) Generation mechanism of a dipole field by a magnetohydrodynamic dynamo. Phys Rev E 55(4):4617–4626

Kageyama, A, Sato T (1997b) Velocity and magnetic field structures in a magnetohydrodynamic dynamo. Phys Plasmas 4(5):1569–1575

de Koker N, Steinle-Neumann G, Vlcek V (2012) Electrical resistivity and thermal conductivity of liquid Fe alloys at high P and T, and heat flux in Earth's core. Proc Natl Acad Sci USA 109(11):4070–4073

Konôpková Z, McWilliams S, Gómez-Pérez N, Goncharov A (2016) Direct measurement of thermal conductivity in solid iron at planetary core conditions. Nature 534:99–101

Kutzner C, Christensen UR (2002) From stable dipolar towards reversing numerical dynamos. Phys Earth Planet Int 131(1):29–45

Lesur V, Whaler K, Wardinski I (2015) Are geomagnetic data consistent with stably stratified flow at the core-mantle boundary? Geophys J Int 201:929–946

Livermore PW, Hollerbach R, Finlay CC (2017) An accelerating high-latitude jet in Earth's core. Nat Geosci 10:62–68. doi:10.1038/ngeo2859
Mininni PD (2011) Scale interactions in magnetohydrodynamic turbulence. Ann Rev Fluid Mech 43:377–397
Moffatt HK (1978) Magnetic field generation in electrically conducting fluids. Cambridge University Press, London
Ohta K, Kuwayama Y, Hirose K, Shimizu K, Ohishi Y (2016) Experimental determination of the electrical resistivity of iron at Earth's core conditions. Nature 534(7605):95–98
Oliveira JS, Langlais B, Pais MA, Amit H (2015) A modified equivalent source dipole method to model partially distributed magnetic field measurements, with application to Mercury. J Geophys Res 120:1075–1094
Olsen N, Mandea M (2008) Rapidly changing flows in the Earth's core. Nature Geosci 1(6):390–394
Olson P, Amit H (2006) Changes in Earth's dipole. Naturwissenschaften 93(11):519–542
Olson P, Aurnou J (1999) A polar vortex in the Earth's core. Nature 402(6758):170–173
Olson P, Christensen UR (2002) The time averaged magnetic field in numerical dynamos with non uniform boundary heat flow. Geophys J Int 151(3):809–823
Olson P, Christensen UR, Glatzmaier GA (1999) Numerical modeling of the geodynamo: mechanisms of field generation and equilibration. J Geophys Res 104(B5):10,383–10,404
Olson P, Sumita I, Aurnou J (2002) Diffusive magnetic images of upwelling patterns in the core. J Geophys Res 107(B12):801–813
O'Rourke JG, Stevenson DJ (2016) Powering Earth's dynamo with magnesium precipitation from the core. Nature 529:387–389
Pais MA, Jault D (2008) Quasi-geostrophic flows responsible for the secular variation of the Earth's magnetic field. Geophys J Int 173(2):421–443
Peña D, Amit H, Pinheiro KJ (2016) Magnetic field stretching at the top of the shell of numerical dynamos. Earth Planets Space 68(1):1–21. doi:10.1186/s40623-016-0453-x
Pozzo M, Davies C, Gubbins D, Alfe D (2012) Thermal and electrical conductivity of iron at Earth's core conditions. Nature 485(7398):355–358
Press WH, Teukolsky SA, Vetterling WT, Flannery BP (1992) Numerical recipes in FORTRAN. Cambridge University Press, Cambridge
Takahashi F, Shimizu H (2012) A detailed analysis of a dynamo mechanism in a rapidly rotating spherical shell. J Fluid Mech 701:228–250
Tarduno J, Watkeys M, Huffman T, Cottrell R, Blackman E, Wendt A, Scribner C, Wagner C (2015) Antiquity of the South Atlantic Anomaly and evidence for top-down control on the geodynamo. Nat Commun 6:7865
Terra-Nova F, Amit H, Hartmann G, Trinidade RIF, Pinheiro KJ (2017) Relating the South Atlantic Anomaly and geomagnetic flux patches. Phys Earth Planet Int 266:39–53
Whaler K (1980) Does the whole of the Earth's core convect? Nature 287(5782):528–530
Wicht J (2002) Inner-core conductivity in numerical dynamo simulations. Phys Earth Planet Inter 132:281–302
Wicht J, Heyner D (2014) Mercury's magnetic field in the messenger era. In: Jin S (ed). Planetary Geodesy and Remote Sensing. CRC Press. pp 223–262

16

Simultaneous measurements of elastic wave velocities and electrical conductivity in a brine-saturated granitic rock under confining pressures and their implication for interpretation of geophysical observations

Tohru Watanabe[1*] and Akiyoshi Higuchi[1,2]

Abstract

Simultaneous measurements of elastic wave velocity and electrical conductivity in a brine-saturated granitic rock were conducted under confining pressures of up to 180 MPa. Contrasting changes in velocity and conductivity were observed. As the confining pressure increased to 50 MPa, compressional and shear wave velocities increased by less than 10 %. On the other hand, electrical conductivity decreased by an order of magnitude. Both changes must be caused by the closure of cracks under pressures. Microstructural examinations showed that most cracks were open grain boundaries. In reality, a crack is composed of many segments with different apertures. If crack segments have a similar length, segments with small apertures are closed at low pressures to greatly reduce conductivity, while those with wide apertures are open even at high pressures. The latter must form an interconnected fluid path to maintain the electrical conduction through fluid. A power law distribution of apertures causes a steep decrease in conductivity at low pressures. An empirical relation between the crack density parameter and normalized conductivity was obtained. The normalized conductivity is the ratio of bulk conductivity to the conductivity of a pore fluid. This relation should be a basis for quantitative interpretation of observed seismic velocity and electrical conductivity.

Keywords: Seismic velocity, Electrical conductivity, Fluid, Crack, Pore

Background

Geophysical mapping of fluids is critical for understanding geodynamic processes including seismic activities. Fluids can significantly reduce the frictional strength of fault zones (e.g., Sibson 2009) and weaken the flow strength of rocks through fluid-assisted processes such as pressure solution (e.g., Rutter 1983). Seismic velocity and electrical resistivity structures have been constructed to study the fluid distribution in the continental crust. Though a lot of studies have suggested the pervasive existence of aqueous fluids in the crust (e.g., Ogawa et al. 2001), the fluid distribution has not been quantitatively constrained.

Observations on seismic velocity and electrical resistivity should be combined to make a quantitative inference on fluid distribution. It is impossible to infer the amount of fluid only from observed seismic velocity. Seismic velocity of a fluid-bearing rock depends on the elastic properties of the solid and fluid phases and the geometry and amount of the fluid (e.g., Takei 2002). Even if we know the elastic property and geometry of the fluid, we cannot estimate the amount of fluid. Since the lithology of a study region is usually unknown, elastic properties of the rock matrix must be assumed. In

* Correspondence: twatnabe@sci.u-toyama.ac.jp
[1]Graduate School of Science and Engineering, University of Toyama, 3190 Gofuku, Toyama 930-8555, Japan
Full list of author information is available at the end of the article

addition, the fluid amount cannot be inferred only from observed resistivity. Electrical resistivity of a fluid-bearing rock depends on the resistivity of solid and fluid phases and the geometry and amount of fluid (e.g., Schmeling 1986). The resistivity of fluid is usually much lower than that of the rock matrix. If the fluid phase forms an interconnected path, the electrical conduction is dominated by the conduction through the fluid phase. The bulk conductivity hardly depends on the conductivity of the rock matrix. The estimation of the amount of fluid requires the knowledge of fluid resistivity, which is usually unknown. Owing to the uncertainty in fluid resistivity, the estimated amount of fluid must thus have large uncertainty. However, the fluid amount estimated from observed resistivity must be identical to that estimated from observed seismic velocity. This can be a constraint to reduce uncertainty in the interpretation of velocity and resistivity.

In order to make a combined interpretation of velocity and resistivity, we must have a thorough understanding of velocity and resistivity in fluid-bearing rocks. Velocity and resistivity in a fluid-bearing rock should be formulated as a function of a structural parameter such as fluid volume fraction. Fluids are mainly situated within cracks in crustal rocks (e.g., O'Connell and Budiansky 1974). Based on the inclusion theory (Eshelby 1957), the influence of fluid-filled cracks on elastic properties of rocks has long been studied. For thin cracks, the effective elastic constants are formulated as a function of the crack density parameter (O'Connell and Budiansky 1974). In their pioneering work, Brace et al. (1965) showed that electrical resistivity of brine-saturated rocks largely increased with increasing confining pressure. The increase in confining pressure closed cracks to squeeze brine out to result in a large increase in resistivity. Later, the observed resistivity change was reasonably reproduced by a percolation model (Johnson and Manning 1986). If we relate the change in resistivity to that in velocity, we can make a combined interpretation of velocity and resistivity. Simultaneous measurements of velocity and resistivity in a fluid-bearing rock are required to give a basis for the combined interpretation.

We have conducted simultaneous measurements of elastic wave velocity and electrical conductivity (the inverse of resistivity) of a brine-saturated granitic rock with changing confining and pore-fluid pressures. In this paper, we report measurements of velocity and conductivity and examined pore spaces. The nature of the conduction path will be discussed with the percolation model devised by Johnson and Manning (1986). Based on the empirical relation between electrical conductivity and crack density parameter, we propose a method for the combined interpretation of seismic velocity and electrical resistivity.

Methods

Samples

A fine-grained (100–500 μm) biotite granite (Aji, Kagawa prefecture, Japan) was used as a rock sample. The rock sample is composed of 52.8 % plagioclase, 36.0 % quartz, 3.0 % K-feldspar, and 8.2 % biotite (Fig. 1a). The apparent density is 2.658–2.668 g/cm^3, and the porosity is 1.9–2.3 % (Table 1). Each sample has a cylindrical shape with dimensions of 26 mm in diameter and 30 mm in length. The apparent density was calculated from the apparent volume and the mass of a cylindrical sample. The porosity was calculated from the apparent density and the density of the solid matrix. The density of the solid matrix was calculated from the volume and mass of a crushed rock sample. The volume was measured with the gas expansion method.

Pores in a rock sample (AJG02) were examined with scanning electron microscopy (SEM) and X-ray computer-aided tomography (CT). An SEM image (Fig. 1b) shows that there are two types of cracks: intragrain cracks (igc) and grain boundary cracks (open grain boundaries, ogb). Numerous round pores are seen on the polished surface. However, in addition to real pores, they include damages from polishing. X-ray CT images (Fig. 1c, d) show that open grain boundaries (arrows) are pervasive in a rock sample and that there are also a number of round pores (rp), which might be traces of fluids and do not seem to form an interconnected path of fluid. X-ray CT observation was conducted by using a CT system (v|tome|x L300, GE Sensing & Inspection Technologies) at Tokyo Metropolitan Industrial Technology Research Institute, Jonan Branch. A sample for X-ray CT ($D = 2$ mm, $L = 6$ mm) was made from Sample AJG02 with ultrasonic machining.

Ultrasonic velocity and strain measurements on dry samples

Ultrasonic velocities were measured in a dry rock sample (AJG02) to study the anisotropy in elasticity of rock samples. Velocity measurements were made in the axial direction and two mutually orthogonal radial directions of the cylindrical sample. One compressional wave velocity and two shear wave velocities were measured in each of three orthogonal directions. Two shear waves propagating in one direction oscillate in mutually orthogonal directions. Measurements were conducted at room temperature and confining pressures of up to 177 MPa with a pressure vessel (Riken, PV-2 M-S6F). Silicone oil (Shin-Etsu Chemical, KF-96-100cs) was used as a pressure medium.

The pulse transmission technique was employed by using Pb(Zr,Ti)O_3 transducers with the resonant frequency of 2 MHz. The method was similar to that described in Watanabe et al. (2011). Transducers were

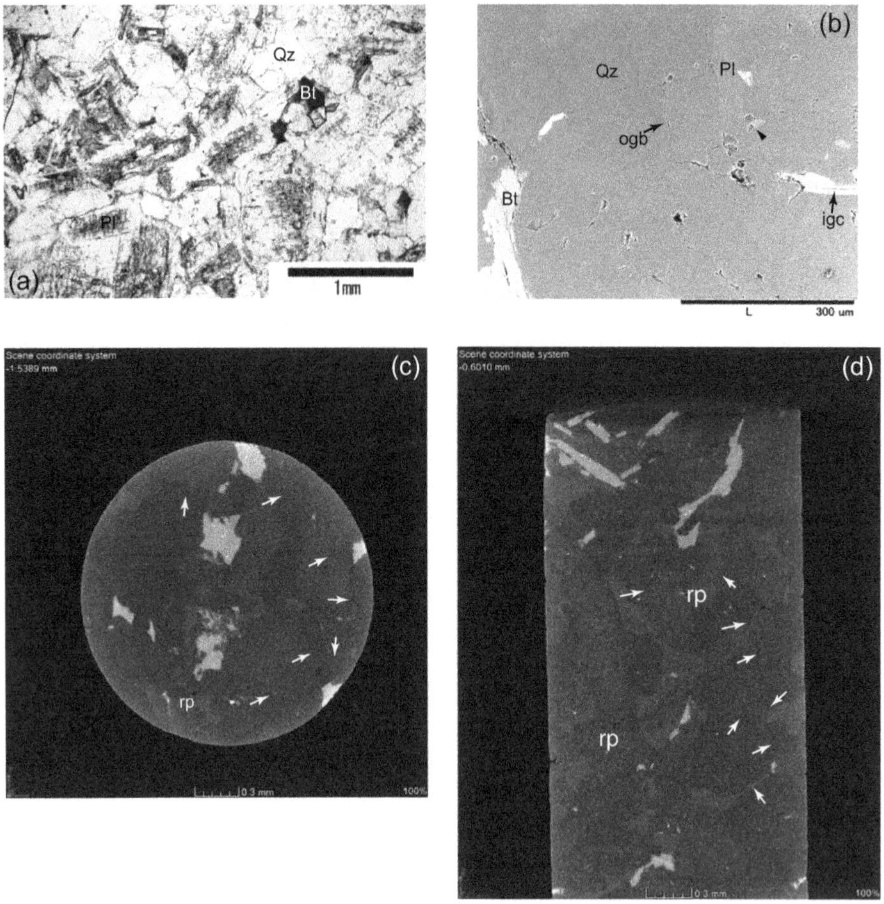

Fig. 1 Microstructures of a granitic rock sample (AJG02). *Pl* plagioclase, *Qz* quartz, *Bt* biotite. **a** A photomicrograph (plane light). The width is parallel to the axis of the cylindrical rock sample. **b** An SEM image. Both intragrain cracks (igc) and intergrain cracks (ogb) can be seen. A *triangle* shows damage from polishing. **c** Radial and **d** axial cross-sectional images of X-ray CT of a cylindrical sample ($D = 2$ mm, $L =$ 6 mm), the axis of which is perpendicular to the thin section. Open grain boundaries (*arrows*) seem to be interconnected. Round pores (rp) are seen on grain boundaries and inside grains

glued to the sample, which was covered with RTV rubber (Shin-Etsu Chemical, KE-45). A function generator (Hewlett Packard, 33120A) applied an electrical rectified pulse to one transducer to excite an elastic wave. The other transducer received the transmitted elastic wave and converted it to an electrical signal, which was digitized and averaged over 1024 times by a digital oscilloscope (Agilent Technologies, 54621A). The sampling interval was 20 ns, and the digitized 8-bit signal was transferred to a computer for analysis. Velocities were calculated from the path length and the travel time.

Linear strains were measured on a dry rock sample (AJG07) under confining pressures to evaluate the volume fraction of cracks. Strains were measured in three mutually orthogonal directions of the cylindrical sample. Measurements were made with a data logger (Tokyo Sokki, TDS-301) and electrical resistance strain gages (Kyowa Electronic Instruments, KFG-1 N-120-C1-11) bonded to the sample surface.

Table 1 List of samples used in this study

Sample	Density (g/cm^3)	Porosity (%)	Measurements
AJG02	2.668(2)	1.9(7)	Vp, Vs (dry)
AJG04	2.658(3)	2.3(7)	Vp, Vs, conductivity (wet)
AJG05	2.664(3)	2.1(7)	Vp, conductivity (wet)
AJG07	2.667(2)	2.0(7)	Vp, strain (dry)

The number inside brackets shows the error in the last digit

Ultrasonic velocity and electrical conductivity measurements on fluid-saturated samples

Ultrasonic velocity and electrical conductivity were measured on fluid-saturated rock samples at room temperature and confining pressures of up to 125 MPa. The confining and pore-fluid pressures were separately controlled by different pumps (Fig. 2a). A pressure vessel (Riken-Seiki, PV-2 M-S14) was equipped with a plastic

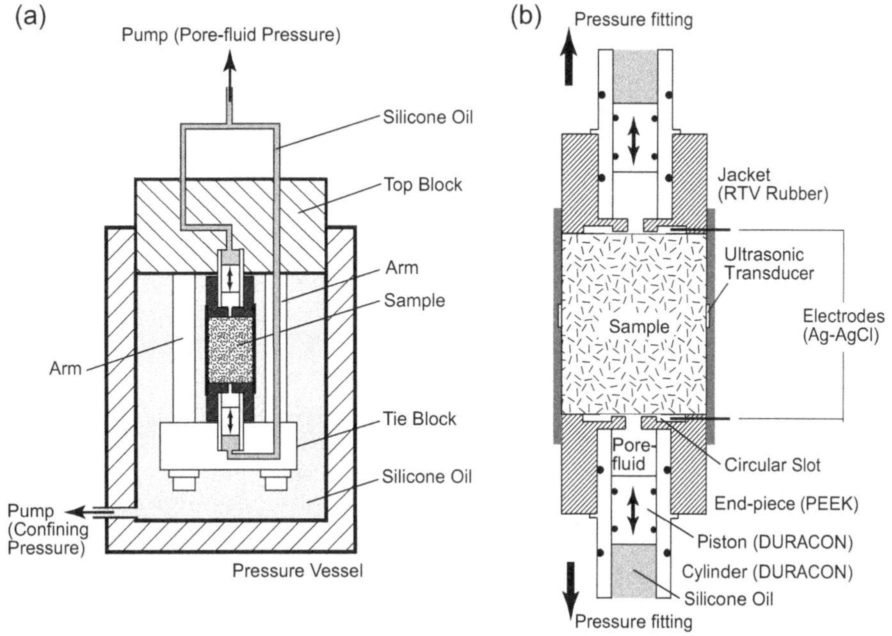

Fig. 2 Schematics of **a** pressure vessel and **b** sample assembly

piston-cylinder system, which was employed for pore-fluid pressure control and electrical isolation (Watanabe and Higuchi 2014). A pump for controlling pore-fluid pressure moves the piston to change the pore-fluid pressure. The aqueous pore fluid is electrically isolated from the metal work of the pressure vessel.

A sample assembly is shown in Fig. 2b. A cylindrical sample was firstly evacuated and saturated with 0.01 mol/L KCl aqueous solution and assembled with end-pieces. A DURACON (acetal copolymer) piston-cylinder system is inserted in a PEEK end-piece. DURACON and PEEK were selected for their high strength. DURACON can minimize the resistance to the piston movement. RTV rubber was used as a jacketing material. The sample assembly was attached to the top block of the pressure vessel with two arms and a tie block (Fig. 2a). Along with the compression in radial directions, the silicone oil between the top block and the upper end-piece and between the tie block and the lower end-piece compresses a sample in the axial direction, causing a hydrostatic compression. The right arm and the tie block are designed to work as a channel for silicone oil (viscosity ~ 0.1 Pa·s) to move the lower piston. The upper piston is moved by the silicone oil in a channel through the top block.

Compressional and shear wave velocities were measured by the pulse transmission technique. Measurements were made in mutually orthogonal radial directions of the cylindrical sample. The shear wave oscillates parallel to the axis of the sample. $Pb(Zr,Ti)O_3$ transducers with the resonant frequency of 2 MHz were bonded to the sample surface which was coated with an epoxy resin for electrical isolation between the pore fluid and transducers. A function generator (Agilent Technologies, 33220A) and a digital oscilloscope (Agilent Technologies, DSO5012A) were used for pulse excitation and data acquisition, respectively.

Electrical impedance was measured in the axial direction by using an LCR meter (NF, ZM2355) with the two-electrode method. Ag-AgCl electrodes were made as in Watanabe and Katagishi (2006). The PEEK end-piece has a hole and a circular slot on the sample side (Fig. 2b). The pore fluid can flow between the hole and the slot. The frequency range was from 40 Hz to 200 kHz. The conductivity was calculated from the sample resistance, the length, and the cross-sectional area of a sample.

Results

Compressional and shear wave velocities in a dry sample

Compressional and shear wave velocities in a dry sample (AJG02) are shown as a function of confining pressure in Fig. 3. Measurements were made during the increase in confining pressure. The *Z*-direction was set parallel to the axis of the cylindrical sample. The *X*- and *Y*-directions are mutually orthogonal and perpendicular to the cylinder axis. Both compressional and shear wave velocities increase with increasing confining pressure. The increase in velocity is larger for compressional waves (~1.5 km/s) than for shear waves (~0.6 km/s). The increase rate of velocity is greatly reduced at pressures higher than 100 MPa. The increase in velocity must thus be caused by the closure of pores. If we consider a

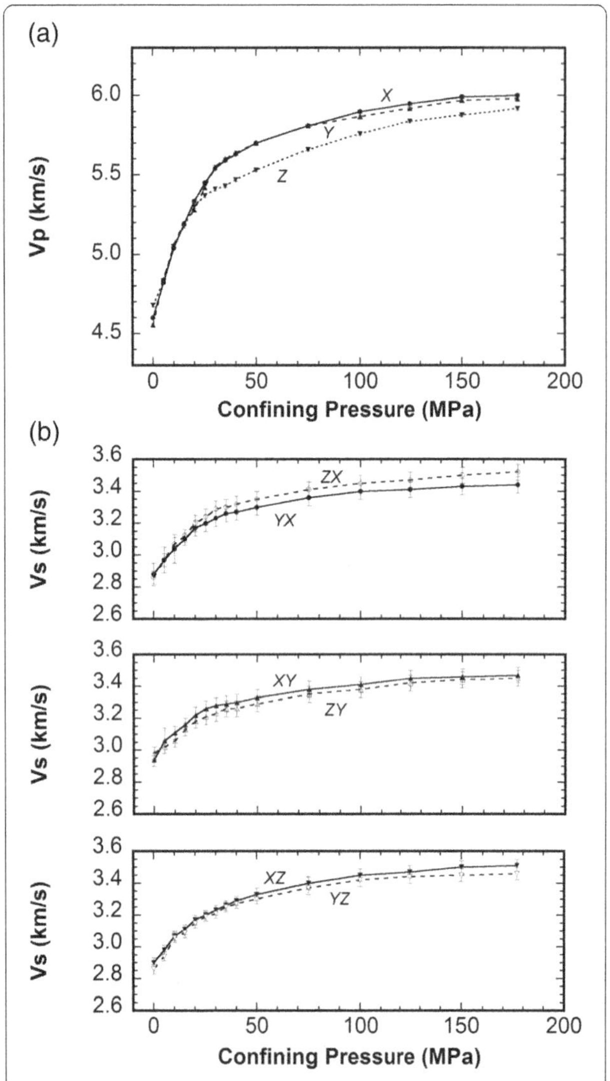

Fig. 3 **a** Compressional and **b** shear wave velocities in a dry rock sample AJG02. Velocities are shown as a function of confining pressure. *Z*-direction is parallel to the axis of the cylindrical sample. *X*- and *Y*-directions are perpendicular to *Z*-direction and mutually orthogonal. The letter for a compressional wave represents the propagation direction. The first letter for a shear wave is the oscillation direction, and the second letter is the propagation direction

spheroidal pore with aspect ratio $\alpha = c/a$, where a and c are the lengths of the semi- major and minor axes, the closure pressure is given by

$$p_c = \frac{\pi E \alpha}{4(1-\nu^2)}, \tag{1}$$

where E and ν are Young's modulus and Poisson's ratio, respectively (Walsh 1965). Pores with smaller aspect ratios thus close at lower pressures.

The sample shows weak velocity anisotropy in the whole range of confining pressure. The anisotropy in compressional and shear wave velocities is quite weak at 177 MPa. The difference between the fastest and slowest compressional wave velocities is only 1.3 %, and no significant difference in shear wave velocity can be seen between two oscillation directions (Fig. 3b). Since the influence of pores is sufficiently suppressed, the elastic wave velocities at 177 MPa are mainly governed by elastic properties of the solid matrix. The solid matrix is thus almost isotropic in elasticity. Although spheroidal pores with large aspect ratios could still be open, their influence on elastic properties is sufficiently small. Elastic wave velocities at atmospheric pressure, which are strongly affected by oblate spheroidal pores, also show weak anisotropy. In consideration of the weak anisotropy in elasticity of the solid matrix, the oblate spheroidal pores, which we call cracks hereafter, must thus be almost randomly oriented.

Volumetric strain in a dry sample

Volumetric strain in a dry rock sample (AJG07) is shown in Fig. 4 as a function of confining pressure. Measurements were made during the increase in confining pressure. The volumetric strain takes positive values for contraction. Linear strains in three mutually orthogonal directions showed that the rock sample deformed isotropically under hydrostatic conditions. The volumetric strain is given by the sum of the three strains. The magnitude of the volumetric strain increases and its increasing rate decreases with increasing confining pressure. The nonlinear increase with pressure is attributed to the closure of cracks. When all cracks are closed, the volumetric strain is caused by the elastic deformation of the solid matrix, and it increases linearly with pressure. The extrapolation of the linear trend to 0 MPa provides an estimate of the crack porosity (e.g., Walsh 1965). However, a linear trend due to elastic deformation of the

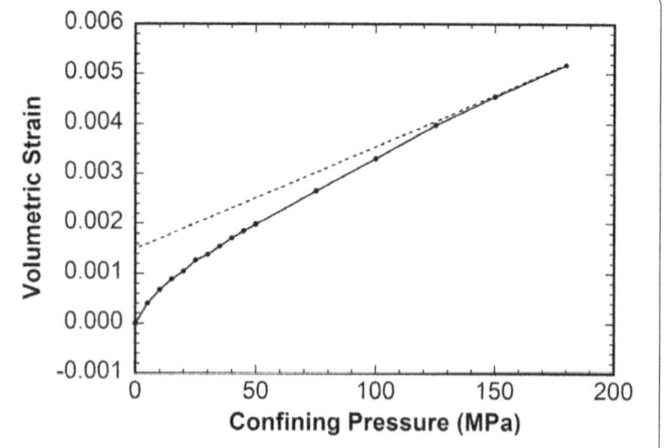

Fig. 4 Volumetric strain of dry rock sample AJG07 as a function of confining pressure. The increasing rate in the volumetric strain decreases with increasing confining pressure. The *broken line* passing two points at 150 and 180 MPa is extrapolated to 0 MPa to give 0.15 % as the lower bound of the crack porosity

solid matrix was not observed in our experiment, since the confining pressure is less than 200 MPa. A line passing two points at 150 and 180 MPa provides an estimate of the crack porosity, which is 0.15 % as the intercept on the volumetric strain axis. It is the lower bound of the crack porosity, since the linear trend due to elastic deformation of the solid matrix has a more gradual slope.

Influence of confining pressure on velocities and conductivity

Simultaneous measurements of velocities and conductivity were made on wet sample AJG04 at confining pressures of up to 125 MPa. The pore-fluid pressure was kept at 0.1 MPa (atmospheric pressure). Compressional and shear wave velocities and electrical conductivity are shown in Fig. 5 as a function of confining pressure. Average compressional and shear wave velocities in dry sample AJG02 are shown for comparison. The confining pressure was first increased to 5 MPa and then increased to 25, 50, 75, 100, and 125 MPa. Velocities and conductivity were allowed to become stationary values before changing the confining pressure. Each pressure condition was kept for 50–100 h. Measurements were made during the increase in confining pressure.

Velocities increased but conductivity decreased with increasing confining pressure. Most changes were observed at confining pressures lower than 50 MPa. The velocities in AJG04 (wet) were higher than those in AJG02 (dry), but the difference between velocities in wet and dry samples decreased with increasing pressure. The changes in velocities and conductivity must thus be attributed to the closure of pores with increasing pressure. The compressional wave velocity increased by 0.34 km/s (6.0 %) from 0.1 to 125 MPa and the shear wave velocity by 0.27 km/s (8.5 %).

Conductivity showed a large change at low pressures in contrast to velocity changes. Conductivity decreased by more than one order of magnitude from 0.1 to 25 MPa but showed no remarkable change at higher pressures. Similar changes in conductivity have been observed in previous studies (Brace et al. 1965; Lockner and Byerlee 1985). Though the porosity was quite low, the observed conductivity clearly demonstrated the interconnection of fluid. The change in conductivity suggests that the connectivity was greatly reduced by the closure of pores.

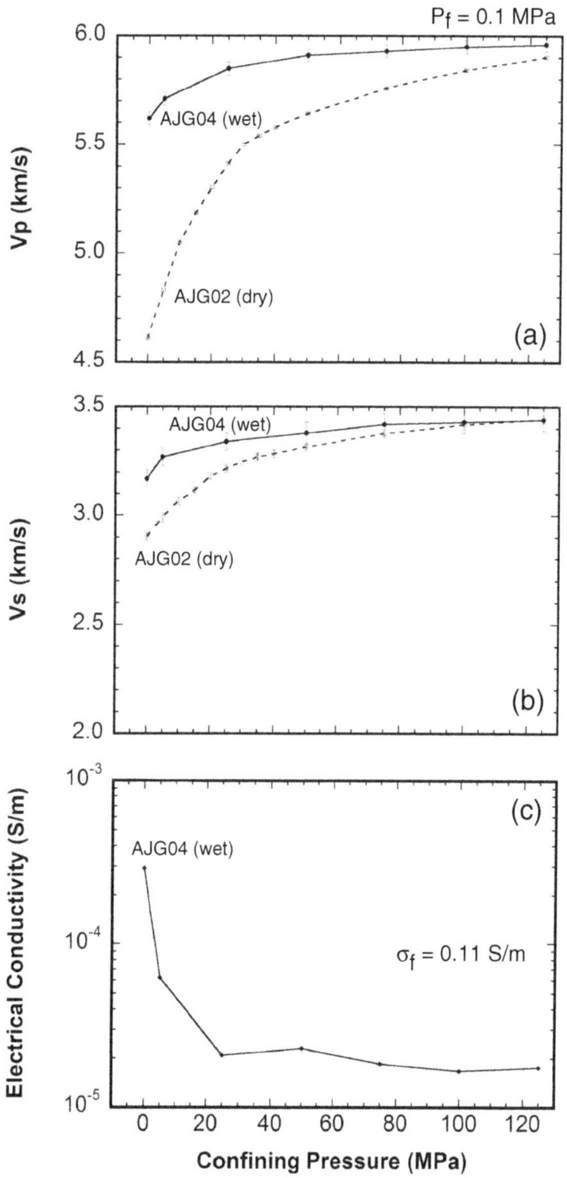

Fig. 5 a Compressional and **b** shear wave velocities and **c** electrical conductivity in brine-saturated sample AJG04. Velocities and conductivity are shown as a function of confining pressure. The pore-fluid pressure was kept at atmospheric pressure (0.1 MPa). Average compressional and shear wave velocities in dry sample AJG02 are also shown for comparison

Influence of pore-fluid pressure on velocities and conductivity

Velocity and conductivity in wet sample AJG06 were measured at various confining and pore-fluid pressures. Compressional and shear wave velocities and electrical conductivity are shown in Fig. 6 as a function of confining and pore-fluid pressures. The confining pressure was first increased to 5, 10, and 15 MPa, while the pore-fluid pressure was kept at 0.1 MPa. The confining pressure was then kept at 15 MPa, and the pore-fluid pressure was increased to 5 and 10 MPa. The confining and pore-fluid pressures were then increased alternately. Each pressure condition was kept for 50–100 h until velocities and conductivity become stationary values. The differential pressure, which is given by

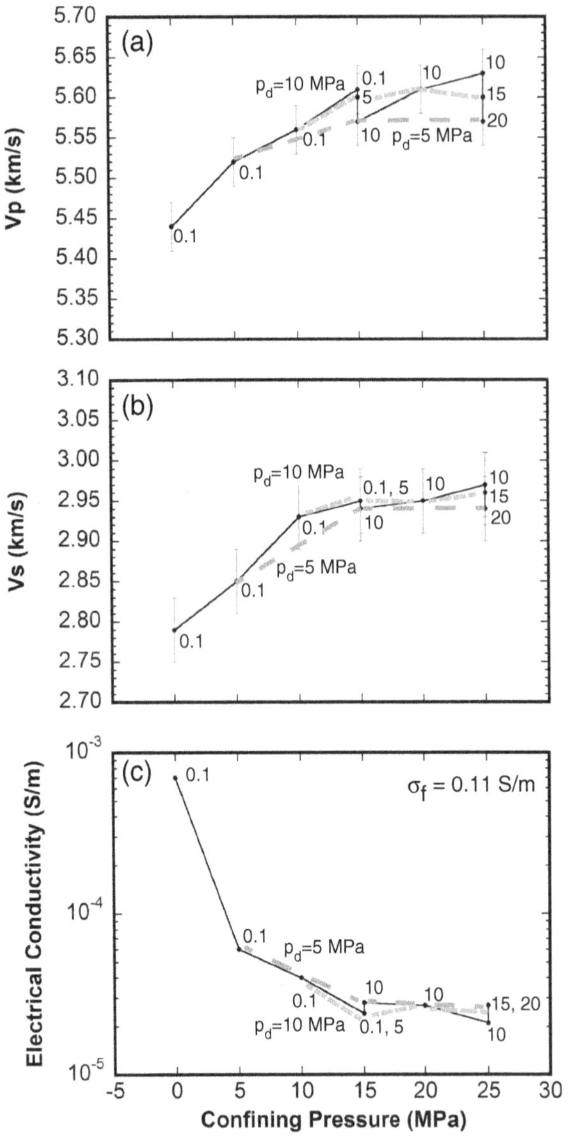

Fig. 6 **a** Compressional and **b** shear wave velocities and **c** conductivity in brine-saturated sample AJG06. Velocities and conductivity are shown as a function of confining pressure at different pore-fluid pressures (in MPa; see labels). Data points for the differential pressures of 5 MPa and 10 MPa are respectively tied with *gray broken lines*

$$p_d = p_c - p_f,$$

is also indicated in Fig. 6.

The pore-fluid pressure and confining pressure have influences on velocities and conductivity in the opposite directions. When the confining pressure was kept constant, velocities decreased but conductivity increased with increasing pore-fluid pressure. The increase in pore-fluid pressure must open pores to decrease velocities and increase conductivity.

Neither velocities nor conductivity was constant at constant differential pressures. Velocities increased but conductivity decreased at a constant differential pressure as the confining pressure was increased. Similar changes in velocity were reported for oceanic basalt and dolerite (Christensen 1984) and Berea sandstone (Christensen and Wang 1985). When confining pressure is increased, the pore-fluid pressure must be increased by an amount greater than the confining pressure to maintain a constant velocity. The theory of fluid-saturated porous materials (Biot and Willis 1957; Geertsma 1957) showed that the bulk volumetric deformation is not governed by the differential pressure but by the effective pressure

$$p_{\text{eff}} = p_c - np_f$$

$$n = 1 - \frac{\beta_s}{\beta},$$

where β_s and β are the compressibility of the solid grain and bulk material, respectively. Christensen and Wang (1985) suggested that a physical property sensitive to bulk volume shows a similar effective pressure law. Both velocity and conductivity are strongly dependent on the confining pressure of less than 50 MPa. They are sensitive to bulk volume and thus expected to depend on the effective pressure. The compressibility of a bulk sample is evaluated to be 4.4×10^{-11} (1/Pa) at the confining pressure of 10 MPa from the volumetric strain of a dry sample (Fig. 4). The compressibility of a plagioclase grain, which is the dominant phase, is 2.0×10^{-11} (1/Pa) (Hearmon 1979). These compressibility values give $n = 0.55$. When confining pressure is increased by Δp, the pore-fluid pressure must be increased by $1.8\,\Delta p$ to maintain velocity and conductivity. This explains our observations.

Discussion

Saturation degree of pores

We firstly consider the shape of pores in granite rock samples on the basis of the confining pressure dependence of velocities. For simplicity, a pore is supposed to have a spheroidal shape. The closure pressure given by Eq. (1) depends on its aspect ratio and elastic properties of the solid matrix. Since the influence of pores on elastic properties is sufficiently suppressed at high pressures, elastic properties of the solid phase are estimated from velocities in a dry rock sample (AJG02) at 177 MPa (Table 2). The pores closed below 125 MPa must thus have aspect ratios less than 2×10^{-3}. Such oblate spheroidal pores can be treated as circular cracks, and their influence on velocities are formulated in terms of the crack density parameter defined by

$$\varepsilon = \frac{1}{V}\sum a^3 = N_c\langle a^3\rangle,$$

Table 2 Elastic properties of solid matrix estimated from velocities in dry sample AJG02 at 177 MPa

Properties	Estimated values
Bulk modulus, K	51.9(5) (GPa)
Shear modulus, G	32.3(4) (GPa)
Young's modulus, E	80.3(8) (GPa)
Poisson's ratio, ν	0.242(3)

The number inside brackets shows the error in the last digit

where a is the radius of a crack and the summation is over all cracks in a volume V (O'Connell and Budiansky 1974). The crack density parameter is also expressed by the average of a^3 and the number of cracks per unit volume, N_c.

The crack density parameter in a dry sample can be evaluated by comparing measured and calculated velocities. Once the crack density parameter is known, velocities can be calculated for various degrees of fluid saturation (O'Connell and Budiansky 1974). The comparison between measured and calculated velocities will give us an estimation of the saturation degree of pores.

The crack density parameter in Sample AJG04 (wet) was assumed to be identical to that in Sample AJG02 (dry). This assumption should be reasonable since the mean compressional wave velocities in Samples AJG02 and AJG04 were almost identical at atmospheric pressure and dry state (AJG02 4.61(1) km/s, AJG04 4.65(1) km/s). Measured velocities in Sample AJG02 were compared with velocities calculated as a function of the crack density parameter. Elastic properties of the solid phase (Table 2) were used in the calculation of velocity. The crack density parameter at a confining pressure was determined to minimize

$$S = \left(V_p^{\text{obs}} - V_p^{\text{calc}}\right)^2 + \left(V_s^{\text{obs}} - V_s^{\text{calc}}\right)^2.$$

The estimated crack density parameter in Sample AJG02 is shown in Fig. 7 as a function of confining pressure. It rapidly decreases with increasing confining pressure. Measured compressional and shear wave velocities in AJG02 were reasonably reproduced by the estimated crack density parameter (Fig. 8).

The saturation degree in Sample AJG04 (wet) was estimated by comparing measured and calculated velocities. By using the estimated crack density parameter, the compressional and shear wave velocities were calculated as a function of confining pressure for the saturation degrees of 80, 90, and 100 % (Fig. 8). The effect of the fluid is to glue the opposing faces of the crack together with respect to relative normal displacement while not inhibiting relative sliding (O'Connell and Budiansky 1974). In the calculation of

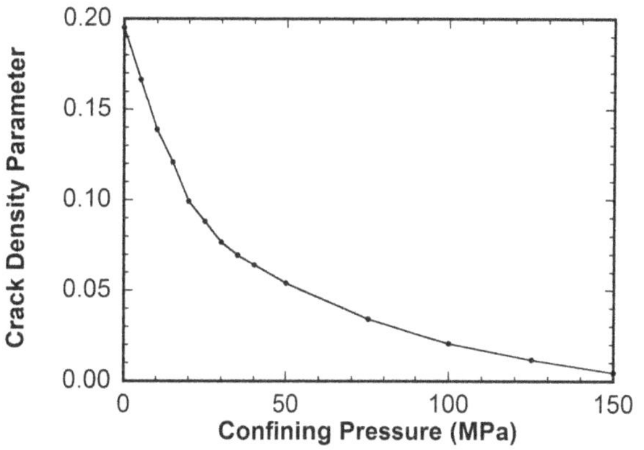

Fig. 7 Crack density parameter in dry sample AJG02 as a function of confining pressure. The crack density parameter was determined to minimize the square sum of differences between measured and calculated velocities

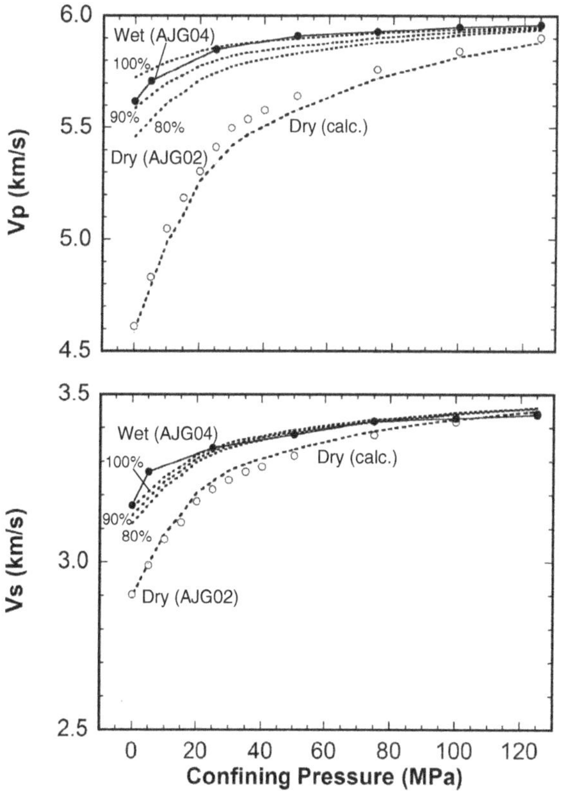

Fig. 8 Measured Vp and Vs in brine-saturated sample AJG04 and velocities for different saturation degrees. The measured velocities in brine-saturated sample AJG04 are shown by *solid circles*, and those in dry sample AJG02 by *open circles*. Calculated velocities for determined crack density parameters (Fig. 7) are shown for comparison (*broken lines*). Velocities for various saturation degrees were calculated by using the crack density parameter in dry sample AJG02. The comparison of measured and calculated velocities suggests that pores are mostly filled with aqueous solution

velocity, all cracks were assumed to be isolated. Even if an elastic wave induces a spatial variation in the fluid pressure, the fluid cannot flow between cracks to reduce the pressure variation. In reality, cracks are interconnected as demonstrated by electrical conductivity. However, the frequency of the elastic wave (2 MHz) is much higher than the characteristic frequency for the flow between cracks,

$$f = \frac{1}{2\pi}\frac{K}{\eta}\alpha^3 \sim 1 \text{ (kHz)},$$

where K, η, and α are the bulk modulus of the solid phase, the viscosity of the fluid, and the aspect ratio of a crack, respectively (O'Connell and Budiansky 1977). Thus, the isolation of cracks is a reasonable assumption. The comparison shows that the degree of saturation must be larger than 90 %. Cracks are then almost completely filled with fluid.

Nature of conduction paths

Conductivity steeply decreases at low pressures and shows small changes at high pressures. Similar changes in conductivity of low-porosity rocks were reported by Brace et al. (1965). Johnson and Manning (1986) devised a percolation model to reasonably reproduce the reported conductivity changes. The percolation model is briefly reviewed, and then it is applied to our results. Based on the comparison between measured and calculated conductivities and microstructural observations, the nature of conduction paths will be discussed.

In the percolation model (Johnson and Manning 1986), the conduction path is modeled as a lattice network with an average coordination number Z and composed of two types of bond: cracks and "pores." Cracks are closed easily with pressure, while "pores" remain open at high pressures. They are randomly distributed on the network. Both cracks and "pores" are assumed to occur as cylindrical tubes with the cross-sectional area of A_c and A_p, which are independent of pressure. The tube length is assumed to be a constant. The occupancy fraction of cracks is denoted by f_c and that of "pores" by f_p. When a crack is closed at a high pressure, it no longer works as a bond. It is treated as a void of the lattice network, and its occupancy fraction is denoted by f_v ($f_c + f_p + f_v = 1$). The effective conductivity, σ_{eff}, of the network is evaluated through the effective medium theory (Kirkpatrick 1973) as

$$\sum_i f_i \left[\frac{\sigma_{\text{eff}} - \sigma_i}{(Z/2 - 1)\sigma_{\text{eff}} + \sigma_i}\right] = 0,$$

where σ_i is the conductivity of bond i ($i = c, p, v$). The conductivity of bonds is given by

$$\sigma_i = \frac{\phi_i}{3f_i}\sigma_f \quad \text{for } i = c, p$$

$$\sigma_v = 0,$$

where ϕ_c and ϕ_p are the crack porosity and "pore" porosity, respectively (Johnson and Manning 1986). The fluid conductivity is denoted by σ_f. Because of their stiffness, the occupancy fraction of "pores", f_p, is independent of pressure. Since all cracks are assumed to have the same geometry, the occupancy fraction of cracks, f_c, is proportional to the crack porosity, which is evaluated from strain measurements. The difference between the broken line and the solid curve in Fig. 4 gives the lower bound of the crack porosity as a function of confining pressure.

The conductivity change in Sample AJG04 is compared with the conductivity calculated with the percolation model (Fig. 9). Both calculated and measured values are normalized by the fluid conductivity. The average coordination number Z is set to be 2.3. Johnson and Manning (1986) showed that the steep decrease in conductivity was reasonably well described with an average coordination number slightly greater than 2, and that the calculated values are insensitive to variations of Z in the range $2 < Z < 2.5$. The percolation model roughly reproduces the change in conductivity, though it is a relatively simple model. The steep decrease in conductivity at low pressures is caused by the closure of cracks, which greatly reduces the connectivity of conducting bonds. The conductivity at high pressures is maintained by interconnected stiff "pores."

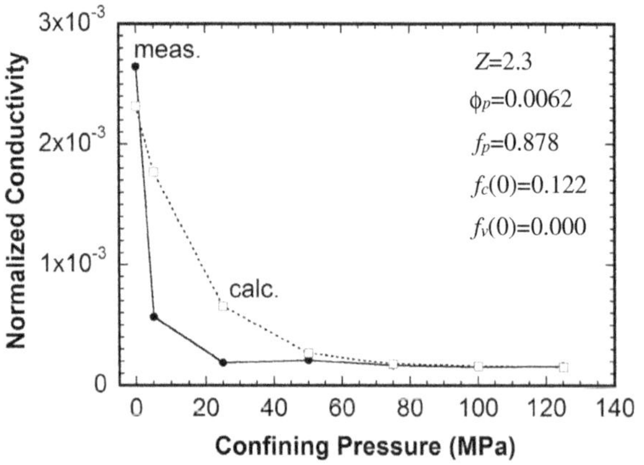

Fig. 9 Measured and calculated conductivities as a function of confining pressure. Conductivities are normalized by the fluid conductivity. The percolation model (Johnson and Manning 1986) was employed for the calculation of conductivity. The crack occupancy fraction is assumed to be proportional to crack porosity. The crack porosity was obtained from the volumetric strain measurement of dry rock sample AJG07

The geometry of cracks should be taken into account for a thorough reproduction of the change in conductivity. Microstructural examination shows that there are three types of pore spaces: intragrain cracks, intergrain cracks (open grain boundaries), and round pores (Fig. 1). Open grain boundaries seem to be the pervasive and the dominant component of the conduction paths at low pressures. In reality, a crack is not an ideal oblate spheroid which closes all at once at a confining pressure. The surface of a crack might be a distribution of asperities which progressively come into contact as the crack closes with pressure (Carlson and Gangi 1985). Wong et al. (1989) measured the crack surface area per unit volume as a function of microcrack aperture and showed that the aperture statistics can be fitted with a power law ($n \sim -1.8$). The crack surface area steeply decreases with its crack aperture. If crack segments have a similar length, segments with smaller apertures close at lower pressures. A power law distribution of aperture thus causes a steep decrease in conductivity at low pressures. On the other hand, crack segments with wide apertures, which are small in area, will be open even at high pressures. The wide segments can work as stiff "pores" and form an interconnected conduction path along with round pores. A quantitative examination of crack aperture should be done on our sample for further understanding.

Implication for geophysical observations

Geophysical observations have revealed contrasting variations of seismic velocity and electrical resistivity (the inverse of conductivity) in the continental crust. The variation of seismic velocity is less than 10 % (e.g., Matsubara et al. 2004), while that of resistivity is several orders of magnitude (e.g., Ogawa et al. 2001). The observed resistivity has suggested that aqueous fluids exist pervasively within the crust. The contrasting variations of velocity and conductivity would be explained by the contrasting dependence of velocity and conductivity on the amount of fluid, which was observed in our experiments. We propose a new method for interpreting observed velocity and resistivity.

We at first assume a rock type of a study region and evaluate the crack density parameter from seismic velocity. Seismic velocity in a fluid-saturated rock depends on elastic moduli of the solid and fluid phases and the shape and amount of the fluid phase. When the fluid phase exists within thin cracks, the impact of fluid on effective elastic moduli is formulated with the crack density parameter (O'Connell and Budiansky 1974). When we assume a rock type and give elastic moduli of the solid phase, we can evaluate the crack density parameter to give the observed seismic velocity.

With the obtained crack density parameter, we then evaluate normalized conductivity and evaluate fluid conductivity. Based on our experiments, we now have a relation between the crack density parameter and normalized conductivity (Fig. 10). Electrical conductivity is normalized by the fluid conductivity. The crack density parameter is now limited to the range of 0.01–0.20. For the crack density parameter of 0–0.01, a large increase in the normalized conductivity is expected. The conductivity of dry crustal rocks is estimated to be less than 10^{-4} S/m (e.g., Kariya and Shankland 1983) and that of crustal fluids 10–100 S/m (Nesbitt 1993). The normalized conductivity is thus expected to be less than 10^{-5} at the crack density parameter of zero. There might be an abrupt increase as the crack density increases to 0.01.

Once we have a relation between the crack density parameter and normalized conductivity in the study region, we can evaluate normalized conductivity from the crack density parameter. The inverse of normalized conductivity gives normalized resistivity: the resistivity normalized by the fluid resistivity. Dividing the observed resistivity by the normalized resistivity leads to the fluid resistivity. Its inverse gives the fluid conductivity. If the obtained fluid conductivity is an unrealistic value, we should change the rock type and evaluate the crack density parameter and normalized conductivity. Since the number of parameters to be estimated is larger than

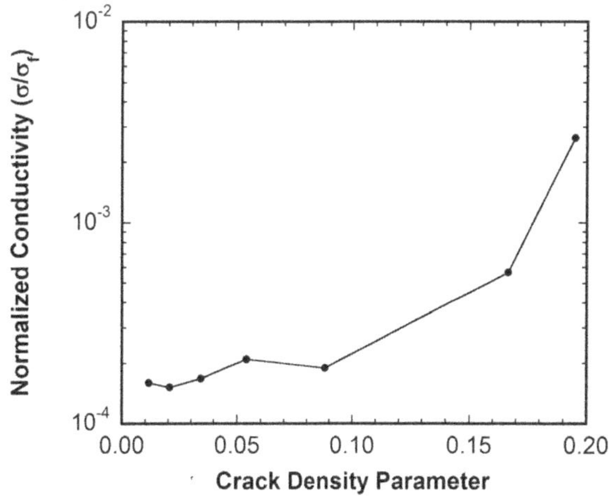

Fig. 10 Relation between normalized conductivity and crack density parameter in brine-saturated granitic rock. The electrical conductivity is normalized by the fluid conductivity. Electrical conductivity was measured in brine-saturated sample AJG04 as a function of confining pressure. The crack density parameter was estimated from compressional and shear wave velocities in dry sample AJG02 as a function of confining pressure. Since Samples AJG02 and AJG04 have similar elastic wave velocities at atmospheric pressure, we assume that the two samples have the same crack density parameter at atmospheric pressure. We thus can relate the conductivity in AJG04 to the crack density parameter in AJG02

that of observables, we will have large uncertainties in the interpretation. The appropriateness of the interpretation should be checked from geological and petrological points of view.

Through the above scheme of interpretation, we can infer rock type, fluid conductivity, and crack density parameter. The estimation of the fluid volume fraction requires the information about the aperture of cracks. If we denote the mean aspect ratio of cracks by α, the fluid volume fraction ϕ is related to the crack density parameter ε as

$$\phi = \frac{4}{3}\pi\alpha\varepsilon.$$

This gives a rough estimate of the fluid volume fraction.

The relation between the crack density parameter and normalized conductivity should be further studied. It is only an empirical relationship, and its applicability to geophysical observations and limitations are not understood. However, as we showed above, once we have such a relation, we can quantitatively interpret seismic velocity and electrical conductivity. The applicability and limitations should be studied both experimentally and theoretically. In experimental studies, the relation between the crack density parameter and normalized conductivity should thus be investigated in a wider range of the crack density parameter in the same rock type. It should also be studied in different rock types. Further theoretical works on the network of grain boundary cracks should give us a basis of the relation between the crack density parameter and normalized conductivity.

Conclusions

Elastic wave velocity and electrical conductivity in a brine-saturated granitic rock were simultaneously measured. Contrasting changes in velocity and conductivity were observed. As the confining pressure increased to 50 MPa, compressional and shear wave velocities increased by less than 10 %. On the other hand, electrical conductivity decreased by an order of magnitude. Both changes must be caused by the closure of cracks under pressures.

Microstructural examinations showed that most cracks were open grain boundaries. In reality, a crack is composed of many segments with different apertures. If crack segments have a similar length, segments with small apertures are closed at low pressures to greatly reduce conductivity, while those with wide apertures are open even at high pressures. The latter must form an interconnected fluid path to maintain the electrical conduction through fluid. A power law distribution of apertures will cause a steep decrease in conductivity at low pressures.

An empirical relation between the crack density parameter and normalized conductivity was obtained. The normalized conductivity is the ratio of bulk conductivity to the conductivity of a pore fluid. This relation should be the basis for a quantitative interpretation of observed seismic velocity and electrical conductivity.

Competing interests
The authors declare that they have no competing interests.

Authors' contributions
TW proposed the topic and designed the study. TW and AH carried out the experimental study and wrote the manuscript. Both authors read and approved the final manuscript.

Acknowledgements
We thank T. Takezawa and A. Monkawa for conducting X-ray CT at Tokyo Metropolitan Industrial Technology Research Institute. We are grateful to A. Yoneda for ultrasonic machining of samples for X-ray CT, F. Maeno for his help in measuring density by the gas expansion method, and Riken-Seiki for designing and building our high-pressure apparatuses. We gratefully appreciate two anonymous reviewers for their careful reading and invaluable comments. This work was supported by JSPS KAKENHI Grant Number 19540444.

Author details
[1]Graduate School of Science and Engineering, University of Toyama, 3190 Gofuku, Toyama 930-8555, Japan. [2]Now at Yachiyo Engineering Co., Ltd., 1-4-70 Siromi, Chuo-ku, Osaka 540-0001, Japan.

References
Biot MA, Willis DG. The elastic coefficients of the theory of consolidation. J Appl Mech. 1957;24:594–601.
Brace WF, Orange AS, Madden TR. The effect of pressure on the electrical resistivity of water-saturated crystalline rocks. J Geophys Res. 1965;70:5669–78.
Carlson RL, Gangi AF. Effect of cracks on the pressure dependence of P wave velocities in crystalline rocks. J Geophys Res. 1985;90:8675–84.
Christensen NI. Pore pressure and oceanic crustal seismic structure. Geophys J Roy Astron Soc. 1984;79:411–23.
Christensen NI, Wang HF. The influence of pore pressure and confining pressure on dynamic elastic properties of Berea sandstone. Geophys. 1985;50:207–13.
Eshelby J. The determination of the elastic field of an ellipsoidal inclusion, and related problems. Proc R Soc Lond A. 1957;241:376–96.
Geertsma J. The effect of fluid pressure decline on volumetric changes of porous rocks. Petrol Trans AIME. 1957;210:331–40.
Hearmon RFS. The elastic constants of crystals and other anisotropic materials. In: Hellwege KH, Hellwege AM, editors. Landolt-Börnstein Tables, III/11. Berlin: Springer-Verlag; 1979. p. 854.
Johnson DL, Manning HJ. Theory of pressure dependent resistivity in crystalline rocks. J Geophys Res. 1986;91:11611–7.
Kariya KA, Shankland TJ. Electrical conductivity of dry lower crustal rocks. Geophys. 1983;48:52–61.
Kirkpatrick S. Percolation and conduction. Rev Modern Phys. 1973;45:574–88.
Lockner DA, Byerlee JD. Complex resistivity measurements of confined rock. J Geophys Res. 1985;90:7837–47.
Matsubara M, Hirata N, Sato H, Sakai S. Lower crustal fluid distribution in the northeastern Japan arc revealed by high-resolution 3D seismic tomography. Tectonophys. 2004;388:33–45.
Nesbitt BE. Electrical resistivities of crustal fluids. J Geophys Res. 1993;98:4301–10.
O'Connell RJ, Budiansky B. Seismic velocity in dry and saturated cracked solids. J Geophys Res. 1974;79:5412–26.
O'Connell RJ, Budiansky B. Viscoelastic properties of fluid-saturated cracked solids. J Geophys Res. 1977;82:5719–35.
Ogawa Y, Mishina M, Goto T, Satoh H, Oshiman N, Kasaya T, et al. Magnetotelluric imaging of fluids in intraplate earthquake zones, NE Japan back arc. Geophys Res Lett. 2001;28:3741–4.
Rutter EH. Pressure solution in nature, theory and experiment. J Geol Soc Lond. 1983;140:725–40.
Schmeling H. Numerical models on the influence of partial melts on elastic, anelastic and electrical properties of rocks. Part II: electrical conductivity. Phys Earth Planet Inter. 1986;43:123–36.

Sibson R. Rupturing in overpressured crust during compressional inversion—the case from NE Honshu, Japan. Tectonophys. 2009;473:404–16.

Takei Y. Effect of pore geometry on V_P/V_S: from equilibrium geometry to crack. J Geophys Res. 2002. doi:10.1029/2001JB000522.

Walsh JB. The effect of cracks on the compressibility of rock. J Geophys Res. 1965;70:381–9.

Watanabe T, Higuchi A. A new apparatus for measuring elastic wave velocity and electrical conductivity of fluid-saturated rocks at various confining and pore-fluid pressures. Geofluids. 2014;14:372–8.

Watanabe T, Katagishi Y. Deviation of linear relation between streaming potential and pore fluid pressure difference in granular material at relatively high Reynolds numbers. Earth Planets Space. 2006;58:1045–51.

Watanabe T, Shirasugi Y, Yano H, Michibayashi K. Seismic velocity in antigorite-bearing serpentinite mylonites. Geol Soc Lond Spec Pub. 2011;360:97–112.

Wong TF, Fredrich JT, Gwanmesia GD. Crack aperture statistics and pore space fractal geometry of Westerly granite and Rutland quartzite: implications for an elastic contact model of rock compressibility. J Geophys Res. 1989;94:10267–78.

Spatio-temporal changes in the seismic velocity induced by the 2011 Tohoku-Oki earthquake and slow slip event revealed from seismic interferometry, using ocean bottom seismometer's records

Miyuu Uemura[1*], Yoshihiro Ito[2], Kazuaki Ohta[2], Ryota Hino[3] and Masanao Shinohara[4]

Abstract

Seismic interferometry is one of the most effective techniques for detecting temporal variations in seismic velocity caused by large earthquakes. Before the 2011 Tohoku-Oki earthquake (M_w9.0) near the Japan Trench, a slow slip event (SSE, M_w7.0) and low-frequency tremors were observed near the trench. Here, we applied a seismic interferometry technique using ambient noise to data from 17 ocean bottom seismometers (OBSs) installed above the focal region before the main shock. We used our technique to detect temporal variations in seismic velocity caused by the main shock, SSE, and low-frequency tremors. In the region above the large coseismic slip area, we detected a 1–2% seismic velocity decrease after the main shock. In addition, we observed very small temporal increases in seismic velocity near the SSE fault during the initial SSE stage. Moreover, for most of the OBSs, we observed temporal variations in the autocorrelation functions (ACFs) during the low-frequency tremors. These may have been caused by temporal variations in the ambient noise source distributions, resulting from low-frequency tremors. These results suggest the possibility of detecting low-frequency tremors using ACF monitoring.

Keywords: Ambient noise, Autocorrelation function, Seismic interferometry, Slow slip event

Introduction

Seismic interferometry is one of the most powerful techniques for obtaining Green's functions. While seismic interferometry has a high temporal resolution (equivalent to analyses using an artificial source), it does not require an artificial source. Wegler et al. (2009) reported a decrease in seismic velocity of 0.5% after the 2004 mid-Niigata earthquake (M_w6.6) based on seismic interferometry applied to ambient noise. Similarly, other studies have reported decreases in seismic velocity accompanied by large earthquakes in several regions, such as Japan and Sumatra (e.g., Nimiya et al. 2017; Sawazaki et al. 2016; Takagi et al. 2012; Xu and Song 2009). A wide variety of techniques have been used in many regions, and most studies have reported variations in the seismic velocity structure accompanying large earthquakes. However, only a few studies have reported variations in the underground structure accompanying slow slip events (SSEs) (Rivet et al. 2011, 2014).

A number of studies have reported variations in seismic velocity of between 0.1 and 5% after a main shock and accompanying SSE. For example, Sawazaki et al. (2016) detected velocity decreases of 3.1% and 1.4% 1 week after the 2014 northern Nagano Prefecture earthquake (M_w6.2), and recovery to velocity decreases of 1.9% and 1.1% 4 weeks after the main shock, using KiK-net data and autocorrelation functions (ACFs). In contrast, Rivet et al. (2011) reported a velocity decrease of only 0.2% during the early part of the 2006 Guerrero SSE, and an almost complete velocity decrease recovery during the later part. Seismic velocity decreases immediately after an earthquake, and velocities recover in proportion to the time elapsed. However, during SSEs, changes are observed in

* Correspondence: uemura.miyuu.25c@st.kyoto-u.ac.jp
[1]Kyoto University, Gokasyo, Uji, Kyoto 611-0011, Japan
Full list of author information is available at the end of the article

seismic velocities, which decrease during the early parts of an SSE period and subsequently recover in the later parts.

The 2011 Tohoku-Oki earthquake (M_w9.0) occurred off the coast of mainland Japan on March 11, 2011, and prior slow earthquakes (M_w7.0) occurred off the coast of mainland Japan nearer the Japan trench about 1 month prior. Prior to the main shock, a clear preslip was not observed near the main shock's epicenter (Hirose 2011; Hino et al. 2013). An episodic SSE and low-frequency tremors were observed beginning 1 month before the earthquake in the region trenchward of the main shock's epicenter (Ito et al. 2013; Ito et al. 2015). The phase velocity of the Rayleigh wave decreased below 0.5%, accompanying the main shock at Tono, Iwate Prefecture (Takagi et al. 2014). Additionally, accompanying a series of earthquakes (including aftershocks), a seismic velocity decrease of approximately 2% was detected in southern Fukushima Prefecture (Minato et al. 2012). Moreover, seismic velocity decreases under the sea floor of 1–5% were reported after the earthquake (Ito and Hino 2013). However, variations in seismic velocity accompanying the preceding SSE have not been reported. Therefore, in this study, we used a seismic interferometry technique using ambient noise to data, and applied it to ocean bottom seismographs (OBSs). We aimed to detect temporal variations in the seismic velocity structure accompanying the SSE.

Methods/Experimental

2011 Tohoku SSE, analysis period, and data

From January 29 to March 9, 2011, an SSE and low-frequency tremors were detected offshore from Miyagi Prefecture, Japan (Ito et al. 2013; Ito et al. 2015; Katakami et al. 2018). Low-frequency tremors were observed during three periods:

Sequence 1. January 24–January 29

Sequence 2. February 16–February 20

Sequence 3. March 5–March 9

We used the vertical components of the continuous records of 17 OBSs from November 2010 to April 2011. These OBSs were installed offshore from Miyagi Prefecture before the SSE (Fig. 1), and some were installed above the SSE fault. The OBSs had a sampling rate of 100 Hz and eigen frequencies of 4.5 Hz. The shallowest OBS was located approximately 300 m below the sea surface, and the deepest was located approximately 4150 m below the sea surface (Table 1). The OBSs recorded wave fields with one vertical and two horizontal components, and almost all OBSs recorded data for more than 6 months. However, we used only the data for the vertical components, because the horizontal orientations at the bottom had not been precisely estimated. Moreover, we have not yet detected significant wave propagations between OBSs using Z–Z cross-correlation function (CCF). Therefore, we analyzed only the ACF in the Z–Z component.

Data analysis

Seismic interferometry is one of the most effective techniques for detecting temporal variations in seismic velocities caused by large earthquakes. Claerbout (1968) showed that a reflected wave (generated at an observation station on the earth's surface, reflected off a boundary under the station, and returned to the station) could be obtained by calculating the ACF from the waveform observed at the station. This technique is only effective if the structure around the site is a horizontal stratified structure. In the field of helioseismology, Duvall et al. (1993) proved that the Green's function between two stations can be obtained by calculating the CCF between the stations. In seismic interferometry, this technique is applied to the field of seismology. Since Campillo and Paul (2003) reported that a CCF constructed from teleseismic coda waves corresponded to the direct wave between two stations in the scattering field, many studies using seismic interferometry have been published (e.g., Hillers and Campillo 2016; Poli et al. 2012).

In seismic interferometry, a Green's function is created by assuming that one station is the hypocenter and another station is an observation station. This is achieved by calculating the CCF between the waveforms of the wave field at each station (e.g., Wapenaar et al. 2010). Assuming a noise source that is uniformly distributed around the stations (both temporally and spatially), the CCF ($r(\tau)$) between the two stations is calculated using the following formula:

$$r(\tau) = \frac{\int_{-\infty}^{\infty} s(t)u(t+\tau)dt}{\sqrt{\int_{-\infty}^{\infty} s(t)^2 dt \int_{-\infty}^{\infty} u(t)^2 dt}} \quad (1)$$

where $s(t)$ and $u(t)$ are the observation waveforms at the two stations. If seismic interferometry is applied to only one observation station (i.e., the ACFs of the station are calculated by replacing $u(t)$ with $s(t)$ in Eq. (1)), then a waveform can be derived by stacking the ACFs. The waveform is generated at the station, reflected by a boundary surface under the station, and returned to the station.

We removed the waveforms of ordinary earthquakes from the analysis, because apparent variations in ambient noise can affect the ACF and CCF values. These ambient noise variations are caused by temporal changes in background seismicity, such as variations in the hypocenter distribution of ambient noise. To exclude the effects of the spatio-temporal variations caused by the temporal changes in background seismicity, we applied weight functions based on a seismic coda wave shape for ordinary earthquakes (Katakami et al. 2017). Generally, regardless of the distance from the hypocenter, after the elapsed time (which is twice the travel time of an S wave), the envelope

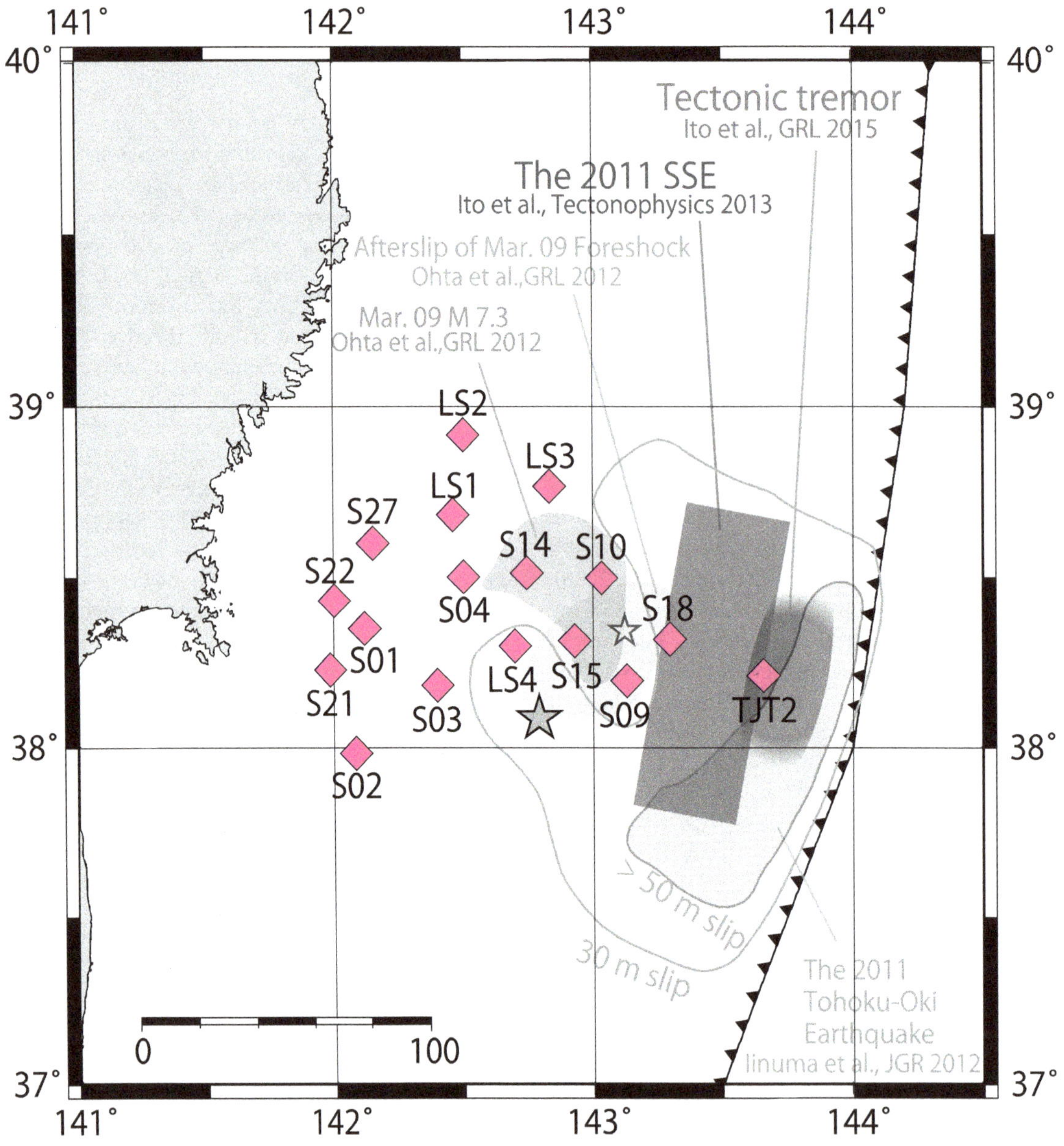

Fig. 1 Locations of OBSs. Location details are presented in Table 1

form $s(t)$ of a seismic coda wave satisfies the following formula:

$$s(t) \propto exp\left[{}^{-\pi ft}\Big/_{Q}\right] \quad (2)$$

where t is the time elapsed since the earthquake, and f and Q are the frequency and attenuation factors within the region, respectively (Rautian and Khalturin 1978). Here, we use a simplified function from Eq. (2) and assume an attenuation factor of 50, which was determined experimentally:

$$s_S(t) = exp\left({}^{-t}\Big/_{50}\right) \quad (3)$$

In addition, we calculated the envelope of the observed seismogram $s_O(t)$ from one vertical and two horizontal components of the continuous record for 1 day:

$$s_O(t) = \sqrt{{}^{\sum_{i=1}^{3} {s_i}^2}\Big/_{3}} \quad (4)$$

Furthermore, we calculated the coefficient $c(t)$ of the cross-correlation between the observed envelope $s_S(t)$ and the calculated envelope $s_O(\tau)$ using a moving time

Table 1 Location and recording period of each OBS. We measured the locations of LS1 and S03 when they were collected, and the locations of other stations when they were set up

Station name	Location of OBSs			Record period
	Longitude (°E)	Latitude (°N)	Depth from sea surface (m)	
LS1 (Additional file 1, Additional file 2)	142.46059	38.68408	1112	November 5, 2010–April 13, 2011
LS2 (Additional file 3, Additional file 4)	142.49997	38.91679	1194	November 5, 2010–October 6, 2011
LS3 (Additional file 5, Additional file 6)	142.83307	38.76621	1403	November 5, 2010–October 6, 2011
LS4 (Additional file 7, Additional file 8)	142.69956	38.29971	1409	November 5, 2010–October 6, 2011
S01 (Additional file 9, Additional file 10)	142.11688	38.35023	524	November 5, 2010–May 21, 2011
S02 (Additional file 11, Additional file 12)	142.08274	37.98356	538	November 5, 2010–May 21, 2011
S03 (Additional file 13, Additional file 14)	142.39969	38.18343	1052	November 5, 2010–April 13, 2011
S04 (Additional file 15, Additional file 16)	142.50041	38.50208	1100	October 1, 2010–July 12, 2011
S09 (Figure 2, Figure 3)	143.13214	38.19742	2041	October 1, 2010–July 12, 2011
S10 (Figure 4, Additional file 17)	143.03415	38.49841	1981	July 1, 2010–March 18, 2011
S14 (Additional file 18, Additional file 19)	142.7457	38.51379	1459	July 1, 2010–May 22, 2011
S15 (Additional file 20, Additional file 21)	142.9276	38.31377	1454	July 1, 2010–May 22, 2011
S18 (Additional file 22, Additional file 23)	143.29608	38.31918	2770	July 1, 2010–March 18, 2011
S21 (Figure 3, Additional file 24)	142.00192	38.4319	358	November 5, 2010–May 21, 2011
S22 (Additional file 25, Additional file 26)	141.98375	38.22922	299	November 5, 2010–May 21, 2011
S27 (Additional file 27, Additional file 28)	142.15009	38.60032	545	November 5, 2010–May 21, 2011
TJT2 (Additional file 29, Additional file 30)	143.65558	38.21309	4147	November 19, 2010–May 23, 2011

window with a length of 300 s. We applied the weight $w(t)$ to each time window for the ACF calculations, according to the cross-correlation coefficient as follows:

$$w(t) = \begin{cases} 0\ (c(t) > 0.5) \\ 1\ (c(t) \leq 0.5) \end{cases} \quad (5)$$

Generally, the coefficient $c(t)$ represents the correlation between the observed and calculated envelopes only after an elapsed time that is twice the travel time of the S wave. In addition, we also applied a weight of 0 before 300 s if the coefficient exceeded 0.5, because the 600 s time window included seismic waves radiating from the ordinary earthquake.

In this study, we investigate spatio-temporal variations of three functions: ACFs of ambient noise, cross-correlation coefficients of the CCFs between a 15-day ACF and a reference ACF, and phase shift based on the CCFs. These values are calculated as shown below.

First, all observation data were corrected for the instrument response before applying the band-pass filter. Then, we applied a band-pass filter at 0.25–2.0 Hz and a one-bit technique to the observed continuous data. These filters were applied to prevent the continuous data from being affected by (1) unexpected signals (such as micro-earthquakes) that could not be removed with the aforementioned weight function, or (2) mechanical or biological noise at the site.

Next, after applying the band-pass filter and one-bit technique at intervals of 0.1 s, we calculated 120 s ACFs using 5 s moving time windows. The coefficient for the time including earthquakes equaled 0 after multiplying the weight $w(t)$ of the time window, and the earthquake effects were removed. The 15-day ACF was defined as the ensemble average of the 120 s ACFs calculated over 15 days. The reference ACF was defined as the ensemble average of 120 s ACFs calculated over a month from November 19 to December 19, the period before the occurrence of SSE and low-frequency tremors.

In addition, we calculated the CCF using a 15 s moving time window between the 15-day ACF and reference ACF with a lag time of − 2 s to 2 s, at intervals of 0.1 s, and estimated the variation in seismic velocity based on the lag time at the maximum CCF. Finally, we objectively evaluated the temporal variations in the 15-day ACFs based on the variations in the cross-correlation coefficient at zero lag time. The shape changes of the 15-day ACFs cause the coefficients to decrease. When a velocity change occurs, the phase of the 15-day ACF delays or progresses, and the shape changes. Even when a velocity change does not occur, it is possible for the shape to change, and a variation in the hypocenter distribution of the ambient noise may explain this shape changes. Therefore, we used the coefficient at zero lag time to detect the temporal variations in the 15-day ACFs caused by, not only velocity change, but also temporal variation of the ambient noise.

Results

Temporal variations in ACF

We detected several temporal variations accompanying the 2011 Tohoku-Oki earthquake, SSE, and low-frequency tremors. The cross-correlation coefficient at zero lag time between the 15-day and reference ACFs during later lapse times decreased after the main shock (Fig. 2b). This decrease was clearer at trenchward stations (e.g., S09 (Fig. 2b) and S10 (Fig. 4b)) than landward stations (e.g., S21 (Fig. 3b)), and continued until the end of the observation period without the full recovery of the pre-main shock coefficient value. At nine stations (TJT2 (Additional file 29(b)), S15 (Additional file 20(b)), S14 (Additional file 18(b)), S09, S04 (Additional file 15(b)), LS4 (Additional file 7(b)), LS3 (Additional file 5(b)), LS2 (Additional file 3(b)), and LS1 (Additional file 1(b))), we detected decreases in the coefficient between the 15-day and reference ACFs after the main shock. We excluded landward stations, such as S21 (Fig. 3b), and stations collected 1 week after the main shock, such as S10 (Fig. 4b).

The cross-correlation coefficient around a lapse time of 10 s decreased after sequence 1 (Fig. 2b). A decrease in the coefficient was detected at most stations installed east of 142.5° E and appeared before sequence 1 at 60% of these stations, such as S10 (Fig. 4b). At all stations, the decreases recovered after 10 days. Similar coefficient variations were detected during sequences 2 and 3, but the decreases were less than that during sequence 1. During SSE, some sustained variations in cross-correlation coefficients were not detected.

Temporal variations in seismic velocity based on variations in the 15-day ACF

We measured phase delays and progressions in 15-day ACFs after the main shock. If the phase of the 15-day ACF was delayed from the phase of the reference ACF, we called this "phase delay," and if the phase of the 15-day ACF proceeded from the phase of the reference ACF, we called this "phase progression." At station S09, the phase delay during the later lapse time was larger than it was during the early lapse time (Fig. 2c). Velocity changes $\frac{dv}{v}$ around the site were calculated from phase changes using the following formula:

$$\frac{dv}{v} = -\frac{dT}{T}, \quad (6)$$

where T is the lapse time and dT is the phase progression at a lapse time of T s (Snieder et al. 2002). For example, the phase delay between a lapse time of 40 s and 50 s was approximately equivalent to a 2% velocity decrease (Fig. 5). This decrease was detected in multiple lapse times. While the velocity recovered slightly over time after the main shock, it did not recover completely by the end of the study period for all lapse times. Similar velocity decreases were detected at five stations near the trench (TJT2 (Additional file 30), S15 (Additional file 21), S04 (Additional file 16), LS1 (Additional file 2), and LS4 (Additional file 8)). These post-main-shock decreases were approximately 1–2% and did not recover completely. An OBS was installed closer to the epicenter of the main shock, and greater velocity decreases were detected.

In addition, we also calculated the velocity changes accompanying the SSE and low-frequency tremors from the phase changes. However, we did not detect obvious changes. Although velocities changed by a few tenths of a percent during the SSE at several stations, such as S09 and S15, the changes were equivalent to those in a normal period when no SSE, no low-frequency tremors, and no large earthquakes occurred (Fig. 5). At all stations, the velocity changes accompanying the SSE and low-frequency tremors were smaller than the velocity fluctuations in the normal period.

Regional characteristics of temporal variations in 15-day ACF

Next, we assessed decreases in the cross-correlation coefficient between the 15-day ACFs and the reference ACF at zero lag time around a lapse time of 10 s during the occurrence of low-frequency tremors. We calculated the average coefficients for the seven periods, shown in Table 2, using a 15 s moving time window from a lapse time of 2.5 s to 17.5 s (Fig. 6).

At the six landward stations (S27, S22, S21, S02, S01, and LS1), the variations for each period were extremely small. This indicates that the cross-correlation coefficients at the six stations were stable in all periods (hereafter region L). However, at the other 11 stations, clear variations were observed for each period. For example, the variation from period II to period III (Fig. 6b) was equivalent to the variation induced by a low-frequency tremor occurrence. A low-frequency tremor did not occur in period II but did occur in period III. At the 11 stations, the variations basically decreased when low-frequency tremors occurred and increased (recovered) after low-frequency occurrences.

The 11 stations were classified into 2 groups. In one group, the coefficient decreased primarily in period II. The eight stations (S15, S14, S10, S04, S03, LS4, LS3, and LS2) located at the center of the OBS array (hereafter Region C) belong to this first group. In another group, the coefficient decreased primarily in period III. The three stations (TJT2, S18, and S09) located near the trench (hereafter Region T) belong to this second group. Regions C and T correspond roughly to the slip area of the largest foreshock and the SSE occurrence area mentioned in Ito et al. (2013), respectively.

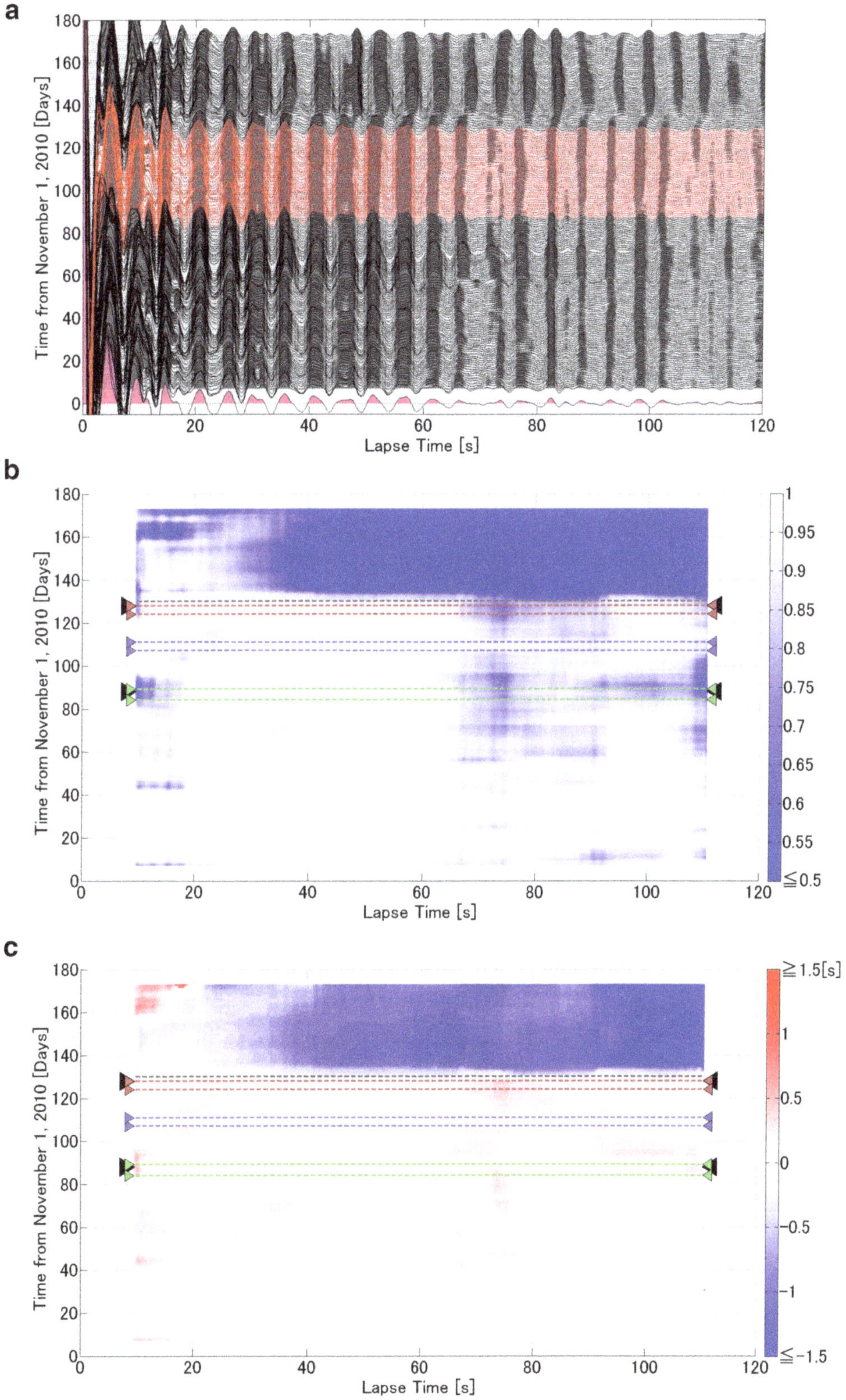

Fig. 2 (See legend on next page.)

(See figure on previous page.)

Fig. 2 Temporal variations in 15-day ACFs at station S09. **a** 15-day ACFs calculated from OBS data over 15 continuous days, including the 7 days before and after the given date. The ACFs plotted in red are in the SSE period, and the bottom ACF is the reference ACF, calculated for the month from November 19 to December 19. **b** The cross-correlation coefficient between the 15-day ACFs and the reference ACF, using a 15-s time window and zero lag time. The period between the black triangles corresponds to the SSE period, and the periods between the red, blue, and green triangles correspond to low-frequency tremor periods in sequences 1, 2, and 3. The magenta broken line represents the main shock. **c** Phase variation in the 15-day ACFs relative to the reference ACF, using a 15-s time window. Warm colors represent phase progressions, and cold colors represent phase delays. The four pairs of triangles and the magenta broken line correspond to each period and the main shock described in (**b**)

Discussion

Temporal variations in seismic velocity and the 15-day ACF

The observed phase delays after the main shock were larger during the later lapse times for the 15-day ACFs than during the earlier lapse times (Fig. 2c). When velocity structures change equally, the travel time increases monotonically for the wave transmitted in a region in proportion with the propagation distance, resulting in an increase in the phase delay during the latter part of the 15-day ACF. In addition, the phase in the latter part of the 15-day ACF carries information regarding waves that have traveled longer distances because of reflection and refraction around the observation station. Moreover, it is possible for the velocity to change equally around a site after the main shock. At S09, the equal velocity decreased after the main shocks were detected for the lapse times of 30–40 s, 40–50 s, 50–60 s, and 60–70 s (Fig. 5). This suggests that a wave reflected many times under S09 reaches this observation station after 60 s. In contrast, the phase delay at TJT2 appeared only for part of the lapse time (Additional file 1), suggesting that the velocity structure did not change equally, but rather in part of the region.

At S09, although phase changes appeared for only a short section around a lapse time of 10 s during sequences 1, 2, and 3 (Fig. 2c), we did not detect a velocity change corresponding to the phase changes (Fig. 5). If the phase changes are caused by an increase in the seismic velocity in part of the region, a phase (or velocity) change several times larger than a phase change at a lapse time of 10 s should appear when the wave transmitted along the same path several times returns to the station. However, such a phase change was not observed. Therefore, the phase changes were not caused by variations in the velocity structure, and it is possible that variations in the hypocenter distribution of ambient noise induced by the low-frequency tremors cause a phase change similar to that caused by a structural change. The phase progression during sequence 1 continued longer after the end of the low-frequency tremor activity and was larger than those during sequences 2 and 3. This phase progression probably included the dummy structural variation accompanying the beginning of the SSE, in addition to the dummy structural variation caused by the variation in the hypocenter distribution accompanying the low-frequency tremors.

When the density of a crack near a hypocenter increases because of the formation of cracks before an earthquake, seismic velocity is expected to decrease around the hypocenter. Several studies using rocks have reported a decrease in the elastic wave velocity accompanying an increase in crack density (e.g., Nishizawa and Kanagawa 2005). According to Lockner et al. (1991), a decrease in elastic wave velocity and decrease in amplitude were detected before a rock sample broke. This was accompanied by the creation of micro-cracks formed by acoustic emissions. Therefore, we speculate that the observed phase delays and decreases in cross-correlation coefficients may have been caused by the creation of micro-cracks. We detected decreases in velocity before the main shock, and noted that these velocity decreases began earlier at the observation station nearer the hypocenter of the main shock. However, we detected no velocity changes before the SSE around the SSE region. This suggests the possibility that the crack density varied little before the SSE, although the density varied enough to change the 15-day ACF before the main shock. Overall, the results suggest that the mechanism of structural change in seismic velocity accompanying an SSE detected by seismic interferometry may differ from that associated with an earthquake.

Spatio-temporal variations in cross-correlation coefficients at each station

We classified the temporal variations in the cross-correlation coefficients around a lapse time of 10 s at all stations (Fig. 6) into three groups: region T, region C, and region L. These three regions corresponded to three slip areas: region L corresponded to the area that slipped only during the main shock (Iinuma et al. 2012), region C corresponded to the slip area of the largest foreshock (Ohta et al. 2012), and region T corresponded to the SSE slip area (Ito et al. 2013). In "Temporal variations in seismic velocity and the 15-day ACF" section, we indicated the possibility that the phase changes around a lapse time of 10 s were caused by the occurrence of low-frequency tremors. The cross-correlation coefficient between the 15-day ACF and the reference ACF at zero lag time decreases if the phase of the 15-day ACF changes. Although the coefficients decreased during sequences 1, 2, and 3, the time when the decrease of sequence 1 began was different at each

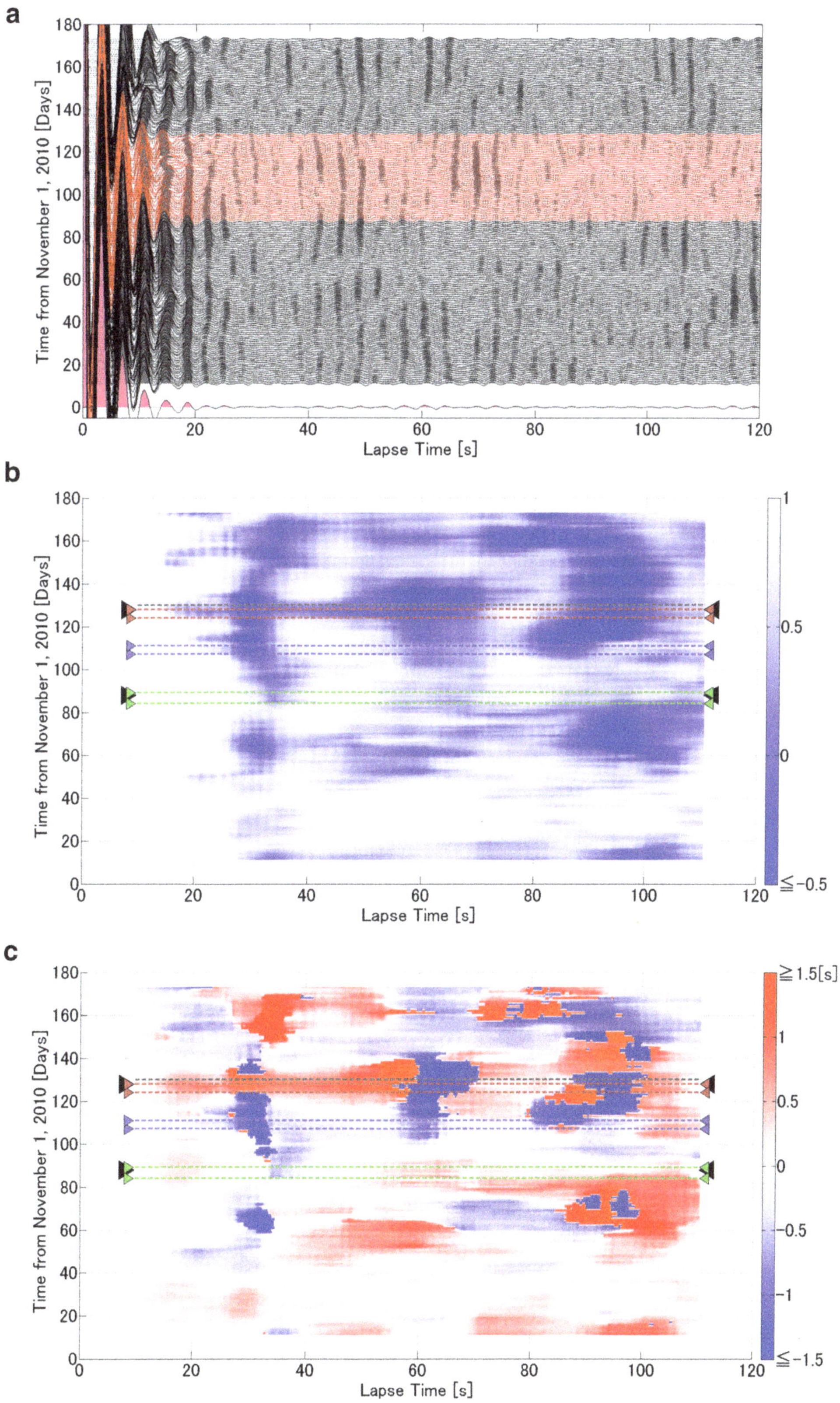

Fig. 3 (See legend on next page.)

(See figure on previous page.)

Fig. 3 Temporal variations in 15-day ACFs at station S21. **a** 15-day ACFs calculated from OBS data over 15 continuous days, including the 7 days before and after the given date. The ACFs plotted in red are in the SSE period, and the bottom ACF is the reference ACF, calculated for the month from November 19 to December 19. **b** The cross-correlation coefficient between the 15-day ACFs and the reference ACF, using a 15-s time window and zero lag time. The period between the black triangles corresponds to the SSE period, and the periods between the red, blue, and green triangles correspond to low-frequency tremor periods in sequences 1, 2, and 3. The magenta broken line represents the main shock. **c** Phase variation in the 15-day ACFs relative to the reference ACF, using a 15-s time window. Warm colors represent phase progressions, and cold colors represent phase delays. The four pairs of triangles and the magenta broken line correspond to each period and the main shock described in (**b**)

observation station. In region C, where coefficient decreases started from period II before sequence 1, it is possible that low-frequency tremors occurred during period II, and that the 15-day ACF changed because of variations in the hypocenter distribution. In agreement, Katakami et al. (2018) reported that micro-low-frequency tremors occurred near region C in the same period. Therefore, the coefficient decreases during period II were caused by the occurrence of a low-frequency tremor.

Meanwhile, the widths of the lapse time section, where the coefficients decreased substantially, were different in during period II and during low-frequency tremor periods (III, V, and VII). The coefficients during period II at the stations installed in region C decreased substantially in a wide section between lapse times of 10 s and 20 s (Fig. 4b). However, the coefficients during periods V and VII at the stations installed in region C decreased substantially in narrow sections around a lapse time of 10 s. The coefficients during period III at stations installed in region T also decreased largely in a narrow range, around a lapse time of 10 s (Fig. 2b). The width of the lapse time section during period II was different from the widths of the lapse-time section during periods V and VII at the same station installed in region C. In addition, although the stations were different, the width of the lapse time section during period II at the station installed in region C was different from the widths of the lapse time section during period III at stations installed in region T. This suggests that the decrease during period II was unique in terms of the width of the lapse time section where the coefficients decreased substantially compared to the coefficient decreases for the section widths during periods III, V, and VII. Therefore, it is possible that the decrease during period II occurred not only because of low-frequency tremors, as detected by Katakami et al. (2018), but also because of an additional event, for example, the preparation process of the 2011 Tohoku-Oki earthquake.

Estimated velocity reductions during 2011 Tohoku-Oki earthquake and SSE

The decrease in velocity under an observation station represents one interpretation of the cause of a phase delay. When a wave reflects off a surface and returns to the station, a phase delay is observed if the seismic wave velocity decreases along the propagation path. The delay can also be explained by an increase in the distance between an observation station and a reflection surface, because the propagation distance of the reflected wave is determined by the distance between an observation station and a reflection surface. Assuming that the distance to the reflection surface did not change, the observed phase delays after the main shock were equivalent to a 2% or lower seismic velocity decrease at six stations (TJT2, S15, S09, S04, LS4, and LS1) near the trench. Ito and Hino (2013) studied the sea floor in the same region as the present study, and detected decreases in seismic velocity of 1–5% after the main shock in the overall analysis region. Moreover, a decrease in seismic velocity of 2% in the southern Fukushima Prefecture (Minato et al. 2012) and 0.1–0.5% in the Iwate Prefecture (Takagi et al. 2012) were reported after the main shock using land data. In the present study, the observed seismic velocity decreases were lower than those of Ito and Hino (2013), but roughly equivalent to those of other regions that shook during the main shock.

In the SSE period, we detected a perturbation in seismic velocity of a few tenths of a percent around the SSE area. At TJT2, S18 (Additional file 23), and S09, which were installed near the SSE region, the velocity changes calculated from the phase changes at lapse times of 20–30 s, 30–40 s, and 40–50 s increased slightly during the initial SSE stage, and did not change after that stage (e.g., Fig. 5). Two studies of the 2006 SSE and 2009 SSE at Guerrero, Mexico, detected decreases and recoveries in seismic velocity during the SSEs using data from stations landward of the SSE slip regions (Rivet et al. 2011, 2014). The researchers proposed that the velocity decreased by 0.8% in the former part of the SSE period and recovered almost completely in the latter part. In this study, increases in velocity were observed in the former part of the SSE period using data from both landward and trenchward stations. However, after that, we did not observe an additional decrease (recovery) during the observation period. We observed a difference between the two Guerrero SSEs and the Tohoku-Oki SSE with respect to whether or not any large earthquakes occurred at the plate boundary immediately after the SSEs. The Guerrero SSEs concluded without any large earthquakes, but the Tohoku-Oki SSE concluded with large earthquakes with foreshocks, a

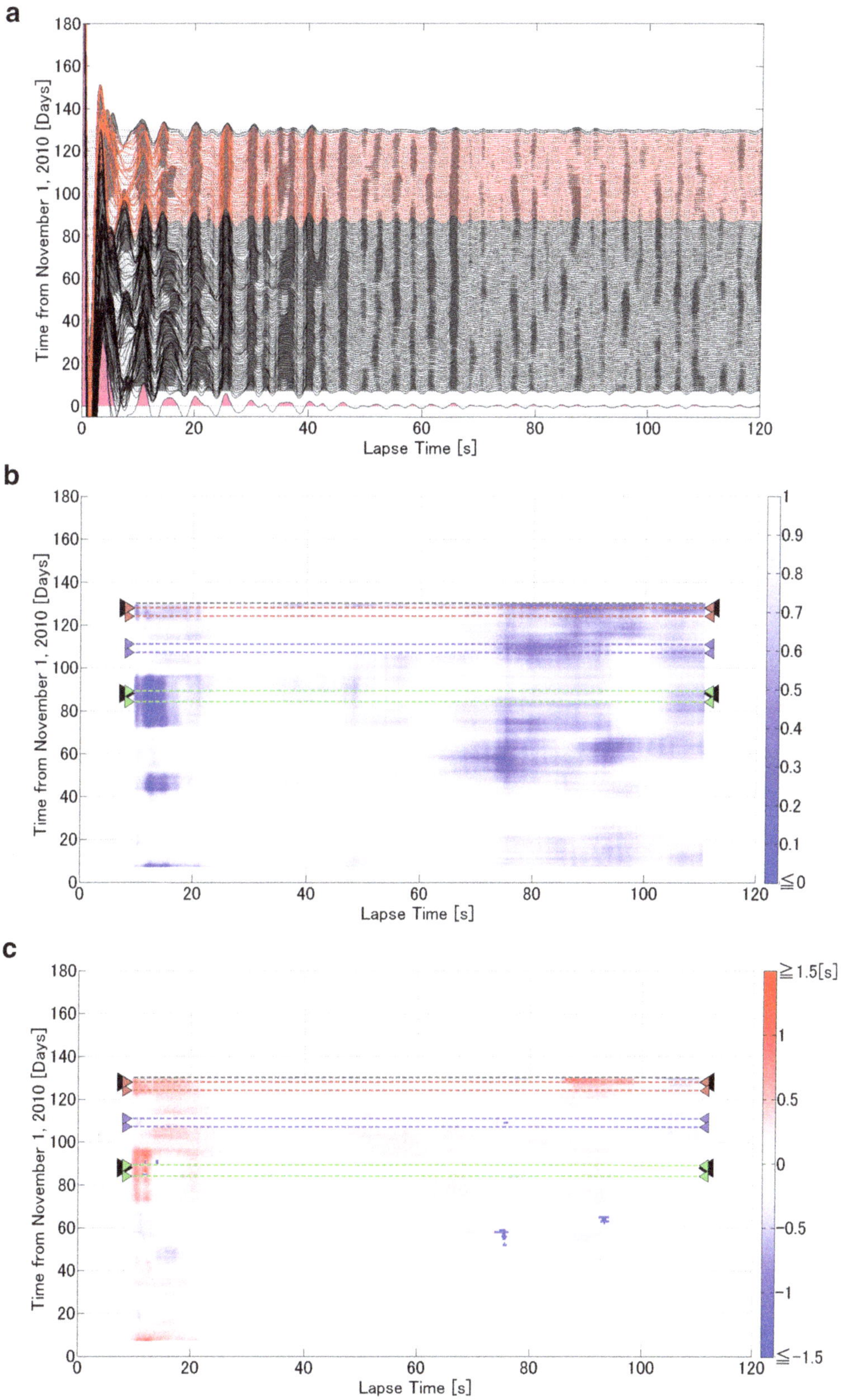

Fig. 4 (See legend on next page.)

(See figure on previous page.)

Fig. 4 Temporal variations in 15-day ACFs at station S10. **a** 15-day ACFs calculated from OBS data over 15 continuous days, including the 7 days before and after the given date. The ACFs plotted in red are in the SSE period, and the bottom ACF is the reference ACF, calculated for the month from November 19 to December 19. **b** The cross-correlation coefficient between the 15-day ACFs and the reference ACF, using a 15-s time window and zero lag time. The period between the black triangles corresponds to the SSE period, and the periods between the red, blue, and green triangles correspond to low-frequency tremor periods in sequences 1, 2, and 3. The magenta broken line represents the main shock. **c** Phase variation in the 15-day ACFs relative to the reference ACF, using a 15-s time window. Warm colors represent phase progressions, and cold colors represent phase delays. The four pairs of triangles and the magenta broken line correspond to each period and the main shock described in (**b**)

mainshock, and aftershocks. Because SSEs were detected by GPS data, if another crustal deformation, such as a large earthquake, occurs during an SSE, it is possible that the timing of the end of the SSE is not detected clearly. Moreover, because earthquakes release stress near the hypocenter, an occurrence of large earthquakes ends SSEs that occur near the hypocenter. It is highly likely that the Guerrero SSEs were observed from beginning to end completely because no large earthquakes occurred during the Guerrero SSEs. On the other hand, it is highly likely that the 2011 Tohoku-Oki earthquake occurred before the Tohoku-Oki SSE finished, and it is possible that the Tohoku-Oki SSE was ended suddenly by the 2011 Tohoku-Oki earthquake. If the 2011 Tohoku-Oki earthquake ended the Tohoku-Oki SSE before the velocity recovery period in SSE, it is not surprising that we could not detect velocity recovery in the latter part of the Tohoku-Oki SSE.

Conclusion

We detected decreases in seismic velocity of no more than 2% at six stations near the hypocenter of the 2011 Tohoku-Oki earthquake. These decreases were detected after the main shock using seismic interferometry with ambient noise. The largest velocity reduction was observed at S09, located between the hypocenter of the main shock and the slip region of the SSE before the main shock.

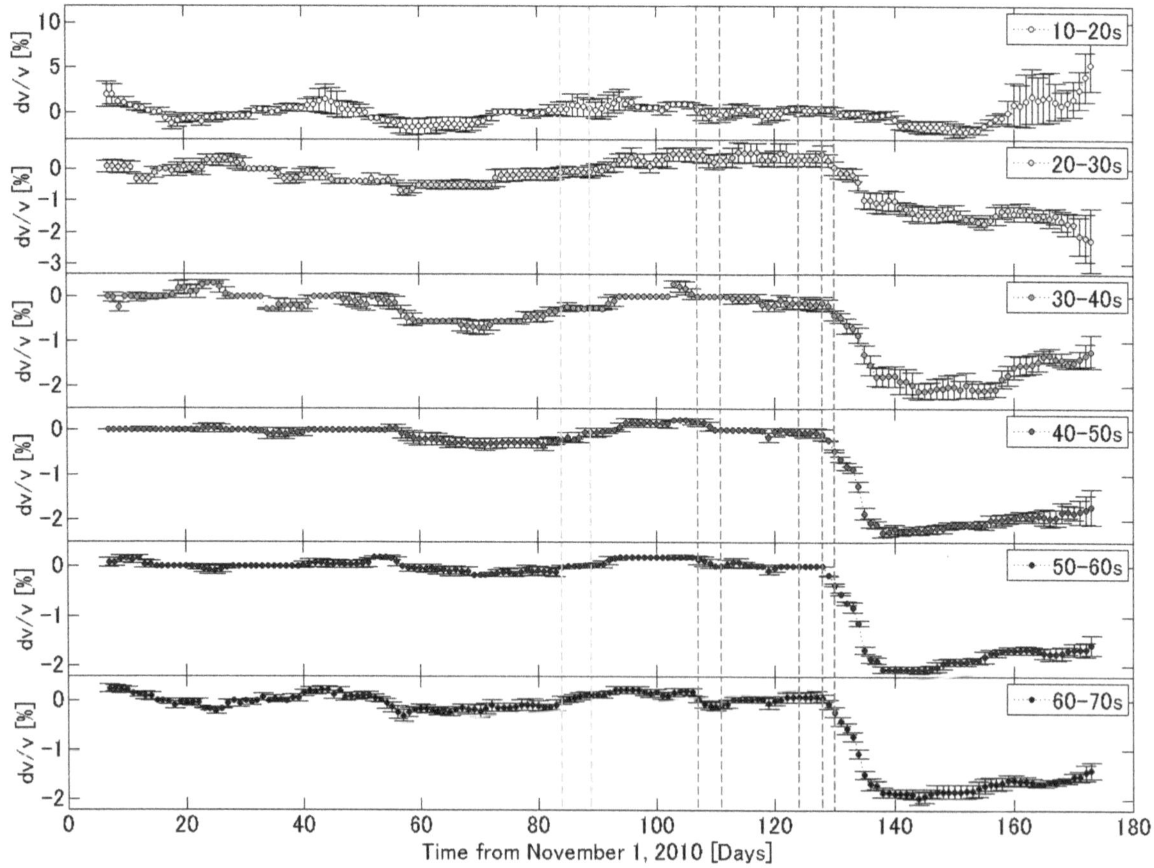

Fig. 5 Temporal variations in seismic velocity at S09. These figures show the temporal velocity changes calculated from the phase delays at each lapse time: from 10–20 s, 20–30 s, 30–40 s, 40–50 s, 50–60 s, and 60–70 s. The four colors of broken lines correspond to the occurrence of each low-frequency tremor and to the main shock

Table 2 Periods used for calculation of variations in cross-correlation coefficients

	Period
I	November 26, 2010–December 14, 2010
II	January 9, 2011–January 23, 2011
III (sequence 1)	January 24, 2011–January 29, 2011
IV	February 6, 2011–February 15, 2011
V (sequence 2)	February 16, 2011–February 20, 2011
VI	February 22, 2011–March 3, 2011
VII (sequence 3)	March 5, 2011–March 9, 2011

In addition, we detected temporal perturbations in the seismic velocity during the SSE at three stations around the SSE region. The velocities increased a few tenths of a percent immediately after the SSE began and did not change after that. Velocity decreases in the early parts of SSEs and recoveries in the latter parts of SSEs were reported in previous studies; however, we did not detect any velocity recoveries in the latter period.

We observed variations in the 15-day ACFs before a lapse time of 15 s at most stations installed east of 142.5° E. These variations were accompanied by low-frequency tremors, although no variations in seismic velocity were detected. Specifically, the 15-day ACFs showed substantial, long-lasting variations during sequence 1. The variations during sequence 1 included not only the effects of the low-frequency tremors but also the effects of the SSE. All stations were classified based on their characteristics when variations in the 15-day ACFs were observed. Variations in the 15-day ACFs at stations installed between 142.5° E and 143° E were observed 15 days before sequence 1, and variations at stations installed east of 143° E near the trench were observed during sequence 1. These regions roughly corresponded to the slip areas of the largest foreshock and SSE. Moreover, the variations in the 15-day ACFs indicated the possibility that low-frequency tremors occurred before sequence 1 between 142.5° E and 143° E.

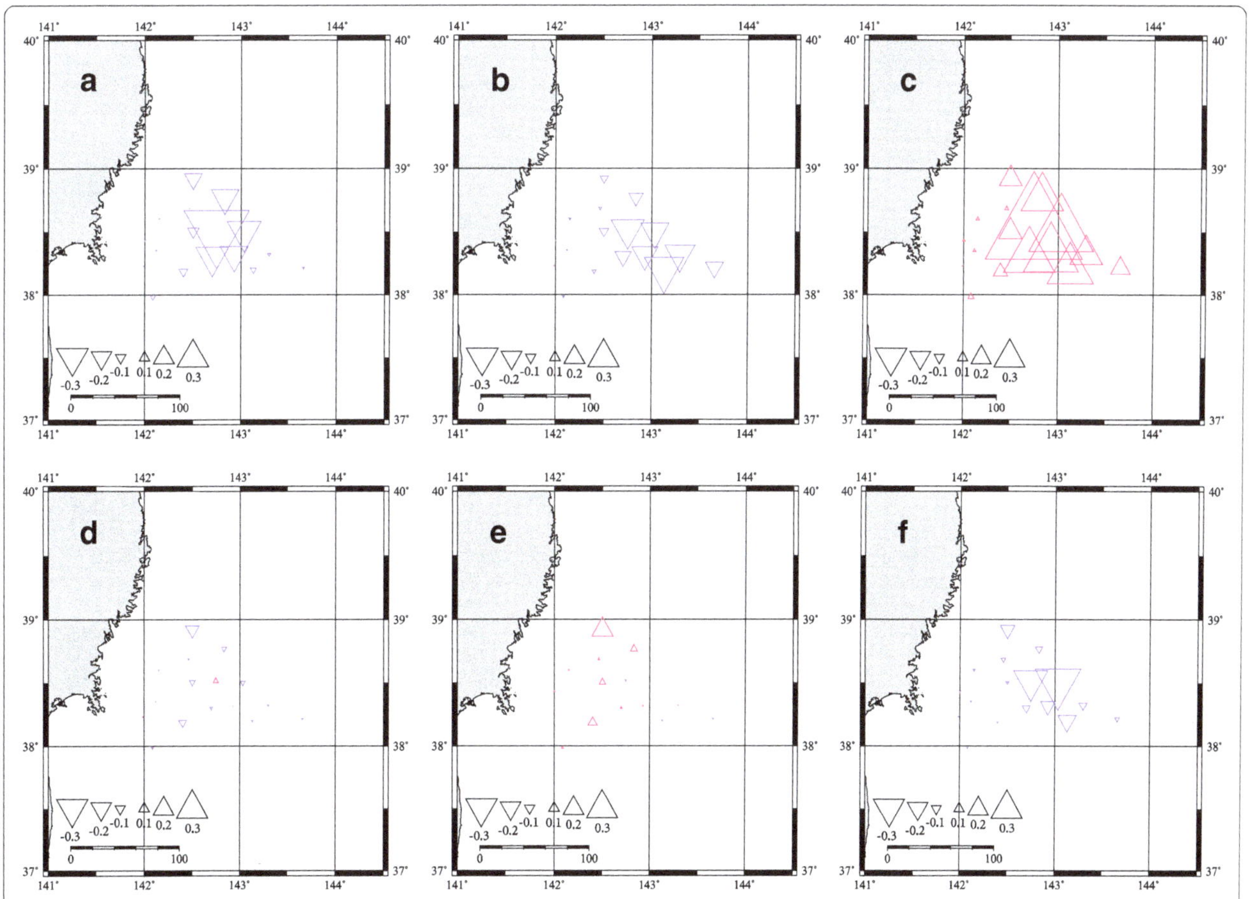

Fig. 6 Variations in cross-correlation coefficient averages for each period at a lag time of zero between the 15-day ACFs and the reference ACF, using a 15-s time window from 2.5 to 17.5 s at each OBS. These figures show variations in the cross-correlation coefficient averages **a** from period I to period II, **b** from period II to period III, **c** from period III to period IV, **d** from period IV to period V, **e** from period V to period VI, and **f** from period VI to period VII. Details of these periods are presented in Table 2. A red upward triangle represents an increase in the cross-correlation coefficient, and a blue downward triangle represents a decrease

Additional files

Additional file 1: Temporal variations in 15-day ACFs at Station LS1. (a) 15-day ACFs calculated from OBS data over 15 continuous days, including the 7 days before and after the given date. The ACFs plotted in red are in the SSE period, and the bottom ACF is the reference ACF, calculated for the month from November 19 to December 19. (b) The cross-correlation coefficient between the 15-day ACFs and the reference ACF, using a 15 s time window and zero lag time. The period between the black triangles corresponds to the SSE period, and the periods between the red, blue, and green triangles correspond to low-frequency tremor periods in Sequences 1, 2, and 3. The magenta broken line represents the main shock. (c) Phase variation in the 15-day ACFs relative to the reference ACF, using a 15 s time window. Warm colors represent phase progressions, and cold colors represent phase delays. The four pairs of triangles and the magenta broken line correspond to each period and the main shock described in (b).

Additional file 2: Temporal variations in seismic velocity at LS1. These figures show the temporal velocity changes calculated from the phase delays at each lapse time: from 10 to 20 s, 20–30 s, 30–40 s, 40–50 s, 50–60 s, and 60–70 s. The four colors of broken lines correspond to the occurrence of each low-frequency tremor and to the main shock.

Additional file 3: Temporal variations in 15-day ACFs at Station LS2. (a) 15-day ACFs calculated from OBS data over 15 continuous days, including the 7 days before and after the given date. The ACFs plotted in red are in the SSE period, and the bottom ACF is the reference ACF, calculated for the month from November 19 to December 19. (b) The cross-correlation coefficient between the 15-day ACFs and the reference ACF, using a 15 s time window and zero lag time. The period between the black triangles corresponds to the SSE period, and the periods between the red, blue, and green triangles correspond to low-frequency tremor periods in Sequences 1, 2, and 3. The magenta broken line represents the main shock. (c) Phase variation in the 15-day ACFs relative to the reference ACF, using a 15 s time window. Warm colors represent phase progressions, and cold colors represent phase delays. The four pairs of triangles and the magenta broken line correspond to each period and the main shock described in (b).

Additional file 4: Temporal variations in seismic velocity at LS2. These figures show the temporal velocity changes calculated from the phase delays at each lapse time: from 10 to 20 s, 20–30 s, 30–40 s, 40–50 s, 50–60 s, and 60–70 s. The four colors of broken lines correspond to the occurrence of each low-frequency tremor and to the main shock.

Additional file 5: Temporal variations in 15-day ACFs at Station LS3. (a) 15-day ACFs calculated from OBS data over 15 continuous days, including the 7 days before and after the given date. The ACFs plotted in red are in the SSE period, and the bottom ACF is the reference ACF, calculated for the month from November 19 to December 19. (b) The cross-correlation coefficient between the 15-day ACFs and the reference ACF, using a 15 s time window and zero lag time. The period between the black triangles corresponds to the SSE period, and the periods between the red, blue, and green triangles correspond to low-frequency tremor periods in Sequences 1, 2, and 3. The magenta broken line represents the main shock. (c) Phase variation in the 15-day ACFs relative to the reference ACF, using a 15 s time window. Warm colors represent phase progressions, and cold colors represent phase delays. The four pairs of triangles and the magenta broken line correspond to each period and the main shock described in (b).

Additional file 6: Temporal variations in seismic velocity at LS3. These figures show the temporal velocity changes calculated from the phase delays at each lapse time: from 10 to 20 s, 20–30 s, 30–40 s, 40–50 s, 50–60 s, and 60–70 s. The four colors of broken lines correspond to the occurrence of each low-frequency tremor and to the main shock.

Additional file 7: Temporal variations in 15-day ACFs at Station LS4. (a) 15-day ACFs calculated from OBS data over 15 continuous days, including the 7 days before and after the given date. The ACFs plotted in red are in the SSE period, and the bottom ACF is the reference ACF, calculated for the month from November 19 to December 19. (b) The cross-correlation coefficient between the 15-day ACFs and the reference ACF, using a 15 s time window and zero lag time. The period between the black triangles corresponds to the SSE period, and the periods between the red, blue, and green triangles correspond to low-frequency tremor periods in Sequences 1, 2, and 3. The magenta broken line represents the main shock. (c) Phase variation in the 15-day ACFs relative to the reference ACF, using a 15 s time window. Warm colors represent phase progressions, and cold colors represent phase delays. The four pairs of triangles and the magenta broken line correspond to each period and the main shock described in (b).

Additional file 8: Temporal variations in seismic velocity at LS4. These figures show the temporal velocity changes calculated from the phase delays at each lapse time: from 10 to 20 s, 20–30 s, 30–40 s, 40–50 s, 50–60 s, and 60–70 s. The four colors of broken lines correspond to the occurrence of each low-frequency tremor and to the main shock.

Additional file 9: Temporal variations in 15-day ACFs at Station S01. (a) 15-day ACFs calculated from OBS data over 15 continuous days, including the 7 days before and after the given date. The ACFs plotted in red are in the SSE period, and the bottom ACF is the reference ACF, calculated for the month from November 19 to December 19. (b) The cross-correlation coefficient between the 15-day ACFs and the reference ACF, using a 15 s time window and zero lag time. The period between the black triangles corresponds to the SSE period, and the periods between the red, blue, and green triangles correspond to low-frequency tremor periods in Sequences 1, 2, and 3. The magenta broken line represents the main shock. (c) Phase variation in the 15-day ACFs relative to the reference ACF, using a 15 s time window. Warm colors represent phase progressions, and cold colors represent phase delays. The four pairs of triangles and the magenta broken line correspond to each period and the main shock described in (b).

Additional file 10: Temporal variations in seismic velocity at S01. These figures show the temporal velocity changes calculated from the phase delays at each lapse time: from 10 to 20 s, 20–30 s, 30–40 s, 40–50 s, 50–60 s, and 60–70 s. The four colors of broken lines correspond to the occurrence of each low-frequency tremor and to the main shock.

Additional file 11: Temporal variations in 15-day ACFs at Station S02. (a) 15-day ACFs calculated from OBS data over 15 continuous days, including the 7 days before and after the given date. The ACFs plotted in red are in the SSE period, and the bottom ACF is the reference ACF, calculated for the month from November 19 to December 19. (b) The cross-correlation coefficient between the 15-day ACFs and the reference ACF, using a 15 s time window and zero lag time. The period between the black triangles corresponds to the SSE period, and the periods between the red, blue, and green triangles correspond to low-frequency tremor periods in Sequences 1, 2, and 3. The magenta broken line represents the main shock. (c) Phase variation in the 15-day ACFs relative to the reference ACF, using a 15 s time window. Warm colors represent phase progressions, and cold colors represent phase delays. The four pairs of triangles and the magenta broken line correspond to each period and the main shock described in (b).

Additional file 12: Temporal variations in seismic velocity at S02. These figures show the temporal velocity changes calculated from the phase delays at each lapse time: from 10 to 20 s, 20–30 s, 30–40 s, 40–50 s, 50–60 s, and 60–70 s. The four colors of broken lines correspond to the occurrence of each low-frequency tremor and to the main shock.

Additional file 13: Temporal variations in 15-day ACFs at Station S03. (a) 15-day ACFs calculated from OBS data over 15 continuous days, including the 7 days before and after the given date. The ACFs plotted in red are in the SSE period, and the bottom ACF is the reference ACF, calculated for the month from November 19 to December 19. (b) The cross-correlation coefficient between the 15-day ACFs and the reference ACF, using a 15 s time window and zero lag time. The period between the black triangles corresponds to the SSE period, and the periods between the red, blue, and green triangles correspond to low-frequency tremor periods in Sequences 1, 2, and 3. The magenta broken line represents the main shock. (c) Phase variation in the 15-day ACFs relative to the reference ACF, using a 15 s time window. Warm colors represent phase progressions, and cold colors represent phase delays. The four pairs of triangles and the magenta broken line correspond to each period and the main shock described in (b).

Additional file 14: Temporal variations in seismic velocity at S03. These figures show the temporal velocity changes calculated from the phase delays at each lapse time: from 10 to 20 s, 20–30 s, 30–40 s, 40–50 s, 50–60 s, and 60–70 s. The four colors of broken lines correspond to the occurrence of each low-frequency tremor and to the main shock.

Additional file 15: Temporal variations in 15-day ACFs at Station S04. (a) 15-day ACFs calculated from OBS data over 15 continuous days, including the 7 days before and after the given date. The ACFs plotted in red are in the SSE period, and the bottom ACF is the reference ACF, calculated for the month from November 19 to December 19. (b) The cross-correlation coefficient between the 15-day ACFs and the reference ACF, using a 15 s time window and zero lag time. The period between the black triangles corresponds to the SSE period, and the periods between the red, blue, and green triangles correspond to low-frequency tremor periods in Sequences 1, 2, and 3. The magenta broken line represents the main shock. (c) Phase variation in the 15-day ACFs relative to the reference ACF, using a 15 s time window. Warm colors represent phase progressions, and cold colors represent phase delays. The four pairs of triangles and the magenta broken line correspond to each period and the main shock described in (b).

Additional file 16: Temporal variations in seismic velocity at S04. These figures show the temporal velocity changes calculated from the phase delays at each lapse time: from 10 to 20 s, 20–30 s, 30–40 s, 40–50 s, 50–60 s, and 60–70 s. The four colors of broken lines correspond to the occurrence of each low-frequency tremor and to the main shock.

Additional file 17: Temporal variations in seismic velocity at S10. These figures show the temporal velocity changes calculated from the phase delays at each lapse time: from 10 to 20 s, 20–30 s, 30–40 s, 40–50 s, 50–60 s, and 60–70 s. The four colors of broken lines correspond to the occurrence of each low-frequency tremor and to the main shock.

Additional file 18: Temporal variations in 15-day ACFs at Station S14. (a) 15-day ACFs calculated from OBS data over 15 continuous days, including the 7 days before and after the given date. The ACFs plotted in red are in the SSE period, and the bottom ACF is the reference ACF, calculated for the month from November 19 to December 19. (b) The cross-correlation coefficient between the 15-day ACFs and the reference ACF, using a 15 s time window and zero lag time. The period between the black triangles corresponds to the SSE period, and the periods between the red, blue, and green triangles correspond to low-frequency tremor periods in Sequences 1, 2, and 3. The magenta broken line represents the main shock. (c) Phase variation in the 15-day ACFs relative to the reference ACF, using a 15 s time window. Warm colors represent phase progressions, and cold colors represent phase delays. The four pairs of triangles and the magenta broken line correspond to each period and the main shock described in (b).

Additional file 19: Temporal variations in seismic velocity at S14. These figures show the temporal velocity changes calculated from the phase delays at each lapse time: from 10 to 20 s, 20–30 s, 30–40 s, 40–50 s, 50–60 s, and 60–70 s. The four colors of broken lines correspond to the occurrence of each low-frequency tremor and to the main shock.

Additional file 20: Temporal variations in 15-day ACFs at Station S15. (a) 15-day ACFs calculated from OBS data over 15 continuous days, including the 7 days before and after the given date. The ACFs plotted in red are in the SSE period, and the bottom ACF is the reference ACF, calculated for the month from November 19 to December 19. (b) The cross-correlation coefficient between the 15-day ACFs and the reference ACF, using a 15 s time window and zero lag time. The period between the black triangles corresponds to the SSE period, and the periods between the red, blue, and green triangles correspond to low-frequency tremor periods in Sequences 1, 2, and 3. The magenta broken line represents the main shock. (c) Phase variation in the 15-day ACFs relative to the reference ACF, using a 15 s time window. Warm colors represent phase progressions, and cold colors represent phase delays. The four pairs of triangles and the magenta broken line correspond to each period and the main shock described in (b).

Additional file 21: Temporal variations in seismic velocity at S15. These figures show the temporal velocity changes calculated from the phase delays at each lapse time: from 10 to 20 s, 20–30 s, 30–40 s, 40–50 s, 50–60 s, and 60–70 s. The four colors of broken lines correspond to the occurrence of each low-frequency tremor and to the main shock.

Additional file 22: Temporal variations in 15-day ACFs at Station S18. (a) 15-day ACFs calculated from OBS data over 15 continuous days, including the 7 days before and after the given date. The ACFs plotted in red are in the SSE period, and the bottom ACF is the reference ACF, calculated for the month from November 19 to December 19. (b) The cross-correlation coefficient between the 15-day ACFs and the reference ACF, using a 15 s time window and zero lag time. The period between the black triangles corresponds to the SSE period, and the periods between the red, blue, and green triangles correspond to low-frequency tremor periods in Sequences 1, 2, and 3. The magenta broken line represents the main shock. (c) Phase variation in the 15-day ACFs relative to the reference ACF, using a 15 s time window. Warm colors represent phase progressions, and cold colors represent phase delays. The four pairs of triangles and the magenta broken line correspond to each period and the main shock described in (b).

Additional file 23: Temporal variations in seismic velocity at S18. These figures show the temporal velocity changes calculated from the phase delays at each lapse time: from 10 to 20 s, 20–30 s, 30–40 s, 40–50 s, 50–60 s, and 60–70 s. The four colors of broken lines correspond to the occurrence of each low-frequency tremor and to the main shock.

Additional file 24: Temporal variations in seismic velocity at S21. These figures show the temporal velocity changes calculated from the phase delays at each lapse time: from 10 to 20 s, 20–30 s, 30–40 s, 40–50 s, 50–60 s, and 60–70 s. The four colors of broken lines correspond to the occurrence of each low-frequency tremor and to the main shock.

Additional file 25: Temporal variations in 15-day ACFs at Station S22. (a) 15-day ACFs calculated from OBS data over 15 continuous days, including the 7 days before and after the given date. The ACFs plotted in red are in the SSE period, and the bottom ACF is the reference ACF, calculated for the month from November 19 to December 19. (b) The cross-correlation coefficient between the 15-day ACFs and the reference ACF, using a 15 s time window and zero lag time. The period between the black triangles corresponds to the SSE period, and the periods between the red, blue, and green triangles correspond to low-frequency tremor periods in Sequences 1, 2, and 3. The magenta broken line represents the main shock. (c) Phase variation in the 15-day ACFs relative to the reference ACF, using a 15 s time window. Warm colors represent phase progressions, and cold colors represent phase delays. The four pairs of triangles and the magenta broken line correspond to each period and the main shock described in (b).

Additional file 26: Temporal variations in seismic velocity at S22. These figures show the temporal velocity changes calculated from the phase delays at each lapse time: from 10 to 20 s, 20–30 s, 30–40 s, 40–50 s, 50–60 s, and 60–70 s. The four colors of broken lines correspond to the occurrence of each low-frequency tremor and to the main shock.

Additional file 27: Temporal variations in 15-day ACFs at Station S27. (a) 15-day ACFs calculated from OBS data over 15 continuous days, including the 7 days before and after the given date. The ACFs plotted in red are in the SSE period, and the bottom ACF is the reference ACF, calculated for the month from November 19 to December 19. (b) The cross-correlation coefficient between the 15-day ACFs and the reference ACF, using a 15 s time window and zero lag time. The period between the black triangles corresponds to the SSE period, and the periods between the red, blue, and green triangles correspond to low-frequency tremor periods in Sequences 1, 2, and 3. The magenta broken line represents the main shock. (c) Phase variation in the 15-day ACFs relative to the reference ACF, using a 15 s time window. Warm colors represent phase progressions, and cold colors represent phase delays. The four pairs of triangles and the magenta broken line correspond to each period and the main shock described in (b).

Additional file 28: Temporal variations in seismic velocity at S27. These figures show the temporal velocity changes calculated from the phase delays at each lapse time: from 10 to 20 s, 20–30 s, 30–40 s, 40–50 s, 50–60 s, and 60–70 s. The four colors of broken lines correspond to the occurrence of each low-frequency tremor and to the main shock.

Additional file 29: Temporal variations in 15-day ACFs at Station TJT2. (a) 15-day ACFs calculated from OBS data over 15 continuous days, including the 7 days before and after the given date. The ACFs plotted in red are in the SSE period, and the bottom ACF is the reference ACF, calculated for the month from November 19 to December 19. (b) The cross-correlation coefficient between the 15-day ACFs and the reference ACF, using a 15 s time window and zero lag time. The period between the black triangles corresponds to the SSE period, and the periods between the red, blue, and green triangles correspond to low-frequency tremor periods in Sequences 1, 2, and 3. The magenta broken line represents the main shock. (c) Phase variation in the 15-day ACFs relative to the reference ACF, using a 15 s time window. Warm colors represent phase progressions, and cold colors represent phase delays. The four pairs of triangles and the magenta broken line correspond to each period and the main shock described in (b).

Additional file 30: Temporal variations in seismic velocity at TJT2. These figures show the temporal velocity changes calculated from the phase delays at each lapse time: from 10 to 20 s, 20–30 s, 30–40 s, 40–50 s, 50–60 s, and 60–70 s. The four colors of broken lines correspond to the occurrence of each low-frequency tremor and to the main shock.

Abbreviations
ACF: Autocorrelation Function; CCF: Cross-correlation function; OBS: Ocean bottom seismographs; SSE: Slow slip event

Acknowledgements
We thank Masatoshi Miyazawa for helpful discussions. We thank Masa Kinoshita, and two anonymous reviewers for their constructive comments. We also thank Editage for the English language revision.

Funding
This work was supported by the Japan Society for the Promotion of Science (JSPS) KAKENHI grant (#26257206) and Japan Science and Technology Agency—Japan International Cooperation Agency Science and Technology Research Partnership for Sustainable Development (JST-JICA SATREPS) (#15543611) grants to M. U. and Y. I., and a JSPS KAKENHI grant (#26000002) grant to R. H., M. S., Y. I., K. O., and M. U.

Authors' contributions
MU conceived the study, analyzed the data, and interpreted the analysis results. YI proposed the topic, collected the data, and helped the interpretation. KO collected the data and helped the interpretation. RH and MS collected the data and collaborated with the corresponding author in the construction of manuscript. All authors read and approved the final manuscript.

Authors' information
M. U. is a student in a doctor's course in division of earth and planetary sciences at Kyoto University. Y. I. is an associate professor at Disaster Prevention Research Institute, Kyoto University. K. O. is an assistant professor at Disaster Prevention Research Institute, Kyoto University. R. H. is a professor at International Research Institute of Disaster Science, Tohoku University. M. S. is a professor at Earthquake Research Institute, The University of Tokyo.

Competing interests
The authors declare that they have no competing interests.

Author details
[1]Kyoto University, Gokasyo, Uji, Kyoto 611-0011, Japan. [2]Disaster Prevention Research Institute, Kyoto University, Gokasyo, Uji, Kyoto 611-0011, Japan. [3]International Research Institute of Disaster Science, Tohoku University, 6-6, Aoba, Aramaki, Aoba-ku, Sendai 980-8578, Japan. [4]Department of Earth and Planetary Science, Tokyo University, 1-1-1, Yayoi, Bunkyo-ku, Tokyo, Japan.

References
Campillo M, Paul A (2003) Long-range correlations in the diffuse seismic coda. Science 299:547–549

Claerbout JF (1968) Synthesis of a layered medium from its acoustic transmission response. Geophysics 33:264–269

Duvall TL, Jefferies SM, Harvey JW, Pomerantz MA (1993) Time-distance helioseismology. Nature 362:430–432

Hillers G, Campillo M (2016) Fault zone reverberations from cross-correlations of earthquake waveforms and seismic noise. Geophys J Int 204:1503–1517

Hino R, Inazu D, Ohta Y, Ito Y, Suzuki S, Iinuma T, Osada Y, Kido M, Fujimoto H, Kaneda Y (2013) Was the 2011 Tohoku-Oki earthquake preceded by aseismic Preslip? Examination of seafloor vertical deformation data near the epicenter. Mar Geophys Res. https://doi.org/10.1007/s11001-013-9208-2

Hirose H (2011) Tilt records prior to the 2011 off the Pacific coast of Tohoku earthquake. Earth Planets Space 63:513–518

Iinuma T, Hino R, Kido M, Inazu D, Osada Y, Ito Y, Ohzono M, Tsushima H, Suzuki S, Fujimoto H, Miura S (2012) Coseismic slip distribution of the 2011 off the Pacific coast of Tohoku earthquake (M9.0) refined by means of seafloor geodetic data. J Geophys Res 117:B07409. https://doi.org/10.1029/2012JB009186

Ito Y, Hino R (2013) Velocity reduction in an offshore region after the 2011 Tohoku-Oki earthquake, revealed from ocean-bottom seismic records, Proceedings of the 11th SEGJ International Symposium, pp 523–526

Ito Y, Hino R, Kido M, Fujimoto H, Osada Y, Inazu D, Ohta Y, Iinuma T, Ohzono M, Mishina M, Suzuki K, Tsuji T, Ashi J (2013) Episodic slow slip events in the Japan subduction zone before the 2011 Tohoku-Oki earthquake. Tectonophysics 600:14–26

Ito Y, Hino R, Suzuki S, Kaneda Y (2015) Episodic tremor and slip near the Japan trench prior to the 2011 Tohoku-Oki earthquake. Geophys Res Lett 42:1725–1731

Katakami S, Ito Y, Ohta K, Hino R, Suzuki S, Shinohara M (2018) Spatiotemporal variation of tectonic tremor activity before the Tohoku-Oki earthquake. J Geophys Res 123. https://doi.org/10.1029/2018JB016651.

Katakami S, Yamashita Y, Yakihara H, Shimizu H, Ito Y, Ohta K (2017) Tidal response in shallow tectonic tremors. Geophys Res Lett 44:9699–9706

Lockner DA, Byerlee JD, Kuksenko V, Ponomarev A, Sidorin A (1991) Quasi-static fault growth and shear fracture energy in granite. Nature 350(6313):39–42

Minato S, Tsuji T, Ohmi S, Matsuoka T (2012) Monitoring seismic velocity change caused by the 2011 Tohoku-Oki earthquake using ambient noise records. Geophys Res Lett 39(9). https://doi.org/10.1029/2012GL051405

Nimiya H, Ikeda T, Tsuji T (2017) Spatial and temporal seismic velocity changes on Kyushu Island during the 2016 Kumamoto earthquake. Sci Adv 3(11). https://doi.org/10.1126/sciadv.1700813

Nishizawa S, Kanagawa K (2005) Seismic wave velocities in rocks II: velocity anisotropy in metamorphic rocks caused by preferred orientations of minerals and cracks. J Geogr 114(6):949–962

Ohta Y, Hino R, Inazu D, Ohzano M, Ito Y, Mishina M, Iinuma T, Nakajima J, Osada Y, Suzuki K, Fujimoto H, Tachibana K, Demachi T, Miura S (2012) Geodetic constraints on Afterslip characteristics following the March 9, 2011, Sanriku-oki earthquake, Japan. Geophys Res Lett 39(16). https://doi.org/10.1029/2012GL052430

Poli P, Pedersen HA, Campillo M, the POLENET/LAPNET Working Group (2012) Emergence of body waves from cross-correlation of short period seismic noise. Geophys J Int 188:549–558

Rautian TG, Khalturin VI (1978) The use of coda for determination of the earthquake source Spectrum. Bull Seismol Soc Am 68(4):923–948

Rivet D, Campillo M, Radiguet M, Zigone D, Cruz-Atienza V, Shapiro NM, Kostoglodov V, Cotte N, Cougoulat G, Walpersdorf A, Daub E (2014) Seismic velocity change, strain rate and non-volcanic tremors during the 2009-2010 slow slip event in Guerrero, Mexico. Geophys J Int 196:447–460

Rivet D, Campillo M, Shapiro NM, Cruz-Atienza V, Radiguet M, Cotte N, Kostoglodov V (2011) Seismic evidence of nonlinear crustal deformation during a large slow slip event in Mexico. Geophys Res Lett 38(8). https://doi.org/10.1029/2011GL047151
Sawazaki K, Saito T, Ueno T, Shiomi K (2016) Estimation of seismic velocity changes at different depths associated with the 2014 northern Nagano prefecture earthquake, Japan (Mw6.2) by joint interferometric analysis of NIED hi-net and KiK-net records. Prog Earth Planet Sci 3:36
Snieder R, Grêt A, Douma H, Scales J (2002) Coda wave interferometry for estimating nonlinear behavior in seismic velocity. Science 295:2253–2255
Takagi R, Nakahara H, Kono T, Okada T (2014) Separating body and Rayleigh waves with cross terms of the cross-correlation tensor of ambient noise. J Geophys Res 119(3):2005–2018
Takagi R, Okada T, Nakahara H, Umino N, Hasegawa A (2012) Coseismic velocity change in and around the focal region of the 2008 Iwate-Miyagi Nairiku earthquake. J Geophys Res 117(B6). https://doi.org/10.1029/2012JB009252
Wapenaar K, Draganov D, Snieder R, Campman X, Verdel A (2010) Tutorial on seismic interferometry: part 1—basic principles and applications. Geophysics 75:A195–A209
Wegler U, Nakahara H, Sens-Schönfelder C, Korn M, Shiomi K (2009) Sudden drop of seismic velocity after the 2004 Mw 6.6 mid-Niigata earthquake, Japan, observed with passive image interferometry. J Geophys Res 114(B6). https://doi.org/10.1029/2008JB005869
Xu ZJ, Song X (2009) Temporal changes of surface wave velocity associated with major Sumatra earthquakes from ambient noise correlation. Proc Natl Acad Sci U S A 106:14207–14212

Permissions

All chapters in this book were first published in PEPS, by Springer; hereby published with permission under the Creative Commons Attribution License or equivalent. Every chapter published in this book has been scrutinized by our experts. Their significance has been extensively debated. The topics covered herein carry significant findings which will fuel the growth of the discipline. They may even be implemented as practical applications or may be referred to as a beginning point for another development.

The contributors of this book come from diverse backgrounds, making this book a truly international effort. This book will bring forth new frontiers with its revolutionizing research information and detailed analysis of the nascent developments around the world.

We would like to thank all the contributing authors for lending their expertise to make the book truly unique. They have played a crucial role in the development of this book. Without their invaluable contributions this book wouldn't have been possible. They have made vital efforts to compile up to date information on the varied aspects of this subject to make this book a valuable addition to the collection of many professionals and students.

This book was conceptualized with the vision of imparting up-to-date information and advanced data in this field. To ensure the same, a matchless editorial board was set up. Every individual on the board went through rigorous rounds of assessment to prove their worth. After which they invested a large part of their time researching and compiling the most relevant data for our readers.

The editorial board has been involved in producing this book since its inception. They have spent rigorous hours researching and exploring the diverse topics which have resulted in the successful publishing of this book. They have passed on their knowledge of decades through this book. To expedite this challenging task, the publisher supported the team at every step. A small team of assistant editors was also appointed to further simplify the editing procedure and attain best results for the readers.

Apart from the editorial board, the designing team has also invested a significant amount of their time in understanding the subject and creating the most relevant covers. They scrutinized every image to scout for the most suitable representation of the subject and create an appropriate cover for the book.

The publishing team has been an ardent support to the editorial, designing and production team. Their endless efforts to recruit the best for this project, has resulted in the accomplishment of this book. They are a veteran in the field of academics and their pool of knowledge is as vast as their experience in printing. Their expertise and guidance has proved useful at every step. Their uncompromising quality standards have made this book an exceptional effort. Their encouragement from time to time has been an inspiration for everyone.

The publisher and the editorial board hope that this book will prove to be a valuable piece of knowledge for researchers, students, practitioners and scholars across the globe.

List of Contributors

Chang-Hwa Chen and Benjamin Fong Chao
Institute of Earth Sciences, Academia Sinica, 128 Academia Road, Section 2, Nankang, Taipei 115, Taiwan

Shogo Komori
Institute of Earth Sciences, Academia Sinica, 128 Academia Road, Section 2, Nankang, Taipei 115, Taiwan
Geological Survey of Japan (GSJ), AIST, Central 7, 1-1-1 Higashi, Tsukuba, Ibaraki 305-8567, Japan

Mitsuru Utsugi, Tsuneomi Kagiyama and Hiroyuki Inoue
Aso Volcanological Laboratory, Kyoto University, Minamiaso, Kumamoto 869-1404, Japan

Hsieh-Tang Chiang
Institute of Oceanography, National Taiwan University, No. 1, Sec. 4, Roosevelt Road, Taipei 10617, Taiwan

Ryokei Yoshimura
Research Center for Earthquake Prediction, Disaster Prevention Research Institute, Kyoto University, Gokasho, Uji, Kyoto 611-0011, Japan

Wataru Kanda
Volcanic Fluid Research Center, Tokyo Institute of Technology, 2-12-1 Ookayama, Meguro, Tokyo 152-8551, Japan

John P. Platt and William Lamborn Schmidt
Department of Earth Sciences, University of Southern California, Los Angeles, CA 90089-0740, USA

Haoran Xia
Department of Earth Sciences, University of Southern California, Los Angeles, CA 90089-0740, USA
Chevron Energy Technology Company, 1500 Louisiana St, Houston, TX 77002, USA

Weijia Kuang
Planetary Geodynamics Laboratory, NASA Goddard Space Flight Center, 8800 Greenbelt Road, Greenbelt, MD 20771, USA

Andrew Tangborn
Joint Center for Earth System Technologies, University of Maryland, Baltimore County, 1000 Hilltop Circle, Baltimore, MD 21250, USA

Hidenori Aiki
Institute for Space-Earth Environmental Research, Nagoya University, Nagoya City, 464-8601 Aichi, Japan
Application Laboratory, Japan Agency for Marine-Earth Science and Technology, Yokohama, Japan

Martin Claus
GEOMAR Helmholtz-Zentrum für Ozeanforschung Kiel, Kiel, Germany

Richard J. Greatbatch
GEOMAR Helmholtz-Zentrum für Ozeanforschung Kiel, Kiel, Germany
Faculty of Mathematics and Natural Sciences, University of Kiel, Kiel, Germany

Edward Biegert, Bernhard Vowinckel, Raphael Ouillon and Eckart Meiburg
Department of Mechanical Engineering, University of California, Santa Barbara, Engineering II, Santa Barbara, CA 93106, USA

Yusuke Yokota, Tadashi Ishikawa, Mariko Sato, Shun-ichi Watanabe, Hiroaki Saito, Naoto Ujihara, Yoshihiro Matsumoto, Shin-ichi Toyama, Masayuki Fujita and Tetsuichiro Yabuki
Hydrographic and Oceanographic Department, Japan Coast Guard, 2-5-18 Aomi, Koto-ku, Tokyo 135-0064, Japan

Masashi Mochizuki
National Research Institute for Earth Science and Disaster Prevention, 3-1 Tennodai, Tsukuba, Ibaraki 305-0006, Japan

Akira Asada
Institute of Industrial Science, University of Tokyo, 4-6-1 Komaba, Meguro-ku, Tokyo 153-8505, Japan

Yasuko Yamagishi and Hide Sakaguchi
Department of Mathematical Science and Advanced Technology, Japan Agency for Marine-Earth Science and Technology, 3173-25, Showa-machi, Kanazawa-ku, Yokohama 236-0001, Japan

Ayako Nakanishi, Seiichi Miura and Shuichi Kodaira
Research and Development Center for Earthquake and Tsunami, Japan Agency for Marine-Earth Science and Technology, 3173-25, Showa-machi, Kanazawa-ku, Yokohama 236-0001, Japan

Shigekazu Kusumoto
Graduate School of Science and Engineering for Research (Science), University of Toyama, 3910 Gofuku, Toyama 930-8555, Japan

Tatsunori Ikeda
International Institute for Carbon-Neutral Energy Research (WPI-I2CNER), Kyushu University, Fukuoka 819-0395, Japan

Takeshi Tsuji
International Institute for Carbon-Neutral Energy Research (WPI-I2CNER), Kyushu University, Fukuoka 819-0395, Japan
Department of Earth Resources Engineering, Kyushu University, Fukuoka 819-0395, Japan

Chie Kusu
Graduate School of Environment and Information Sciences, Yokohama National University, 79-7 Tokiwadai, Hodogaya-ku, Yokohama 240-8501, Japan

Makoto Okada
College of Science, Ibaraki University, 2-1-1 Bunkyo, Mito, Ibaraki 310-8512, Japan

Atsushi Nozaki
Hiratsuka City Museum, 12-41 Sengencho, Hiratsuka, Kanagawa 254-0041, Japan

Ryuichi Majima
Faculty of Environment and Information Sciences, Yokohama National University, 79-7 Tokiwadai, Hodogaya-ku, Yokohama 240-8501, Japan

Hideki Wada
Faculty of Science, Shizuoka University, 836 Ohya, Suruga-ku, Shizuoka 422-8529, Japan

Virginia Strati
Department of Physics and Earth Sciences, University of Ferrara, Via Saragat 1, 44121 Ferrara, Italy
Legnaro National Laboratories, INFN, Viale dell'Università, 2, 35020 Legnaro, Italy

Marica Baldoncini, Fabio Mantovani and Barbara Ricci
Department of Physics and Earth Sciences, University of Ferrara, Via Saragat 1, 44121 Ferrara, Italy
Ferrara Section, INFN, Via Saragat 1, 44121 Ferrara, Italy

Ivan Callegari and Gerti Xhixha
Legnaro National Laboratories, INFN, Viale dell'Università, 2, 35020 Legnaro, Italy

William F McDonough
Department of Geology, University of Maryland, 237 Regents Drive, College Park, MD 20742, USA

Kento Iio
Department of Natural History Sciences, Graduate School of Science, Hokkaido University, N10W8, Kita-ku, Sapporo 060-0810, Japan
Geospatial Information Authority of Japan, Kitasato 1, Tsukuba 305-0811, Japan

Masato Furuya
Department of Earth and Planetary Sciences, Faculty of Science, Hokkaido University, N10W8, Kita-ku, Sapporo 060-0810, Japan

Hiroyuki A. Shimizu, Takehiro Koyaguchi and Yujiro J. Suzuki
Earthquake Research Institute, The University of Tokyo, 1-1-1 Yayoi, 113-0032 Bunkyo-ku, Tokyo, Japan

Satoru Tanaka
Department of Deep Earth Structure and Dynamics Research, Japan Agency for Marine-Earth Science and Technology, Yokosuka 237-0061, Japan

Hrvoje Tkalčić
Research School of Earth Sciences, The Australian National University, Canberra ACT 2601, Australia

Diego Peña and Katia J. Pinheiro
Geophysics Department, Observatório Nacional, Rio de Janeiro 20921-400, Brazil
CNRS, Université de Nantes, Nantes Atlantiques Universités, UMR CNRS 6112, Laboratoire de Planétologie et de Géodynamique, 2 rue de la Houssinière, 44000 Nantes, France

Hagay Amit
CNRS, Université de Nantes, Nantes Atlantiques Universités, UMR CNRS 6112, Laboratoire de Planétologie et de Géodynamique, 2 rue de la Houssinière, 44000 Nantes, France

Tohru Watanabe
Graduate School of Science and Engineering, University of Toyama, 3190 Gofuku, Toyama 930-8555, Japan

Akiyoshi Higuchi
Graduate School of Science and Engineering, University of Toyama, 3190 Gofuku, Toyama 930-8555, Japan
Now at Yachiyo Engineering Co., Ltd., 1-4-70 Siromi, Chuo-ku, Osaka 540-0001, Japan

Miyuu Uemura
Kyoto University, Gokasyo, Uji, Kyoto 611-0011, Japan

Yoshihiro Ito and Kazuaki Ohta
Disaster Prevention Research Institute, Kyoto University, Gokasyo, Uji, Kyoto 611-0011, Japan

Ryota Hino
Intemational Research Institute of Disaster Science, Tohoku University, 6-6, Aoba, Aramaki, Aoba-ku, Sendai 980-8578, Japan

Masanao Shinohara
Department of Earth and Planetary Science, Tokyo University, 1-1-1, Yayoi, Bunkyo-ku, Tokyo, Japan

Index

CPSIA information can be obtained
at www.ICGtesting.com
Printed in the USA
BVHW010414300822
645595BV00042B/696

9 781639 890187